Modeling and Optimization in Science and Technologies

Volume 16

The book series Modeling and Optimization in Science and Technologies (MOST) publishes basic principles as well as novel theories and methods in the fast-evolving field of modeling and optimization. Topics of interest include, but are not limited to: methods for analysis, design and control of complex systems, networks and machines; methods for analysis, visualization and management of large data sets; use of supercomputers for modeling complex systems; digital signal processing; molecular modeling; and tools and software solutions for different scientific and technological purposes. Special emphasis is given to publications discussing novel theories and practical solutions that, by overcoming the limitations of traditional methods, may successfully address modern scientific challenges, thus promoting scientific and technological progress. The series publishes monographs, contributed volumes and conference proceedings, as well as advanced textbooks. The main targets of the series are graduate students, researchers and professionals working at the forefront of their fields.

Indexed by SCOPUS. The books of the series are submitted for indexing to Web of Science.

More information about this series at http://www.springer.com/series/10577

Fouad Bennis • Rajib Kumar Bhattacharjya
Editors

Nature-Inspired Methods for Metaheuristics Optimization

Algorithms and Applications in Science and Engineering

Springer

Editors
Fouad Bennis
Ecole Central de Nantes
Nantes cedex 3, France

Rajib Kumar Bhattacharjya
Department of Civil Engineering
Indian Institute of Technology Guwahati
Guwahati, Assam, India

ISSN 2196-7326 ISSN 2196-7334 (electronic)
Modeling and Optimization in Science and Technologies
ISBN 978-3-030-26457-4 ISBN 978-3-030-26458-1 (eBook)
https://doi.org/10.1007/978-3-030-26458-1

This Springer imprint is published by the registered company Springer Nature Switzerland AG.
The registered company address is: Gewerbestrasse 11, 6330 Cham, Switzerland

Preface

Optimization is a powerful technique for obtaining the best possible solution of various engineering design and planning problems. Over the last two to three decades, several metaheuristic algorithms have been developed by various researchers which have the capability to solve more complex engineering optimization problems. While the gradient-based classical optimization algorithms are not capable of handling the non-linear nonconvex as well as multimodal problems, the metaheuristic algorithms can handle these types of problems efficiently. As a result, these algorithms are now used in various fields of engineering and technology. This book contains the metaheuristic optimization algorithms along with their engineering applications. The algorithms have been explained using some standard benchmark problems of various complexities. The applicability of these algorithms in some complex engineering problem has also been shown.

The book is divided into two sections. The first one, consisting of 14 chapters, deals with the principles of the metaheuristic optimization algorithms. The algorithms discussed in this section are genetic algorithms, artificial bee colony algorithm, differential evolution, bacterial foraging algorithm, firefly algorithm, biogeography-based optimization, invasive weed growth optimization, shuffled frog-leaping algorithm, teaching-learning-based optimization algorithm and particle swarm optimization.

Genetic algorithm, one of the relatively older metaheuristic optimization algorithms, was developed based on Charles Darwin's theory of natural evolution. Chapters 1, 2 and 3 deal with the theory of genetic algorithm. Chapter 2 explains genetic algorithm using the social and cultural behaviour of human beings. On the other hand, in Chap. 3, an interactive genetic algorithm is used to interpret the user's choices in deriving the fitness function of the problem. Differential evolution, an evolutionary algorithm from the field of evolutionary computation, is presented in Chap. 4. The ability to communicate mutually is a fundamental social behaviour of all animals, birds, insects, etc. Inspired by the social behaviour of animals, birds and insects, various optimization algorithms have been developed. Some of them are artificial bee colony algorithm, firefly algorithm, shuffled frog-leaping algorithm, particle swarm optimization, etc. Chapters 5, 6, 7 and 8 present artificial bee colony algorithm, firefly algorithm, shuffled frog-leaping algorithm and particle swarm optimization, respectively. Bacterial foraging optimization, another algorithm inspired by the group foraging behaviour of bacteria, is presented in Chap. 9. The other metaheuristic optimization algorithm presented in this section are artificial immune system (Chap. 10), inspired by the principles and processes of the vertebrate immune system; biogeography-based optimization (Chap. 11), inspired by the migration of species between habitats; invasive weed growth optimization (Chap. 12), inspired by the ecological behaviour of weeds in colonizing and finding a suitable place for growth; teaching-learning-based optimization (Chap. 13), which is based on the effect of a teacher on a learner; and the multi-agent system (Chap. 14).

The second part of the book consists of 12 chapters dealing mainly with the applications of the metaheuristic algorithms to different planning and design problems. They cover solutions to real-world problems in areas such as Civil Engineering, Mechanical Engineering, Electrical Engineering, Chemical Engineering, Robotics, Environmental Science, etc. Since their introduction by Professor John Holland, genetic algorithms have been widely applied for solving various engineering optimization problems. Due to the ability to avoid the local optimal solution(s) of a problem, this class of algorithms has gradually received more and more attention from researchers of different fields of engineering. Chapter 15 shows the application of genetic algorithms for humanoid robot gait generation. The application of genetic algorithm for obtaining optimal river bank protection measures for the river Brahmaputra has been presented in Chap. 16. Chapters 17 and 18 show how the genetic algorithm can be applied in finding unknown groundwater pollution sources and critical slip circle of an earthen embankment, respectively. The application of genetic algorithm in corridor allocation problem and facility layout problem has been presented in Chaps. 19 and 20.

Non-dominated sorting genetic algorithm (NSGA) is currently one of the most robust algorithms for solving multi-objective optimization problems. The algorithm is based on non-dominated sorting and has the ability to obtain the true Pareto-optimal font while maintaining a good diversity in the population. Chapters 21, 22 and 23 show some applications of NSGA-II algorithm. Further applications addressed in this second part include job scheduling using teaching-learning-based optimization (Chap. 24), branch and bound optimization (Chap. 25) for optimizing radar search pattern and optimization of the DRASTIC methodology for assessing groundwater vulnerability (Chap. 26).

Despite some unavoidable limitations, we expect this book to provide readers with the necessary theoretical knowledge to understand and use the set of metaheuristic optimization algorithms presented here in a number of engineering applications and to be an important source of inspiration for future research and developments.

Nantes cedex 3, France Fouad Bennis
Guwahati, Assam, India Rajib Kumar Bhattacharjya

Contents

Contributors

Bahriye Akay Department of Computer Engineering, Erciyes University, Melikgazi, Kayseri, Turkey

Rustu Akay Department of Mechatronics Engineering, Erciyes University, Melikgazi, Kayseri, Turkey

Frederic Barbaresco Thales Air Systems, Voie Pierre-Gilles de Gennes, Limours, France

Fouad Bennis Laboratoire des Sciences du Numérique de Nantes, UMR CNRS 6004, Nantes, France

Rajib Kumar Bhattacharjya Department of Civil Engineering, Indian Institute of Technology Guwahati, Guwahati, Assam, India

D. Blumenthal AgroParistech, Massy, France

Yann Briheche Thales Air Systems, Voie Pierre-Gilles de Gennes, Limours, France
Laboratoire des Sciences du Numérique de Nantes, UMR CNRS 6004, Nantes, France

Damien Chablat Laboratoire des Sciences du Numérique de Nantes, UMR CNRS 6004, Nantes, France

Sandeep Singh Chauhan Department of Chemical Engineering, Indian Institute of Technology Guwahati, Guwahati, Assam, India

Dilip Datta Department of Mechanical Engineering, Tezpur University, Tezpur, Assam, India

Abhijit Deka Department of Mechanical Engineering, School of Engineering, Tezpur University, Tezpur, Assam, India

Anirban Dhar Department of Civil Engineering, Indian Institute of Technology Kharagpur, Kharagpur, West Bengal, India

Xavier Fischer ESTIA Engineering School, Technopôle Izarbel, Bidart, France
IMC-I2M, UMR CNRS 5295, Université de Bordeaux, Bordeaux, France

B. G. Rajeev Gandhi Department of Civil Engineering, Indian Institute of Technology Guwahati, Guwahati, Assam, India

Sahajpreet Kaur Garewal Department of Civil Engineering, Visvesvaraya National Institute of Technology, Nagpur, Maharashtra, India

Nithin V. George Department of Electrical Engineering, Indian Institute of Technology Gandhinagar, Gandhinagar, Gujarat, India

Aniruddha D. Ghare Department of Civil Engineering, Visvesvaraya National Institute of Technology, Nagpur, Maharashtra, India

Sirajul Islam Department of Civil Engineering, Assam Engineering College, Guwahati, India

Hriday Mani Kalita Department of Civil Engineering, National Institute of Technology Meghalaya, Shillong, Meghalaya, India

Zahnupriya Kalita Department of Mechanical Engineering, Tezpur University, Tezpur, Assam, India

Konstantinos L. Katsifarakis Department of Civil Engineering, Aristotle University of Thessaloniki, Thessaloniki, Greece

Remya Kommadath Department of Chemical Engineering, Indian Institute of Technology Guwahati, Guwahati, Assam, India

Yiannis N. Kontos Department of Civil Engineering, Aristotle University of Thessaloniki, Thessaloniki, Greece

Prakash Kotecha Department of Chemical Engineering, Indian Institute of Technology Guwahati, Guwahati, Assam, India

Dilip Kumar Department of Civil Engineering, G B Pant Engineering College, Pauri, Uttarakhand, India

Komal Kumari Department of Civil Engineering, Indian Institute of Technology Kharagpur, Kharagpur, West Bengal, India

Ricardo Mejía-Gutiérrez Design Engineering Research Group (GRID), Universidad EAFIT, Medellín, Colombia

Selva B. Munusamy Department of Civil Engineering, Indian Institute of Technology Kharagpur, Kharagpur, West Bengal, India

Vivek Patel Department of Mechanical Engineering, Pandit Deendayal Petroleum University, Gandhinagar, Gujarat, India

Rohan Patidar Department of Electrical Engineering, Indian Institute of Technology Gandhinagar, Gandhinagar, Gujarat, India

Apoorv P. Patwardhan Department of Electrical Engineering, Indian Institute of Technology Gandhinagar, Gandhinagar, Gujarat, India

J.-F. Petiot LS2N, Ecole Centrale de Nantes, Nantes, France

E. Poirson LS2N, Ecole Centrale de Nantes, Nantes, France

N. Rajasekar Department of Energy and Power Electronics, School of Electrical Engineering, Vellore Institute of Technology (VIT) - Vellore, Vellore, India

J. Prasanth Ram New Horizon College of Engineering (NHCE), Bengaluru, Karnataka

Arup Kumar Sarma Civil Engineering Department, Indian Institute of Technology Guwahati, Guwahati, Assam, India

Vimal Savsani Department of Mechanical Engineering, Pandit Deendayal Petroleum University, Gandhinagar, Gujarat, India

Leichombam Sophia College of Food Technology, Central Agricultural University, Imphal, India

Bipul Talukdar Department of Civil Engineering, Assam Engineering College, Guwahati, India

Mohamed Tawhid Department of Mathematics and Statistics, Thompson Rivers University, Kamloops, BC, Canada

Avinash D. Vasudeo Department of Civil Engineering, Visvesvaraya National Institute of Technology, Nagpur, Maharashtra, India

Teresa Zielinska Faculty of Power and Aerospace Engineering, Warsaw University of Technology, Warsaw, Poland

Part I
Algorithms

Chapter 1
Genetic Algorithms: A Mature Bio-inspired Optimization Technique for Difficult Problems

Konstantinos L. Katsifarakis and Yiannis N. Kontos

Abstract This chapter is dedicated to the method of genetic algorithms. Aiming at offering the essentials and at encouraging prospective users, it includes: (a) Description of the basic idea and the respective terminology, (b) Presentation of the basic genetic operators (selection, crossover, mutation), together with some additional ones, (c) Investigation of the values of basic parameters (crossover and mutation probability), (d) Outline of techniques for handling constraints and of conditions for the termination of the optimization process, and (e) Discussion on advantages and disadvantages of genetic algorithms. Moreover, the relationship between overall accuracy and optimization process accuracy is discussed, and some hints regarding teaching course modules on genetic algorithms are presented.

Keywords Optimization · Genetic algorithms · Genetic operators · Constraint handling · Accuracy · Terminology

1.1 Introduction

Genetic algorithms (GAs) are the older sister, if not the mother, of bio-inspired optimization techniques. They were initially introduced by Holland [11] and they have evolved into a very efficient mathematical tool for solving difficult optimization problems, especially when the respective objective functions exhibit many local optima or discontinuous derivatives. Despite successful use of GAs in most scientific fields, they should not be considered as a panacea. As the "no free lunch" theorem states, no method is superior to all others for every problem (e.g. Wolpert and Macready [30]).

The scientific literature on the theoretical background and the applications of genetic algorithms is already quite rich. It is difficult to compile even a list of books,

---

K. L. Katsifarakis (✉) · Y. N. Kontos
Department of Civil Engineering, Aristotle University of Thessaloniki, Thessaloniki, Greece
e-mail: klkats@civil.auth.gr

F. Bennis, R. K. Bhattacharjya (eds.), *Nature-Inspired Methods for Metaheuristics Optimization*, Modeling and Optimization in Science and Technologies 16,
https://doi.org/10.1007/978-3-030-26458-1_1

exclusively dedicated, or offering a lion's share to their study. Rather indicatively, we record the following: Goldberg [8], Rawlins [25], Michalewicz [23], Dasgupta and Michalewicz [3], Reeves and Raw [26], Sivanandam and Deepa [27], Yu and Gen [31]. A question that might arise, then, is the purpose, or the contribution of one more book chapter on genetic algorithms. Our answer is as follows:

This chapter aims at serving as a practical guide for those who would like to use this powerful optimization tool efficiently, and even adapt existing codes to their needs, without going deep into theory. It also aims at providing teaching material for university courses dealing with advanced optimization techniques. Inevitably, it starts with a description of the basic GA features. Our presentation (in Sects. 1.2 and 1.3) follows that of Katsifarakis and Karpouzos [15] and makes use of the same notation.

1.2 The Basic Idea and the Terminology

Genetic Algorithms are essentially a simplified mathematical imitation of the biological process of evolution of species. Biology is the source of the respective terminology, too. The population, which evolves in the mathematical GA framework, consists of potential solutions of the examined problem, namely of sets of the respective decision variables; each solution is considered as an individual, consisting of one chromosome. For this reason, the terms "individual" and "chromosome" are practically equivalent and they are used interchangeably.

To apply the genetic operators, namely to proceed with simulation of the evolution of species process, chromosomes should be properly encoded. In most early applications and in the majority of recent ones binary encoding is used; first, each decision variable is expressed as a binary (or Gray) number, and then these numbers are concatenated, forming a binary string. Each digit of that string is called gene (or character) and the different values that it can take are called allele. Some authors, though, define a gene as a subset of the chromosome that represents a decision variable.

The encoded form of the chromosomes is called genotype, while the decoded set of decision variables is called phenotype. Both of these puzzling terms are loans from Biology, with Greek origin. The origins of "phenotype", for instance, are traced in the Greek words φαίνομαι (phenomai = appear, look like, cf. phenomenon) and τύπος (typos = type).

Most early applications of Genetic algorithms have adopted binary encoding. The main reasons were: (a) easy construction of genetic operators and (b) the belief in the theory of schemata or similarity templates (e.g. [8, 11]). In recent years, though, real-coded genetic algorithms, appear more and more often. An important criterion is chromosome length (CL), namely the number of its genes. CL depends on the number of decision variables, the range of their values and the required accuracy. If, for instance, a parameter can vary between the real values a_1 and b_1 and an accuracy of d decimal points is sought, the required number of binary digits is the smallest integer m that fulfils the following inequality (e.g. [15]):

$$2^{m} - 1 \geq (b_1 - a_1)\, 10^{d} \tag{1.1}$$

When the decision variables exhibit very large ranges of values, or when very high accuracy is required, binary chromosome length may become excessive. In such cases, real-coded genetic algorithms, where each variable is represented by one gene, expressed as a real number, are preferred. They are also a better choice for some problems, such as the well-known "Travelling salesman problem" (e.g. [23]), where use of binary genetic operators is not convenient. Octal, hexadecimal and other encodings have also been used.

GAs are a population-based technique, namely its application starts with a set of chromosomes, which constitute the initial population, or the first generation. In most cases, initial population is randomly generated. Nevertheless, one could take advantage of any information on optimal solution and its location in the search space (namely the set of all possible solutions). The population size (PS), namely the number of initial solutions, is defined by the user.

Construction of the first-generation chromosomes is followed by their evaluation. The evaluation process depends entirely on the specific application of genetic algorithms and ends up with attributing a fitness value FV(I) to each chromosome I. Then the next generation is produced, applying genetic operators, which are also based on biological patterns, usually in a two-step process. The overall procedure, namely chromosome evaluation-application of genetic operators in order to produce new sets of solutions (generations), is repeated for a number of times, explicitly or implicitly defined by the user (see Sect. 1.4). It should be mentioned that in most cases, the population size PS is the same in all generations. Nevertheless, genetic algorithms with variable PS have been reported in the literature, claiming better performance in some problems (e.g. [20]).

1.3 Genetic Operators

Genetic operators could be defined as distinct processes, which are used to produce new chromosomes from the existing ones. In the "main stream", generation-based GAs, a two-step procedure is followed, involving an "intermediate" population between successive generations. The main genetic operators, which are described in the following paragraphs, are: (a) selection (b) crossover and (c) mutation. Many other operators have been also proposed and used. Some of them are briefly outlined, as well.

1.3.1 Selection

The basic idea of any GA selection process is borrowed from the core of Darwin's theory, namely survival of the fittest. In the generation-based GA framework, selection is used to produce the intermediate population, which consists of copies of current generation chromosomes. The number of each chromosome's copies depends on its fitness, which has been calculated during the evaluation process. Determinism is moderated, though, through introducing some element of randomness, offering a chance of "survival" to weak chromosomes. The most common selection processes are: (a) The biased roulette wheel and (b) The tournament.

1.3.1.1 The Biased Roulette Wheel

This process, which was initially conceived for maximization problems, is based on the idea of a roulette wheel with PS unequal slots, with sizes proportional to the fitness of the respective chromosomes. Then, each time the roulette wheel is "spinned", the selection probability p(I) of a chromosome I is equal to:

$$\mathrm{p\,(I)} = \frac{\mathrm{FV\,(I)}}{\mathrm{SFV}} \tag{1.2}$$

where SFV is the sum of the fitness values of all chromosomes of current generation. Roulette wheel spinning is repeated PS times, to select PS chromosome copies. The resulting intermediate population will include (statistically) more than one copies of the fittest chromosomes, at the expense of some weak ones. Nevertheless, even the worst chromosome has a tiny chance to "survive", namely to be copied to the intermediate population. A simple example follows.

Example 1.1 The biased roulette wheel process is used in solving a maximization problem. The population size PS = 25. The fitness values of current generation chromosomes are shown in Table 1.1. Then, SFV is equal to 1104. The respective p(I) values can be calculated using Eq. (1.2), and the resulting biased roulette-wheel is shown in Fig. 1.1.

The p(I) values of the best and worst chromosome are p(1) = 10/1104 and p(5) = 110/1104 respectively. The probability PT(I), of having at least one copy to the intermediate population can be calculated by the following formula:

$$\mathrm{PT(I)} = 1 - (1 - \mathrm{p(I)})^{\mathrm{PS}} \tag{1.3}$$

Table 1.1 Chromosome fitness values (Example 1.1)

I	1	2	3	4	5	6	7	8	9	10	11	12	13
FV(I)	10	90	52	12	110	24	42	14	48	37	69	15	85
I	14	15	16	17	18	19	20	21	22	23	24	25	
FV(I)	11	29	54	17	23	92	11	65	56	43	24	71	

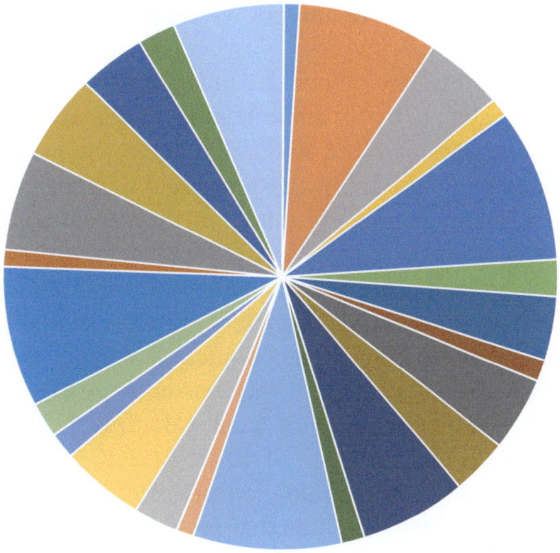

Fig. 1.1 Biased roulette wheel (PS = 25, fitness values of Table 1.1)

It results that PT(1) = 0.203 and PT(5) = 0.927. One can see that the total survival probability of the worst chromosome is substantial, as it slightly exceeds 20%. Moreover, there is 7.3% probability that the intermediate population will not include copies of the best chromosome.

The computational application of the biased roulette wheel process is quite simple [15]: At a preliminary step, the fitness value FV(J) of chromosome J is replaced by the sum of the FV(I) values for I = 1 to J, while, at the same time, SFV is calculated. Then, the following procedure is repeated PS times: A number X_R between 0 and SFV is randomly selected. If FV(K-1) < X_R < FV(K), a copy of chromosome K is added to the intermediate population.

It is desirable to have good solutions in the initial population, or to find some, in the first generations. Nevertheless, if few chromosomes of a generation have very large fitness values compared to the rest, biased roulette wheel selection may lead to an intermediate population dominated by their copies. This, in turn, will shrink further search around the respective solutions and may trap the whole process to local optima. To avoid this phenomenon, which is called "premature convergence", some authors suggest scaling down the range of fitness values in the first generations. Another option is to use the ranking of the solutions, instead of their fitness values, in order to construct the slots of the roulette wheel.

Roulette wheel can be used in minimization problems, too, since most of them can be stated as maximization ones, simply by considering the inverse of the evaluation function. Another option is subtraction of each FV(J) from a properly selected number Vmax (certainly larger than the maximum FV(J)), and use of the remainder in subsequent calculations. To avoid suppressing the actual differences between FV(J) values, Vmax should not be too large, with respect to them.

1.3.1.2 The Tournament Method

The tournament method depends exclusively on chromosome ranking and is directly applicable both to minimization and maximization problems. Its idea is based on contests between KK individuals, where KK a predefined number. In the GA framework, the following procedure is repeated PS times: KK chromosomes of the current generation are randomly selected, and their fitness values are compared to each other. A copy of the chromosome with the best (largest or smallest) fitness value is added to the intermediate population.

The value of KK, which is called selection constant, usually ranges between 3 and 5. It depends on the population size and the desired selective pressure, namely the degree to which best chromosomes are favored. Increase of KK value reduces the "survival" probability of weak solutions, namely it increases selective pressure. In some versions, survival of the worst (KK-1) chromosomes is completely impossible.

A comparative evaluation of selection processes can be found in Goldberg and Deb [9]. They conclude that ranking based processes (e.g. tournament) have some advantages over proportionate ones (biased roulette wheel). We tend to prefer the tournament method, mainly due to its direct application to minimization problems.

1.3.1.3 The Elitist Approach

The selection processes, which have been outlined, do not guarantee that a copy of the best chromosome of one generation will be included in the next one, too. To overcome this weakness, many codes use the so-called elitist approach, namely they include an additional procedure of passing at least one copy of the best chromosome to the next generation. To our opinion, it is better to pass two copies to the intermediate population: one to continue unaltered to the next generation, and one to freely participate in the evolutionary process, achieved by applying the other genetic operators.

1.3.2 Reproduction Operators

Selection does not introduce any new solutions. This task is left to "reproduction" operators, which apply to randomly selected members of the intermediate population. Not-selected members pass unaltered to the next generation, together with the new solutions.

We would like to mention that in some papers, particularly early ones, the term "reproduction" is associated with selection. In this chapter, we use it to denote processes that produce new solutions, following the majority of recent works.

Due to the versatility of genetic algorithms, many reproduction operators have been conceived and used. Some of them are problem-specific, while others are of general use. A few of them are outlined in the following sections. Application to binary GAs is discussed first, and then, when applicable, to real-coded GAs, as well.

We start with the two main operators, namely crossover and mutation.

1.3.2.1 Crossover

Crossover is a gene recombination process. In its basic form, it applies to pairs of chromosomes, which are named parents. Each member of the intermediate population can participate to this "mating process", with a "crossover probability" CRP (usually larger than 0.5). The degree of randomness in pair-forming varies. In some crossover versions parents can be successive chromosomes only, while in others no such restrictions exist. An integer number J_C, between 1 and (CL-1), is randomly selected, as well. Then, each parent binary string is cut into two pieces, immediately after gene J_C. The first piece of each parent is combined with the second piece of the other, as shown in Fig. 1.2a. In this way two new chromosomes, called offspring, are formed, which substitute their parents in the next generation. If each chromosome represents a set of N concatenated variables, crossover is in some versions restricted to the N-1 points that separate these variables.

Gene recombination, induced by crossover, does not guarantee that any offspring will be fitter than both parents. Let us consider, for instance, that the parent chromosomes A and B of Fig. 1.2 represent solutions of a maximization problem with one decision variable. Their fitness values are 810,360 and 473,709, respectively, while those of the two offspring, which resulted from one-point crossover, are 810,093 and 473,976. So, both are less fit than parent A, although they are fitter than parent B. To cope with such cases, in some crossover versions, offspring replace parents, only if they have better fitness values. While this modification looks plausible, it does not guarantee better overall performance, in particular when each chromosome represents a set of N concatenated variables, as is usually the case.

Direct extensions of the basic crossover process are two and multiple-point crossover. As their name implies, parent chromosomes are cut in more than two pieces, which they exchange, in order to produce the offspring. Two-point crossover (after the second and the fourteenth gene) is shown in Fig. 1.2b. The offspring fitness values are 998,008 and 286,061 respectively, namely one of them is fitter than both parents. A further extension is uniform crossover, which introduces additional randomness; the parents are cut in N pieces, but offspring formation is

Parent A	**11000101110/101111000**	**11/000101110101/111000**
Parent B	***01110011101/001101101***	***01/110011101001/101101***
Offspring A	**11000101110/*001101101***	**11/*110011101001/*111000**
Offspring B	***01110011101/*101111000**	***01/*000101110101/*101101***
	(a)	**(b)**

Fig. 1.2 (**a**) One-point crossover (**b**) Two-point crossover (slashes indicate crossover points)

achieved through selection between respective parental pieces with a predetermined probability (usually, but not necessarily, equal to 0.5). There is no guarantee, though, that any of these more complicated crossover versions perform better than the basic one.

Another extension is three or multi-parent crossover. As the names imply, genes of three or more parent chromosomes are recombined to form each offspring (e.g. [27]). Operators with more than two offspring per parent couple have been also designed and used. In this case, the best two make it to the next generation.

Simple and multi-point crossover applies in a similar way to real-coded genetic algorithms, as well. Another option is to equate each gene of an offspring to the arithmetic or to the geometric mean of the respective parent genes. An interesting extension is blend crossover (BLX-α), introduced by Eshelman and Schaffer [5], where offspring are produced in the following way: A real positive constant α is selected ($\alpha = 0.5$ is a common choice). Let gene i of parent chromosomes A and B be equal to A_i and B_i respectively. The quantity $c = |A_i - B_i|$ is calculated and a number d, ranging in the interval $[\min(A_i, B_i) - \alpha \times c, \max(A_i, B_i) + \alpha \times c]$ is randomly selected. Finally, gene i of the offspring is set equal to d. This process is repeated for all genes of the parent chromosomes.

Another interesting variant is parent-centric crossover, where a "male" and a "female" parent are defined. The probability to create offspring is larger in the search area around the female parent, while the male one is used only to define the extent of this area (e.g. [7]). Moreover, multi-parent crossover is used quite often with real-coded GAs (e.g. [4]). A review of crossover operators can be found in Herrera et al. [10].

1.3.2.2 Mutation

Mutation applies to genes. It introduces new "genetic" material, and its aim is twofold: (a) extension of search to more areas of the solution space, where better solutions could be found (mainly in the first generations), and (b) local refinement of good solutions (mainly in the last generations). Nevertheless, as with crossover, mutation can lead to better or to worse solutions.

Application to binary genetic algorithms, is quite simple: The gene that is selected for mutation is changed from 0 to 1 and vice versa. The effect of this change to the phenotype, namely to the decoded value of the respective decision variable, depends on the gene position in the chromosome. Let us consider, for instance, that the 10-digit binary string [1001001111] represents one decision variable DV. The respective decoded value is 591. If the first digit is mutated from 1 to 0, DV is reduced to 79; if the last digit is selected for mutation, DV changes only slightly (to 590).

In every generation, the mutation probability MP is the same for all genes of all chromosomes. It is generally much smaller than the crossover probability, but there is no consensus on the choice of its value. This issue is discussed in more detail in Sect. 1.3.2.3.

In real-coded genetic algorithms, mutation is performed by: (a) multiplying the selected gene by a number close to 1 or (b) by adding or subtracting a small (compared to the gene value) and randomly generated number from it. Replacement by a random number (in the range of variation of the selected gene) has been proposed, too. In "non-uniform" mutation, the magnitude of the added (or subtracted) amount is gradually reduced from generation to generation (e.g. [23]).

1.3.2.3 Mutation and Crossover Probability

The choice of basic parameters, such as crossover and mutation probability, plays an important role in GA performance. MP is much smaller than CRP, because it applies to genes, not to chromosomes. In many applications (e.g. [13]) a value close to 1/CL, CL being the chromosome length, has been used. The idea is to have, statistically, one mutation per chromosome. In any case, too small MP values may fail to hinder premature convergence, while too large ones render the search random.

The practical rule of setting MP = 1/CL was further investigated by Kontos and Katsifarakis [18, 19], at least as far as the combination of GAs and simplified groundwater flow and mass transport models is concerned. The theoretical problem studied, refers to the optimal management of polluted aquifers, bearing few fractures of known geometrical characteristics, namely to the cost minimization of Hydraulic Control (HC) or Pump-And-Treat (PAT) pollution control techniques. The aim was to maintain water quality and total flow rate of existing production wells, using additional wells for hydrodynamic control of pollutant plumes. The objective function of the respective optimization problem included the three main cost items, namely pumping cost, pipe network amortization cost and pumped polluted water treatment cost.

Steady-state flow in an infinite, confined, homogeneous and isotropic aquifer (with thickness a = 50 m, hydraulic conductivity $K = 10^{-4}$ m/s, porosity n = 0.2), was assumed. Two circular pollutant plumes (with radii of 50 and 60 m, respectively) were taken into account, that would eventually contaminate production wells 1 and 2 (pumping a total flow rate of 250 l/s), if no measures were taken. Additional wells (max flow rate of 120 l/s each) could be used to control or pump the plumes (Fig. 1.3). All wells were assumed to have a radius of 0.25 m. Regarding the fractures, one was almost perpendicular to the expected flow lines between plume 1 and well 1 (70 m long) and one almost parallel to the flow lines between plume 2 and well 2 (50 m long). Each fracture was simulated as an one-dimensional high speed runway for water and pollutant particles, not affecting hydraulic head distribution. To further alleviate the computational load, a two dimensional groundwater flow model was implemented, combined with a particle tracking code for advective mass transport simulation.

Advective pollutant transport was simulated by calculating the trajectories of particles with infinitesimal mass, starting from 16 checkpoints, symmetrically placed on the circular perimeter of each plume. The study period was assumed to be equal to the hypothetical pollutant deactivation period (1000 days = 100 time-

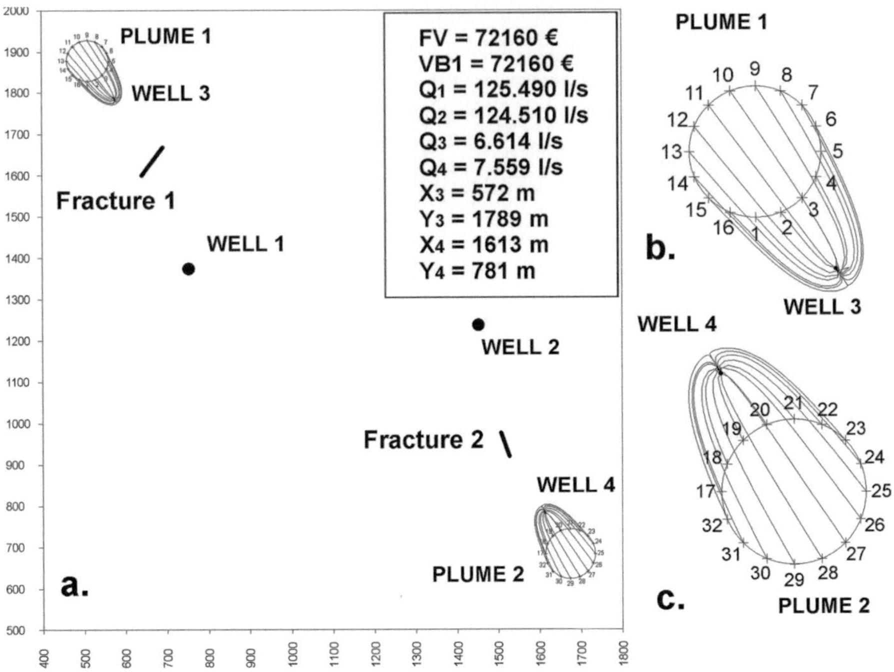

Fig. 1.3 Theoretical flow field with checkpoints' trajectories of best solution of the simplified problem, aiming at minimization of pumping cost only. Magnified view of each plume on the right [19]

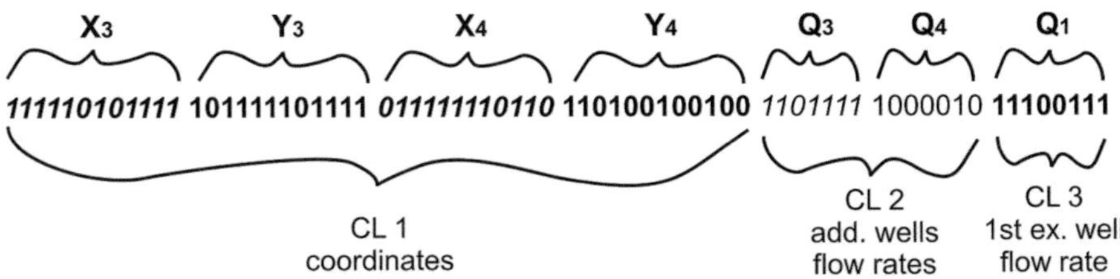

Fig. 1.4 Typical 70-digit binary chromosome of the pumping cost minimization problem of Kontos and Katsifarakis [19]

steps of 10 days each). The "pollution arrival" zone of a well W was considered proportional to the radius of the aquifer's cylindrical volume, which contains the water pumped by W during one time-step.

Simple binary GAs were used with 70-digit long chromosomes (CL = 70) representing the coordinates of the new wells and the flow rates of all wells (Fig. 1.4). The genetic operators used were tournament selection (including elitist approach), one-point crossover and mutation/antimetathesis. The following GA parameter values were used: population size PS = 60, number of generations NG = 1500 and selection constant KK = 3.

In the simplest form of the problem the optimization entailed that the fitness value FV to be minimized equals only the pumping cost Pcost, which was calculated as:

$$P\cos t = A \cdot \sum_{I=1}^{TNW} Q_I s_I \tag{1.4}$$

where A is a pumping cost coefficient, that depends on the density of pumped fluid, the electricity cost per kWh, pump efficiency and the pumping duration (here A = 6.48), TNW the total number of wells, Q_I the flow rate of well I (l/s) and s_I is the hydraulic head level drawdown at well I (m).

A penalty function (PEN) was added to Pcost, in cases of pollution of an existing well. PEN was proportional to the number of violated constraints (number of pollution particles arriving at existing wells) and to the magnitude of the violation (time step of pollutant arrival at existing wells):

$$PEN = \sum_{I=1}^{NP1} [P_C + P_V \cdot (TP - t_I)] \tag{1.5}$$

where NP1 is the number of checkpoints that pollute an existing well, P_C the constant part of the Penalty function (for each polluting checkpoint), P_V the coefficient of the variable part of the Penalty function (P_C/P_V = 100/10), TP the number of time steps (100) and t_I is the time step during which checkpoint I arrives at the well.

For the described minimization problem, an extended set of test runs was implemented in order to investigate the best CRP and MP values, namely most likely to find the optimal solution. The process lead to the proposed as optimal pumping scheme of Fig. 1.3 and to interesting conclusions, regarding the best choice of MP values. With 10 test runs for every CRP-MP combination (in order to produce a satisfactory statistical sample of solutions), CRP ranging from 0 to 0.6 (step 0.02), and MP ranging from 0.01 to 0.028 (step 0.002), the total number of runs reached 3100, entailing enormous computational volume (the evaluation function was calculated 279 million times and total computational time reached 156 days using Intel Core i5 CPU 660 at 3.33 GHz, 4GB of RAM, OS Windows 7, Visual Basic 6.0).

Figure 1.5 presents the relation of Min or Mean FV with CRP (Fig. 1.5a) and MP (Fig. 1.5b). Min FV refers to the best solution out of all 100 runs for a certain CRP value in Fig. 1.5a, and to the best solution out of all 310 runs for a certain MP value in Fig. 1.5b. Mean FV refers to the average value of FV out of all the respective runs. It is obvious that CRP variation does not substantially affect Min or Mean FV values, but there is definitely a correlation between FV and MP. The obvious trend is that Mean FV values are lower as MP increases, hence it is safe to assume that MP values quite higher than the empirically proposed 1/CL = 0.014 are more likely to reach the ideal global optimum. Figure 1.6 presents the lower 5% percentile of FV

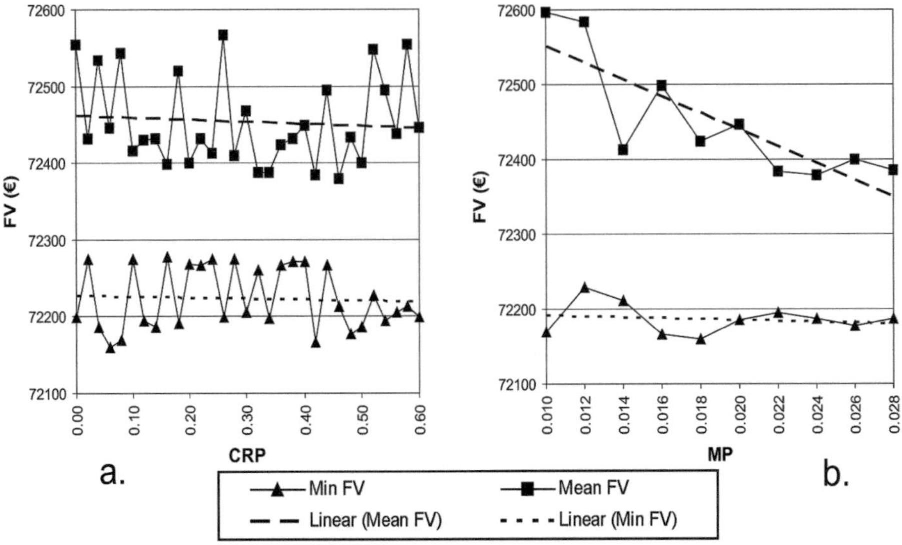

Fig. 1.5 Minimum fitness values (**a**) in relation to CRP and (**b**) in relation to MP for the minimization problem of Kontos [18] and Kontos and Katsifarakis [19]

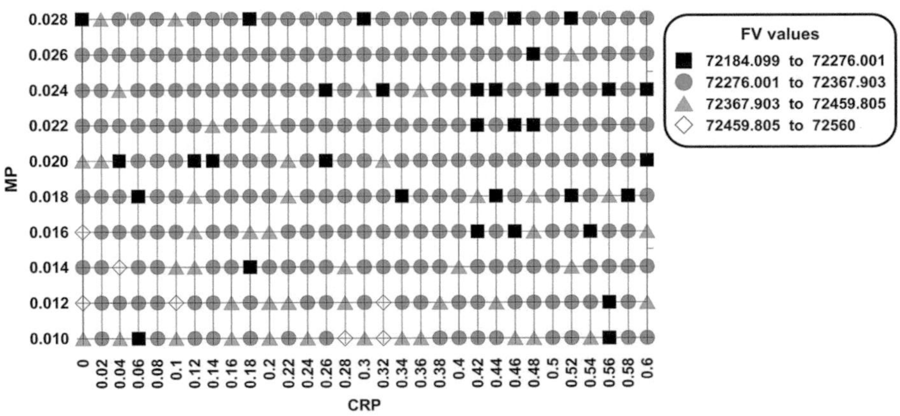

Fig. 1.6 $FV_{5\%\ perc.}$ in relation to CRP and MP values for the minimization problem of Kontos [18] and Kontos and Katsifarakis [19]. Each value's symbol and color opacity denote the quarter it belongs to ($FV_{5\%\ perc.}$ is the lower 5% percentile of Min FV)

for all CRP-MP tests. The higher statistical probability for the algorithm to produce lower FV solutions by CRP-MP values coming from the upper right area of the diagram is apparent.

Despite the fact that the optimal solution, due to the stochastic nature of GAs, resulted from a CRP-MP combination of 0.06–0.01, overall, CRP = 0.42 demonstrated higher probability of finding lower FV values, while as far as MP is concerned, the respective value was 0.024 ($\approx 1.7 \cdot CL^{-1} \approx 0.014$).

Concluding, this research offered some evidence that CRP values larger than 0.40 are more likely to achieve optimal performance for genetic algorithms, at least in similar groundwater management problems. Regarding mutation probability, instead of the empirically proposed MP value of 1/CL, values in the range of 1.5/CL to 2/CL proved to be more likely to direct the algorithm to the global optimum. Nevertheless, more research is needed, including investigation of a possible correlation between the chromosome structure and the best MP values that can lead to the global optimum. There are some indications, that the number of the decision variables that are included in the chromosome, is quite important.

1.3.2.4 Niche, Speciation, Sharing, Crowding, Migration

These techniques are useful in multimodal problems, namely with many maxima or minima, when we are interested in locating some or all of them. The basic idea is to divide the total population in independent or semi-independent subpopulations (called species), which evolve separately in niches, namely in distinct sub-domains of the solution space. To sustain different subpopulations, concentration of most chromosomes to one niche should be discouraged. One technique to achieve this goal, is "sharing". The idea is to reduce a chromosome's fitness, if many individuals are similar to it. It is performed in the following way: First, a sharing function Sim(I,J) is defined for each pair of chromosomes (I,J), based on their "distance" (at the genotypic or at the phenotypic level). The value of Sim(I,J) ranges between 0 and 1 with Sim(I,I) = 1 and Sim(I,J) = 0 for distance larger than a threshold value. Then, for each chromosome I the share factor, namely the sum of the respective Sim(I,J) values, is calculated, and its fitness value FV(I) is divided by this sum.

Niching is also favoured by using a crossover version, where each offspring replaces a member of the intermediate population that is most similar to it. This replacement process is called crowding.

To allow information exchange between subpopulations, a "migration rate" can be introduced, allowing crossover between chromosomes that have evolved in different niches. Migration is allowed to occur with a predefined schedule, e.g. every 10 generations, or if evolution of subpopulations is stalled.

Niching requires large total population size, depending on the number of the subpopulations, and, generally, entails an increase of computational load.

1.3.2.5 Antimetathesis

Antimetathesis is a mutation-like operator, which has been introduced by Katsifarakis and Karpouzos [14]. It applies to pairs of successive genes. Any gene of a chromosome (except for the last one) can be selected, with equal probability AP. If it is 1, it is changed to 0, and the following gene is set to 1. The opposite happens if the selected gene is 0. More explicitly, the following happen, with regard to gene pairs: (a) [11] becomes [01] (b) [00] becomes [10] (c) [10] becomes [01] and (d)

[01] becomes [10]. In the first two cases, antimetathesis is equivalent to mutation at the selected position. In the last two though, it is equivalent to mutation of both genes.

It is not recommended to use antimetathesis instead of mutation, but to use the two operators interchangeably (e.g. in the even and odd generations respectively). The merits of the combination in refining solutions can be seen through the following example [15]:

A genetic algorithm is used to find the optimum value of function F(x), x being an integer from 0 to 1000. Let's assume that this optimum occurs for x = 82, and that a good approach, i.e. x = 81, has been obtained. In binary form we have: $82 = [0001010010]_2$ and $81 = [0001010001]_2$.

Comparing the two chromosomes, we can see that mutation alone cannot improve the solution. Antimetathesis, though, can find the optimum, if applied to the ninth position of the chromosome. On the contrary, mutation can lead to the optimum, starting from x = 83. To put it in a more general way, use of antimetathesis could alleviate the "Hamming cliff" problem, which is inherent to use of binary numbers. This problem is due to the fact that successive binary numbers may have more than one different digits (or "Hamming distance" larger than 1), e.g. when the largest of them is equal to a power of 2 (e.g. $15 = [01111]_2$, while $16 = [10000]_2$).

Antimetathesis is also complementary to mutation in leading search to different solutions, for the following reason: The jump, or change caused by mutation, is always equal to a power of 2. The change introduced by antimetathesis, applied at position i, is equal to $2^i–2^{i-1}$. Thus, the solution space can be searched more thoroughly, if the two operators are used interchangeably.

1.4 Termination of the Optimization Procedure

Ideally, the optimization procedure should stop, as soon as the optimal solution is found. This criterion cannot be used with GAs, because in most cases, there is no guarantee that the optimum will be found, and even if it is found, there is no way to verify the achievement. Moreover, available computer resources are often restricted, while, in some cases, we may be satisfied even with good solutions, fulfilling certain standards (e.g. cost lower than a certain threshold). For these reasons, termination of the optimization procedure, should be defined by the user. The obvious and simplest way is to define the number of generations. Some other termination criteria, which have been summarized in many previous works (e.g. [15]), are:

- No better solution is found over a predetermined number of generations.
- The variation of the average fitness value of the population is below a threshold over a predetermined number of generations.
- A predetermined desired value of the objective function is reached by the best individual.

- The median fitness of the population is better than a predetermined value. This approach will lead to more than one satisfactory solutions.

1.5 Constraint Handling

Optimization targets, expressed in the GA framework by the chromosome evaluation process, are in most cases subject to constraints, which have usually the form of lower and/or upper bounds. For instance, availability of machinery could be translated in upper bound constraints, in optimizing work construction schedules.

In some heavily constrained problems, it might be difficult to find enough feasible solutions, in order to construct the initial population. Even when constraints are "light", though, reproduction operators may result in chromosomes, which represent infeasible solutions. To overcome this difficulty, many constraint-handling approaches have been developed. The most common are:

(a) Rejection of "infeasible" chromosomes, namely chromosomes that violate constraints.
(b) Chromosome repair, i.e. application of algorithms that correct infeasible chromosomes.
(c) Modification of the reproduction operators, in order to produce only "feasible" chromosomes.
(d) Inclusion of penalties in the evaluation process, to reduce the fitness of "infeasible" chromosomes

The first approach, which is equivalent to imposing the death penalty, even for small crimes, is not efficient, especially when the percentage of infeasible chromosomes is large. Applicability of chromosome repair and reproduction operator modification techniques depends on the required additional computational load. The most usual approach is inclusion of penalties in the chromosome evaluation process. Penalty form and magnitude, though, are problem specific, and should be decided by the user.

Regarding their form, penalties can be: (a) constant (b) proportional to the number of violated constraints and (c) increasing (linearly, quadratically, logarithmically etc.) with the degree of violation of each constraint. The latter are usually the most efficient (e.g. [16]). Regarding their magnitude, penalties should be large enough to guarantee observance of the constraints. Excessive magnitude, though, may level out differences in the evaluation function and may hinder the overall optimization process. Moreover, in search spaces with disjoint feasible regions, it would hinder moving from one to another, "stepping" on infeasible solutions. For these reasons, many researchers follow the minimum penalty rule, namely imposing the smallest penalties, which guarantee that the optimal solution is feasible (e.g. [21]). Others, though, have opted to use penalties, which render the worst feasible solution better than the best infeasible one (e.g. [24]).

The magnitude and the form of penalty functions depend also on the elasticity of the constraints. If small violations can be tolerated, the penalty could combine a rather small constant term, with strong dependence on the magnitude of constraint violation. Moreover, some variants include penalty adaptation with the evolution of the optimization procedure, such as penalty increase if the best solution is infeasible and penalty reduction if the best solution is feasible for a predetermined number of generations.

It should not be concluded, from the above discussion, that some form of penalty is always better than some other constraint-handling technique. An example follows:

Example 1.2 Suppose that we seek to minimize cost of pumping a given total flow rate Qtot from N wells. If Q(J) is the flow rate of well J, each chromosome represents a combination of N Q(J) values, while the evaluation function expresses the respective pumping cost. The constraint has the following form:

$$\sum_{J=1}^{N} Q(J) = Qtot \tag{1.6}$$

This constraint is practically never fulfilled, as each of the N flow rate values can get any value between 0 and a predetermined upper bound (e.g. Qtot), resulting in a sum QS between 0 and N×Qtot. Observance of the constraint can be efficiently imposed by chromosome repair, i.e. by multiplying each Q(J) by the factor Qtot/QS (e.g. [14]). Use of a penalty for QS exceeding Qtot, would be less efficient.

More ways to handle constraints can be found in the literature, such as introduction of an artificial immune system [2].

1.6 Steady State Genetic Algorithms

Steady state genetic algorithms do not follow the generation pattern, which has been described so far. Instead, new chromosomes resulting from reproduction operators replace immediately an equal number of existing ones. Chromosomes that are replaced are usually randomly selected, e.g. using a tournament process. Other options that have been used, include replacement of: (a) The "oldest" member of the population, (b) The member of the existing population, which is the most similar with the new one (crowding approach, as discussed in Sect. 1.3.2.4) and (c) The weakest individual. This option, though, might result in quick reduction of population diversity and in premature convergence.

1.7 Selection of Optimization Technique-Advantages and Disadvantages of Genetic Algorithms

Optimization problems arise in almost every scientific field (e.g. [6]), sometimes without even mentioning the term. The problem variety is reflected to the large number of optimization methods. Choosing the best method is sometimes a difficult task; as Reeves and Raw [26] warns us, "there is no royal road to optimization".

Experience (when not reduced to habit) and intuition may facilitate the choice, and use of more than one optimization approaches may be sometimes necessary. Moreover, use of hybrid techniques, e.g. combining genetic algorithms with a local search method, is an interesting option (e.g. [12, 22]).

Genetic algorithms are usually a very good choice for "difficult" optimization problems, namely non-linear, with "recalcitrant" objective functions (e.g. non-differentiable or with discontinuous derivatives), discontinuous search space, etc. Their main comparative advantages have been summarized in most relevant books and book chapters. A short list follows:

- Contrary to traditional methods, search for the global optimum starts from many points of the search space. Of course, GAs share this advantage with other population based techniques.
- GAs use data from the objective function only. For this reason, their application is not restricted by continuity or differentiability requirements, as already stated.
- There is no need to linearize namely to simplify (or oversimplify) the investigated problem (see also Sect. 1.8).
- They are very flexible, regarding the construction of the evaluation function.
- They do not involve complex mathematics, inviting users to modify genetic operators or even to create new ones, tailored to the investigated problems.
- They are naturally adapted to parallel processing.
- They can be easily combined with other optimization methods and computational techniques.
- Many times, they end up with more than one good solutions of the investigated problem. This point deserves some further discussion. It has already been mentioned is Sect. 1.3.2.4, that specific techniques, like niching, have been developed to locate many optima. This is quite understandable when studying mathematical functions; but in real world problems, when one seeks the lowest operational cost, for instance, the advantage is not that obvious. Even in such problems, though, finding a number of good solutions of similar fitness is an asset, because it offers the possibility of selecting the most suitable one, based on [15]:

    (a) Additional criteria, which have not been included in the evaluation process, because their expression in mathematical terms is difficult (e.g. intangible parameters, like aesthetics) or because their inclusion leads to disproportionate increase of the computational load.
    (b) Experience or scientific intuition, which could be interpreted as additional criteria, not explicitly stated.
    (c) Solution stability to changes of the values of input parameters; this is very important, when these values are not accurately known, as in many real-world problems. This stability could be checked by means of sensitivity analysis.

Genetic algorithms have some weak points, too. The most important are the following:

- In most cases, there is no guarantee that the global optimum has been reached (as in linear programming, for instance).
- The fitness value has to be calculated for every chromosome of each generation. With rather conservative values for the respective parameters, such as population size PS = 40 and number of generations NG = 100, 4000 chromosome evaluations will be performed. For this reason, the total computational volume increases rapidly with the complexity of the chromosome evaluation process, the number of generations and the population size.
- GAs are rather slow in local search. This disadvantage can be overcome, if necessary, by combining GAs with local search techniques.
- The efficiency of the method depends on the form of the evaluation function. With step functions for instance, the GA performance is mediocre.

1.8 Overall Accuracy vs Accuracy of the Optimization Procedure

Quite often, engineering problems have the form of predicting response of natural or man-made systems to certain human activities (e.g. hydraulic head level drawdown due to pumping) or of guaranteeing a desirable system's response to given external pressures (proper construction of a bridge, in order to "survive" a "design" earthquake with minor damage). The first step towards the solution is to construct the simulation model, namely to describe mathematically the studied systems and/or processes with adequate accuracy, and the second to solve this model, either analytically or numerically.

Optimization, on the other hand, could be defined as a procedure of finding the best (or, in a broader sense, a sufficiently good solution) of a problem, based on predefined criteria and under certain constraints. In the GA framework, the optimization criteria determine the chromosome evaluation process. This process may include solution of a relevant engineering problem. For instance, in the pumping cost minimization problem of Example 1.2, the respective hydraulic head level drawdown should be calculated, namely a groundwater flow problem should be solved. Detailed simulation models may result in high computational volume per chromosome evaluation. When the total computational volume becomes a restricting factor, one has two options: (a) Use smaller number of generations and/or population size, thus reducing the efficiency of the optimization process and (b) Use simplified simulation models, or even surrogate models (e.g. [1, 28]), which eventually compromise the accuracy of the description of the physical problem.

In any case, accuracy of the optimization process does not guarantee overall accuracy. To clarify the difference, let us study, from this point of view, the following problem, which has been presented by Tselepidou and Katsifarakis [29], and further discussed by Katsifarakis and Tselepidou [17]:

A geothermal aquifer consisting of 2 zones of different transmissivities (T_1 and T_2) and 4 zones of different temperatures (θ_1, θ_2, θ_3 and θ_4) is considered. The central zone has the highest temperature. A square area, with dimensions 3 km x 3 km, is available for geothermal development (Fig. 1.7). Its borders are shown with grey line. Zones of different transmissivities are indicated with different tones of grey, while their interface is shown as a grey line. Boundaries of thermal zones are also indicated by means of black lines.

The required total flow rate Qtot depends on the average temperature θ_{av} of pumped water and varies from 500 to 600 l/s (for θ_{av} varying from 80 °C to 60 °C respectively). To pump Qtot, 4 existing and 6 new wells will be used. The former are shown in Fig. 1.8 as small black squares. Geothermal water is collected to a central water tank at the border of the geothermal area, which appears as a larger black

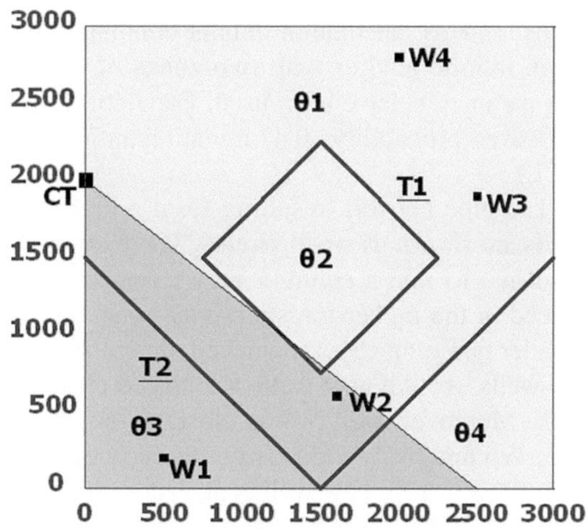

Fig. 1.7 External and internal boundaries of geothermal area and layout of the existing wells [17]

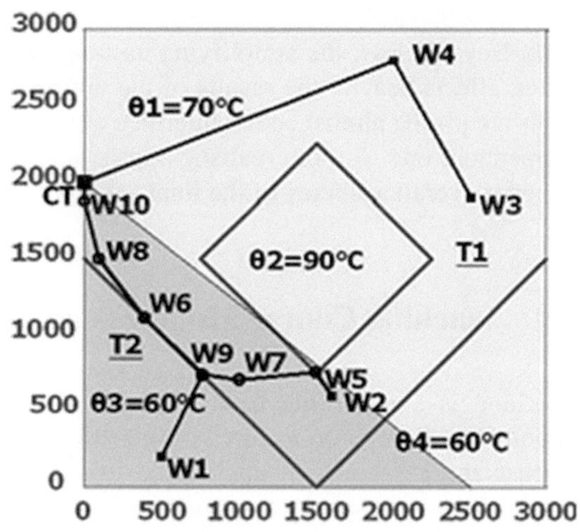

Fig. 1.8 Optimal layout of wells and the respective pipe network [17]

square (indicated as CT). The optimization procedure aims at finding the distribution of Qtot to the 10 wells, together with the locations (coordinates) of the 6 new ones that minimize the sum of two major cost components, namely: (a) annual pumping (operation) cost and (b) amortization of construction cost of the pipe network that carries the hot water from the wells to a central water tank.

A basic simplifying assumption is that the temperature of water, pumped from a well J, is equal to the temperature θ_I of the thermal zone, to which the well belongs. Out of the many data sets, discussed in the aforementioned references, we focus on the following one:

$$\mathrm{T}_1 = 0.001\mathrm{m}^2/\mathrm{s}, \mathrm{T}_2 = 0.01\mathrm{m}^2/s, \theta_1 = 70^{\circ}\mathrm{C}, \theta_2 = 90^{\circ}\mathrm{C}, \theta_3 = 60^{\circ}\mathrm{C}\theta_4 = 60^{\circ}\mathrm{C}.$$

Each chromosome represents a combination of 10 well flow rates and 6 pairs of coordinates. The respective fitness value is equal to the sum of the two cost items, and its calculation entails computation of hydraulic head level drawdown in an infinite aquifer with two zones of different transmissivities. The following GA parameters have been used: Population size: 60, number of generations: 2000, crossover probability: 0.4, mutation/antimetathesis probability: 0.0042 (close to 1/CL).

The pipe network resulting from a typical run appears in Fig. 1.8, where new wells are shown as small circles. There are many indications that the GA code has managed to find a solution very close to the optimal one: (a) All new wells are placed in the higher transmissivity zone, although its area is smaller. In this way, smaller pumping cost is achieved, due to smaller hydraulic head level drawdown at the wells. (b) All new wells are placed closer than the existing ones to the central tank. Moreover, well W9 is close to the Fermat point of the triangle defined by W1, W6 and W7. In this way, the pipe network length (and amortization cost) is rather low. (c) No new well is placed in the lowest temperature zone. Moreover, the new well W5 is placed in the very small area, which is characterized by the higher transmissivity and the highest temperature. Thus, θ_{av} is large, resulting in smaller Qtot. Nevertheless, the simplifying assumption regarding temperature of pumped water, affects heavily the results of the optimization process, since 3 out of 6 new wells are placed almost on the interface of two thermal zones, just inside the higher temperature one. A more realistic approximation of that temperature, would result in better overall accuracy of the final solution.

1.9 Teaching Course Modules on Genetic Algorithms

Teaching is a rewarding task, to those who enjoy it. The first author had the opportunity to include lectures on genetic algorithms in graduate courses and to discuss and explain their application to students, working on their Diploma and Doctoral Theses. Moreover, he has gained some insight into the difficulties of the respective notions (and the weak points of his own presentations, as well), from students' repeated mistakes in relevant tests.

One important issue is clarifying the difference between chromosome and evaluation function, as many students mistake one for the other. Another issue is stating all problem constraints and understanding their relationship with the chromosome evaluation process.

As there is no guarantee that GAs will end up with the globally optimal solution, it is very important to develop qualitative criteria, in order to estimate how good the derived "best" solution is. Such criteria have been presented in Sect. 1.8, for the problem discussed there. Moreover, restrictions both on overall accuracy and on the accuracy of the optimization procedure should be made clear to the students, as many tend to consider computer results as some kind of divine truth.

Regarding "technical" details, some students have difficulties with binary encoding. Nevertheless, it is worth teaching it, as it promotes better understanding of the decimal system, too.

Testing student knowledge is not easy, either. To help teachers, two exercises, one rather simple and one more sophisticated, are presented below:

Exercise 1: In order to minimize cost of pumping a total flow rate Qtot from 9 wells, you are going to use the method of binary genetic algorithms. Calculate the chromosome length (number of genes), if each of the 9 well flow rates can be up to 130 l/s and the required accuracy is 1 l/s.

Exercise 2: You aim at optimizing the operation of a reservoir, through determining the level of stored water at weekly basis. The optimization targets are: (a) Maximization of total revenue from electricity production (during the whole year) and (b) Elimination of flood damage during the high flood risk period of the year (by storing temporarily excess water quantities and restricting outflows to safe levels).

Available data include: (a) The height of the reservoir's dam (b) the stored water volume as a function of the water level at the reservoir (c) The required storage volume (per week), in order to eliminate flood damage and (d) The capacity (largest flow rate) that can be carried by the pipe, which connects the reservoir with the electricity production station. The water level at the reservoir, which feeds the electricity production station, and the respective flow rate towards the electricity station are considered as constant each week (and equal to the respective weekly average value).

In order to solve the problem, you are going to use the method of genetic algorithms. (a) Are you going to formulate a maximization or a minimization problem? (b) Which (and how many) variables are you going to include at each chromosome? (c) What are you going to include at the optimization function? (d) Which constraints could you state?

Moreover, the cost minimization problem, discussed in Sect. 1.8, can serve as an exercise basis.

1.10 Concluding Remarks

Genetic algorithms are a very strong tool for tackling difficult optimization problems. Although the respective terminology, originating from Biology, may sound strange to engineers, application of the method to most engineering problems is easy, due to its versatility. What is even more important, GAs promote creativity of the users, if their basic principles are understood. This is manifested by the variety of existing genetic operators and it is probably the best feature of the method, allowing not only its survival, but its continuous thriving, in a very competitive "environment", namely that of modern optimization techniques.

Understanding the basic principles of genetic algorithms, paves the way for understanding most nature-inspired and population-based techniques, which are currently available, allowing researchers a wide choice, based on the features of the investigated problem, or on their personal scientific background and taste.

While no method should be regarded as a panacea for optimization problems, we believe that genetic algorithms are at least as important as linear programming to practicing engineers, and to researchers, as well. For this reason, they should enjoy a fair share in optimization courses of engineering departments' curricula.

References

1. Bhattacharjya RK, Datta B (2005) Optimal management of coastal aquifers using linked simulation optimization approach. Water Resour Manag 19:295–320
2. Coello CAC, Cortés NC (2004) Hybridizing a genetic algorithm with an artificial immune system for global optimization. Eng Optim 36(5):607–634
3. Dasgupta D, Michalewicz Z (eds) (1997) Evolutionary algorithms on engineering applications. Springer, Berlin/Heidelberg/New York
4. Deb K, Anand A, Joshi D (2002) A computationally efficient evolutionary algorithm for real-parameter evolution. Evol Comput 10(4):371–395
5. Eshelman LJ, Schaffer JD (1993) Real-coded genetic algorithms and interval schemata. In: Whitley DL (ed) Foundation of genetic algorithms II. Morgan Kaufmann, San Mateo, pp 187–202
6. Floudas CA, Pardalos PM (eds) (2008) Encyclopedia of optimization, 2nd edn. Springer, New York
7. García-Martínez C, Lozano M, Herrera F, Molina D, Sánchez AM (2008) Global and local real-coded genetic algorithms based on parent-centric crossover operators. Eur J Oper Res 185:1088–1113
8. Goldberg DE (1989) Genetic algorithms in search, optimization and machine learning. Addison-Wesley Publishing Company, Reading
9. Goldberg DE, Deb K (1995) A comparative analysis of selection schemes used in genetic algorithms. In: Rawlins GJE (ed) Foundations of genetic algorithms. Morgan Kaufmann, San Mateo, pp 69–93
10. Herrera F, Lozano M, Sánchez AM (2005) Hybrid crossover operators for real-coded genetic algorithms: an experimental study. Soft Comput 9:280–298
11. Holland JH (1975) Adaptation in natural and artificial systems. University of Michigan Press, Ann Arbor

12. Huang WC, Yuan LC, Lee CM (2002) Linking genetic algorithms with stochastic dynamic programming to the long-term operation of a multireservoir system. Water Resour Res 38(12):1304–1312
13. Karpouzos DK, Katsifarakis KL (2013) A set of new benchmark optimization problems for water resources management. Water Resour Manag 27(9):3333–3348
14. Katsifarakis KL, Karpouzos DK (1998) Minimization of pumping cost in zoned aquifers by means of genetic algorithms. In: Katsifarakis KL, Korfiatis GP, Mylopoulos YA, Demetracopoulos AC (eds) Proceedings of an international conference on protection and restoration of the environment IV, Sani Greece, pp 61–68
15. Katsifarakis KL, Karpouzos DK (2012) Genetic algorithms and water resources management: an established, yet evolving, relationship. In: Katsifarakis KL (ed) Hydrology, hydraulics and water resources management: a heuristic optimisation approach. WIT Press, Southampton/-Boston, pp 7–37. ISBN 978-1-84564-664-6
16. Katsifarakis KL, Petala Z (2006) Combining genetic algorithms and boundary elements to optimize coastal aquifers' management. J Hydrol 327(1–2):200–207
17. Katsifarakis KL, Tselepidou K (2015) Optimizing design and operation of low enthalpy geothermal systems. In: Chandra Sharma U, Prasad R, Sivakumar S (eds) Energy science and technology. Vol. 9: Geothermal and Ocean energy, Studium Press. ISBN: 1-62699-070-0, 190-213
18. Kontos YN (2013) Optimal management of fractured coastal aquifers with pollution problems (in Greek), PhD thesis, Department of Civil Engineering, Aristotle University of Thessaloniki, Thessaloniki, Greece, p 465
19. Kontos YN, Katsifarakis KL (2012) Optimization of management of polluted fractured aquifers using genetic algorithms. Eur Water 40:31–42
20. Koumousis VK, Katsaras CP (2006) A saw-tooth genetic algorithm combining the effects of variable population size and Reinitialization to enhance performance. IEEE Trans Evol Comput 10(1):19–28
21. Li X, Zang G (2015) Minimum penalty for constrained evolutionary optimization. Comput Optim Appl 60(2):513–544
22. Mahinthakumar GK, Sayeed M (2005) Hybrid genetic algorithm—local search methods for solving groundwater source identification inverse problems. J Water Resour Plan Manag 131(1):45–57
23. Michalewicz Z (1996) Genetic algorithms + data structures = evolution programs. Springer, Berlin/Heidelberg
24. Michalewicz Z, Xiao J (1995) Evaluation of paths in evolutionary planner/navigator. In: Proceedings of the 1995 international workshop on biologically inspired evolutionary systems, Tokyo, Japan, pp 45–52
25. Rawlins GJE (1991) Foundations of genetic algorithms. Morgan Kaufmann Publishers, San Francisco, p 1991
26. Reeves CR, Raw JE (2003) Genetic algorithms-principles and perspectives. Kluwer Academic Publishers, Boston
27. Sivanandam SN, Deepa SN (2008) Introduction to genetic algorithms. Springer, Berlin/Heidelberg
28. Sreekanth J, Datta B (2010) Multi-objective management of saltwater intrusion in coastal aquifers using genetic programming and modular neural network based surrogate models. J Hydrol 393(3–4):245–256
29. Tselepidou K, Katsifarakis KL (2010) Optimization of the exploitation system of a low enthalpy geothermal aquifer with zones of different transmissivities and temperatures. Renew Energy 35:1408–1413
30. Wolpert DH, Macready WG (1997) No free lunch theorems for optimization. IEEE Trans Evol Comput 1(1):67–82
31. Yu X, Gen M (2010) Introduction to evolutionary algorithms. Springer, London/Dordrecht/Heidelberg/New York

Chapter 2
Introduction to Genetic Algorithm with a Simple Analogy

Arup Kumar Sarma

Abstract Genetic Algorithm (GA), which is now considered as a well-established and one of the most widely applied optimization techniques, started its journey when Von Neumann first forwarded the theory of self-reproducing automata during fifties (Fellenius W: Calculation of the stability of earth dams. Trans, of 2nd congress on Large Dams, vol 4, pp 445–459, 1936). However, implementation of this ideas came into application during eighties (Baker R: Determination of the critical slip surface in slope stability computations. Int J Numer Anal Methods Geomech 4:333–359, 1980; Bishop AW: The use of slip circle in the stability analysis of slopes. Geotechnique London 5:7–17, 1955; Chen Z-Y, Shao C-M: Evolution of minimum factor of safety in slope stability analysis. Canadian Geotech J Ottawa 25:735–748, 1988; Celestino TB, Duncan JM: Simplified search for noncircular slip surface. In: Proceedings of the 10th international conference on SMFE, pp 391–394, 1981). Because of the obvious advantage of using GA in optimizing even complex non-linear functions, it has now been used by many researchers for solving varieties of optimization problems. Basically GA is a computerized search optimization algorithms based on the principles of survival of the fittest, first laid down by Charles Darwin. Concept of GA has so far been presented by different researchers in different forms, mostly in a mathematical framework. The terms like gene, chromosome, cross over, mutation and generation, which are basically derived from biological origin, many a time, become confusing to readers of other disciplines and also it become difficult to appreciate how these processes will lead to an optimal solution. Therefore, in this article, the basic philosophy of GA is presented by highlighting how it works and how the natural biological processes can help in obtaining the most optimal or at least near optimal solution through a simple analogy.

A. K. Sarma (✉)
Civil Engineering Department, Indian Institute of Technology Guwahati, Guwahati, Assam, India
e-mail: aks@iitg.ernet.in

F. Bennis, R. K. Bhattacharjya (eds.), *Nature-Inspired Methods for Metaheuristics Optimization*, Modeling and Optimization in Science and Technologies 16,
https://doi.org/10.1007/978-3-030-26458-1_2

2.1 Introduction

Genetic Algorithm (GA), which is now considered as a well-established and one of the most widely applied optimization techniques, started its journey when Von Neumann first forwarded the theory of self-reproducing automata during fifties [5]. However, implementation of this ideas came into application during eighties [1–4]. Because of the obvious advantage of using GA in optimizing even complex non-linear functions, it has now been used by many researchers for solving varieties of optimization problems. Basically GA is a computerized search optimization algorithms based on the mechanics of natural genetics and natural selection. The basic concept of GAs is designed to simulate processes in natural system necessary for evolution, specifically those that follow the principles of survival of the fittest, first laid down by Charles Darwin.

Concept of GA has so far been presented by different researchers in different forms, mostly in a mathematical framework [2]. The terms like gene, chromosome, cross over, mutation and generation, which are basically derived from biological origin, many a time, become confusing to readers of other disciplines and also it become difficult to appreciate how these processes will lead to an optimal solution. Therefore, in this article, first of all I am trying to present the basic philosophy of GA highlighting how it works and how the natural biological processes can help in obtaining the most optimal or at least near optimal solution through a simple analogy. I hope this analogy will help a GA beginner from any discipline to appreciate the idea of Genetic Algorithm and its techniques.

2.2 A Simple Analogy to GA

Let me place a question: if we want to find the highest possible intelligence level of a society in any time domain, can we really have an answer to it? Please note that the problem in hand is not to find the most intelligent person of a society among the present generation, rather it is to find the highest possible intelligence level of a society in any time domain, and the society is not a static entity rather it's a dynamic one and depending on various factors intelligence level of the next generation may become higher than the present generation and thus with progress of generation, the level of intelligence in a society is also expected to increase. If the problem was to find the most intelligent person of the society, then one could do it easily by setting a questionnaire objectively to test their intelligence level and then identifying the best one based on the mark they score in the test. As the problem is to find the highest possible intelligence level at any time domain, we will have to imagine about the future time as well, and can visualize how we can proceed to arrive at the highest possible intelligence level. In the discussion we will be drawing analogy to all typical terms of GA, so that reader can relate the biological process for moving towards optimal with the GA techniques.

Readers may consider themselves to be ruler of the society under consideration and imagine the following steps presented below:

1. As we do not know the intelligence level of any individual, we will have to consider that every member of the society has the potential to have the highest possible intelligence level and therefore, each one is a potential solution. To start with, let us, through some test, select a total of say 100 intelligent persons from the society comprising male and female, so that they can also play the role of good parents in the present generation. We can device different methods for selection of these 100 persons. In this case, the objective is to find the highest possible intelligence level. To move in a systematic manner, we can make 100 groups in the society and can conduct a test of intelligence in each of these groups to select the best from each group to make a total of 100 intelligent persons. To evaluate intelligence level, we can prepare a questionnaire for evaluating different attributes that determine intelligence level. For example, learning ability, inquisitiveness, alertness, analytical ability, far-sightedness, reflection to a situation, logical reasoning and intuitions are some of the genetic characteristics or variables that can be tested to determine fitness of a person for considering him/her as an intelligent person. While these genetic characteristics, which can vary from person to person are equivalent to variables of a problem, questionnaire set to check fitness of an individual based on value of these characteristics can be said as ***fitness function*** of GA. Combination of these genetic properties (***gene***) when placed together in sequence in the form of a string is referred as ***chromosome*** in GA as in bioscience genes are located on a chromosome. Thus in GA a chromosome is nothing but a number set formed by the value of the variables (gene); which may either be written in binary form or in decimal form. The procedure of selecting the proposed 100 best fit person can be regarded as the first step in GA, which is referred as ***selection of initial population*** of potential solutions. The process of dividing total population of the society into 100 arbitrary groups and selecting the best from each group to form a sample of initial 100 potential solutions from the current generation is analogues to the ***Roulette wheel*** selection technique. Figure 2.1 depicts the selection of initial 100 populations.
2. The above step has given us 100 intelligent persons, and we could of course select the top scorer among these 100 persons as the most intelligent person. However, our objective is to find the highest possible intelligence level in any time domain. This means we need to experiment on how the intelligence level can increase in the coming future generation. To form a new generation, let us proceed with the following steps.
3. Let us arrange marriage among male and female of these 100 intelligent persons from present generation. However, all male and female may not get involve into marriage and so we can assume that about 70% to 80% of these intelligent persons will get involve in marriage (Fig. 2.2). Because of mating of an intelligent male and an intelligent female there is possibility that the children born will inherit good genetic properties of both mother and father and will become more

Fig. 2.1 Selection of intelligent persons

Fig. 2.2 Marriage of intelligent male and female representing cross over operation of GA

Fig. 2.3 Birth of children with new genetic material

intelligent. However, if some of the children acquire only the bad properties of the parents then their intelligence level may become even inferior to their parents. Also, some time, while one child becomes intelligent the other may not become equally intelligent. Marriage of male and female to give birth to children and exchange of genetic properties between parents to form child of new generation is analogues to the ***cross over*** operation of genetic algorithm. However, as shown in Fig. 2.3, in GA we restrict the number of children to 2(two) from one pair of male and female (parents).

4. Although the children generally derive the genetic characteristics of parents, they can derive some new characteristics that were not there in their parents. In biological terms, such arbitrary change in some of the characteristics is called ***mutation***. While most of the children possess their parent's characteristic, only limited numbers of children in a society (say 2–3%) undergo changes in some of his/her characteristic through mutation. Thus mutation brings to a child some new characteristics that are not in her/his parents. Therefore, mutation probability, though generally very low, yet extremely important, as it can introduce new genetic characteristics to a person, which increases the chance of having a person of higher intelligence level than that of the parents. This happen naturally in biological system, and therefore, the children born from their parents have the chance to become more intelligent than their parents. This principle is followed in genetic algorithm as well.
5. Once the children are born, we need to check if they are more intelligent than their parents by the same procedure of determining intelligence level objectively. If the children are more intelligent, then their parents are replaced by the children (Fig. 2.4). Otherwise, to ensure that the intelligence level of the next generation never goes below the present generation, intelligent, and hence smart parents are allowed to continue as active member of the society in the next generation to help building a better intelligent society. Thus as member, the new generation will contain some intelligent smart persons from the previous generation and some intelligent children (Fig. 2.5).

Fig. 2.4 New generation replacing the parents

Fig. 2.5 Intelligent parents and children of new generation

6. Intelligence level of this new generation can now be tested by the same questionnaire and highest intelligence level of this generation can be evaluated. To have a better group, we may decide to select only those from this generation who bear a minimum level of intelligence. However, this will reduce the number of person in the current generation to produce the next generation. For example if only 80 out of these 100 intelligent person satisfy the minimum intelligence criteria then we will be having only 80 persons to get involved into the reproduction process and hence to generate the next generation. This limitation can be overcome by generating 20 additional clones (Fig. 2.6) of intelligent person in the current generation which is equivalent to allowing some intelligent persons to participate in more than one marriage (cross over). Thus our population size will remain the same.
7. The process of reproduction and evaluation as indicated from step 2 through 6 can be repeated for several generations until we observe that the increase in highest intelligence level of the society is no more significant with increase in generation. This can be considered as the highest possible intelligence level of that society. Interestingly, as we are repeatedly generating new generation of better intelligence level by changing genetic characteristics of each of the participating persons, it is expected that the all persons (potential solution) from the final generation will have combinations of genetic properties those will make their intelligence level very competitive and will be very close to the highest intelligence level. Thus one will have many possible alternate solutions.

Fig. 2.6 Formation of clones

2.3 Conclusion

Genetic Algorithm is now being used by researchers from different field for solving variety of optimization problems. In the analogy presented, we are testing intelligence level of a person based on the mark scored by her/him (potential solution) in the questions set for testing their learning ability, inquisitiveness, alertness, analytical ability, far-sightedness, reflection to a situation, logical reasoning and intuitions etc., which are considered as their genetic characteristics and values of these variables determine the fitness. It is worth mentioning that, once we get a value for each of these characteristics, calculation of intelligence level (Fitness value) does not introduce any mathematical complexities, irrespective of whatever weightage or whatever complex functions we use for calculating the fitness based on the value of these parameters (variables). Thus GA can be applied easily to any complex objective function or fitness function. Also in the ultimate solution, we get several potential solutions, which is another advantage of GA. With the increase in computational capability GA can now be used for handling large constraint optimization problem. Only limitation of GA is that to have quick conversion one should have some idea about the range of values of these variables. I hope the above explanation of Genetic Algorithm, elaborating how different procedures and GA techniques work in sequence to lead towards an optimal solution, will be helpful in appreciating the working principles of genetic algorithm in an easier way.

References

1. Davis L (ed) (1987) Genetic algorithms and simulated annealing. Pitman, London
2. Dejong KA (1985) Genetic algorithm: a 10 year perspective, in Grefenstette, 1985
3. Goldberg DE (1989) Genetic algorithm in search, optimization and machine learning. Addison Wesley, Reading
4. Holland J (1975) Adaptation in natural and artificial systems. University of Michigan Press, Ann Arbor
5. J Von Neumann (1966) Theory of self-reproducing automata, AW Burks – IEEE transactions on neural networks

Chapter 3
Interactive Genetic Algorithm to Collect User Perceptions. Application to the Design of Stemmed Glasses

E. Poirson, J.-F. Petiot, and D. Blumenthal

Abstract To avoid flops, the control of the risks in product innovation and the reduction of the innovation cycles require valid and fast customer's assessments. A methodology must be proposed to the designer to take into account the perceptions of the user. The method presented is based on an iterative process of user selection of representative CAD models of the product. An IGA is used to interpret the user's choices and introduce new products. In the center of this methodology, the user who, thanks to his decisions, will guide the evolution of the algorithm and its convergence. After a description of the IGA, a study on the convergence of the IGA is presented, according to the tuning parameters of the algorithm and the size of the design problem. An experiment was carried out with a set of 20 users on the application case proposed a steemed glass. The implementation of the perceptive tests and the analysis of the results, using Hierarchical Ascendant Classification (HAC) is described. The main contributions of the paper are proposals of (1) an interactive product optimization methodology; (2) a procedure for parameterizing interactive genetic algorithms; (3) a detection of perceptive trends that characterize customer expectations; (4) an experimental application on a real life product.

Keywords Interactive genetic algorithms · Shape design · Convergence · Perceptive tests · Design methodology · Genetic operators

---

E. Poirson (✉) · J.-F. Petiot
LS2N, Ecole Centrale de Nantes, Nantes, France
e-mail: Emilie.poirson@ec-nantes.fr; jean-francois.petiot@ec-nantes.fr

D. Blumenthal
AgroParistech, Massy, France
e-mail: david.blumenthal@agroparistech.fr

F. Bennis, R. K. Bhattacharjya (eds.), *Nature-Inspired Methods for Metaheuristics Optimization*, Modeling and Optimization in Science and Technologies 16,
https://doi.org/10.1007/978-3-030-26458-1_3

3.1 Introduction

To allow the companies to remain competitive, the current society imposes times of development of product/service shorter and shorter, products cheaper and cheaper and always better quality. The leitmotif of the Quality Function Deployment methodology is for illustration "the right product the first time and each time". QFD is a method of introducing quality at the design stage to satisfy the customer. The method helps product designers to explicitly identify the needs of consumers, correlate them with the technical characteristics given by the engineers, and evaluate the potential characteristics of the product compared to those already existing in the market [1]. The QFD is adapted for performance indications and product usages study. For example, [25] presents a methodology to mathematically calculate the relation between the technical customer requirements and the technical product characteristics. This methodology is used to fill the house of quality tool of the quality function deployment method during the product design and development.

This methodology is part of the user-oriented design research topic. One of the major points of this orientation is taking into account not only the expected performances but also the perceptions or emotions that the client wishes to experience. To understand and integrate them throughout the product development cycle, verbalization is regularly used. The language causes problems whether in differences of translation or definitions of the verbatim between qualifications of participants [18]. Indeed, on hedonic studies for which a population of naive subjects is questioned, the precision in the vocabulary misses accuracy compared to the expert subjects who used a combination of both descriptions and hedonic terms when describing a product (wine in [11]), indicating that they are better at communicating and describing what they like. It is particularly difficult to verbalize certain perceptions, and even more emotions. To overcome this difficulty, pictorial representations are used, as PrEmo [2], tools that uses illustrated characters represented the 5 basics emotions. These methods are part of the current Kansei Engineering, a method founded by Mr. Nagamachi at Hiroshima University about 30 years ago. Kansei engineering is used to quantify people's perceptions and to translate them into the design elements. Sensory tests, statistical methods, and AI techniques have been applied to formalize these relationships. Various modeling methods are proposed to provide useful design rules or trend prediction [17, 29].

Another branch of the literature of user oriented design is human-computer interactions where an algorithm gradually refines the propositions made to the users, based on their previous assessments. An example of this is interactive evolutionary computation (IEC), where the user is used as an evaluator in an evolutionary process [27]. In classical EC, a mathematical function evaluates the fitness (adaptation of the population to the environment). In IEC, this function is implicit and is provided by the assessments of the user. Since the user evaluates the fitness, there is no need for a prior mathematical function. The intervention of humans to replace the mathematical evaluation of fitness has been applied in many fields (music, writing, education, food industry, etc.) involving different sensory modalities. This IEC method has also been used in a number of fashion applications [15].

Let's firstly present the family of algorithms chosen for the study: the genetic algorithm.

3.2 Background on Genetic Algorithms

3.2.1 Definition

Genetic algorithms are evolutionary optimization methods developed originally by Holland [9]. They are based on iterative generations of population of individuals, converging step-by-step toward solutions. The term individual refers here to a specific set of values for the design variables. In other words, an individual is a specific candidate design. A metaheuristic is used to improve the current population of solutions. Based on the principle of Darwins natural evolution theory [3], the algorithm proceeds to a selection of parents, which spread their genetic dominant heritage in the next generation. The general principle of a genetic algorithm can be decomposed in 3 steps. (1) Consider an initial population of individuals randomly created, arbitrarily chosen by the designer, or generated by another calculation process. (2) This population iss evaluated according to the constraints and objectives formulated in the optimization problem. If the stopping criteria of the algorithm are satisfied, the algorithm stops; otherwise (3) the genetic operators are applied to this population in order to change it to a new population of the individuals that best meet the requirements. The steps 2–3 are repeated till satisfaction of the mathematical fitness.

Genetic algorithms are often used in the literature to explore design space, encourage creativity [23] or to help innovation [19].

3.2.2 Encoding of the Design Variables

Each chromosome represents a particular design. For each design, we can define design factor. Each factor can thus take different values called levels. A chromosome containing one level for each variable represents a particular combination, thus design. A binary string represents a variable where the length of the string depends on the number of allowed levels for the variables. For example, a product defined by 5 factors on 4 levels will be encoded by a vector of 5 patterns of 2 bits (4 combinations).

3.2.3 The Genetic Operators

Genetic operators have two objectives: to converge the current population of individuals to a set of optimal solutions (process efficiency) and to explore the largest proportion of the design space (process diversification). These operators are selection, crossover, and mutation. Each operator is performed randomly on individuals.

The selection is a recopy of an individual. In the implementation here, the crossover is a single point crossover: from two individuals, a point of crossing is selected and the two headpieces are swapped. One of the two children is randomly selected. The mutation is a change of one randomly selected factor value. The crossover rate (Rc), mutation (Rm), and selection (Rs) are real values chosen between 0 and 1 such that $Rc + Rm + Rs = 1$. Each selected chromosome will therefore undergo one of these 3 operations.

One of the main points of the genetic algorithm is the fitness function that allows the evaluation of the adaptability of the design in the environment. We have already discussed the difficulty of objectifying a function when studying perceptions. One solution is to replace the fitness evaluation by a human one, thus to let the system be interactive.

3.3 Interactive Genetic Algorithm

3.3.1 Synoptic of the IGA Process

Classically, the fitness evaluation of the individuals concerned is calculated numerically with a mathematical function known beforehand. A particular category of GA, Interactive Genetic Algorithms (IGA), introduces the user in the optimization loop to assess the fitness. During each iteration of the process, the user selects solutions that he/she considers to be the closest to the desired objective. After a number of iterations, the method may converge toward one or several solutions that fulfill the users objective. These algorithms are used to explore design spaces [13]. Since the user decides the individuals fitness, there is no need for a prior and unique formulation of the fitness function. For some applications, such as finding the semantic dimensions of a product, this advantage is crucial.

The general process of the IGA can be seen in Fig. 3.1. The IGA creates an initial population of designs by generating the chromosomes, and presents them to a user as CAD drawings. The number of individuals in a population is chosen by the experimenter according to the number of variables and levels and the size of the screen to represent the designs if visual experiment.

Based on personal criteria the user then selects a subset of these individuals. This number is also a part of the setting up. Selection consists in choosing,

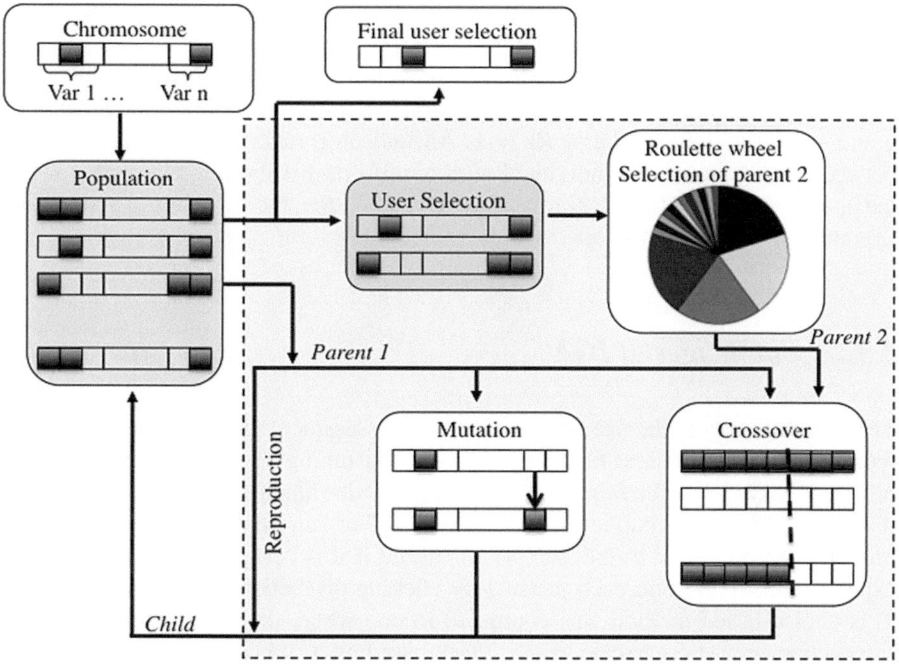

Fig. 3.1 Synoptic of the IGA process [14]

among the individuals of a given population, some solutions that seem better according to the constraints and objectives defined by the instruction of the test. This selected product are favored in the randomly election for the genetic operators. The most popular method of election is the roulette method, where individuals with better performances (selected by the user) have a higher probability of being selected (illustration in Fig. 3.1). The crossover operator consists in crossing several individuals with each other in order to generate better individuals. The mutation operator consists of modifying certain variables of a given individual in order to create a new, different individual and thus ensure that the search for solutions runs through the biggest design space as possible. Each generic operator has its own settings and the effectiveness of the generic algorithm strongly depends on these settings. Several studies are devoted to the influence of these parameters on the performance of the genetic algorithm [10, 12, 30].

By definition, in the MOGA-II algorithm [22], each chromosome of the population is a parent of the following generation, which gives robustness to the algorithm. A new population is thus created and evaluated, comparing each child to its parents. This iterative method runs until the program has reached the maximum of generations chosen by the experimenter or until the user estimates that the target has been reached.

The efficiency of the IGA is ruled by its 3 operators: crossover, mutation, and selection, chosen randomly for each individual of the population. The rate at which these operations occur is determined by the value of their weights. The crossover rate (*Rc*), mutation rate (*Rm*) and selection rate (*Rs*) are real values chosen between 0 and 1 such that $Rc + Rm + Rs = 1$. An indicator, rand(i), is randomly chosen between 0 and 1 for each individual with a uniform distribution. (1) rand(i) < Rc, the operation is a crossover. (2) $Rc \leq \text{rand(i)} \leq Rc + Rm$, the operation is a mutation. (3) rand(i) > $Rc + Rm$, the operation is a selection.

3.3.2 Challenges of IGA

The first difficulty in the use of IGA lies in the assessment of its fitness by the user: requiring the user to assess fitness can be time consuming or tiresome [26]. Gong [4] proposed to calculate the fitness by observing the time taken by the user to determine if an individual in a group is either "satisfactory" or "unsatisfactory". For example, products are presented to the user who is asked if it is "elegant". The users had to express their opinion on each product by clicking on "satisfied" or "not satisfied." This choice based method was compared to one where the users could rank their satisfaction on a sliding scale [6]. The task based on a slide required the evaluation of the entire population for each generation, which was tedious for the user. An individual's level of concentration influences their decision time, which can distort the results. Another way to assess the fitness is to use a choice-based scheme. In [24] the author's study a choice task for preference elicitation. The results showed that choices are practical to represent preference with a tolerable interaction effort and that algorithms based on binary choices are able to converge in a limited number of iterations.

Another difficulty in the use of IGA in user's tests lies in the convergence of the algorithm: an algorithm that is too slow may imply user's fatigued. To speed up the convergence, numbers of tools exist: krigging metamodel [16], chaining of neural network models and IGA [28] or cooperative interactive genetic algorithm based on the ratings of previous users [5]. If the acceleration of convergence is significant, it is limited to case studies with a reduced number of variables, because the choices can only be made on the products presented.

It is therefore, for our study, to find the right settings of the algorithm to have a reasonable convergence, that is to say, that give the user enough time to allow him to reach his target but not too much to tire him. To avoid the risk of premature convergence because of the selection operator, the mutation operator plays the role of "noise", jumping randomly in another part of the design space. Crossover will also explore new combinations. The difficulty is to balance their ratio.

3.3.3 Set up of the Genetic Algorithms

To determine the most adapted crossover rate (*Rc*), mutation rate (*Rm*) and selection rate (*Rs*), we define an automatic mode. A virtual user simulates to evaluate products. It is assumed that the user has an idea in mind and the more he advances in the experience, the closer he gets to it. We therefore suppose coherence in his reasoning and his choices. The goal for the user is to ensure that the proposed products are getting closer and closer to the target he has in mind. To simulate the choices of a virtual user, a target product was arbitrarily chosen in the design space: $t = [t_1, t_2, \ldots, T_{n\,var}]$. We calculate the distance to this target for each of the designs presented. The "selection" of the user is therefore logically composed of products that are the closest to the target. Calling this algorithm IGA is a misnomer since the algorithm is no longer interactive but for better understanding, automatic mode for IGA will be kept.

For each individual *j*, the distance *d(j,t)* to the target was defined by the Euclidian distance function:

$$d(j,t) = \sqrt{\sum_{i=1}^{n\,var} \left(rankx_{ij} - rankt_i\right)^2} \quad (3.1)$$

with: $rankt_i$: rank of the level of the variable t_i
$rankx_{ij}$: rank of the level of the variable x_i for individual j

By assumption, the Euclidian distance function used was considered as representative of the perceptual distance of the user (tests with other distance functions showed that the IGA parameters were not too sensitive to the nature of the distance function used). The automatic mode was used to tune the parameters of the IGA [20]. This was done in order to be able to launch several simulations in the same conditions, and to have an average estimate of the convergence rates. Knowing the convergence rates allows designers to limit the design space (the number of variables and their levels) to a reasonable size so that the number of generations required for convergence can be decreased. It was estimated that a subject could process 30 generations in manual mode before becoming fatigued.

The objective is to assess, for a fixed number of generations, the average quality of the solutions provided by the IGA. The IGA is used in *automatic* mode (the fitness is simulated by the computation of the Euclidian distance to a given target). For the optimal set of IGA parameters, the IGA process was simulated *N* times (Monte Carlo method). The quality of the solutions is assessed by two criteria: the average distance to the target of the best individual of the last generation *g*, and the average distance to the target of all the individuals of the last generation. The average distance to the target of the best individual of the last generation d_{best} is given by:

$$d_{best} = \frac{1}{N}\sum_{i=1}^{N} d^g\left(j^i_{best}, t\right) \tag{3.2}$$

$$j^i_{best} = \underset{j\in\{1,\ldots popsize\}}{\text{argmin}} \left(d^g(j, t)\right) \tag{3.3}$$

with: j^i_{best}: the individual minimizing the distance to the target for simulation i

For each iteration of the IGA, the two individuals which minimize this distance are automatically selected. These individuals are considered as the closest individuals to the target. According to this selection, the IGA generates a new population of individuals.

We arbitrarily consider that the IGA converges and then stops the process if the distance between an individual, generated by the IGA, and the target is lower than 2 (difference of 2 ranks). The parameters in input of the Automatic IGA are (1) the number of variables, (2) the number of level for each variable (the same for all variables), (3) the "roulette" wheel rate (weight of a selected parent to create the next generation), (4) the mutation and crossover rates (R_m and R_c). Each one has coherent intervals of variation. For each set of IGA parameters, the IGA process is simulated $N = 10$ times and the number of generations is measured. The mean value of this number of generation is computed, associated to the set of parameters. We estimated around 20 the number of generations which can be supported by a subject before fatigue. Consequently, we decided to select a size of problem defined by 5 variables and 5 levels for each variable for the rest of the study (the IGA, in automatic mode, needs 10 generations to converge for this size of problem). The best set of IGA parameters, for the size of problem, defined by 5 variables and 5 levels, with a Wheelrate of 16, $R_m = 0.15$ and $R_c = 0.8$.

The results show a global improvement of the fitness of the population according to the n° of generations: the average distance $d_{average}$ of a population to the target decreases with the number of generation and the quality of the population globally increases. The distance $d_{average}$ reaches a plateau above 60 generations (this distance is not null because the population is not homogeneous). The distance to the target of the best individual d_{best} also decreases, and is null above 60 generations. It signifies that above 60 generations, the method always converged toward the target for all the simulations ($N = 100$). For 30 generations, the distance d_{best} is approximately equal to 0.2. A study of the distribution of the distance $d^g\left(j^i_{best}, t\right)$ for all the simulations ($N = 100$) showed that 60% of the simulations converged perfectly toward the target, 30% converged toward the target with a difference of one level for one variable, and 10% converged toward the target with a difference of one level for two variables.

All these results make us confident in the ability of the IGA to converge when real subjects are used.

3.4 Application Case: Protocol and Results

3.4.1 Goal-Seeking Task

To provide an example of the methodology, a study was conducted on a simple product common to every day French life; a wine glass. The digital mock-up of the glass (Fig. 3.2) was defined by 5 dimensional variables (V_1 to V_5) with 5 levels.

Two tests were proposed to a panel of 20 subjects (age between 22 and 25), students at Ecole Centrale de Nantes: a goal seeking task and the research of the most "elegant" glass.

For both tests, the following conditions were set:

- The values of the IGA parameters: $Rw = 16$, $Rc = 0.8$, $Rm = 0.15$.
- The IGA were allowed to run through a maximum of 20 generations.
- If the subjects estimated that the task was fulfilled and the design selected satisfied, they could stop the test before the 20th generation.
- At the end of the entire process, the subject had to give the selected product a score from 0 to 10, corresponding to the quality of the solution, with respect to the task.

The type of perceptive test was a choice of maximum 2 (0, 1 or 2) products on 8 presented. It was dimensioned because of the size of the screen (3 lines of 3 products), and the fact that the target stay visible during all the test thus, take the ninth place.

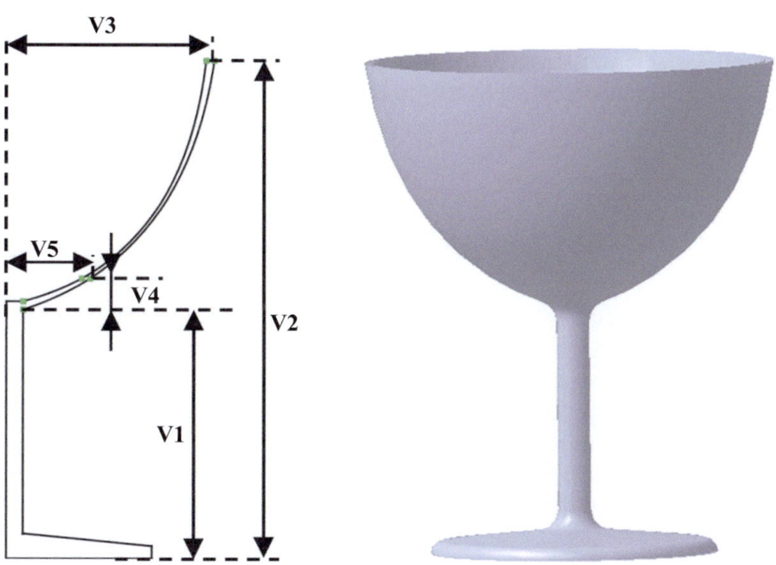

Fig. 3.2 Definition of the variables (V_1 to V_5) to parameterize the geometry of the glass

The test had three goals:

1. Warming up the users, familiarize them with the interface,
2. Confirming the performance of the method for the experimenter testing whether subjects were able to make the IGA populations converge toward a given goal,
3. Giving a feedback of the choice of the variables on this particular product.

Goal 1 and 2: Familiarization with the Task and Convergence of the Algorithm
The test lasted an average of around 8 min per user. The target needed an average of 16.75 generations to appear. We estimated that it was a reasonable duration for a perceptive test. Around half of the subjects (Pop 1) stopped the algorithm before the last generation, meaning that they thought they had reached the target. This information confirms the hypothesis of our study.

Goal 3
To estimate the discrepancy between the target and the choices of the subjects, for each variable of the 5 variables, the standard deviation is calculated on the rank of the variable to avoid a scale effect. For example, V1 could take $i = 5$ levels: $l_1 = 3.5$, $l_2 = 5$, $l_3 = 7$, $l_4 = 9$, $l_5 = 10$. The values themselves l_i are not used but only the level i are compared.

This standard deviation σ_j estimates, for each variable, the degree of agreement of the subjects' choices with the target (e.g. $\sigma_j = 0$ if all the subjects chose the same level than the target for the variable *Vj*). The results show that the 5 variables chosen do not have the same influence on the perceived shape of the glass. In Pop 1, everybody chose a target with the exact level for V1 and V3. The conclusion was that the variations of these variables on the form were easy for the user to perceive. In contrast, the choices were more heterogeneous on V4 (the height of the point controlling the form of the glass). The differences were not perceptible in the same way for all variables, at least for the representation mode chosen. All the variables don't have the same perceptive influence on the form of the product. This result is useful for the designer to distinguish his product on more impacting parameters.

3.4.2 *Free Task on "Elegant" Glass: Protocol*

In this test, the subjects were not given a target glass, but were told to select the most *elegant* wine glass, according to their personal opinions. We assume that each subject has his/her own target in mind, and makes consistent choices during the test (no change of target). The objective function to minimize corresponds to the distance between the target in mind and the products presented. For example, if the ideal of elegance is a kind of champagne flute, the subject will be attracted by glasses with a long stem, small diameter and elongated, rejecting the glass of Fig. 3.2 for example.

Figure 3.3 describes the process of IGA, showing the exchange of data between the subject and the algorithm. A population of 8 glasses was firstly generated. We

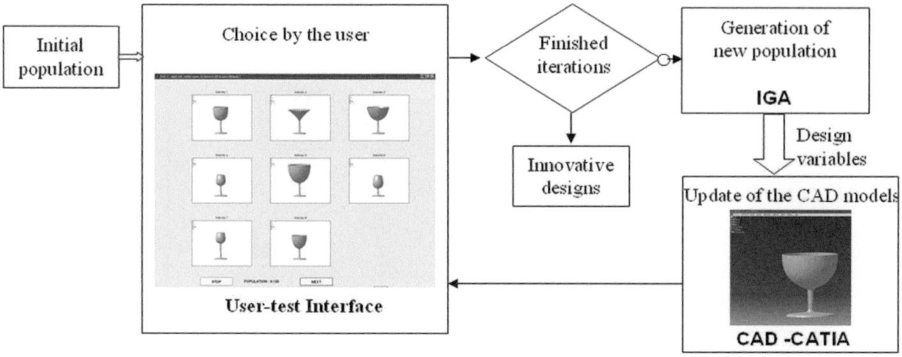

Fig. 3.3 Framework of the iterative user-test

keep 8 products, even if the 9th place dedicated to the target in the previous study is free, not to disturb the subject, to avoid being always considered as a target. This population is modeled in CAD tool (Catia V5 in our application) and presented to the subject via an interface (Matlab). Based on the choices done, a new population is generated and represented by new CAD modeling. The use of an Interactive Genetic Algorithm requires a constant exchange of data between the interface, the algorithm, the tool for product representation.

At the end of the process, the subject had to select one glass, and rate it on a scale from 0 (not at all elegant) to 10 (perfectly elegant). As in the previous test, the subject could stop the process and make their final choice before the 20th iteration (Fig. 3.3: "Finished iterations"? = yes).

The aims of the 2nd test were:

- To know to which extent the process converges toward a satisfying solution for the subject in the case of a simple product.
- To know if design trends concerning the elegance of a glass can be extracted from the results of the test. In other words, the problem was to uncover relevant information about the elegance concerning the design variables of the form.
- To estimate the variability of the results concerning the elegant semantic dimension, and to show how this information can be taken into account for product design.

3.4.3 *Free Task on "Elegant" Glass: Results*

Among the 20 subjects, 8 stopped the test before the 20th generation, giving a score of 10/10 to the selected product. These 8 subjects were considered as perfectly satisfied with their choice. For the 12 other subjects, the average satisfaction score of their chosen glass was 8.3/10. This relatively high score and the low standard

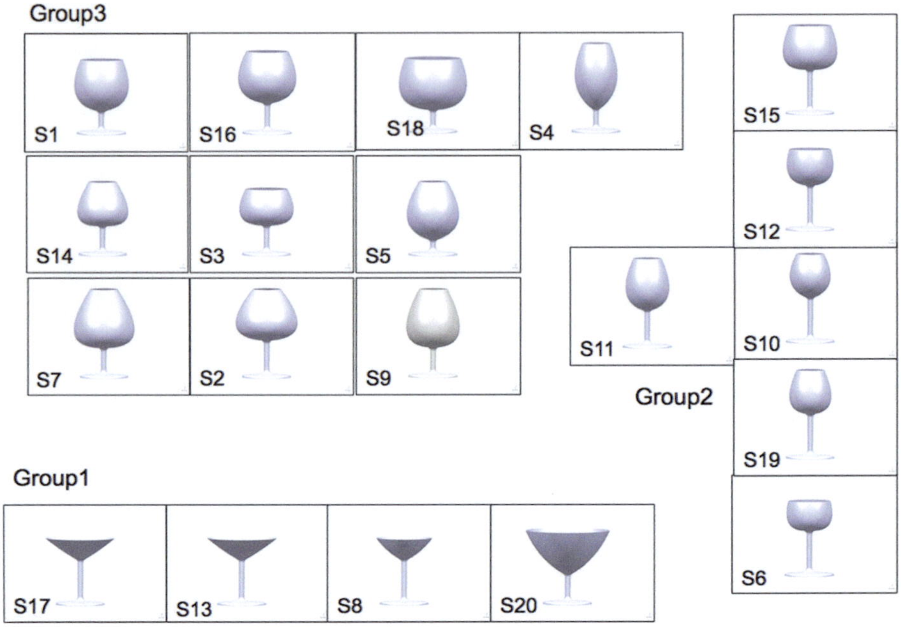

Fig. 3.4 Forms of the different glasses in the three groups

deviation imply that the subjects found satisfying enough glasses, and that 20 generations were sufficient to converge toward the desired product.

The first analyze considered the global population (20 subjects). Concerning the final glass chosen by the subjects, the results showed that only 2 subjects chose exactly the same glass. A great diversity in the final glass was noticed. The standard deviation of a subject compared to the 19 others was computed on each of the 5 levels. The results show that no variable was subjected to a great consensus concerning the elegance of the glass: the variability was of the same order for each variable, and no particular level could be considered as representative of the elegance. The conclusion led us to study more precisely the panel, searching for groups of subjects with similar image of elegance.

In order to provide a partition of the glasses and to define groups of targets similar from a perceptual point of view, a hierarchical ascendant classification (HAC) [7] has been performed. Details on this analyze can be found in [21]. To illustrate the potential of identifying homogeneous groups of subjects, three groups (HAC) were considered. The composition of the three groups is given in Fig. 3.4. While some of the users selected similar products, only S13 and S17 chose the exact same product.

To find a unique representation, a optimization was performed inside a group, aiming to minimize the dissatisfaction of the subject with the chosen representation. The results are presented Fig. 3.5.

The main result here is that the semantic dimension of elegance is subjective. There were at least 3 elegance types among only 20 students and on a total

Target 1 (Group 1)	Target 2 (Group 2)	Target 3 (Group 3)

Fig. 3.5 Forms of the 3 proposed glasses, minimizing the dissatisfaction

population of 5^5 glasses. A company could decide what type of elegance it wants for its glasses. That decision cannot be taken by the IGA methodology, as it is a strategic choice relating to the company's target values. One can also observe that the users have to be representative of the company's target user base. For example, Target 1 does not match a "traditional" wine glass form, which could signify that the users are not wine amateurs.

This test had to be considered as a pilot study, as the number of subjects was weak, and the product was relatively simple. This test allowed the description of the method and of the statistical data analysis, which could then used on a more substantial study with a more complex product.

3.4.4 Conclusions on the Two Tests on the Glasses

A goal-seeking test was carried out to give an estimation of the convergence abilities of the IGA, and revealed the problem of just noticeable differences in perception of forms. In Test 1, within a small number of generations, 80% of subjects either found a glass similar to the target, or one that they perceived to be the target (as they stopped the task before the end) in a less than 10 min. Perceptual tests with 20 subjects were conducted to demonstrate the validity of IGA and to conduct an analysis of the results. The perceptual tests on the semantic dimension – elegance – of a glass were performed to show how the results can be analyzed and how to deal with inter-individual differences among users.

3.5 Synthesis and Perspectives on the Use of IGA for Design

The interest of IGA was previously demonstrated underlining the opportunities for design. If the case studied was simple, the methodology is adaptable to very different and more complicated product or system. The different steps must be broken down according to the study.

Preliminary step: product and representation.

The first step is, the product/system defined, to choose which sensorial organ will be activate for the perceptive tests. In the example above, the visual sense was used but hearing can also be studied (concerning touch, smell and taste, the difficulty of generating new samples is for the moment an obstacle to the use of IGAs). The combination of both senses can be also imagined. Each representation has to be coherent with the goal of the designer. For example, a screenshot of a CAD model can be if needed completed in Virtual Reality Modeling Language (VRML), or a control of the sound level can be given to the subject. The interest of this method is that the representation of the product is independent of the algorithm (Fig. 3.3). The interface of this study is functional, and the computation time for updating the model is reasonable; in the wine glass test, the time between generations (time to built the 8 CAD-models) was less than 8 s. The interface with the CAD software is a significant advantage. Replacing the glass by another product of 5 variables on 5 levels is immediate. To increase or decrease some of these values, a modification of the set up of the algorithm is simply needed.

Step 1: Variables and levels

Not totally independent of the previous step, the product must be configurable. The IGA can help the designer to reduce the complexity of his model. In the goal-seeking task, the subjects showed the convergence and divergence of results, meaning that their comprehension of the variables were different. We could therefore consider some parameters as not influential for the perceived quality of a product, which allows the designer to reduce his design space.

Step 2: parameterization of the algorithm and protocol

IGA implies the evaluation of the subject as fitness function. The human fatigue must be taken into account in the protocol, designing a tool limiting the duration and difficulty of the task. For the duration, the convergence must be studied, depending on the different parameters of the algorithm (mutation, crossover rates, wheelrate in the case of Moga II). The targeted task is an adapted tool for this step. Concerning the difficulty of the task, the choice made is a selection of 2 (maximum) on 8 products. If it seems affordable for a naïve subject, the quantity of information is not very important, especially increasing the number of variables, thus the design space. Other interventions of the user can be considered in the algorithm, not only on the evaluation of the fitness but also on the set up of the algorithm. For example, the subject could stop, even momentarily, the evolution of a criterion (variable) or reduce its scale and accelerate the convergence. He can also be asked the to score at the end of each step the product(s) he/she chose and one or two randomly chosen product from this generation. Hybrid versions of IGA, mixing evaluation of the fitness by or the machine or the human (not exclusively human), are also studied in this way.

Step 3: analysis of results

A phenomenal amount of information is generated by such perceptive tasks. In the previous study, the distances between the selected product and the target or between the final population and the target is computed and studied. The IGA makes it possible to investigate the inter-individual differences for a group of users in the relationship between the semantics and the design features. It is possible to separate which attributes are common for the entire group, which attributes are specific to a subgroup of subjects, and which attributes are typical of a particular user. Other results can be exploited as the not chosen products. If a common characteristic appears in all the choices of the subject, he could reject some values of parameters, that is also very useful for the designer. The choice process could be redefine by asking the subject to select 3 categories of products: desired products, neutral products, rejected products. Modeling the user's path in the presented database can help to predict its next choices and thus accelerate convergence.

3.6 Conclusion

The general process for detecting design trends using Interactive Genetic Algorithms and user-tests was described. These tests are proposed to enhance innovation and to understand which design features are representative of a given semantic dimension. Applied to a simple example, it is adapted to other product and system (for example, layout problem [8]).

A limitation of the IGA algorithm used concerns the relative small size of the design space. To tackle bigger design spaces, several perspectives can be drawn. Concerning the algorithm, we could fit a model of the response of the subject (for example with a neural network) during the assessments, and optimize this response with classical GA. The tools of machine learning can be used to enrich the perceptive tests and modelize the behavior of the subject.

References

1. Aungst S, Barton R, Wilson D (2003) The virtual integrated design method. Qual Eng 15:565–579
2. Desmet P (2003) Measuring emotion: development and application of an instrument to measure emotional responses to products, Human-computer interaction series, 3. Springer, Dordrecht
3. Goldberg DE (1989) Genetic algorithms in search, optimisation and machine learning. Addison Wesley, Reading
4. Gong DW, Pan FP (2003) Theory and applications of adaptive genetic algorithms. China University of Mining and Technology, Xuzhou
5. Gong D, Zhou Y, Li T (2005) Cooperative interactive genetic algorithm based on user's preference. Int J Inf Technol 11:1–10

6. Gong DW, Guo GS (2007) Interactive genetic algorithms with interval fitness of evolutionary individuals, dynamics of continuous, discrete and impulsive systems, series B: complex systems and applications-modeling. Control Simul 14(s2):446–450
7. Hair JF, Tatham RL, Anderson RE, Black W (1998) Multivariate data analysis, 5th edn. Prentice Hall, Upper Saddle River
8. Hasda RK, Bhattacharjya RK, Bennis F (2017) Modified genetic algorithms for solving facility layout problems. Int J Interact Des Manuf (IJIDeM) 11(3):713–725
9. Holland JH (1992) Adaptation in natural and artificial systems: an introductory analysis with applications to biology, control and artificial intelligence. MIT Press, Cambridge, MA
10. Hong TP, Wang H-S, Lin W-Y, Lee W-Y (2002) Evolution of appropriate crossover and mutation operators in a genetic process. Appl Intell 16(1):7–17
11. Hopfer H, Heymann H (2014) Judging wine quality: do we need experts, consumers or trained panelists? Food Qual Prefer 36:1–2
12. Jilkova J, Raida Z (2008) Influence of multiple crossover and mutation to the convergence of genetic optimization. MIKON 2008, XVII international conference on microwaves, radar and wireless communications in Poland
13. Kelly JC, Wakefield GH, Papalambros PY (2011) Evidence for using interactive genetic algorithms in shape preference assessment. Int J Prod Dev 13(2):168–184
14. Kelly J, Papalambros PY, Seifert CM (2008) Interactive genetic algorithms for use as creativity enhancement tools. In: Proceedings of the AAAI spring symposium, Stanford, CA, pp 34–39
15. Kim HS, Cho SB (2006) Application of interactive genetic algorithm to fashion design. Eng Des 38:224–237
16. Li M, Li G, Azarm S (2008) A kriging Metamodel assisted multi- objective genetic algorithm for design optimization. ASME J Mech Des 130(3):031401
17. Nagamachi M (1995) Kansei engineering: a new ergonomic consumer-oriented technology for product development. Int J Ind Ergon 15:3–11
18. Poirson E, Petiot J-F, Richard F (2010a) A method for perceptual evaluation of products by naive subjects: application to car engine sounds. Int J Ergon 40(5):504–516
19. Poirson E, Petiot J-F, Aliouat E, Boivin L, Blumenthal D (2010b) Interactive user tests to enhance innovation; application to car dasboard design. International conference on kansei engineering and emotion research KEER 2010
20. Poirson E, Petiot J-F, Aliouat E, Boivin L, Blumenthal D (2010c) Study of the convergence of Interactive Genetic Algorithm in iterative user's tests: application to car dashboard design. In: Proceedings of IDMME – virtual concept 2010 Bordeaux, France
21. Poirson E, Petiot JF, Boivin L, Blumenthal D (2013) Eliciting user perceptions using assessment tests based on an interactive genetic algorithm. J Mech De Am Soc Mech Eng 135(3):1–16
22. Poles S, Rigoni E, Robic T (2004) MOGA-II performance on noisy optimization problems. In: Proceedings of the International conference on bioinspired optimization methods and their applications, BIOMA2004, 11–12 October 2004, Ljubljana, Slovenia, pp 51–62
23. Qian L, Ben-Arieh D (2009) Joint pricing and platform configuration in product family design with genetic algorithm. In: Proceedings of IDETC/CIE 2009, San Diego, CA, USA
24. Ren Y, Papalambros PY (2011) A design preference elicitation query as an optimization process. J Mech Des 133(1):111004
25. Shabestari SS, Bender B (2017) Enhanced integrated sensitivity analysis in model based QFD method. In: Proceedings of the 21st international conference on engineering design (ICED 17), Vancouver, Canada, 4, pp 317–326
26. Swait J, Adamowicz W (2001) The influence of task complexity on consumer choice: a latent class model of decision strategy switching. J Consum Res 28:135–148
27. Takagi H (2001) Interactive evolutionary computation: fusion of the capabilities of EC optimization and human evaluation. Proc IEEE 89(9):1275–1296

28. Tseng I, Cagan J,Kotovsky K (2011) Learning stylistic desires and generating preferred designs of consumers using neural networks and genetic algorithms. DETC2011-48642, ASME IDETC – design automation conference, Washington, DC
29. Yoshida S, Aoyama H (2008) Basic study on trend prediction for style design. ASME International Design engineering technical conferences, Brooklyn, New York, USA
30. Zhang J, Chung HSH, Zhong J (2005) Adaptive crossover and mutation in genetic algorithms based on clustering technique. In: Proceedings of the 7th annual conference on genetic and evolutionary computation, GECCO'05, Washington DC, USA – June 25–29. ACM, New York, pp 1577–1578. ISBN:1-59593-010-8, https://doi.org/10.1145/1068009.1068267

Chapter 4
Differential Evolution and Its Application in Identification of Virus Release Location in a Sewer Line

B. G. Rajeev Gandhi and R. K. Bhattacharjya

Abstract Differential Evolution is a stochastic, population-based optimization algorithm for solving the nonlinear optimization problems. The algorithm was introduced by Storn and Price (Minimizing the real functions of the ICEC 1996 contest by differential evolution. In: Proceedings of IEEE international conference on evolutionary computation, pp 842–844). Later, the algorithm was explored, and problem specific modifications have been done by many researchers. The beauty of this meta-heuristic optimization algorithm is that the optimization search can be made to be bounded, global as well as local search by changing the parameters of the algorithm. This chapter gives the basic idea behind the origin of this technique and its recent advancements. The algorithm is then applied for identification of unknown virus release locations in an underground sewer line.

Keywords Virus transport · Leaking sewers · Simulation-optimization · Source identification

4.1 Introduction

Many natural processes have optimization at their core. The Darwinian evolution is based on the optimization principle as the statement "survival of the fittest" states the same. Nature has the perfect balance between the number of resources available and the lives that depend upon these resources. If one increases that decreases the other eventually. This creates a necessity for optimization in all the natural processes. The observation and careful understanding of natural processes lead to a completely different approach of optimization which is known as nature-inspired metaheuristic optimization techniques. The gradient based classical optimization algorithms are

---

B. G. R. Gandhi (✉) · R. K. Bhattacharjya
Department of Civil Engineering, Indian Institute of Technology Guwahati, Guwahati, Assam, India
e-mail: b.rajeev@iitg.ac.in; rkbc@iitg.ac.in

F. Bennis, R. K. Bhattacharjya (eds.), *Nature-Inspired Methods for Metaheuristics Optimization*, Modeling and Optimization in Science and Technologies 16,
https://doi.org/10.1007/978-3-030-26458-1_4

problem dependent and thus works for a specific problem only. As such, it may not be suitable for all types of problems. On the other hand, the metaheuristic approaches are problem independent and work well for any type of problem [4].

There are many algorithms which aim at different processes happening in nature. Some of the algorithms concentrate on the genetic part of the natural evolution called Genetic Algorithms (GA). Some of the algorithms are inspired from the different processes that happen in a particular species such as bird flocks giving Particle Swarm Optimization (PSO), fireflies (Fire Fly Algorithm), leaping of frogs for food (Shuffled Frog Leaping Algorithm) and building of ant colonies (Ant Colony Optimization) *etc*. One more algorithm coming from the family of evolutionary algorithms is the Differential Evolution (DE) that has emerged in the mid-1990s. The DE algorithm was first presented as a technical report by R. Storn and K.V. Price in 1995 [8]. The DE algorithm was demonstrated at the First International Contest on Evolutionary Optimization in May 1996 [9]. DE received the third price on the first ICEO held at Nagoya, Japan and it was given the first price on the second ICEO [7]. By the next decade after DE was proposed, many variants and improvements were published such as Self-Adaptive DE [1], Opposition based DE [13], DE with global and local neighborhood [10], JADE [5], *etc*.

The Differential Evolution comes under the same category of the Evolutionary Strategies with the scaled difference between the parent vectors or genomes from the previous generation as input. Thus, this algorithm resembles mostly of the Controlled Random Search algorithm. There are many advantages for researchers to use DE as an optimization tool. One of the advantages is the algorithm is very simple to understand and also to be coded, which allows a wide range of applications in many fields of engineering and sciences. It can be programmed easily with minimal programming skill. The DE algorithm also uses a smaller number of parameters (F and Cr), which can be easily handled. The significance of the parameters is well understood and can be explored by running the program under different conditions. The section "Parameters and Sensitivity" gives much more details on their significance in the convergence of DE towards the optimal solution. In this chapter, DE is presented starting with the basic concepts, different structures and the applications of DE on a large-scale optimization problem.

4.2 Structure of the Algorithm

The algorithm DE consists of four basic steps – initialization of the population, mutation using difference vectors, recombination, and selection. Once the initialization of the population is made, the rest of the three steps continue until the termination criterion is not reached. This whole process is explained in the flowchart shown in Fig. 4.1. Each step of the algorithm is explained in the subsequent subsections.

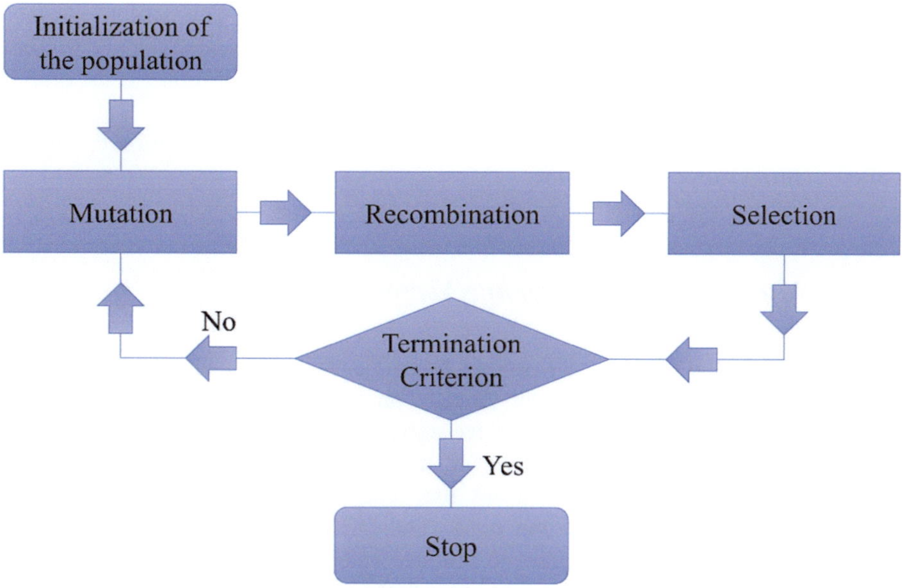

Fig. 4.1 Differential evolution algorithm

4.2.1 *Initialization of the Population*

The initial population of all nature inspired meta-heuristic optimization algorithms are created randomly in the feasible search space according to the problem. Let the search space be a region '***R***' of ***D***-dimensional space, where '***D***' is the number of decision variables. The number of decision variables and the bounds of the variables are decided according to the problem to be solved. '$\boldsymbol{X_d}$' is the vector containing the decision variables. The population decided for the problem contains a set of these vectors equal to the total number of population (M). Therefore, the whole population for each of the generation '***G***' can be represented as shown in Eq. 4.1.

$$\left\{X_{d,p}^{G}\right\} = \left\{x_{1,p}^{G}, x_{2,p}^{G}, x_{3,p}^{G}, \ldots, x_{D,p}^{G}\right\} \tag{4.1}$$

Here 'p' extends from 1 to P (total population) and 'G' extends from 1 to Gen (total generations) respectively.

If the search space is bounded, the minimum and maximum of each of the decision variables exists which can be represented the same way as vectors '$\boldsymbol{X_{min}}$' and '$\boldsymbol{X_{max}}$' respectively. The initial population can be generated for the specified bound of the variables as shown in Eq. 4.2.

$$x_{d,p}^{1} = x_{d,\min} + rand_{d,p}(0,1) \times \left(x_{d,\max} - x_{d,\min}\right) \tag{4.2}$$

'$rand_{d,p}$' is a random number generated between 0 and 1. The random number distribution is assumed to be a uniform distribution. This generates a set of vectors for each of the population assuming the search space to be continuous. If the search space is discrete (say integers only), then an integer random number is generated between the minimum and the maximum of the bounds. The population created is called a chromosome or a genome.

4.2.2 Mutation with Difference Vectors

The mutation in the evolutionary process resembles the sudden change in the gene, instructing the gene to do different things than it is supposed to perform. These mutations can cause the gene to survive extreme conditions, which can help the chromosome to survive longer. However, the mutations are not always effective, because too much of mutation causes the gene to behave completely different and leads to the extinction of the gene. The DE adopts the mutation property from evolution to change the genes to produce better population than the previous generation. The mutation is generally carried out with the scaled difference of the vectors in interest. There are 5 types of mutations with the difference vectors [12]. The best and mostly used one is the random mutation, which can be executed using Eq. 4.3. The random mutation is started by selecting three random vectors '*r1*, *r2* and *r3*' from the population. The scaled difference of these vectors '*r2* and *r3*' with the scale factor '*F*' (which is a positive number less than 1) is taken and added to the initial random vector '*r1*'. The random vector '*r2* and *r3*' give the direction to which the search should perform, and '*F*' gives the size of step length. The random mutation changes the population towards the optimal solution over a number of generations.

$$v_{d,p}^{G} = x_{d,r1}^{G} + F\left(x_{d,r2}^{G} - x_{d,r3}^{G}\right) \tag{4.3}$$

There are other types of mutations with the difference vectors used by different researchers. Two of them are listed in Eqs. 4.4 and 4.5. The mutation process that creates a vector close to the best vector in the current generation is shown in the Eq. 4.4 [6]. The mutation that takes the current chromosome of interest to the best in the current generation is shown in Eq. 4.5 [6].

$$v_{d,p}^{G} = x_{d,best}^{G} + F\left(x_{d,r1}^{G} - x_{d,r2}^{G}\right) \tag{4.4}$$

$$v_{d,p}^{G} = x_{d,p}^{G} + F\left(x_{d,best}^{G} - x_{d,p}^{G}\right) + F\left(x_{d,r1}^{G} - x_{d,r2}^{G}\right) \tag{4.5}$$

Here only two random vectors are generated choosing two integers generated randomly between 1 and P (the total population). There are also other types of mutations, which are mostly a combination of one or more than one of the above three listed mutations. Some require 5 random populations; some require 4 and some require 6 or 7 populations. However, the basic idea of the mutation is to create a better vector to replace the weaker solutions in the population. As long as the mutated population is better, any type of mutation can be adopted.

4.2.3 Recombination or Crossover

The recombination allows the mutated vector, called the target vector to get mixed with the population to form an offspring vector denoted by 'U'. The recombination is also called as 'crossover' because the vectors from parent generations are used to produce the offspring. There are two common types of crossover used by the researchers. One of which is a simple binomial crossover and the other is the exponential crossover. The simple binomial crossover is given in Eq. 4.6. It uses the mutated vector on one hand and the actual vector on the other hand combined with a crossover rate and a random number as shown in Eq. 4.6.

$$u_{d,p}^{G} = \begin{cases} v_{d,p}^{G} \ \ if \ \ \text{rand}\,(0,1) <= C_r \ \ or \ \ d = d_{rand} \\ \qquad x_{d,p}^{G} \ \ otherwise \end{cases} \tag{4.6}$$

Here, 'd_{rand}' is a random number generated between 1 and D (number of decision variables). 'C_r' is the crossover rate chosen between 0 and 1. This condition for the crossover rate allows the certain number of mutated population to enter the population from the previous generation. The condition that the chosen decision variable is equal to the randomly generated decision variable ensures that not all the decision variables are changed in the current population. This condition also allows the population to try a different combination of decision variables to solve the problem. This is very helpful when there are a very large number of decision variables.

The other type of crossover that is quite useful for large scaled optimization problems is the exponential crossover. In this crossover, an integer 'n' between 1 and D is chosen at random. The nth vector is the starting point to search for a better solution. Another integer 'L' is chosen between 1 and D, but this time with a choosing probability according to Eq. 4.7.

$$P\,(L = \nu) = (C_r)\,\nu - 1 \tag{4.7}$$

Here, 'C_r' represents the crossover rate which can be any fraction between 0 and 1. Whereas 'ν' is any integer between 1 and D. This probability can be executed in the way as given in the pseudo code below [11].

L = 0;

While (rand(0,1) <= C_r and L<= D)

L = L+1;

After choosing the random integers '*n*' and '*L*', the vector for the next generation can be chosen by recombination as shown in Eq. 4.8 where $\langle n\rangle_D$ represents modulo function with modulus D.

$$u_{d,p}^{G}=\begin{cases} v_{d,p}^{G} \;\; if \;\; j=\langle n\rangle_D, \langle n-1\rangle_D, \langle n-2\rangle_D, \ldots, \langle n-L+1\rangle_D \\ x_{d,p}^{G} \;\; otherwise \end{cases} \tag{4.8}$$

4.2.4 Selection

The selection decides whether the offspring vector is suitable for the population in the next generation. The condition for survival of the offspring is explained in the Eq. 4.9. The fitter vector will replace the original one using the following rules.

$$X_p^{G+1}=\begin{cases} U_p^G \;\; if \;\; f\left(U_p^G\right)\le f\left(X_p^G\right) \\ X_p^G \;\; if \;\; f\left(U_p^G\right)> f\left(X_p^G\right) \end{cases} \tag{4.9}$$

Here, the less than symbol ensures the population to converge towards the best solution and the equal to symbol ensures the migration of population in between the vectors corresponding to flat fitness. After the selection of the offspring for the next generation, the termination criterion is checked and the whole cycle continues until the optimal solution is reached.

4.3 Parameters and Sensitivity

Differential Evolution is very easy to code because of the simple nature of the algorithm. Also, a few parameters are required to be adjusted. There are only two parameters that make DE one of the most successful metaheuristic algorithms. The mutation with difference vectors takes care of the wide range of search space with the parameter '*F*'. The crossover rate 'C_r' takes care not to replace all the population by the new population created by the mutation operator. The combination of mutation, recombination and selection always assures population to converge and reach the optimal solution of the problem. But, to assure convergence, the user should select the best combination of these two parameters. To analyze the best combination of the parameters, a two variable objective function is chosen (the sphere function) with only one optimum at (0, 0). The mutation method used is random mutation and the crossover is the binomial crossover. Three values of F are chosen with three crossover rates each giving rise to nine combinations of choices for the chosen values of F and C_r.

The rates of C_r are chosen to be 0, 0.5 and 1. The C_r value of 0 will almost not allow the new vectors to enter the population in the next generation. The rate 0.5 allows the new vectors to be in the population of next generation by almost 50% and C_r value of 1 will allow 100% replacement of the old population. The values of F are also chosen to be 0, 0.5 and 1. Here, the value of 0 tells that the new vector is nothing but the randomly chosen vector or the best vector, according to the mutation method used. The value of 0.5 indicates that the difference of two random vectors is scaled to be half of the actual difference, which gives the wide range for the search by creating the difference vector not too far when the difference is high enough. The value of 1 indicates that there is no scaling of the difference vectors used to get the new population. Figure 4.2 shows all the nine different combinations of the parameters F and C_r.

It can be observed from the figures that as the crossover increases, the randomness in the population is also increasing. The crossover rate of 0.5 shows fast convergence of the solutions. Although the crossover rate of 0 shows good results towards convergence, the population gets stagnated more in the same place, as this allows the population to change very little in the generations next to come. The only replacement in the population is due to the other condition where the chosen random number is equal to the variable to be replaced. That is the reason for the '+' pattern observed for the lower rate of crossover. As the crossover rate increases, the pattern slowly increases towards randomness. From the Fig. 4.2, it can also be observed that the value of F has a lesser impact on the behavior of the population, as it is just a scaling factor for the difference vector that can either stay or not stay in the population of next generation based on the condition for crossover. However, a higher value of F ensures the required widening of the search space by scaling the difference vectors. An optimal search operation requires faster convergence and a smaller degree of randomness in the population. This ensures to check all the local optimal solutions and to decide on the global optimal based on the information from all the local optima. Thus, the value of F and Cr can be chosen to be greater than 0.5 for better results.

Some researchers also prefer to alter the values of C_r and F as the DE operation goes on [11]. These variable parameters improve the population and help to achieve the optimal solution faster. For example, at the beginning of the search, a larger value of F can be chosen to explore the entire search space. But as the solution convergence towards optima, the value of F can be reduced iteration after iteration to obtain the exact optimal solution of the problem. This helps the algorithm to get the solution faster. Similarly, the crossover rate should be increased as the solution reaches the actual optimal. This helps more population to get replaced from the previous population allowing the algorithm to search for more solutions in lesser time. When it comes to solving large scale optimization problems or problems with a larger number of decision variables, it is almost always necessary to change the parameters as the algorithm progresses.

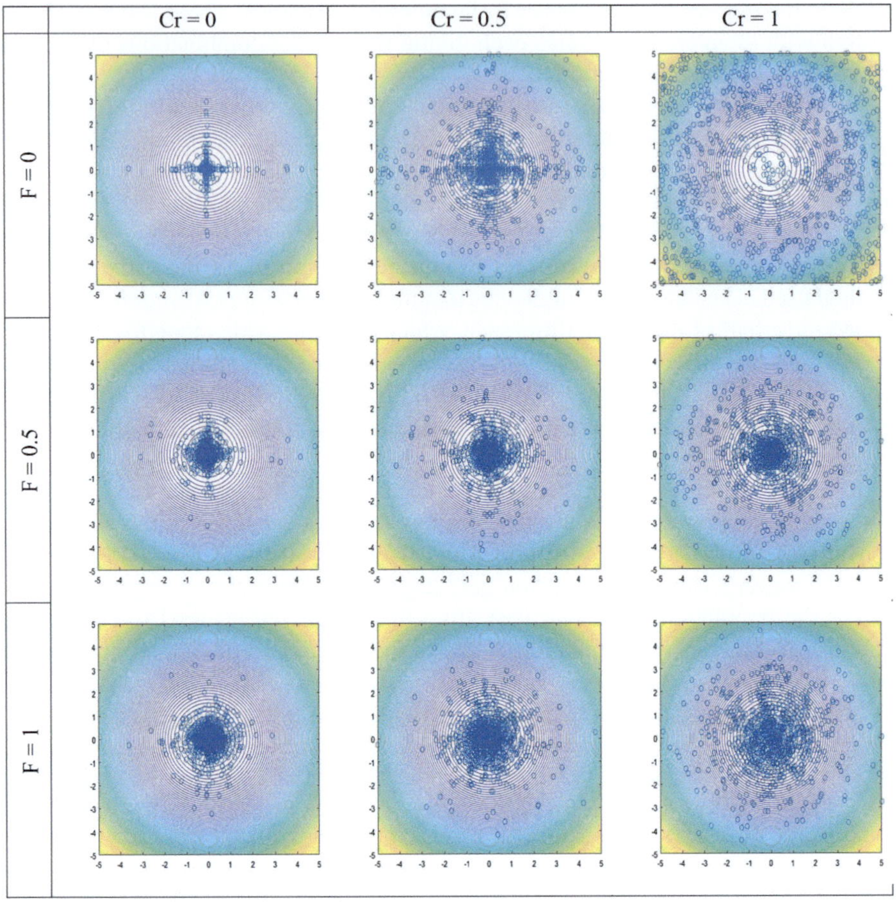

Fig. 4.2 Nine combinations of the parameters of DE on sphere function on 10th iteration

4.4 Differential Evolution on Mathematical Functions

In the previous subsection, the sensitivity of the model parameters on the sphere function has been discussed. The combination of the parameters and their effect on the convergence have also been discussed. In this subsection, the performance of DE on some of the popular mathematical functions is discussed. The mathematical functions are chosen in such a way that DE can be tested on different kinds of problem such as problems having alternate optimal solutions, problems having local and global optimal solutions, and the problems with single optimal solution. The functions considered are Cross-in-tray function, Rastrigin function and the Goldstein-Price function.

4.4.1 Cross-in-Tray Function

The cross-in-tray function is a multi-optimal function with four optimal points in the domain −10 to 10 in both the directions. The function is given in Eq. 4.10 and the minima are given in Eq. 4.11.

$$f(x, y) = -0.001\left[\left|\sin x \sin y \times e^{\left(\left|100-\frac{\sqrt{x^2+y^2}}{\pi}\right|\right)}\right| + 1\right]^{0.1} \tag{4.10}$$

$$\text{Min} = \begin{cases} f(1.34941, -1.34941) & = -2.06261 \\ f(1.34941, 1.34941) & = -2.06261 \\ f(-1.34941, 1.34941) & = -2.06261 \\ f(-1.34941, -1.34941) & = -2.06261 \end{cases} \tag{4.11}$$

The contour and the surface plots of the cross-in-tray function are given in the Fig. 4.3a, b. It can be observed in the Fig. 4.3b that the function represents a cross in the tray containing egg holder like shapes. The four minima can be seen in the contour plot in the Fig. 4.3a. The function is solved by using the differential evolution taking the F and Cr values to be 0.5 and 0.7 respectively. The population size is 10 and the model is run for 500 generations. The variation of the function values with the generation for 20 runs is given in the Fig. 4.4.

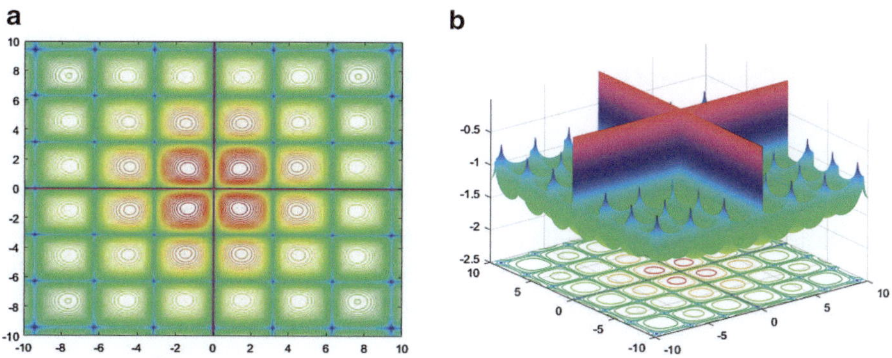

Fig. 4.3 (**a**) Contour plot of Cross-in-tray function. (**b**) Isometric view of Cross-in-tray function

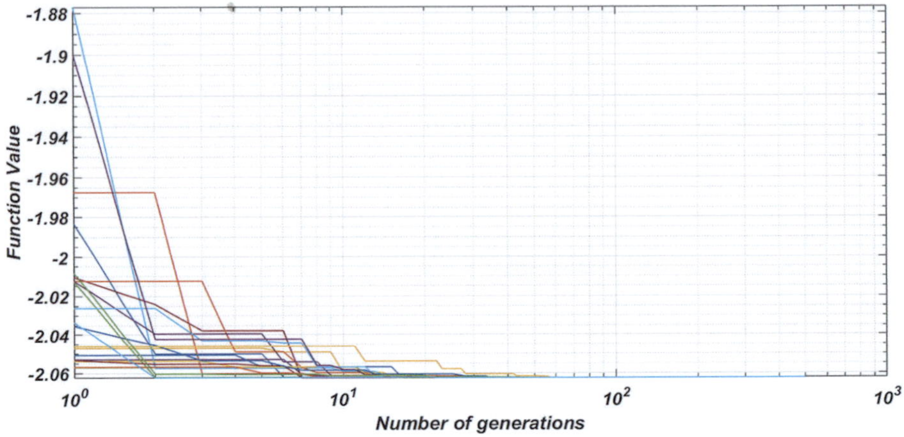

Fig. 4.4 Logarithmic plot of the function value with the number of generations over 20 runs

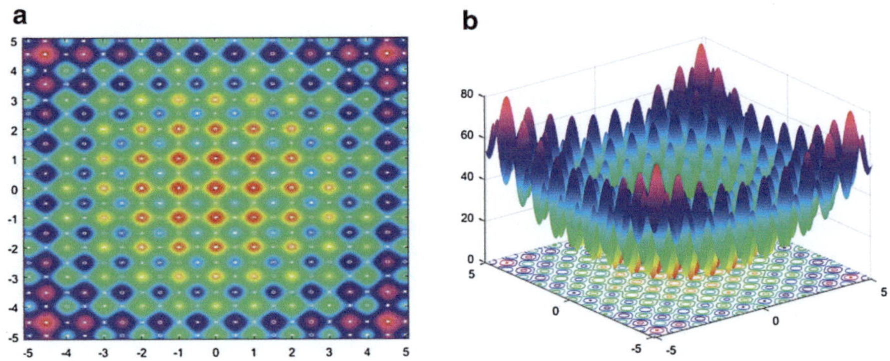

Fig. 4.5 (**a**) Contour plot of Rastrigin function. (**b**) Surface plot of Rastrigin function

4.4.2 Rastrigin Function

The Rastrigin function is a single global optimal function with many local optima. The domain is taken to be $-5.12 <= x, y <= 5.12$. The equation representing the function is given in the Eq. 4.12 and the minimum of the function is zero at (0, 0).

$$f(x, y) = 2A + x^2 + y^2 - A\cos(2\pi x) - A\cos(2\pi y) \tag{4.12}$$

The contour plot and the surface of the function as a plot in 3D are given in the Fig. 4.5a, b respectively. The optimization to find the minima is done using differential evolution with Crossover rate 0.7 and F value at 0.5. The population size and the number of generations are taken to be 10 and 500 respectively. Figure 4.6 shows the function value with the generations over 20 runs.

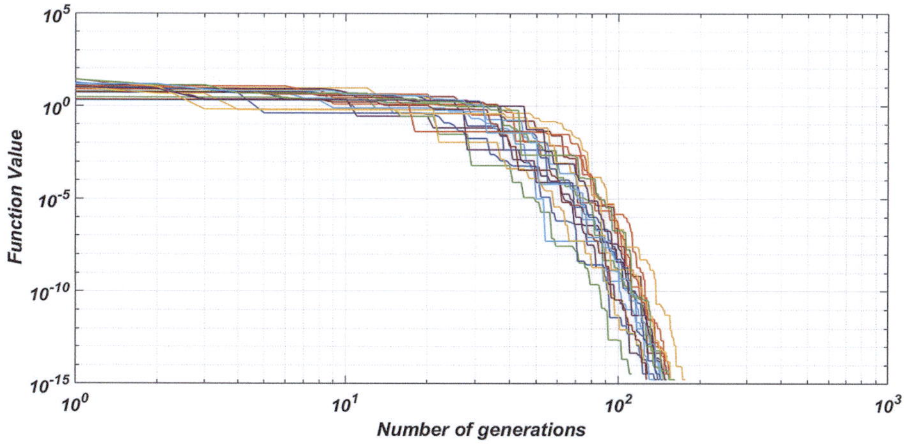

Fig. 4.6 Logarithmic plot of the function value with the number of generations over 20 runs

4.4.3 Goldstein-Price Function

The Goldstein-Price function is a single optimal function with a lesser number of local optima. The domain is between −2 and 2 for both the variables. The equation of the function is given in Eq. 4.13. The minimum of the function is at (0, −1) with a minimum value of 3.

$$f(x, y) = \left[1 + (x + y + 1)^2 \left(19 - 14x + 3x^2 - 14y + 6xy + 3y^2\right)\right] \times \left[30 + (2x - 3y)^2 \left(18 - 32x + 12x^2 + 48y - 36xy + 27y^2\right)\right] \tag{4.13}$$

The contour and the surface plot for the Goldstein-Price function are shown in the Fig. 4.7a, b respectively. The optimization is done using DE with Cr of 0.7 and F value of 0.5 respectively. The population size and the number of generations are taken to be 10 and 500. The optimization is carried out for 20 runs. The function value with respect to the number of generations over the 20 runs is shown in the Fig. 4.8.

Example Problem
For evaluating the performance of the DE on a large-scale optimization problem, the problem considered by Gandhi et al. [3] for identifying the leakage of a sewer line is considered. The number of leaks and the strength of the leaks are not known. This kind of problems is called a source identification problem, where the source of the contaminant (virus in this case) is identified based on the temporal concentration information collected from the observation wells. Figure 4.9 describes the study area, sewage line, the location of the monitoring wells and the pumping wells. The dimension of the study area is 200 × 200 × 100 m.

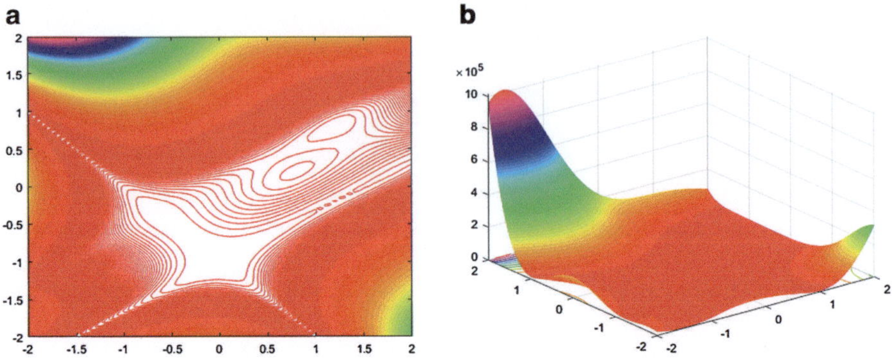

Fig. 4.7 (**a**) Contour plot of Goldstein-Price function. (**b**) Surface plot of Goldstein-Price function

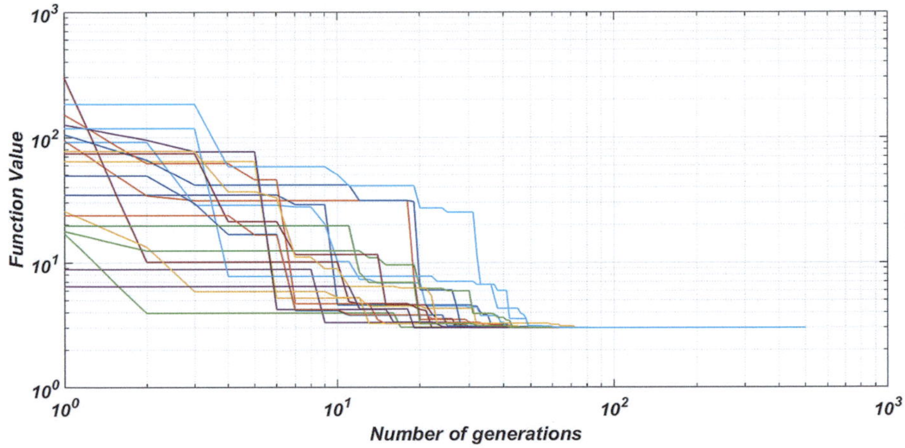

Fig. 4.8 Logarithmic plot of the function value with the number of generations over 20 runs

For solving the source identification problem, the aquifer simulation models, *i.e.* flow and transport models must be incorporated with the optimization model [3]. Here, the groundwater flow model is developed using MODFLOW and the virus transport model is developed using MT3DMS modules available in the Groundwater Modelling Software (GMS). The virus transport is solved considering that sorption occurs only in equilibrium phase with a first order irreversible reaction and following a liner relation between the aqueous and sorbed phase of the concentrations. Gandhi et al. [3] considered two cases, however in this chapter one of the most complicated mixed integer problems is attempted using DE. This sewer line has 20 observation wells that allow collecting the samples from the wells. The samples are tested for viruses in groundwater. There are records for the tests conducted for 150 days. Based on this dataset, the number and location of sources along with the strength of sources are to be identified. Table 4.1 gives

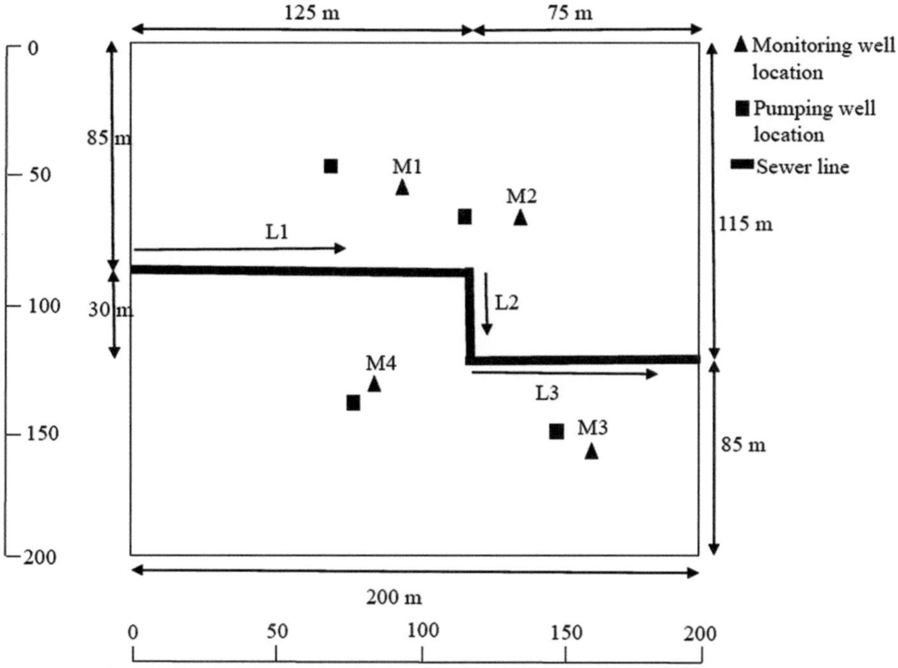

Fig. 4.9 Sewer line that has leaks somewhere along its length. [3]

Table 4.1 Aquifer flow and transport parameters for virus

Flow parameters	Hydrodynamic dispersion parameters	Virus and sorption parameters
$K_{xx} = 9.0 \times 10^{-5}$ ms^{-1}	$\alpha_L = 2$ m	$\lambda = 0.03$ d^{-1}
$K_{yy} = 9.0 \times 10^{-5}$ ms^{-1}	$\alpha_{TH} = 0.2$ m	$\lambda^* = 0.02$ d^{-1}
$K_{zz} = 9.0 \times 10^{-5}$ ms^{-1}	$\alpha_{TV} = 0.1$ m	$K_{eq} = 10^{-4}$ m^{3}kg^{-1}
$S_s = 0.001$ m^{-1}	$D^* = 6 \times 10^{-10}$ m^2 s^{-}	$\rho_b = 1800$ kg m^{-3}

the parameters of the aquifer for solving both the flow and transport models. The virus gets inactivated most of the time according to a first order irreversible reaction as discussed above. The rates of inactivation of the viruses λ and λ^* are the rates in aqueous and in sorbed phase. The slope of the relation in the concentration of virus in the sorbed phase to the concentration of virus in the aqueous phase is the distribution coefficient 'K_{eq}'. The hydraulic conductivity is given by K_{xx}, K_{yy} and K_{zz} representing the hydraulic conductivities in all the three principal directions. S_s represents the specific storage in the aquifer. The dispersivity in longitudinal, transverse horizontal and transverse vertical directions are given by α_L, α_{TH} and α_{TV}. D^* represents the molecular diffusion coefficient and ρ_b represents the bulk density of the porous medium.

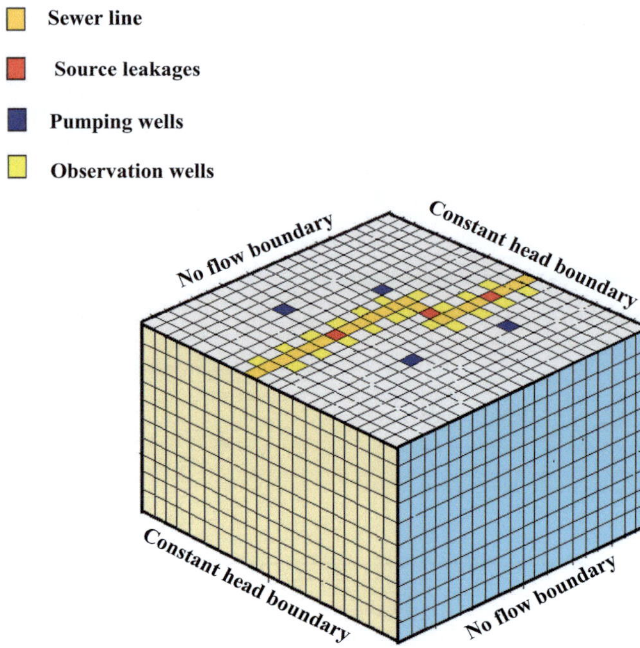

Fig. 4.10 Discretization of the study area and the location of the observation wells

The locations of the observation wells in the aquifer are given in Fig. 4.10. There are 20 observation wells in the study area.

The partial differential equations for solving the flow and the transport equation are given in Eq. 4.14 and 4.15.

$$\frac{\partial}{\partial x}\left(K_{xx}\frac{\partial \phi}{\partial x}\right)+\frac{\partial}{\partial y}\left(K_{yy}\frac{\partial \phi}{\partial y}\right)+\frac{\partial}{\partial z}\left(K_{zz}\frac{\partial \phi}{\partial z}\right)+W=S_s\frac{\partial \phi}{\partial t} \tag{4.14}$$

Where, K is the hydraulic conductivity tensor. ϕ is the piezometric head. W is the volumetric flux per unit volume flowing in and out of the control volume representing the source and sink terms (T^{-1}). This equation is taken as the governing equation for flow through porous media [2].

$$\begin{aligned} n\frac{\partial C}{\partial t}+\rho_b k_{eq}\frac{\partial C}{\partial t} &= \frac{\partial}{\partial x_i}\left(D_{ij}\frac{\partial C}{\partial x_j}\right)-\frac{\partial}{\partial x_i}(q_i C) \\ &\quad - n\mu_l C-\rho_b K_{eq}\mu_{s,eq}C-q'_s C+q_s C_s \end{aligned} \tag{4.15}$$

Here, C is the mass of free viruses per unit volume in the aqueous phase (ML^{-3}). In short, we refer to it as the free virus concentration. The adsorbed virus concentration is given in terms of the mass of viruses per unit mass of the soil (MM^{-1}). We refer to it as the attached virus concentrations. Further, n is the porosity (−); D is the hydrodynamic dispersion tensor ($L^2\ T^{-1}$); q is the Darcy's flow velocity vector (LT^{-1}); μ_l is the inactivation rate coefficient for free viruses (T^{-1}); $\mu_{s,eq}$ is the inactivation rate coefficient for attached viruses to equilibrium sites (T^{-1}). q'_s is the rate of change of transient ground water storage; q_s is the pumping at the source or sinks and C_s is the concentration of the source or sink.

The simulation model consists of the solution of these two equations for given inputs. The optimization model minimizes the sum of the absolute difference between the observed concentration and the simulated concentration at all the observation wells. The objective function for the optimization model is given in Eq. 4.16.

$$\text{Minimize } f = \sum_{k=1}^{k=sk} \sum_{k=1}^{k=oc} \left(OC_o^k - SC_o^k\right)^2 . W_0 \tag{4.16}$$

Here, OC_o^k is the observed concentration at the well location o for time step$k$; $SC_o^k$is the simulated concentration at the well location o for time step$k$; oc is the total number of observation well locations; $sk$is the total number of observation time steps and W_0 is the weightage of corresponding observed well location. Here, the weightage was taken as$\left(1 / \left(OC_o^k + 1\right)\right)^2$. The weightage is selected so as to normalize the concentration at each location. If at all the observed concentration happens to be zero at some observation location, the effect of such concentrations will shoot up the objective function to a very large value. So, the additive coefficient 1 is taken for minimizing this effect. Additionally, it will not have much effect for wells with higher concentrations.

The observed concentrations can be obtained from the field. However, for the hypothetical problem considered here, the aquifer simulation model is used with known source locations and strengths of sources to numerically simulate the observed concentrations at observation locations at different times. Then the optimization model is run for arbitrary location and leakage to find out the actual source strengths and locations by minimizing the objective function. The flowchart of the algorithm is given in Fig. 4.11. The number of leaks in the sewer line are unknown, so 5 leakages are assumed with continuous leakages (i.e. the leakage is at the same rate in all the time steps). Therefore, the total number of variables are 10. Five of the variables are integer type variables. The area is discretized into rectangular grids in each layer. The grids are numbered from the origin of the coordinates counting 1–20 in the first row of the first layer, 21–40 in the second row of the first layer and so on. The sewer line starts from 161, continues till 173 and goes column wise up to 233 and then again runs along the row from 233 to 240. Therefore, the potential locations are 161, 162, 163, 164, 165, 166, 167, 168, 169, 170, 171, 172, 173, 193, 213, 233, 234, 235, 236, 237, 238, 239 and 240. Thus, from these 23 locations, 5 suspected locations are the actual leakages. The source strengths have no upper limit, but it cannot be less than zero.

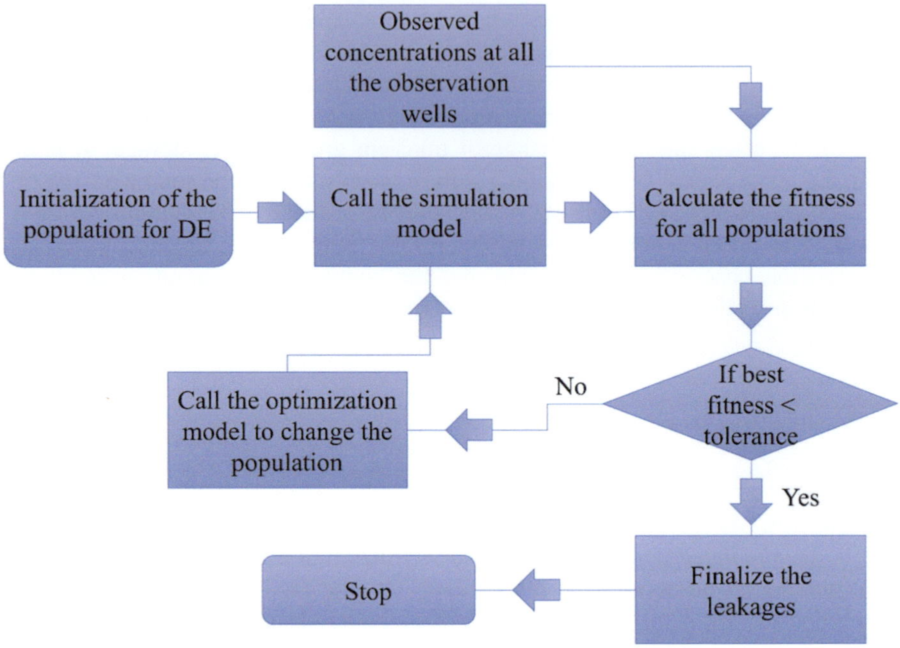

Fig. 4.11 Flow chart describing the procedure followed for optimization

The differential evolution method is used to solve the optimization problem. The objective function is the error function and the minimum value is always close to zero. As such, the tolerance for the fitness function is taken as 10^{-6}. The values of F and C_r are taken to be varying according to the fitness and the generation. The Eqs. 4.17 and 4.18 describes the function used for mutation scaling factor and the crossover rate. The mutation scaling factor decreases as the generations increases from 1 to a value slightly greater than zero by the end of the optimization. The infeasible and the deviating populations created in each generation are always eliminated by the selection operator. A population size of 20 is considered.

$$F_p = \begin{cases} f_p \Big/ \sum\limits_{p=1}^{p=P} f_p & if \ \text{rand}\,(1,2) = 1 \\ 1 - {}^{i-1}\!\Big/_{gen} & if \ \text{rand}\,(1,2) = 2 \end{cases} \tag{4.17}$$

$$C_r = \begin{cases} 1 \ if \ \ mean\,(f_p) \le 100 \\ \quad 0.7 \ otherwise \end{cases} \tag{4.18}$$

Table 4.2 Results from the DE, Actual and Gandhi et al. [3]

Type of variable	Value	Location and flux (Differential evolution)
Location	167	167
	213	167
	237	213
	N.A. (Dummy)	213
	N.A. (Dummy)	237
Flux	2000.000	1991.839
	800.000	8.176
	1000.000	1.023
	0.000	798.978
	0.000	1000.000

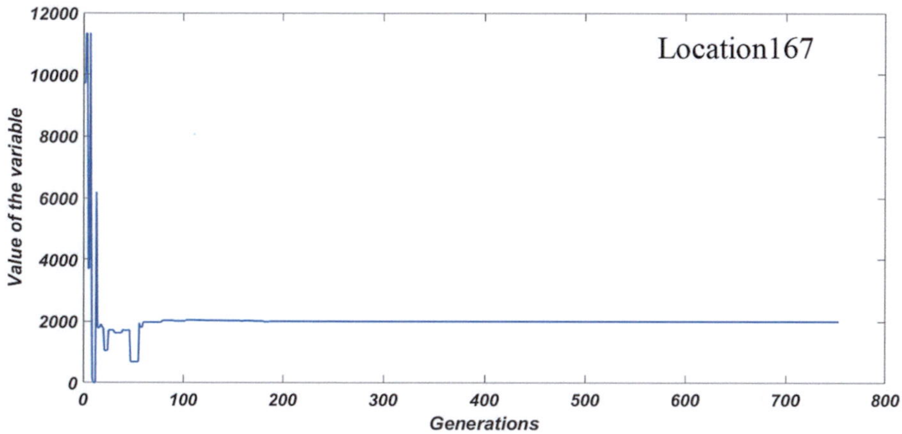

Fig. 4.12 Combination of the strengths at location 167 with generations

While solving such mixed integer problems, a minor modification for DE is necessary. At all the generations, the populations should be checked with the decision variables to be rounded off to the nearest integer in the integer decision variables. The rest of DE where there are continuous variables can run as it is. The problem took 753 generations to reach the optimal solution. The optimal solution is shown in the Table 4.2.

The results show that values of the source leakage are almost closely identified with a minute relative error of 0.001% which can be considered almost negligible. The locations are exactly identified by DE. Here, two of the location variables are assumed to be dummy. But DE has not recognized them as dummy and they are recognized as the actual locations sharing the total value of the source strength, which is totally valid. Figures 4.12, 4.13 and 4.14 shows the best population for the variable corresponding to locations 167, 213 and 237 respectively. Figure 4.15 shows the total population converging towards the best fitness over the generations. The line with higher thickness represents the best population.

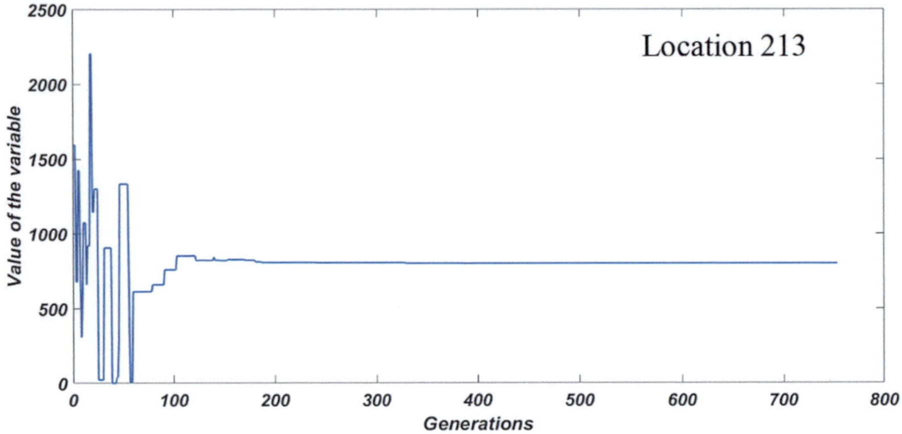

Fig. 4.13 Combination of the strengths at location 213 with generations

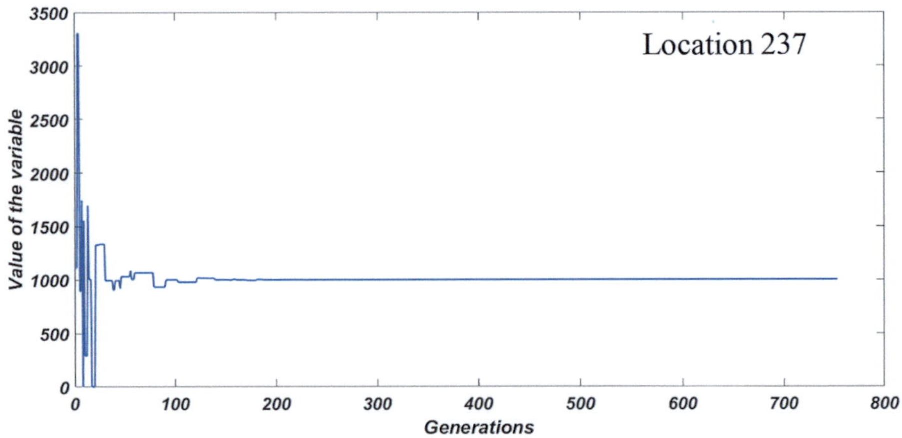

Fig. 4.14 Combination of the strengths at location 237 with generations

The Fig. 4.15 clearly shows that there is still some difference in the fitness value of the population. This indicates that the optimal solution of the problem has not reached, and the population will continue to converge to the actual optimal solution of the problem. The objective function value at termination is 3.9239×10^{-7}. In this problem, the DE took 15,060 function evaluations to converge to the optimal solution. DE is advantageous in many aspects of finding the global solution easily. Based on the observations from the previous research and the problem solved using DE, the following conclusions were made.

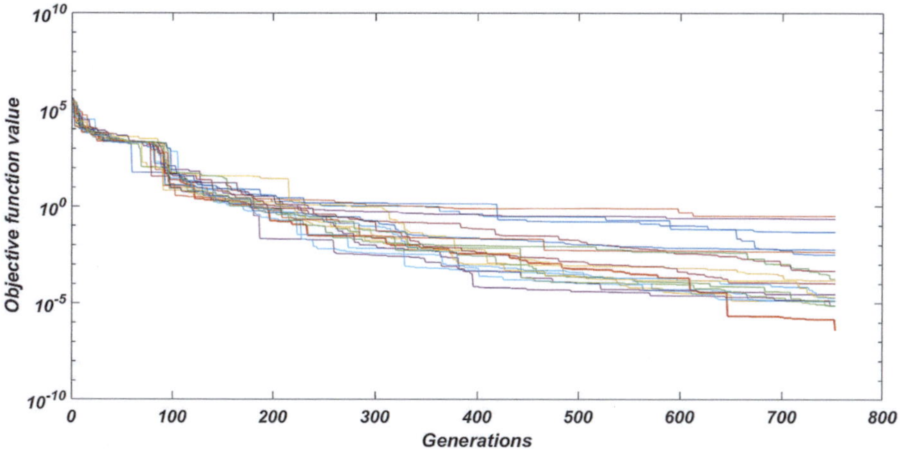

Fig. 4.15 All the populations converging towards optima

4.5 Conclusions

The differential evolution algorithm is easy to understand and one of the most basic theories of natural evolution. The evolution of the population is through the difference in the initial population and as the convergence occurs, the population tend to converge at optimal point in the search space. The rate of crossover and the mutation scale factor decide the number of function evaluations required to reach the optimal solution. These parameters of the algorithm can be decided based on their behaviors in various combinations. Dynamic parameters (*i.e.* parameters changing with generations) gives best result compared to the parameters that are more static or fixed. The nature and complexity of the problem plays a key role in selecting the parameters. For the non-convex problems or the problems having a larger number of decision variables, dynamic parameter setting is necessary. Whereas the problems that are convex and simple or having lesser number of variables can be solved using any combination of the parameters. Overall, DE is one of the most robust algorithms which has the capability to handle the complex engineering optimization problems. DE also rated as one of the easiest algorithms to program and handle as well as can be guaranteed to derive the optimal solution.

References

1. Qin AK, Huang VL, Suganthan PN (2009) Differential evolution algorithm with strategy adaptation for global numerical optimization. IEEE Trans Evol Comput 13(2):398–417
2. Bear J (1988) Dynamics of fluids in porous media. Dover Publications, New York

3. Gandhi BGR, Bhattacharjya RK, Satish MG (2016) Simulation–optimization-based virus source identification model for 3D unconfined aquifer considering source locations and number as variable. J Hazard Toxic Radioact Waste 21(2):04016019. http://ascelibrary.org/doi/abs/10.1061/(ASCE)HZ.2153-5515.0000334
4. Blum C, Roli A (2003) Metaheuristics in combinatorial optimization: overview and conceptual comparison. ACM Comput Surv 35(3):268–308
5. Zhang J, Sanderson AC (2009) JADE: adaptive differential evolution with optional external archive. IEEE Trans Evol Comput 13(5):945–958
6. Price K, Storn R, Lampinen J (2005) Differential evolution—a practical approach to global optimization. Springer, Berlin
7. Price KV (1997) Differential evolution vs. the functions of the 2nd ICEO. In: Proceedings of IEEE international conference on evolutionary computation, Indianapolis, Indiana, USA. IEEE Press, New York, pp 153–157
8. Storn R, Price KV (1995) Differential evolution: a simple and efficient adaptive scheme for global optimization over continuous spaces," ICSI, USA, Tech. Rep. TR-95-012 [Online]. Available: http://icsi.berkeley.edu/~storn/litera.html
9. Storn R, Price KV (1996) Minimizing the real functions of the ICEC 1996 contest by differential evolution. In: Proceedings of IEEE international conference on evolutionary computation, pp 842–844
10. Das S, Abraham A, Chakraborty UK, Konar A (2009) Differential evolution using a neighborhood based mutation operator. IEEE Trans Evol Comput 13(3):526–553
11. Das S, Suganthan PN (2011) Differential evolution – a survey of the state of the art. IEEE Trans Evol Comput 15(1):4–31
12. Das S, Mullick SS, Suganthan PN (2016) Recent advances in differential evolution – an updated survey. Swarm Evol Comput 27:1–30
13. Rahnamayan S, Tizhoosh HR, Salama MMA (Feb. 2008) Opposition-based differential evolution. IEEE Trans Evol Comput 12(1):64–79

Chapter 5
Artificial Bee Colony Algorithm and an Application to Software Defect Prediction

Rustu Akay and Bahriye Akay

Abstract The term swarm refers to any restrained collection of interacting agents or individuals. For survival, creatures need to live or perform some tasks collectively such as defending against predators, foraging, mating, etc. An intelligent swarm optimizes a goal or task by responding adaptively to the local and/or global environmental changes in a collective manner. Swarm intelligence research field deals with designing algorithms inspired by the collective behavior of social creatures. Task division and self-organization abilities lead swarm intelligence to occur in a colony and the self-organization is characterized by positive-feedback, negative feedback, fluctuation and multiple interactions in order to use the local information to form a global pattern without a supervision. Ants, termites, birds, and fishes are some examples of social animals that have swarm intelligence and inspire researchers to design problem solving techniques. Bees also are a typical example of creatures performing tasks collectively in nest site selection, nest building, mating, and foraging. Among these activities, foraging might be the most crucial one for the survival of a bee colony. The swarm intelligence in foraging of a honey bee colony inspired Karaboga to design an optimization algorithm, Artificial Bee Colony (ABC), in which the search is guided by the bees' exploration and exploitation mechanisms to maximize the quality of the honey within the hive. In this chapter, first, the foraging behavior of real honey bees is summarized and then, the details of Artificial Bee Colony algorithm about how it mimics the foraging behavior is provided. In the third section, an application of ABC algorithm is carried out on a software engineering problem, software defect prediction.

R. Akay
Department of Mechatronics Engineering, Erciyes University, Melikgazi, Kayseri, Turkey
e-mail: akay@erciyes.edu.tr

B. Akay (✉)
Department of Computer Engineering, Erciyes University, Melikgazi, Kayseri, Turkey
e-mail: bahriye@erciyes.edu.tr

F. Bennis, R. K. Bhattacharjya (eds.), *Nature-Inspired Methods for Metaheuristics Optimization*, Modeling and Optimization in Science and Technologies 16,
https://doi.org/10.1007/978-3-030-26458-1_5

Keywords Meta-heuristics · Swarm intelligence · Artificial bee colony algorithm · Software defect prediction · Classification · Neural network training

5.1 Introduction

Swarm intelligence deals with achieving a goal based on collective behavior patterns of individual creatures without a central supervision. There are tasks in a swarm performed by specialized agents (division of tasks). The role of an individual is identified by the current state of the swarm and it may change the role or its state based on the information obtained from the others or the environment, which is called self-organization. Self-organization arises from multiple interactions, positive feedback, negative feedback and fluctuation properties in the swarm. Information held by the agents is distributed through the swarm by local interactions of individuals via a kind of communication (multiple interactions). An individual may repeat a promising behavior (positive feedback) or may change its behavior to a more profitable behavior after interaction, or may abandon a behavior that is no longer profitable (negative feedback). Some agents in the swarm try to find out undiscovered behavior patterns which are supposed to bring innovation to the swarm (fluctuation).

Ants, termites, bees, birds, and fishes are example creatures that exhibit swarm intelligence properties in their vital activities. Researchers have been inspired by the swarm intelligence coming from nature and proposed problem solving techniques. Ant colony optimization by Dorigo et al. [12] and Particle swarm optimization by Kennedy and Eberhart [19] are popular problem solving tools that simulate the swarm intelligence in an ant colony and fish schools, respectively.

The bees have many intelligent behavior patterns such as task division in the nest building, mating, navigation, nest site selection, and foraging in addition to their inherent features such as photographic memories, space-age sensory, etc [16]. Foraging is also a collective activity of honey bees and it is very crucial to ensure the survival of the colony. There is a division of labor between bees in the foraging task. The colony is divided into three categories for foraging: employed bees, onlooker bees and scout bees. The role of the bees and task division may change adaptively depending on hive's internal and environmental conditions or the nectar amount of the food sources they are exploiting. The foraging process is fired by a forager bee to search for a food source. When a source is discovered, the bee exploits the nectar, keeps the exploited nectar in her stomach and enzymes are secreted on the nectar. The bee turns back to the hive, unloads the nectar in her stomach to the honeycomb cells and cells are closed by some special antibacterial and anti-fermentative substances. Later on, the bee can return-back to the flower to exploit more or before turning back, she can dance in the hive to recruit the other bees to the profitable source. Without a central supervision, bees communicate between each other and an information flow network is sustained in the colony. The communication is performed by dancing which includes information about the spatial location and the quality of the food source discovered. Therefore, the onlooker bees are informed about the high quality sources and a positive feedback effect occurs towards good solutions.

Artificial bee colony (ABC) algorithm developed by Karaboga [14] is a swarm intelligence algorithm that mimics the foraging behaviour of honey bees. The search space is composed of the points that correspond to the food source locations. An initial population of randomly generated locations in the search space goes through the phases of the ABC algorithm, which are employed bee, onlooker bee and scout bee phases. As in real bees, the foraging process starts with exploration which means sending scouts to the sources. Once a source is chosen by the scout bee in the initialization, this forager bee becomes an employed bee. The employed bee memorizes the source and starts to exploit the source. While exploration brings global search capability, exploitation introduces local search ability. Onlooker bees chose food source based on the information gathered from the employed bees. This is simulated in ABC algorithm by a probabilistic selection scheme. If a source can not be improved through a certain number of periods, this source is assumed to be exhausted and its bee becomes a scout bee. ABC algorithm is a successful swarm based meta-heuristic algorithm and applied to solve unconstrained and constrained, single-objective and multi-objective, and continuous and combinatorial design problems [2, 16, 18].

This chapter gives a comprehensive explanation about the ABC algorithm and its application to a software engineering problem. In the second section, the details of the ABC algorithm is presented, its application to a software engineering problem is provided in the third chapter and finally it is concluded.

5.2 ABC Algorithm

Artificial Bee Colony (ABC) Algorithm developed by Karaboga [14] is a swarm intelligence algorithm which simulates the behaviors of employed bees, onlooker bees and scout bees in foraging process. As a swarm algorithm, ABC algorithm has self-organization and division of labor properties. Division of labor in foraging process manifests as assigning different roles and responsibilities to the agents in the swarm such that the bees are categorized as employed, onlooker and scout bee. The employed bees are assigned to exploitation task and also responsible from informing the onlooker bees about the location and quality of the food source by dancing. The onlooker bees watch the dances of the employed bees and decide a food source to fly and exploit. The scout bees are assigned to explore new sources in the environment depending on an internal motivation or based on possible external clues. These roles can change intelligently based on the conditions of the hive. A scout bee becomes an employed bee when she finds a new source and the employed bee becomes a scout bee when her source is exhausted. An onlooker bee waits in the hive and watches the dances in the dance area and she behaves as employed bee when she decides a source about which she has gathered information.

The other characteristic of a swarm is self-organization which is based on positive feedback, negative feedback, fluctuation and multiple-interactions properties. Dancing of an employed bee attracts the other bees and recruits them to the source pointed by the information in dancing. As more bees select a rich source, more bees dance about the rich source and as more bees dance about the rich source, the number of the bees recruited to a specific source increases, which is the positive feedback. Because the positive feedback may cause all the population repeating the same behavior pattern, it should be balanced inversely when it is needed. The negative feedback breaks the excessive accumulation effect of positive feedback. For example, a bee abandonees her source when its nectar is exhausted and she tries to find new sources. The fluctuation introduces new patterns to bring innovation into the behavior patterns. In a bee colony, random flights performed by the scout bees are examples of the fluctuation. The multiple interactions are established based on the communication of the agents and a network is constructed among them. Spread of the information among bees is carried out by their dances.

In the algorithm, a food source location corresponds to a possible solution to the optimization problem. Therefore, the number of decision variables in the design problem corresponds to the dimension of the food source (dimension of the solution). For numeric optimization, each dimension of the solution is allowed to take values within a parameter-specific range. The surfaces constructed by all values of the design parameters within the ranges establish a search space ($\mathbb{R}^n$) in which the optimum solution is explored. An initial food source population is generated and the nectar amount of each source is evaluated. Nectar amount calculation corresponds to fitness calculation of a solution. The ABC algorithm searches for the optimum solution in the search space by means of the employed, onlooker and scout bee phases until maximum number of cycles is reached. Main steps of the ABC algorithm are given in Algorithm 1 [15, 17].

```
Data: Assign values for the control parameters;
CS: Number of Food Sources,
MCN: Maximum Cycle Number,
limit: A control parameter to decide whether a source is exhausted
begin
    Initialize the food source locations and evaluate them;
    cycle = 1;
    while cycle < MCN do
        Employed Bees' Phase;
        Onlooker Bees' Phase;
        Memorize the Best Solution;
        Scout Bee Phase;
        cycle ++;
    end
end
```

Algorithm 1: Basic ABC algorithm

ABC algorithm has three control parameters of which the values are decided before running the algorithm. Two of them are common in all swarm and population based algorithms: swarm size (CS) and maximum number of cycles (MCN). The other control parameter called limit is the number of exploitations to decide whether a source is exhausted or not. Once their values are assigned, an initial food source population is generated by using Eq. 5.1:

$$x_{ij} = x_j^{min} + rand(0,1)(x_j^{max} - x_j^{min}) \tag{5.1}$$

where $i = 1 \ldots CS$, $j = 1 \ldots D$, CS is the number of food sources, D is the dimension of the problem, problem-specific x_j^{min} and x_j^{max} are lower and upper boundary of jth dimension, respectively.

After the initial population is generated, the phases of the ABC algorithm are iterated until the termination criteria is satisfied.

In the employed bee phase the ABC algorithm (Algorithm 2), an analogy of the food source exploitation is conducted by a local search defined by Eq. 5.2:

$$\upsilon_{ij} = x_{ij} + \phi_{ij}(x_{ij} - x_{kj}) \tag{5.2}$$

where i is the solution currently being exploited, k is a randomly chosen neighbor solution and ϕ_{ij} is a real random number within the range [−1,1] drawn from uniform distribution.

In the local search defined by Eq. 5.2, only one randomly chosen dimension of the current solution (parameter j) is changed. In some modified versions of the ABC algorithm, the local search changes more than one parameter of the current solution (Eq. 5.3) [1].

$$\upsilon_{ij} = \begin{cases} x_{ij} + \phi_{ij}(x_{ij} - x_{kj}), & if \;\; R_{ij} < MR \\ x_{ij} \qquad\qquad , & \text{otherwise} \end{cases} \tag{5.3}$$

where R_{ij} is a real random number drawn from the uniform distribution within (0,1) and MR, is the modification rate. If $R_{ij} < MR$ is satisfied, then, υ_{ij} is taken from the local search, otherwise taken from x_{ij}.

When a new solution (υ_i) is generated, a greedy selection is applied between the current solution (x_i) and the new solution produced by the local search (υ_i). The better one is kept and the other one is discarded from the population. If the new solution is better, because it is a just discovered solution, the number of exploitations on this solution is zero. Otherwise, if the current solution is better, its counter holding its number of exploitations is increased by 1.

An employed bee unloads the nectar and then gives information to onlookers about the quality and the location of her source. High quality solutions have high chance to be selected but the solutions with low quality can also be selected by the onlookers. The probability of each solution (p_i) can be calculated proportionately to its fitness value:

```
Data: Food Source Population;
begin
    foreach food source x_i do
        New Solution x' ⟵ produced by Eq. 5.2;
        f(x') ⟵ evaluate new solution;
        if f(x') < f(x_i) then
            x_i ⟵ x'; exploit(x_i) ⟵ ∅;
        else
            exploit(x_i) = exploit(x_i) + 1;
        end
    end
end
```

Algorithm 2: Employed Bees' Phase of the ABC algorithm

$$p_i = \frac{fitness_i}{\sum_{i=1}^{SN} fitness_i} \tag{5.4}$$

In the onlooker bees' phase of the ABC algorithm (Algorithm 3), the onlooker bees select food sources to fly using a stochastic selection scheme based on the probability values of the food sources. The multiple interactions phenomena in ABC is achieved using the probability values which establishes a communication network between the bees about the sources. When an onlooker bee selects a source, the bee performs a local search in the vicinity of that solution. As in the employed bees' phase, a greedy selection is applied and the exploitation counters are updated. The difference between the employed bees' and onlooker bees' is that the onlooker bees choose the sources based on their quality. Solutions with better quality are likely to be selected. This shows the positive feedback property in the ABC algorithm.

The scout bees' phase of the ABC algorithm is given in Algorithm 4. In real bees, a source is exhausted by the exploitations of its nectar. In the ABC algorithm, if the number of exploitations related to a source exceeds the control parameter, limit, the source is assumed to be exhausted. The ABC algorithm has ability to forget about the patterns that make the convergence supersaturated. This is the negative feedback of ABC algorithm. Instead of the exhausted solutions, the scout bees in ABC algorithm make global search and try to explore undiscovered sources randomly, which is the fluctuation characteristic of the ABC algorithm.

5.3 An Engineering Application: Software Defect Prediction

Software testing is checking whether the output specifications are satisfied when the input specifications are provided [24] and defect prediction deals with estimating number of defects or faults in the design or source code of a software in order to improve product quality and performance capability.

```
Data: Food Source Population; Probability of each solution
begin
    foreach food source x_i do
        p_i ⟵ assign probability by (5.4);
    end
    i ⟵ 0 ;
    t ⟵ 0 ;
    while t < CS do
        r ⟵ rand(0, 1);
        if r < p(i) then
            t ⟵ t + 1;
            x' ⟵ a new solution produced by Eq. 5.2;
            f(x') ⟵ evaluate new solution;
            if f(x') < f(x_i) then
                x_i ⟵ x'; exploit(x_i) ⟵ ∅;
            else
                exploit(x_i) = exploit(x_i) + 1;
            end
        end
        i ⟵ (i + 1) mod (CS − 1);
    end
end
```

Algorithm 3: Onlooker Bees' Phase of the ABC algorithm

```
Data: Food Source Population, Exploitation Counters;
begin
    si = {i : exploit(i) = max(exploit)};
    if exploit(si) > limit then
        x_si ⟵ random solution by Eq. 5.1;
        exploit(si) ⟵ ∅;
    end
end
```

Algorithm 4: Scout Bees' Phase of the ABC algorithm

A testing team plans and manages the test execution activities to achieve the quality and performance metrics and to ensure all defects are found and fixed. Better plans and managing activities are the goal of software development organizations by predicting the defects because huge amount of budget and resource is allocated to testing phase to discover the defects. Detecting defects in later stages of the software life cycle causes high relative cost of fixing a fault [5]. Predicting the software defects in advance is one of the major issues in software engineering field. Predicting the defects and the fault-prone modules that needs refactoring in earlier stages of software production process helps to produce more robust and reliable software systems, increases software quality and makes the maintenance easier. By a good defect estimation model, more effort and time can be assigned to the modules that are fault-prone and this also have positive impact on customer satisfaction, suppressing the regression effects in software and minimizing the schedule variance.

Software defect prediction problem can be considered as a classification problem or symbolic regression problem. Regression techniques aim to predict the quantity and density of the defects. Classification techniques can classify the modules into fault-prone module and defect-free module groups based on the software-metrics collected [9]. They learn from the data in previous releases of the current project or the data of other similar projects to forecast the software defects in the projects.

5.3.1 Artificial Neural Networks for Predicting Software Defects

Several models have been proposed based on machine learning and statistical techniques for classification of software defects [8]. Menzies et al. [22] reported 71% success rate of defect predictors which is higher than some industrial methods including manual code reviews.

In traditional statistical classification procedures, classification is performed based on a probability model which calculates the posterior probability. A major disadvantage of the statistical models is that they make assumptions on the underlying probability model and they require a knowledge about both data and the model [26].

An alternative machine learning technique for classification is Artificial Neural Network (ANN). ANNs which simulate the biological neural networks, do not make assumptions on the data and the underlying model. ANNs have been applied to a variety of real world classification tasks in industry, business and science [25]. In an ANN, a response is generated when a signal is transmitted in the network by means of interconnecting neurons through synapses. Each neuron produces an output by processing the signals coming from the neurons in the previous layer (Fig. 5.1). Output y_i of the ith neuron can be described by Eq. 5.5:

$$y_i = f_i(\sum_{j=1}^{N_i} \omega_{ij} x_j + \theta_i) \tag{5.5}$$

where N_i denotes the number of connections to ith neuron, so it is the number of neurons in the previous layer, x_{ij} is the input signal to ith neuron coming from jth neuron, ω_{ij} is the connection weight between ith neuron and jth neuron, θ_i is the threshold (or bias) of ith neuron, and f_i is the transfer function of ith neuron.

A neural network for a classification problem can be assumed as a mapping function from d-dimensional input-space to the M-dimensional class-space, $E : R^d \rightarrow R^M$ [26]. The aim is to minimize an overall error measure (Eq. 5.6):

$$\min_{\omega,\theta \in \mathbb{R}} E(\omega, \theta) \tag{5.6}$$

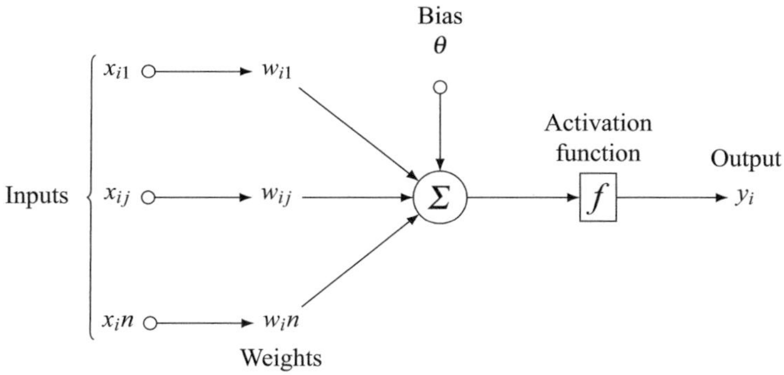

Fig. 5.1 A neuron as a processing node

where $E(\omega, \theta)$ can be defined by (5.7):

$$E(\omega, \theta) = \frac{1}{N_p} \sum_{j=1}^{N_p} \sum_{i=1}^{N_o} (d_i - y_i)^2 \quad (5.7)$$

where d_i and y_i are the desired and the actual values of the ith neuron at the output layer, respectively. N_o is the number of neurons in the output layer and N_p is the number of patterns in the training phase.

5.3.2 ABC Algorithm in Training an ANN for Software Defect Prediction

In the implementation of an optimization algorithm in the training of an ANN, most focus should be given to the encoding scheme and the objective function. The encoding scheme and perturbation operators should be handled carefully so that they work compatibly.

Encoding schemes in the optimization of neural networks fall in to two groups: direct methods and indirect methods. In the direct methods, connection weight matrices are converted a vector and they are concatenated in an individual. In the indirect encoding schemes, the genotype of an individual requires to be interpreted since a compressed description of the network is evolved in a smaller genotype. While the indirect schemes are scalable and lead to the faster convergence, the direct schemes are preferable in small architectures [4, 7].

In this study, since it is easy to apply and the architectures in the study are not so large, we used the direct encoding without compressing. Weight values and bias values of the network are encoded in an individual represented by a vector. In the evaluation of an individual by the cost function, elements of the individual are decomposed into weight matrices and bias vectors to be applied to the neurons in the network.

Another issue that we should give emphasis in designing a neural network is the objective function which guides the search to minimize the output error between the neural network and the actual system. Generalization ability of the neural network, having the least number of local minima, representing the prediction error of the neural net without any additional curvature should be considered in deciding the objective function [6].

In the study, we have used root mean square error (RMSE) as the objective function defined by Eq. 5.8 in the training.

$$\text{RMSE} = \sqrt{E(\omega, \theta)} \tag{5.8}$$

Classification error percentage (CEP) can also be used as the objective function. However, because different individuals with different RMSE error may have the same CEP value [3], RMSE value can guide the search to the optimum regions better. After training NN based on RMSE, we also evaluated the classifier performance based on the classification metrics given in next section.

Different neighborhood mechanisms has been proposed for ABC algorithm as mentioned in Sect. 5.2. In this study, the local search scheme in the basic algorithm (Eq. 5.2) has been used.

5.3.3 Experiments

5.3.3.1 Data Set and Metrics

In the experiments, we used the software data presented by D'Ambros et al. [11] for the purpose of software defect prediction. The number of classes, version information and next release bugs of the datasets used in the experiments are given in Table 5.1.

Each data set has Chidamber and Kemerer (CK) [10] and Object Oriented (OO) metrics which are the static code features used as inputs by our prediction system. These attributes are presented in Table 5.2.

CK and OO metrics are easy to compute, have good predictive power and do not require historical information. OO metrics also have the edge in explanative power. Using CK and OO metrics together in prediction provides an improvement in prediction performance [11]. The preprocessed data sets has 17 attributes plus one target attribute. Target attribute in the data set is the error count related to each system.

Table 5.1 Dataset used in the experiments [11]

System	Prediction release	Time period	#Classes	#Versions	#Transactions	#Post-rel defects
Eclipse JDT Core www.eclipse.org/jdt/core/	3.4	1.1.2005–6.17.2008	997	91	9135	463
Eclipse PDE UI www.eclipse.org/pde/pde-ui/	3.4.1	1.1.2005–9.11.2008	1562	97	5026	401
Equinox framework www.eclipse.org/equinox/	3.4	1.1.2005–6.25.2008	439	91	1616	279
Apache Lucene www.lucene.apache.org	2.4.0	1.1.2005–10.8.2008	691	99	1715	103

Table 5.2 Class level source code metrics [11]

Type	Metric	
CK	WMC	Weighted Method Count
CK	DIT	Depth of Inheritance Tree
CK	RFC	Response For Class
CK	NOC	Number Of Children
CK	CBO	Coupling Between Objects
CK	LCOM	Lack of Cohesion in Methods
OO	FanIn	Number of other classes that reference the class
OO	FanOut	Number of other classes referenced by the class
OO	NOA	Number of attributes
OO	NOPA	Number of public attributes
OO	NOPRA	Number of private attributes
OO	NOAI	Number of attributes inherited
OO	LOC	Number of lines of code
OO	NOM	Number of methods
OO	NOPM	Number of public methods
OO	NOPRM	Number of private methods
OO	NOMI	Number of methods inherited

5.3.3.2 Prediction Performance Evaluation

In the experiments, the performance of ANN models trained by Bayesian Regularization (BR) [13, 21], Levenberg-Mardquart (LM) [20], Scaled Conjugate Gradient (SCG) [23] and ABC algorithms are compared to each other in training and test phases.

Levenberg-Marquardt training algorithm approximates the second-order Hessian matrix using $H = J^T J$ where J is the Jacobian matrix that contains the first derivatives of the network errors with respect to the weights and biases. The Levenberg-Marquardt updates the parameters using Newton-like update defined by Eq. 5.9

$$x_{k+1} = x_k - [J^T J + \mu I]^{-1} J^T e \tag{5.9}$$

When μ is zero, the updating rule works as Newton's method, based on the approximated Hessian matrix. When μ is large, it works as gradient descent algorithm. μ is decreased if a reduction in the objective function is achieved and is increased when the objective function is increased.

BR training algorithm uses LM updating rule and minimizes a linear combination of squared errors and weights, and then determines the correct combination.

SCG training algorithm is a kind of conjugate gradient (CG) algorithm. CG algorithm updates the weights by Eq. 5.10 but chooses the search direction and the step size more carefully by using the second order information (Eq. 5.11).

$$\omega_{k+1} = \omega_k + \alpha_k \cdot p_k \tag{5.10}$$

$$s_k = E''(\omega_k) \cdot p_k \tag{5.11}$$

In SCG algorithm s_k is approximated by Eq. 5.12:

$$s_k \approx \frac{E'(\omega_k + \sigma_k \cdot p_k) - E'(\omega_k)}{\sigma_k}, 0 < \sigma_k \ll 1 \tag{5.12}$$

In order to evaluate the training performance, we used the error count as the output of the ANNs. After the model is trained, in order to the calculate the classification percentage, the error count column was converted into a Boolean attribute based on whether the module has defect or not.

For each problem, 85% and 15% of the data is taken as train and test data, respectively. All the algorithms were run 30 times and in each run, different 85% and 15% portions of the data are assigned to train and test data to make the results independent of the data distribution. Average of the RMSE values obtained from 30 runs of the algorithms are presented in Table 5.3. The effect of the number of the neurons in the hidden layer is also investigated.

On the training of the Eclipse problem, the best RMS error is produced by the ABC algorithm when the number of neurons in hidden layer is 8. The smallest standard deviations are produced by the ABC algorithm. On the testing of ANN models trained by the algorithms for Eclipse problem, the ANN with 16 neurons in hidden layer and trained by ABC algorithm produced the best result. Again, the smallest standard deviations are reported for the ANN trained by ABC algorithm.

On the training of the ANN for Equinox problem, the best RMSE is produced by the ANN model with 32 neurons in hidden layer and trained by ABC algorithm. On the testing for Equinox problem, the ANN model with eight neurons in hidden layer and trained by ABC algorithm is superior to the other cases in the table.

On the training of ANN for Lucene problem, ANN model trained by BR algorithm and with eight neurons in the hidden layer, produced the best RMSE value while on the testing of ANN for this problem, SCG algorithm produced the smallest RMSE. However, on this problem ABC algorithm is very close to BR on training and very close to SCG on testing when the number of neurons is 16.

On the training and testing of ANN for Mylene problem, the best results are produced by the ANN models trained by ABC algorithm and with 16 neurons and 32 neurons, respectively.

For Pde problem, ANN model with 16 neurons and trained by BR algorithm is the best in training while ANN with eight neurons and trained by ABC algorithm is the best in test results.

In addition to the training performance metric, RMSE, the performance of a classifier is evaluated based upon some evaluation indices such as confusion matrix (or error matrix), Precision, Recall, etc.

Confusion matrix (Table 5.4) describes the performance of a supervised classifier upon a set of test data where TP (hit) means that defected module is correctly

Table 5.3 Comparison of the average RMS error values produced by 30 runs of BR, LM, SCG and ABC algorithms

			Train data				Test data			
			#Neurons in hidden Layer				#Neurons in hidden Layer			
Problem	Algorithm		4	8	16	32	4	8	16	32
Eclipse	BR	Mean	0.0899	0.0909	0.0896	0.0917	0.1122	0.1075	0.1160	0.1122
		Std.	0.0066	0.0081	0.0085	0.0097	0.0297	0.0248	0.0260	0.0209
	LM	Mean	0.0952	0.0976	0.0920	0.0999	0.1158	0.1196	0.1152	0.1260
		Std.	0.0132	0.0163	0.0143	0.0195	0.0218	0.0271	0.0325	0.0331
	SCG	Mean	0.0940	0.0915	0.0893	0.0906	0.1090	0.1120	0.1106	0.1067
		Std.	0.0087	0.0131	0.0127	0.0161	0.0256	0.0241	0.0256	0.0186
	ABC	Mean	0.0957	**0.0890**	0.0898	0.0883	0.0937	0.0973	**0.0906**	0.0999
		Std.	0.0028	0.0030	0.0020	0.0025	0.0151	0.0165	0.0136	0.0143
Equinox	BR	Mean	0.0624	0.0651	0.0647	0.0650	0.0864	0.0795	0.0867	0.0845
		Std.	0.0106	0.0124	0.0099	0.0133	0.0279	0.0167	0.0202	0.0285
	LM	Mean	0.0861	0.0781	0.0735	0.0853	0.0984	0.1046	0.1065	0.1118
		Std.	0.0234	0.0199	0.0185	0.0200	0.0308	0.0422	0.0474	0.0445
	SCG	Mean	0.0706	0.0691	0.0667	0.0709	0.0873	0.0897	0.0879	0.0885
		Std.	0.0169	0.0152	0.0143	0.0169	0.0313	0.0307	0.0255	0.0321
	ABC	Mean	0.0740	0.0631	0.0610	**0.0595**	0.0726	**0.0722**	0.0731	0.0758
		Std.	0.0029	0.0033	0.0025	0.0022	0.0178	0.0183	0.0175	0.0144
Lucene	BR	Mean	0.0523	**0.0512**	0.0518	0.0538	0.0673	0.0730	0.0720	0.0678
		Std.	0.0059	0.0067	0.0051	0.0059	0.0225	0.0258	0.0229	0.0262
	LM	Mean	0.0602	0.0602	0.0556	0.0538	0.0769	0.0729	0.0737	0.0737
		Std.	0.0121	0.0126	0.0076	0.0070	0.0319	0.0473	0.0286	0.0285
	SCG	Mean	0.0543	0.0532	0.0536	0.0517	0.0710	**0.0505**	0.0640	0.0655
		Std.	0.0072	0.0040	0.0065	0.0046	0.0242	0.0159	0.0250	0.0231
	ABC	Mean	0.0571	0.0517	0.0513	0.0517	0.0582	0.0564	0.0506	0.0543
		Std.	0.0031	0.0024	0.0021	0.0018	0.0182	0.0163	0.0144	0.0189

Mylyn	BR	Mean	0.0458	0.0453	0.0458	0.0471	0.0537	0.0581	0.0560	0.0611
		Std.	0.0043	0.0042	0.0042	0.0048	0.0108	0.0145	0.0146	0.0188
	LM	Mean	0.0475	0.0469	0.0466	0.0512	0.0531	0.0543	0.0513	0.0558
		Std.	0.0067	0.0072	0.0046	0.0074	0.0153	0.0157	0.0130	0.0160
	SCG	Mean	0.0465	0.0467	0.0469	0.0468	0.0543	0.0533	0.0499	0.0491
		Std.	0.0031	0.0042	0.0030	0.0043	0.0177	0.0160	0.0149	0.0132
	ABC	Mean	0.0468	0.0471	**0.0446**	0.0449	0.0443	0.0496	0.0494	**0.0470**
		Std.	0.0020	0.0036	0.0021	0.0019	0.0103	0.0135	0.0108	0.0101
Pde	BR	Mean	0.0225	0.0231	**0.0214**	0.0247	0.0316	0.0331	0.0314	0.0295
		Std.	0.0046	0.0067	0.0038	0.0056	0.0167	0.0191	0.0157	0.0166
	LM	Mean	0.0290	0.0265	0.0283	0.0273	0.0280	0.0399	0.0326	0.0350
		Std.	0.0080	0.0077	0.0117	0.0077	0.0123	0.0222	0.0187	0.0199
	SCG	Mean	0.0259	0.0259	0.0245	0.0228	0.0267	0.0356	0.0282	0.0350
		Std.	0.0078	0.0069	0.0061	0.0053	0.0102	0.0210	0.0123	0.0168
	ABC	Mean	0.0320	0.0283	0.0243	0.0229	0.0304	**0.0251**	0.0273	0.0287
		Std.	0.0500	0.0026	0.0017	0.0017	0.0179	0.0125	0.0118	0.0173

Table 5.4 Confusion matrix, TP:True positive, TN:True negative, FN:False negative, FP:False positive

		Prediction		
		Positive	Negative	
Actual	Condition Positive	TP	FN	$\sum P$
	Condition Negative	FP	TN	$\sum N$

classified as in actual data, TN (correct rejection) means that defect-free module is classified as defect-free, FP (false alarm, type I error) means that defect-free module is incorrectly classified as defected, FN (miss, type II error) means that defected module is incorrectly classified as defect-free.

The confusion matrices of the ANN classifiers trained by BR, LM, SCG and ABC algorithms are drawn for training and test phases, separately in Fig. 5.2. From the confusion matrix, ANN classifier trained by ABC algorithm shows a better performance in terms of TP, TN and their overall summation. The results of SCG algorithm are close to ABC algorithm. ABC algorithm is more successful on predicting defected modules while SCG algorithm is better on predicting defect-free modules.

Accuracy defined by Eq. 5.13 measures the ratio of the number correctly classified defected and defect-free modules to the total number of predictions. Precision given by Eq. 5.14 is the ratio of the number of correctly classified defected modules to the number of defected modules in the prediction. Recall defined by Eq. 5.15 is the ratio of the number of correctly classified defected modules to the number of defected modules in actual data. Recall and Precision counterbalances each other. Because precision is a measure of correctly classified defected modules with respect to the prediction data and the recall is a measure of correctly classified defected modules with respect to actual data, another metric which is harmonic mean of precision and recall is defined by Eq. 5.16.

$$Accuracy = \frac{TP + TN}{TP + FN + FP + TN} = \frac{TP + TN}{P + N} \tag{5.13}$$

$$Precision = \frac{TP}{TP + FP} \tag{5.14}$$

$$Recall = \frac{TP}{TP + FN} = \frac{TP}{P} \tag{5.15}$$

$$F_1 = \frac{2 * TP}{2 * TP + FP + FN} \tag{5.16}$$

Values of Accuracy, Precision, Recall and F1 metrics produced by the ANN classifiers trained by BR, SCG, LM and ABC algorithms and have 16 neurons in the hidden layer are presented in Table 5.5. On different problems, the algorithms have varying performances. In terms of overall performances considering all problems together, ABC algorithm is the best in terms of the accuracy metric, BR algorithm

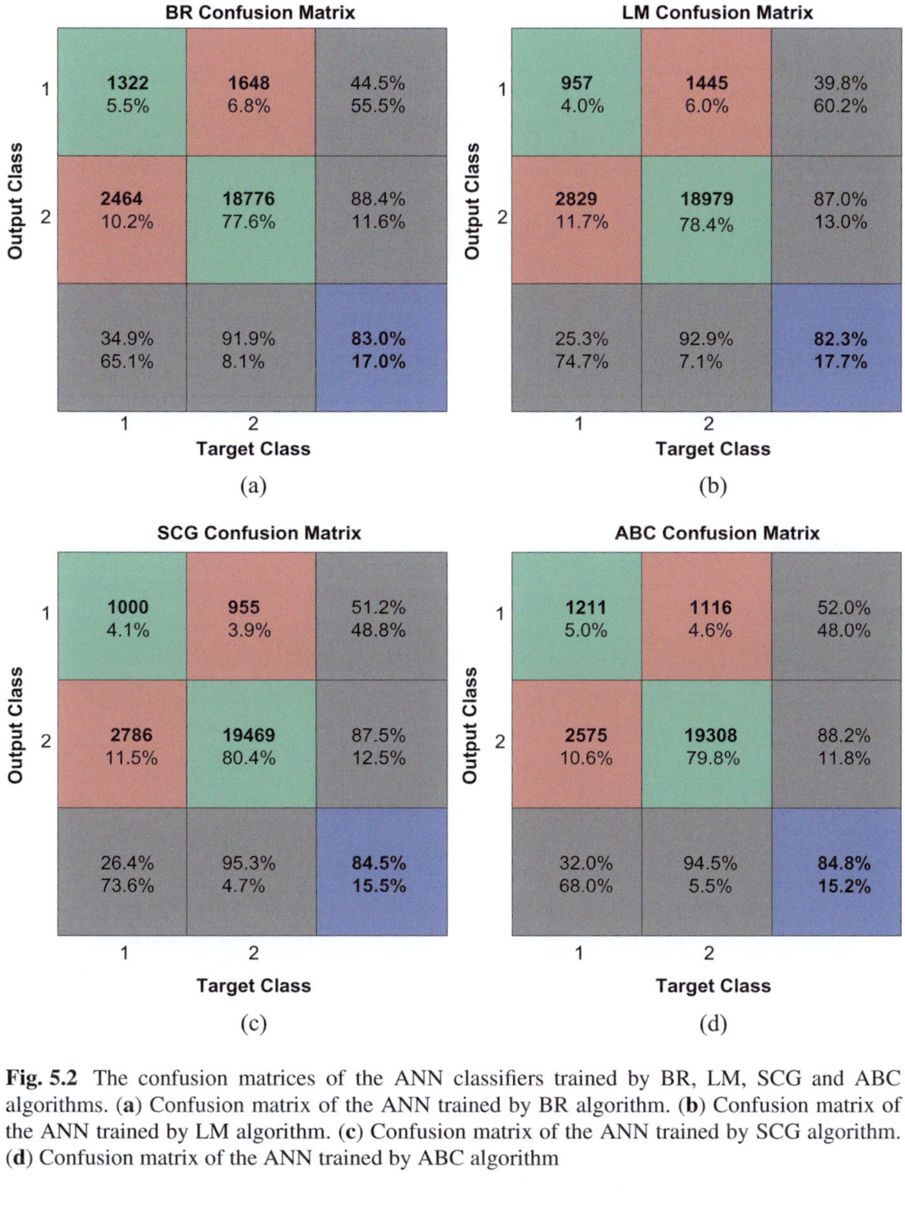

Fig. 5.2 The confusion matrices of the ANN classifiers trained by BR, LM, SCG and ABC algorithms. (**a**) Confusion matrix of the ANN trained by BR algorithm. (**b**) Confusion matrix of the ANN trained by LM algorithm. (**c**) Confusion matrix of the ANN trained by SCG algorithm. (**d**) Confusion matrix of the ANN trained by ABC algorithm

is the best in terms of the precision and the F1 metric and ABC algorithm is the second best. SCG algorithm is the best in terms of the recall metric on the train set and while ABC is the best on test results. On the test results, in terms of F1 metric, which balances Precision and Recall by harmonic mean, ABC algorithm is the best in overall.

Table 5.5 Accuracy, precision, recall and F1 metrics of the ANN classifiers trained by BR, LM, SCG and ABC algorithms

		Train data				Test data			
Problem	Algorithm	Accuracy	Precision	Recall	F1	Accuracy	Precision	Recall	F1
Eclipse	BR	0.7950	0.5067	0.5026	0.5046	0.7862	0.4831	0.4903	0.4867
	LM	0.8063	0.3386	0.5486	0.4188	0.7967	0.3051	0.5265	0.3863
	SCG	0.8390	0.4106	0.6817	0.5125	0.8340	0.3941	0.6801	0.4990
	ABC	0.8348	0.4775	0.6308	0.5435	0.8336	0.4767	0.6383	0.5458
Equinox	BR	0.6942	0.6698	0.6056	0.6361	0.6728	0.6159	0.5789	0.5968
	LM	0.6999	0.6206	0.6248	0.6227	0.6544	0.5329	0.5641	0.5480
	SCG	0.7010	0.5194	0.6590	0.5809	0.6762	0.4689	0.6159	0.5324
	ABC	0.7219	0.6549	0.6506	0.6527	0.7020	0.6176	0.6220	0.6198
Lucene	BR	0.8965	0.2288	0.3977	0.2905	0.8846	0.1724	0.2941	0.2174
	LM	0.8943	0.1914	0.3649	0.2511	0.8721	0.0828	0.1529	0.1074
	SCG	0.9108	0.1337	0.5798	0.2173	0.9029	0.0793	0.3898	0.1318
	ABC	0.9101	0.1387	0.5580	0.2221	0.9006	0.0897	0.3611	0.1436
Mylyn	BR	0.8738	0.2460	0.5567	0.3412	0.8637	0.1823	0.3967	0.2498
	LM	0.8456	0.1877	0.3492	0.2441	0.8412	0.1286	0.2414	0.1678
	SCG	0.8686	0.1454	0.5193	0.2271	0.8711	0.1161	0.4337	0.1832
	ABC	0.8654	0.1504	0.4786	0.2289	0.8699	0.1430	0.4319	0.2149
Pde	BR	0.8371	0.3072	0.3944	0.3454	0.8270	0.2897	0.3479	0.3162
	LM	0.8480	0.2516	0.4265	0.3165	0.8336	0.2178	0.3400	0.2655
	SCG	0.8353	0.2310	0.3614	0.2818	0.8317	0.2285	0.3381	0.2727
	ABC	0.8388	0.2469	0.3820	0.3000	0.8363	0.2457	0.3629	0.2930
All	BR	0.8411	**0.3863**	0.5012	**0.4363**	0.8302	**0.3492**	0.4451	0.3914
	LM	0.8365	0.3052	0.4789	0.3728	0.8235	0.2528	0.3984	0.3093
	SCG	0.8492	0.2856	**0.5510**	0.3762	0.8455	0.2641	0.5115	0.3484
	ABC	**0.8494**	0.3279	0.5451	0.4095	**0.8475**	0.3199	**0.5204**	**0.3962**

From the RMSE values, and the other evaluation metrics, ABC algorithm can be said a good training algorithm for ANNs established for software prediction. Besides, it is a robust algorithm when the standard deviations are considered.

5.4 Conclusion

Researchers have been inspired by the ants, termites, bees, birds, and fishes which have collective intelligence to survive and to adapt the changes in their environment. New problems solving techniques are proposed to the literature based on their swarm intelligence characteristics.

Artificial Bee Colony algorithm is a successful swarm based meta-heuristic algorithm and has been applied to solve may kind of problems in the research field including unconstrained, constrained, linear, nonlinear, single-objective, multi-objective, continuous and combinatorial optimization problems. In this chapter, a comprehensive explanation about the ABC algorithm and swarm intelligence has been provided.

ABC algorithm has also been applied to a software engineering problem: software defect prediction. Essentially, defect prediction is a supervised classification problem and in this study, artificial neural network is employed as a classifier. ABC algorithm is used in training of the classifier for software defect prediction. The performance of the classifier trained by ABC algorithm is compared to those of the ANN classifiers trained by BR, LM, and SCG algorithms. From the results, ANN classifier trained by ABC algorithm is a successful, alternative tool for prediction problems.

Artificial Bee Colony algorithm provides an efficient optimization framework which has a balanced exploration and exploitation capability. There are studies which focus on improving local search ability of the algorithm. Besides, when the discrete representation is adopted for combinatorial problems, designing operators for the algorithm can be suggested as one of the challenging and important gaps to fill in the existing literature.

References

1. Akay B, Karaboga D (2012) A modified artificial bee colony algorithm for real-parameter optimization. Inf Sci 192:120–142
2. Akay B, Karaboga D (2015) A survey on the applications of artificial bee colony in signal, image, and video processing. Signal Image Video Process 9(4):967–990
3. Alba E, Chicano JF (2006) Training neural networks with GA hybrid algorithms. In: Alba E, Martí R (eds) Metaheuristic procedures for training neural networks. Springer, Heidelberg, p 118
4. Azzini A (2006) A new genetic approach for neural network design and optimization. Ph.D. thesis, Universita Degli Studi Di Milano

5. Boehm BW (1984) Software engineering economics. IEEE Trans Softw Eng SE-10(1):4–21
6. Boozarjomehry RB (1997) Application of artincial intelligence in feedback linearization. Ph.D. thesis, The University of Calgary
7. Boozarjomehry RB, Svrcek W (2001) Automatic design of neural network structures. Comput Chem Eng 25:1075–1088
8. Catal C (2011) Software fault prediction: a literature review and current trends. Expert Syst Appl 38(4):4626–4636
9. Catal C, Banarjee S (2012) Application of artificial immune systems paradigm for developing software fault prediction models. In: Khosrow-Pour M (ed) Machine learning: concepts, methodologies, tools and applications. IGI Global, Pennsylvania, pp 371–387
10. Chidamber SR, Kemerer CF (1994) A metrics suite for object oriented design. IEEE Trans Softw Eng 20(6):476–493
11. D'Ambros M, Lanza M, Robbes R (2010) An extensive comparison of bug prediction approaches. In: 2010 7th IEEE working conference on mining software repositories (MSR 2010), Cape Town, South Africa, pp 31–41
12. Dorigo M, Maniezzo V, Colorni A (1991) Positive feedback as a search strategy. Technical report 91-016, Politecnico di Milano, Italy
13. Foresee FD, Hagan MT (1997) Gauss-newton approximation to bayesian regularization. In: Proceedings of the 1997 international joint conference on neural networks, Houston, pp 1930–1935
14. Karaboga D (2005) An idea based on honey bee swarm for numerical optimization. Technical report TR06, Erciyes University, Engineering Faculty, Computer Engineering Department
15. Karaboga D (2010) Artificial bee colony algorithm. Scholarpedia 5(3):6915 revision #91003
16. Karaboga D, Akay B (2009) A survey: algorithms simulating bee swarm intelligence. Artif Intell Rev 31(1–4):61–85
17. Karaboga D, Basturk B (2007) A powerful and efficient algorithm for numerical function optimization: artificial bee colony (abc) algorithm. J Glob Optim 39(3):459–471
18. Karaboga D, Gorkemli B, Ozturk C, Karaboga N (2012) A comprehensive survey: artificial bee colony (ABC) algorithm and applications. Artif Intell Rev 42(1):21–57
19. Kennedy J, Eberhart R (1995) Particle swarm optimization. In: IEEE international conference on neural networks, Piscataway, pp 1942–1948
20. Levenberg K (1944) A method for the solution of certain non-linear problems in least squares. Q J Appl Math II(2):164–168
21. MacKay DJC (1992) Bayesian interpolation. Neural Comput 4(3):415–447
22. Menzies T, Greenwald J, Frank A (2007) Data mining static code attributes to learn defect predictors. IEEE Trans Softw Eng 33(1):2–13
23. Møller MF (1993) A scaled conjugate gradient algorithm for fast supervised learning. Neural Netw 6(4):525–533
24. Schach SR (2008) Object-oriented software engineering. McGraw-Hill, New York
25. Widrow B, Rumelhart DE, Lehr MA (1994) Neural networks: applications in industry, business and science. Commun ACM 37(3):93–105
26. Zhang GP (2000) Neural networks for classification: a survey. Trans Sys Man Cyber Part C 30(4):451–462

Chapter 6
Firefly Algorithm and Its Applications in Engineering Optimization

Dilip Kumar, B. G. Rajeev Gandhi, and Rajib Kumar Bhattacharjya

Abstract Metaheuristic algorithms are optimization algorithms which attempt to enhance the degree of resolution of the solution space iteratively. This is performed by utilizing guided search methods along with some randomness properties. These algorithms are motivated by biological phenomena or the social behavior of the species. While the deterministic optimization methods depend on the nature of the optimization problem, the metaheuristic algorithms are generally problem independent. Due to their specific advantages over the classical methods, these algorithms have been used extensively in solving the different problems in the fields of science and engineering. One such metaheuristic algorithm is the firefly algorithm. It is inspired by the flashing behavior of fireflies and widely used for solving nonlinear- nonconvex optimization problems. This chapter describes the firefly algorithm and its recent modifications. The sensitivity of the parameters affecting the firefly algorithm along with the solution to optimization problems are discussed in this chapter.

Keywords Metaheuristic algorithm · Nonconvex problems · Swarm intelligence

6.1 Introduction

The optimization methods are a mathematical procedure for obtaining the best possible solution of a problem under a given set of conditions. There are different classical methods that can be used for solving an optimization problem. However,

---

D. Kumar (✉)
Department of Civil Engineering, G B Pant Engineering College, Pauri, Uttarakhand, India

B. G. R. Gandhi · R. K. Bhattacharjya
Department of Civil Engineering, Indian Institute of Technology Guwahati, Guwahati, Assam, India
e-mail: b.rajeev@iitg.ac.in; rkbc@iitg.ac.in

F. Bennis, R. K. Bhattacharjya (eds.), *Nature-Inspired Methods for Metaheuristics Optimization*, Modeling and Optimization in Science and Technologies 16,
https://doi.org/10.1007/978-3-030-26458-1_6

for the nonlinear nonconvex problem, the classical methods are not suitable and may provide only the local optimal solution of the problem. In such a situation [1], the global optimization algorithms have to be used for obtaining the best possible solution of the problem. Global optimization is important in many applications, such as water resources management, various structural design, image processing analysis, and network related problems.

The optimization methods available so far can be broadly classified into two main sections: the deterministic algorithms and the stochastic algorithms. In case of deterministic algorithm, we will reach at the same solution every time if we start the search process from the same initial solution. The benefit of using such type of algorithms is that they always converge at the same optima wherever the search begins in a convex space. This might not be applicable for all the problems, because not every problem has a convex search space. In such cases, the algorithm leads to different results based on the location of the initial point. On the other hand, stochastic algorithms use some randomness in their approaches as the name 'stochastic' suggests. Due to this, the stochastic algorithm has the capability to come out of the local optima and can explore the entire search space to obtain the global optimal solution of the problem. Due to this characteristic, they always produce unrepeatable path in each individual run. This assures the algorithm to search all the space and converge to the global optima, if the algorithm is allowed to go with more and more iterations. Most of the stochastic algorithms fall under the category of meta-heuristic algorithms. These algorithms are influenced by the different biological processes as well as social behavior of different species. The algorithms are simple in nature but more powerful and efficient in solving complex engineering optimization problems [6, 7]. Some of them are Genetic Algorithm (GA), Ant Colony Optimization (ACO), Particle Swarm Optimization (PSO), Artificial Bee Colony (ABC), Firefly Algorithm (FA), etc.

Besides their nature in solving complex engineering problems, these algorithms also have some disadvantages. One of the common disadvantages is that one has to properly set the parameters of the algorithm in an effective manner. This can be converted to an advantage if we know how to handle the parameters and understand their importance over the result. The parameters of the algorithm are capable to increase the diversity of the search or to intensely search a local region. These two characteristics of diversification and intensification are the major traits of the algorithm [18, 19]. The first trait of diversification searches the solution globally and hence has a better chance to get to the global optima and at the same time it can take a very large amount of time to converge. The second trait of intensification might give the solution fast enough but has the problem of giving a local optimal solution [18]. If we can balance between these two components and find a way to exactly use these two properties, the solution can be made global optima in a minimum amount of time and computational effort [5, 18, 19]. In this chapter, we have presented the Firefly (FA) algorithm, sensitivity of its parameters and the application of the algorithm.

6.2 Firefly Algorithm

6.2.1 Philosophy of the Algorithm

As discussed earlier, Firefly is a metaheuristic global optimization algorithms motivated by the flashing behavior of fireflies. It was initially proposed by Xin-She Yang in 2008. He explained how the flashing behavior of fireflies can be used as an optimization algorithm. Fireflies are bioluminescent and thus produce flashes of light in a rhythm. They flash either to attract a potential partner or to protect themselves from a predator. Therefore, the fireflies move towards each other based on the intensity of the light. The intensity of light of firefly decreases as the distance in between the fireflies increase [19]. Firefly algorithm is thus followed by these three idealized rules:

- Fireflies are attracted towards each other, based on the intensity of light and regardless of gender.
- The attractiveness and brightness of each firefly are correlated and the firefly with lesser attraction moves towards the firefly with more attraction.
- The firefly brightness is correlated to the objective function [19].

6.2.2 Mathematical Background for the Algorithm

The light intensity and the attractiveness between the fireflies are considered as two important variables in the Firefly algorithm. Every firefly gets attracted towards the other firefly which has a greater brightness than its own flash. In other words, the attractiveness of any firefly is directly proportional to its intensity of light in the flash. The light intensity is known to be inversely proportional to the distance from which the intensity is measured. Therefore, the attractiveness decreases as the distance '*r*' increases. Attractiveness can be defined as the light intensity as observed by another firefly, thus the attractiveness is defined as β as given in Eq. 6.1. Here β_0 is the attractiveness when the distance is zero, γ is the light absorption coefficient and the distance between the fireflies is given as r.

$$\beta = \beta_0 e^{-\gamma r^2} \tag{6.1}$$

The fireflies are assigned the Cartesian coordinates and thus the distance is given in Eq. 6.2 where d is the number of dimensions.

$$r_{ij} = \left|x_i - x_j\right| = \sqrt{\sum_{k=1}^{d} \left(x_{k,i} - x_{k,j}\right)^2} \tag{6.2}$$

Fireflies are attracted towards each other and the movement of the fireflies as a result of the attraction can be calculated using Eq. 6.3. Here the firefly i is attracted to j and moves in its direction with some randomness coefficient α.

$$x_i^{t+1} = x_i^t + \beta_0 e^{-\gamma r_{ij}^2} \left(x_j^t - x_i^t\right) + \alpha \varepsilon_i \\ 0 < \varepsilon_i < 1 \tag{6.3}$$

Equation 6.3 addresses the most key part in the whole algorithm, the movement of the fireflies towards the best solution. The light absorption coefficient γ can have a small value or a large one. If the light absorption coefficient is chosen to be zero, the attractiveness of the fireflies becomes the same as that of the case in which the distance between the fireflies is zero. This means that the firefly with such brightness can be seen from any position making it the global best. If the coefficient γ becomes too large, the movement of the fireflies becomes random due to the introduction of the randomness parameter α. By considering these two asymptotic behaviors, the firefly algorithm can be implemented.

6.2.3 Modified Firefly Algorithm

In case of standard Firefly algorithm, the global best solution is considered as the current best solution. As also seen that the movement of the fireflies are dominated by some randomness in each iteration. Because of these random movements, the attractiveness towards the best firefly might be lost after some iterations which will eventually influence the performance of the algorithm [11, 16, 19]. To manage the randomness in the algorithm, Tilahun and Ong [16] proposed that the fireflies can move in an orderly manner instead of the random movements. The movement should be ordered towards the best solution. His proposed modification has been implemented by generation of a unit vector in the direction of improvement of brightness and to move the firefly in the direction that leads the increase in brightness of the firefly [14, 15]. If no such improvement is found, the current brightest firefly will stay at the current position. Wang et al. [17] also presented some modification to the algorithm to add custom flight for improvement in term of localized searching for a closer solution. Yu et al. [20], Olamaei et al. [12], Farahani et al. [8] proposed an adaptive formulation for the randomization value α. They proposed that when the value of α is large, it is better for the fireflies to explore the unknown search space like global optimization. For smaller values of α, the fireflies will perform the local search.

6.2.4 *Advantages of FA*

Following are some of the advantages of FA algorithms:

- FA can deal with highly non- linear, nonconvex optimization problems efficiently.
- The convergence of FA is fast compared to the other classical methods in achieving the global optima.
- Hybrid tools can be formulated by combining FA with other optimization algorithms [9, 10, 13].
- A good initial solution is not necessary to start the optimization process i.e. it leads to the same optima irrespective of the initial solution.

6.3 Parameters of the Algorithm and Their Sensitivity

The parameters of the firefly algorithm are the light absorption coefficient 'γ', the maximum attractiveness 'β_0' and the randomness parameter 'α'. The algorithm is sensitive to these three parameters. As it was mentioned earlier, the firefly algorithm is a meta-heuristic algorithm which gives the best results when the problem is nonlinear and nonconvex. Hence the 'Ackley' function is chosen to test the parameters and their sensitivity. Each of the coefficients are tested on a wide range of their variation and the sensitivity of these parameters on the solution are presented. The population size is taken as 25. The tolerance of the function is chosen to be 10^{-6} whereas the best value of the function is 0. The maximum number of iterations are taken to be 500. As the algorithm has some randomness as an inherent property, the solutions are run over 5 runs and the results along with the mean are presented.

6.3.1 *Light Absorption Coefficient 'γ'*

The coefficient 'γ' can have a smaller value to a very large value. However, it is not possible to search the limiting situation and hence the range of 0–50 is chosen for the evaluation of the algorithm. The other two parameters β_0 and α are kept at 2 and 0.2 respectively. Figure 6.1 shows the variation of the best function value with the gamma values and Fig. 6.2 shows the variation of number of iterations before convergence of the function with the gamma values.

From the figures, it can be concluded that the convergence of the algorithm is not very sensitive to gamma value. The experiment shows that the algorithm provides the best result when gamma value is kept in the range of 1 and 2. Therefore the light absorption coefficient is chosen to be 1 for the study.

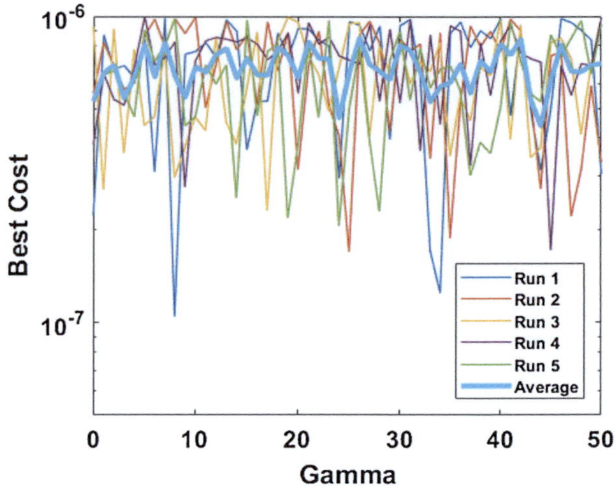

Fig. 6.1 Best cost variation with the gamma values from 0 to 50

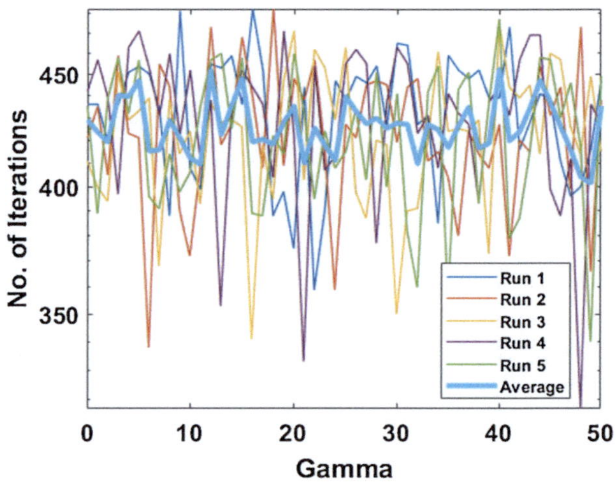

Fig. 6.2 No of iterations with gamma values from 0 to 50

6.3.2 *Maximum Attractiveness 'β_0'*

The coefficient 'β_0' can only be in the range of the variable space. The minimum and the maximum value of the variable space are chosen to be −10 and 10 respectively. Therefore, the value of the maximum attractiveness can only range from 0 to 10. The values of γ and α are fixed at 1 and 0.2 respectively. Figure 6.3 shows the variation of the best function value with the maximum attractiveness and Fig. 6.4 shows the maximum iterations before convergence at the optimal solution.

From the figures, it can be seen that the absolute zero value of the parameter leads to no solution and the values from 0.5 to 10 leads to the solution almost all the time.

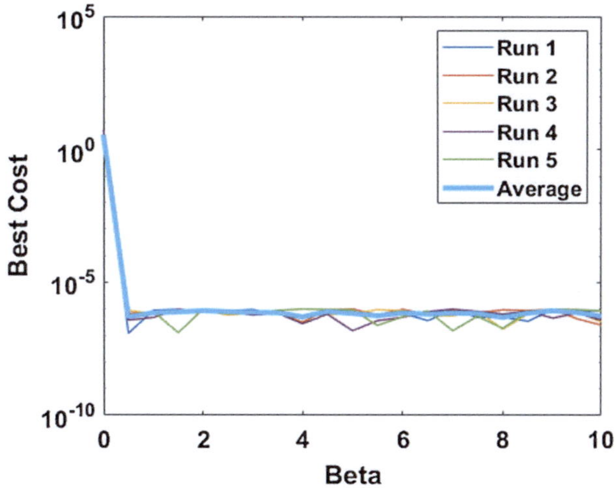

Fig. 6.3 Best cost variation with the β_0 values from 0 to 10

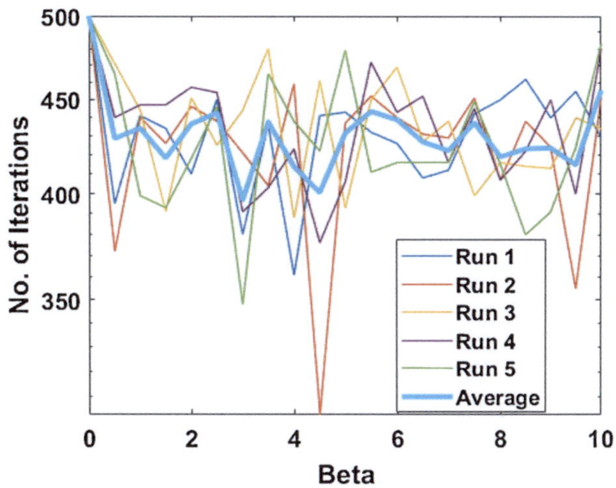

Fig. 6.4 No of iterations with β_0 values from 0 to 10

Although, from Fig. 6.4, it can be observed that the parameter when set between 2 and 4 leads to the best results. As such, any value within the range should work fine.

6.3.3 Randomness Parameter 'α'

The coefficient 'α' can also be only in the range of the whole variable space. The minimum and the maximum value of the variable space are chosen to be −10 and 10 respectively. Therefore the value of the maximum attractiveness can only range

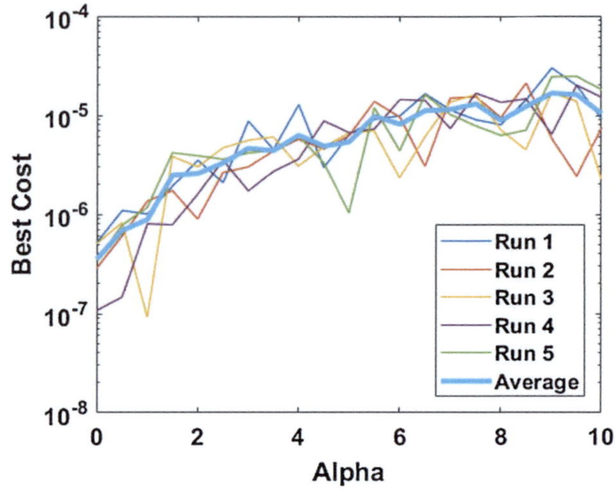

Fig. 6.5 Best cost variation with α values from 0 to 10

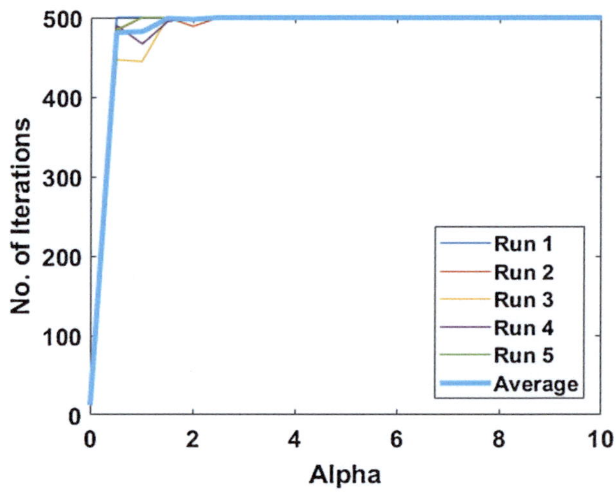

Fig. 6.6 No of iterations with α values from 0 to 10

from 0 to 10. The values of γ and β_0 are fixed at 1 and 2 respectively. Figure 6.5 shows the variation of the best function value with the randomness parameter and Fig. 6.6 shows the maximum iterations before convergence with the randomness parameter.

The parameter α has a considerable amount of variation with different values. Both the figures ensure that the value of the randomness when in a very small range yields the best results. Figure 6.6 shows that the convergence only occurs when the value of the randomness coefficient is between 0 and 0.5. Therefore for all the other calculations, the parameters value of 0.2 is taken.

The parameter sensitivity study concludes that the values of the coefficient α should be between 0 and 0.5, β_0 to be within the range of the solution space and γ to be in between 1 and 2 for better results.

6.4 Firefly Algorithm Applied to a Mathematical Function

The "Ackley' function which has multiple local optima and a single global optimal solution has been considered in this chapter. This is a very difficult problem as it has several local minimum solutions and the algorithm can be trapped in one of the minima.

The mathematical form of the "Ackley' function is given in Eq. 6.4.

$$f(x, y) = -20e^{-0.2\sqrt{0.5(x^2+y^2)}} - e^{0.5(\cos 2\pi x+\cos 2\pi y)} + e + 20 \tag{6.4}$$

The optimal value of the function is 0 at the point 0,0. The representation of the function as a surface and contour plot is given in Fig. 6.7.

The function is solved using the firefly algorithm, with population size of 25, γ value of 1, β_o value of 2 and α value of 0.2. The randomization parameter α is dampened by a factor of 0.98 every iteration. This means by the end of the 500th iteration the value of α will reach the value 8.2×10^{-6}. This decreases the randomization in every iteration and helps the solution to converge by the end of all the iterations.

By applying these parameters, the optimization is run for twenty different iterations and the results over the 20 iterations are all given in Fig. 6.8. It can be observed that the algorithm is capable of finding the global optimal solution of the problem in every run of the algorithm.

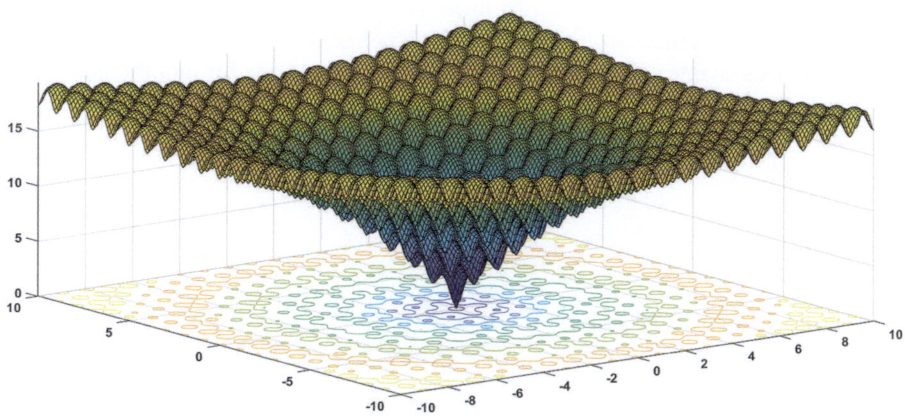

Fig. 6.7 Surface and contour plot of Ackley function

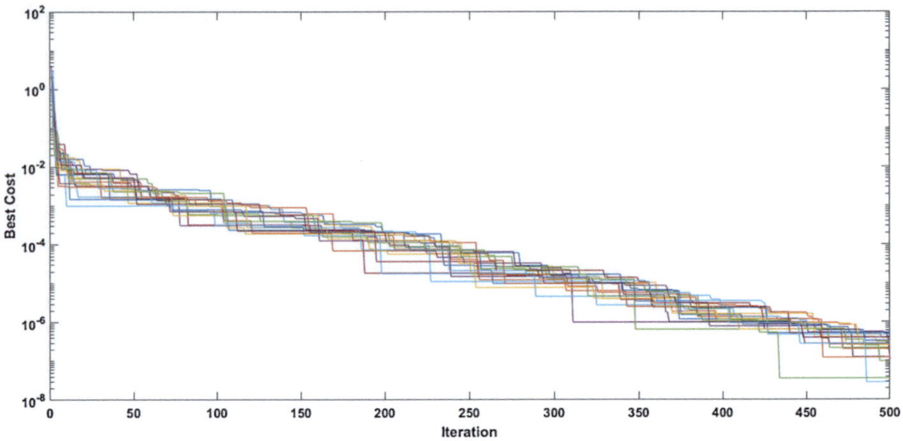

Fig. 6.8 Best cost of Ackley function with iterations over 20 runs

6.5 Conclusion

Firefly algorithm is considered to be a new algorithm in the swarm intelligence family. Despite the fact that it is new, the firefly algorithm have been used in the various types of problems. This algorithm have already proved that it is superior as compared to the other algorithms of swarm intelligence. Even though the firefly algorithm has proven to be superior compared to the previous swarm intelligence, there is a possibility of modification to improve the local search as well as global search [4]. This is to ensure that the solution obtained is the global optimum and not a premature local optimal solution. Firefly algorithm is a suitable metaheuristic algorithm that can be used for multidimensional and nonlinear problems [2, 3]. Since its proposal, the firefly algorithm has undergone some major changes in its structure. The modified algorithm to improve the position of the fireflies and the introduction of the randomization parameter to perform local search were discussed in detail in this chapter. The algorithm has much larger scope for modification with the future studies and it remains one of the most applicative and variant model in the field of science and engineering.

References

1. Abdullah A, Deris S, Mohamad M, Hashim S (2012) A new hybrid firefly algorithm for complex and nonlinear problem. In: Distributed computing and artificial intelligence. Springer, Berlin/Heidelberg, pp 673–680
2. Apostolopoulos T, Vlachos A (2010) Application of the firefly algorithm for solving the economic emissions load dispatch problem. Int J Comb 2011:1–23

3. Apostolopoulos T, Vlachos A (2011) Application of the firefly algorithm for solving the economic emissions load dispatch problem. Int J Comb 2011:1–23
4. Bhattacharjya RK (2006) Optimal design of open channel section incorporating critical flow condition. J Irrig Drain Eng (ASCE) 132(5):513–518. https://doi.org/10.1061/(ASCE)0733-9437(2006)132:5(513)
5. Blum C, Roli A (2003) Metaheuristics in combinatorial optimization: overview and conceptual comparison. ACM Comput Surv (CSUR) 35(3):268–308
6. Breedam A (2001) Comparing descent heuristics and metaheuristics for the vehicle routing problem. Comput Oper Res 28(4):289–315
7. El-Sawy A, Zaki E, Rizk-Allah R (2012) A novel hybrid ant Colony optimization and firefly algorithm for solving constrained engineering design problems. J Nat Sci Math 6(1):1–22
8. Farahani S, Abshouri A, Nasiri B, Meybodi M (2011) A Gaussian firefly algorithm. Int J Mach Learn Comput 1(5):448–453
9. Farahani S, Abshouri A, Nasiri B, Meybodi M (2012) Some hybrid models to improve firefly algorithm performance. Int J Artif Intell 8(S12):97–117
10. Farook S, Raju P (2013) Evolutionary hybrid genetic-firefly algorithm for global optimization. Int J Comput Eng Manag 16(3):37–45
11. Hassanzadeh T, Meybodi M (2012) A new hybrid algorithm based on firefly algorithm and cellular learning automata. Electrical engineering, (ICEE) 2012, 20th Iranian conference on. IEEE, pp 628–633
12. Olamaei J, Moradi M, Kaboodi T (2013) A new adaptive modified firefly algorithm to solve optimal capacitor placement problem 18th Electric Power Distribution Conference, pp 1–6
13. Rizk-Allah R, Zaki E, El-Sawy A (2013) Hybridizing ant colony optimization with firefly algorithm for unconstrained optimization problems. Appl Math Comput 224(3):473–483
14. Srivatsava P, Mallikarjun B, Yang X (2013) Optimal test sequence generation using firefly algorithm. Swarm Evol Comput 8:44–53
15. Talbi E (2009) Metaheuristics, 1st edn. Wiley, Hoboken
16. Tilahun S, Ong H (2012) Modified firefly algorithm. J Appl Math 2012:1–12
17. Wang G, Guo L, Duan H, Liu L, Wang H (2012) A modified firefly algorithm for UCAV path planning. Int J Hybrid Inf Technol 5(3):123–144
18. Yang X and Karamanoglu M. (2013). Swarm Intelligence and Bio-Inspired Computation: An Overview. https://doi.org/10.1016/B978-0-12-405163-8.00001-6.
19. Yang X (2010) Nature-inspired Metaheuristic algorithms, 2nd edn. Luniver Press, Frome
20. Yu S, Yang S, Su S (2013) Self-adaptive step firefly algorithm. J Appl Math 2013:1–8

Chapter 7
Introduction to Shuffled Frog Leaping Algorithm and Its Sensitivity to the Parameters of the Algorithm

B. G. Rajeev Gandhi and R. K. Bhattacharjya

Abstract The optimization techniques is an essential tool in many fields of science and engineering. The evolution of humans and several other species teaches us many techniques which can be replicated to solve the nonlinear nonconvex optimization problems in an efficient and faster way. One such evolutionary trait from frogs can be thought of as an efficient optimization algorithm. This algorithm is based on the food search by an army of frogs. An army of frogs in a swamp search for food on the rocks floating around by leaping onto the nearest possible rock and also communicating with the rest of the frogs for improvising the search process. Each individual frog tries to get maximum food as fast as possible and thus developing a strategy to time their leaps in the best possible way. The algorithm developed replicating this process is called the Shuffled Frog Leaping Algorithm. This chapter discusses the basic principle of Shuffled Frog Leaping Algorithm and its efficiency using common benchmark optimization functions.

Keywords Memetic algorithm · Metaheuristic optimization · Non-convex optimization

7.1 Introduction

Shuffled frog leaping algorithm is a population-based metaheuristic algorithm and a memetic algorithm. Memetic algorithm approaches the solution based on cultural information transfer from one individual to another. Memetic is the term first mentioned by Dawkins [1] in his book 'The Selfish Gene'. A 'meme' is simply some cultural or intellectual information that survives long enough and passes the knowledge from mind to mind by imitation or other non-genetic means.

B. G. R. Gandhi (✉) · R. K. Bhattacharjya
Department of Civil Engineering, Indian Institute of Technology Guwahati, Guwahati, Assam, India
e-mail: b.rajeev@iitg.ac.in; rkbc@iitg.ac.in

F. Bennis, R. K. Bhattacharjya (eds.), *Nature-Inspired Methods for Metaheuristics Optimization*, Modeling and Optimization in Science and Technologies 16,
https://doi.org/10.1007/978-3-030-26458-1_7

Memetic passage of information is faster than the genetic passage, as the transfer of information is in the space. However, the genetic passage takes several generations for the same [7]. The principle and philosophy of shuffled frog leaping algorithm is a combination of Shuffled Complex Evolution (SCE) developed originally by Duan et al. [2] and Particle Swarm Optimization (PSO) developed by Eberhart and Kennedy [3]. SCE algorithm is based on the global process of natural evolution where the population is divided into several complexes and evolves independently. After certain evolutions, they are forced to mix which is called shuffling. The local evolution or the evolution at each sub-complex are driven by the philosophy of PSO [6] in the Shuffled Frog Leaping Algorithm (SFLA). This is almost similar to the teams of researches and conference, where the researchers are allowed to work with their own way of approaching a problem along with their team members. The corresponding teams discuss their ideas in the conference and implement the best idea in the actual applications [7].

7.2 Methodology for SFLA

The SFLA has two steps as mentioned in the introduction. One is the shuffling global search and the other is localized search in each of the memeplexes. The global search and shuffling are done after a specific number of evolutions of all the memeplexes. The frogs are initially placed at positions selected randomly in the search space. The population of all the frogs is then divided into memeplexes where the frogs within the memeplex are allowed to evolve independently. The idea of memetic evolution is that some of the frogs in each of the memeplex can be selected and infected with the idea of the best meme in the memeplex. Then those frogs travel towards the best position. If they cannot get through with the best solution, then the frogs are infected with the best meme in all of the memeplexes (global best). For a significant improvement, the frogs with better memes should dominate the infection rather than the frogs with poor ideas (memes). Triangular probability distribution helps in achieving such a goal. After evolution in memeplexes, the frogs are forced to return for the reshuffle as well as to carry forward the global search [5]. This cycle of improving each memeplex and shuffling gives the whole philosophy of Shuffled Frog Leaping Algorithm (SFLA). The Figs. 7.1 and 7.2 represent the flowcharts for the whole process of shuffling and frog leaping algorithm respectively.

The FLA described in the flowchart is the actual portion where the frogs in the memeplexes evolve. The flowchart given in Fig. 7.2 explains the evolution in the memeplexes.

The shuffled frog leaping algorithm is explained below. The first part of the algorithm is Shuffling and global search. The subset of this algorithm (the second part) is the Frog Leaping Algorithm.

Step 1: Initialize the number of memeplex (n_m) and the no of frogs in each memeplex (n_f). The total population of frogs in the swamp will be $P_f = n_{m*}n_f$.

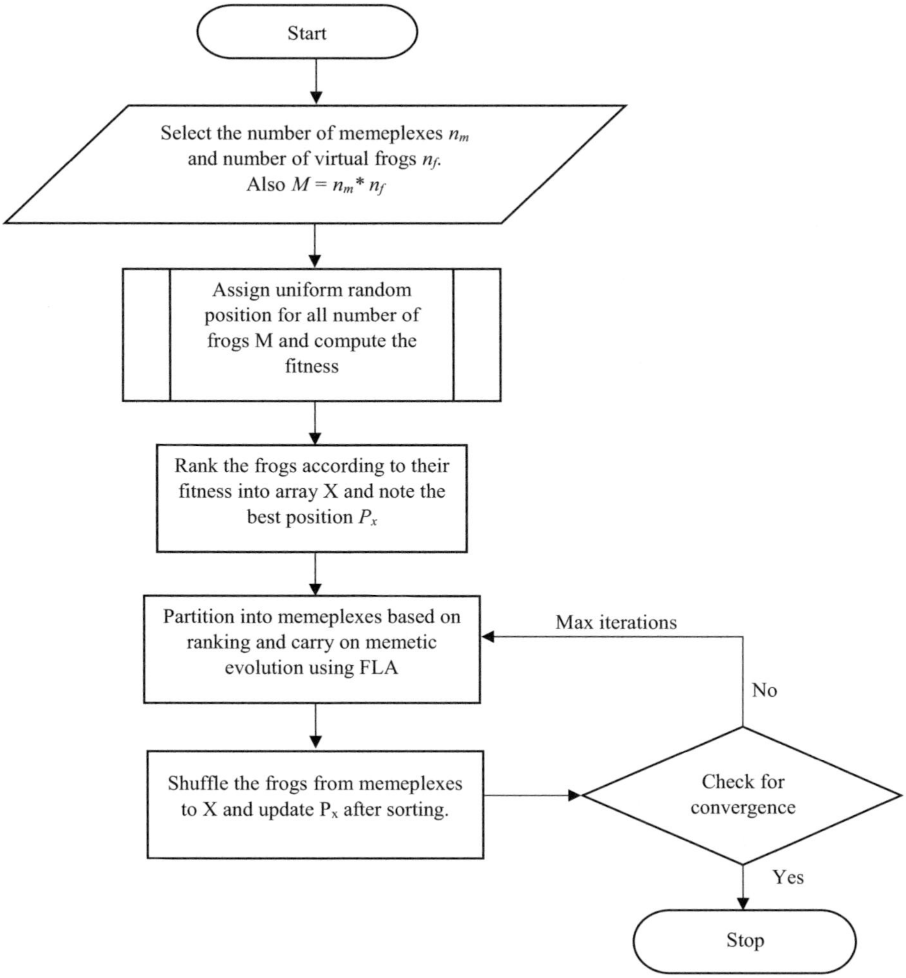

Fig. 7.1 Shuffled frog leaping algorithm on the large scale

Step 2: Select the decision variables for each of the frogs called memes for the virtual population of frogs and compute their performances based on the objective function or the fitness function. Each meme carries a virtual frog consisting of all the decision variables in the feasible space. $M_i = [M_i\ (1)\ M_i\ (2)\ M_i\ (3)\ \ldots\ M_i\ (d)]$ where $i = 1 \ldots .\ P_f$.

Step 3: Rank the frogs according to their performances from best to worst in descending order and store in the vector X. Note the best position in the whole population as P_x.

Step 4: Partition the sorted frogs into 'n_m' different memeplexes with 'n_f' frogs each such that the frogs sort themselves into each memeplex in order. Vectors in

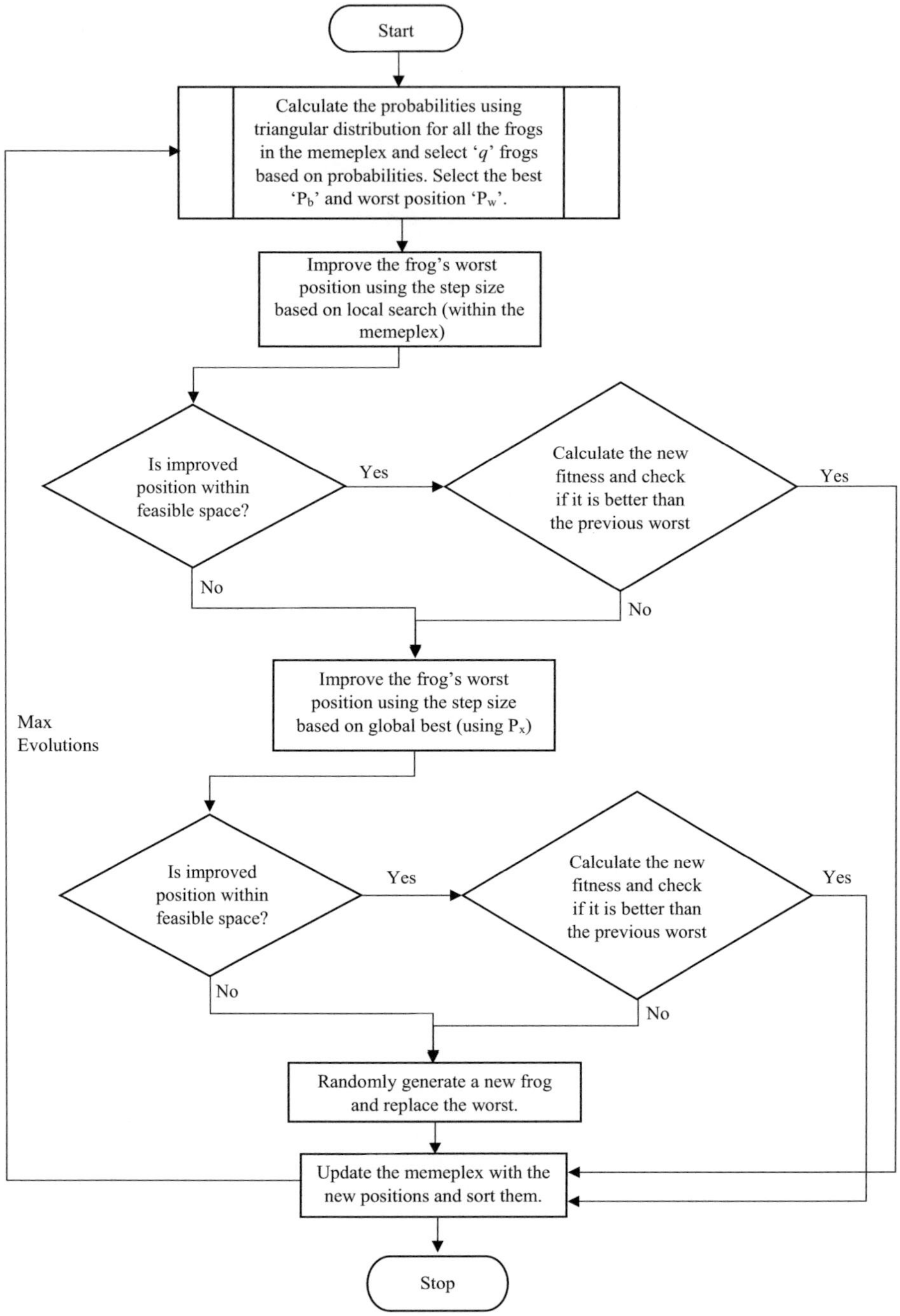

Fig. 7.2 Frog Leaping Algorithm (detailed)

each memeplex are called Y each containing n_f frogs. For example if $n_m = 5$. The frog with rank 1,2,3,4 and 5 would go to memeplex 1. The frogs with ranks 6,7,8,9 and 10 to memeplex 2 and so on.

Step 5: Memetic evolution within each of the memeplexes using Frog Leaping Algorithm (FLA). Y_{old} to Y_{new}.

Step 6: After a certain number of memetic evolutions, the frogs are shuffled into memeplexes again replacing the previous X with Y_{new}. The ranking of frogs are again done and the new best position is updated in P_x.

Step 7: If convergence criterion for the fitness function is satisfied, the algorithm stops. Otherwise, the algorithm resumes to step 4 into division onto memeplexes.

7.2.1 Frog Leaping Algorithm

Step 1: It is not always the best way to involve all the frogs for the local search because the frogs will tend to incline towards the local optima. The best way to overcome this is to construct a submemeplex based on probabilities of the frogs in the memeplex according to their fitness in a triangular distribution defined by.

$$p_j = \frac{2\left(n_f + 1 - j\right)}{n_f\left(n_f + 1\right)} \quad \text{Where } j = 1 \ldots . n_f \tag{7.1}$$

Step 2: Construct a sub memeplex with 'q' frogs based on the probability 'p_j'. The best and the worst position are noted to be P_B and P_w.

Step 3: Improve the worst frog's position based on the best position in submemeplex. The step size is chosen based on the philosophy of PSO as described below.

Step size S $= minimum\ of\ \{int\,[rand \text{ x } (P_B - P_w)]\,, S_{max}\}\ for\ positive\ step$

$Maximum\ of\ \{int\,[rand \text{ x } (P_B - P_w)]\,, -S_{max}\}\ for\ negative\ step$

New position is given by $M_q = P_w + S$

If the new position is feasible then the function value is calculated and the worst frog is replaced with the new frog and go to step 6. Otherwise step 4 is continued with.

Step 4: Improve the worst frog's position based on the best position globally P_x. The step size is chosen based on the philosophy of PSO as described below.

Step size S $= minimum\ of\ \{int\,[rand \text{ x } (P_x - P_w)]\,, S_{max}\}\quad for\ positive\ step$

$Maximum\ of\ \{int\,[rand \text{ x } (P_x - P_w)]\,, -S_{max}\}\ for\ negative\ step$

New position is given by $M_q = P_w + S$

If the new position is feasible then the function value is calculated and the worst frog is replaced with the new frog and to step 6. Otherwise step 5 is continued with.

Step 5: Censorship is done by randomly generating a frog in the feasible space and replacing the worst position. This step is taken only if the local influence over the frog and the global influence both failed to improve the frog's position.
$M_q = r$ and $f(q)=f(r)$.

Step 6: Upgrade the memeplex with the original locations of new solutions added and sort the memeplex

Step 7: Check whether all the memeplexes are upgraded for the maximum number of evolutions. If they are then return to global search for shuffling. Else continue with step 1.

7.2.2 Parameters and Sensitivity

Selection of parameters effects the SFLA performance in many ways. There are major five parameters namely the number of memeplexes (n_m), number of virtual frogs in each memeplex (n_f), number of frogs selected for evolution (q), number of evolutions before reshuffling (N_e) and the maximum step size allowed in an evolution (S_{max}). Each of the factors has an effect individually and the selection of some combination also effects the solution in possible ways.

n_m: The number of memeplexes selected provides the diversity for the frogs initially to settle for more localized searches that can provide the global optima.

n_f: The number of frogs in each of memeplex effects the local memetic strategy if the number of frogs is too small. More number of frogs on the other hand increases the computational burden.

q: The number of frogs to be selected in each of the submemeplex effects the solution times. If too many frogs are selected, all the frogs are infected with unwanted ideas which eventually leads to censorship and increases the search period. On the other hand, if a lesser number of frogs are selected the information transfer will be too slow which will eventually increase the time to reach the optima.

N_e: The number of evolutions should be greater than 1. If the value is less there will be a frequent shuffling of frogs which will eventually reduce the exchange of ideas on a local scale. If the value is larger each memeplex shrinks to local optima resulting in a delay for the global search.

S_{max}: The maximum step size depends on the boundaries of the search space. A low value of maximum step size increases the probability to converge at a local optimal solution. On the other hand, a higher value of maximum step size may affect the global optimal exploration by missing the optima.

The combinations of all these parameters affect the optimization algorithm [4]. For example, the product of the number of memeplexes and the number of frogs actually effect the complexity of the problem. The product plays a key role in addressing the complexity, thus allowing the user to select one of the values based on the other. Number of total frogs in all the memeplexes and the number of evolutions directly affect the computational burden. If both values are high, it will take many function evaluations to eventually reach the optima. However, there is no specific way of selecting the parameters applicable to any function. They mostly depend upon the complexity of the function and number of decision variables.

The behavior of the virtual frogs as the loop proceeds on the Himmelblau function is shown in Fig. 7.3. This function has four optimal solutions with identical function value. In the Fig. 7.3, it can be observed that the frogs initially start at some

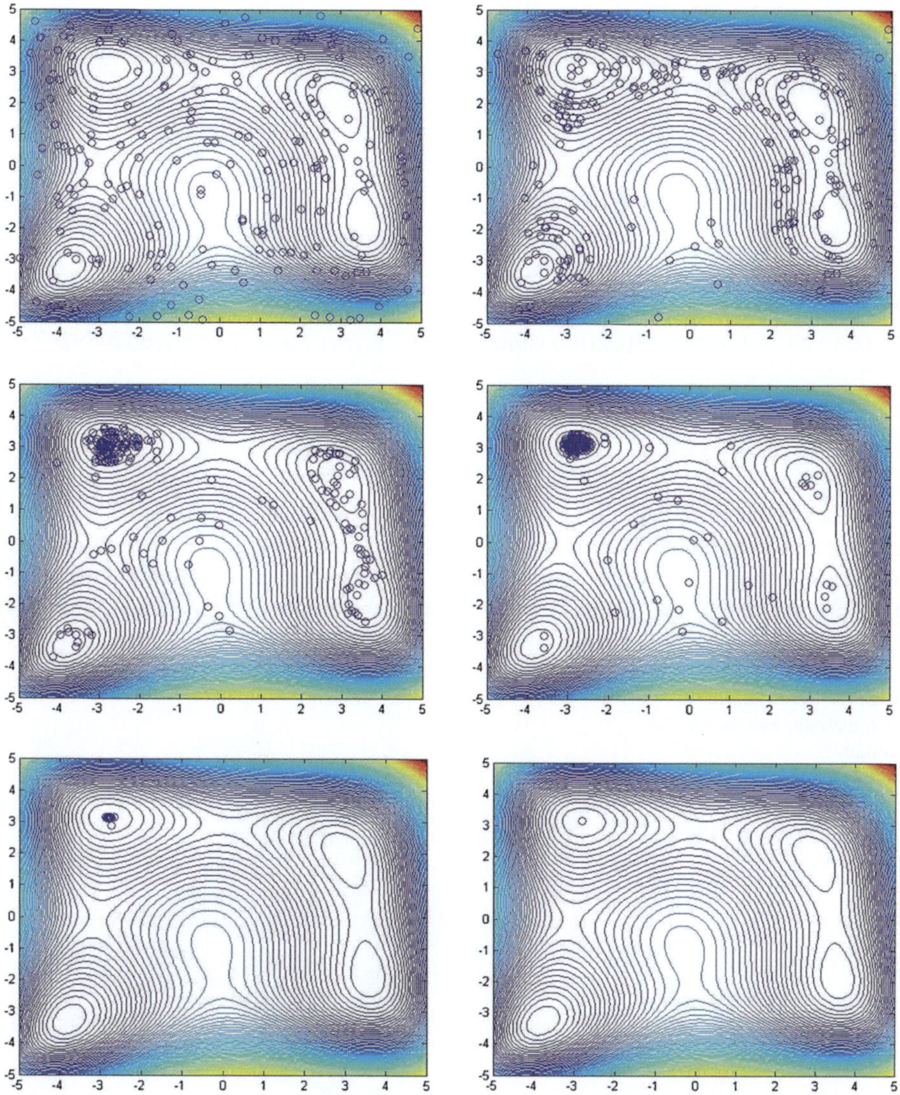

Fig. 7.3 Virtual frogs in movement at the start (top left), 2nd, 4th, 6th 10th and 20th iterations

random locations and leap towards their respective local optima according to their memeplexes.

All the local optimal points are checked and the final optimal solution is decided on the basis of the maximum number of frogs gathered at a single point. It can also be observed in Fig. 7.3 that all the optimal solutions are checked by the algorithm.

7.3 SFLA on Mathematical Functions

The performance of the algorithm on different functions can be tested and the results are described below.

7.3.1 *Himmelblau Function*

Himmelblau function is defined by Eq. 7.2, which has four identical minimum at (3,2), (−2.805118,3.131312), (−3.779310,−3.283186) and (3.584428,−1.848126) as (*x, y*) pairs where the function value is zero. The variation can be seen in Fig. 7.4.

$$f(x, y) = \left(x^2 + y - 11\right)^2 + \left(x + y^2 - 7\right)^2 \tag{7.2}$$

SFLA is run for 20 iterations under the conditions that n_m is 5, n_f is 10, q is 50% of the frogs in memeplex, maximum step S_{max} is 1 and the number of evolutions N_e is 5 before shuffling is done again. The mean of the function evaluations to reach the optima is 2790. The graph in Fig. 7.5 shows the number of function evaluations

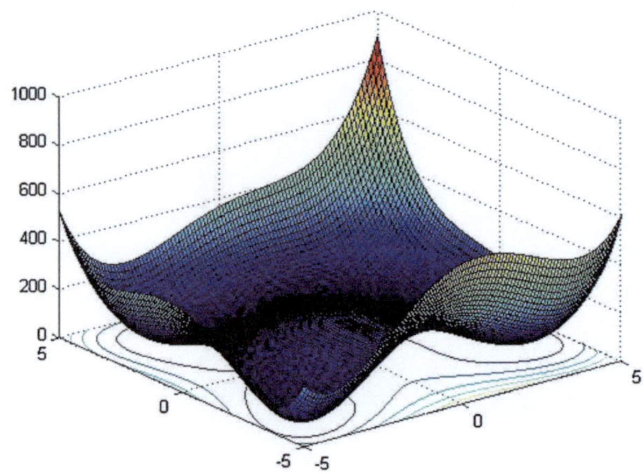

Fig. 7.4 Variation of Himmelblau function in the domain

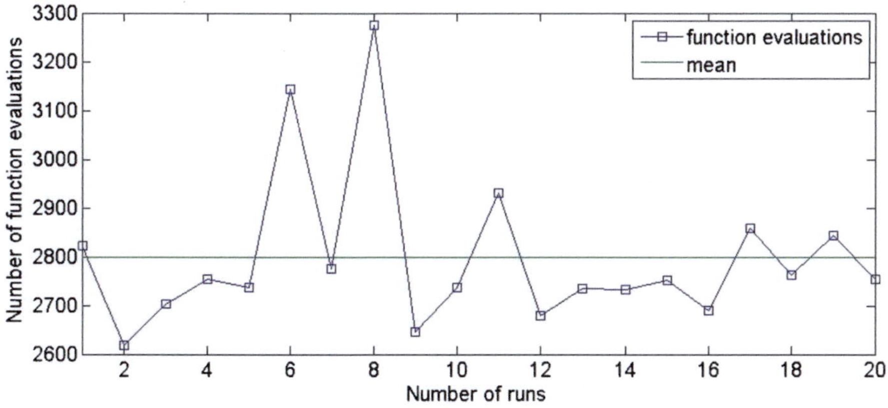

Fig. 7.5 Function evaluations varying with each run for 20 runs

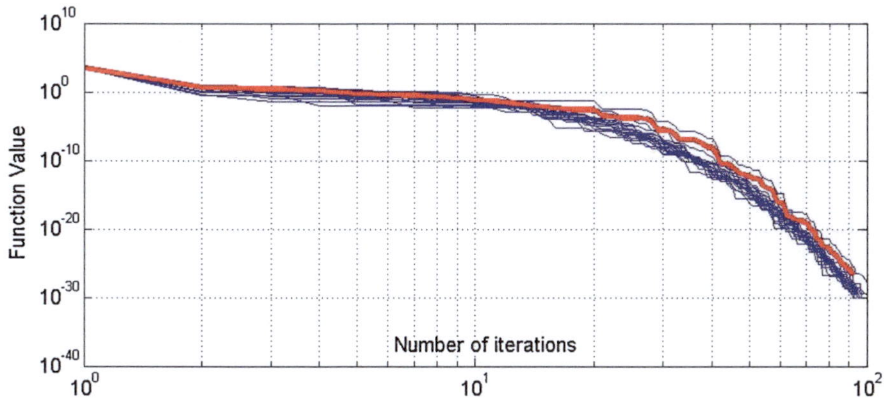

Fig. 7.6 Function value with the number of iterations in log-log scale

with number of runs. Figure 7.6 shows all the function values for 20 runs on one plot along with the mean (in thick red line) with the number of iterations.

7.3.2 Rosenbrock Function

Rosenbrock function is defined in the Eq. 7.3, with two variables and a minimum value of 0 at (1,1). The variation of the function in the domain is given in Fig. 7.7.

$$f(x, y) = 100\left(x_1^2 - x_2\right)^2 + (1 - x_1)^2 \tag{7.3}$$

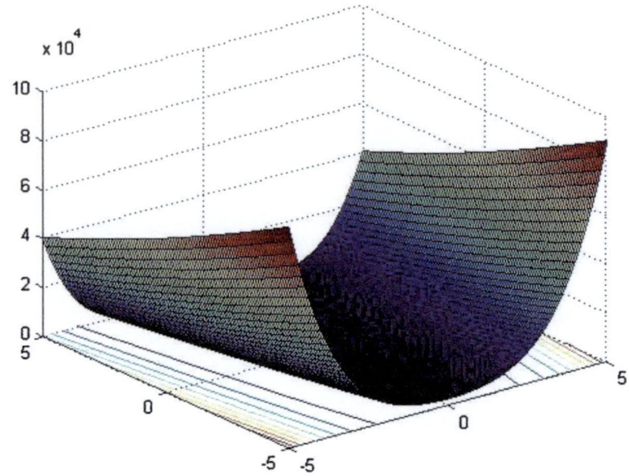

Fig. 7.7 Rosenbrock function in the search space

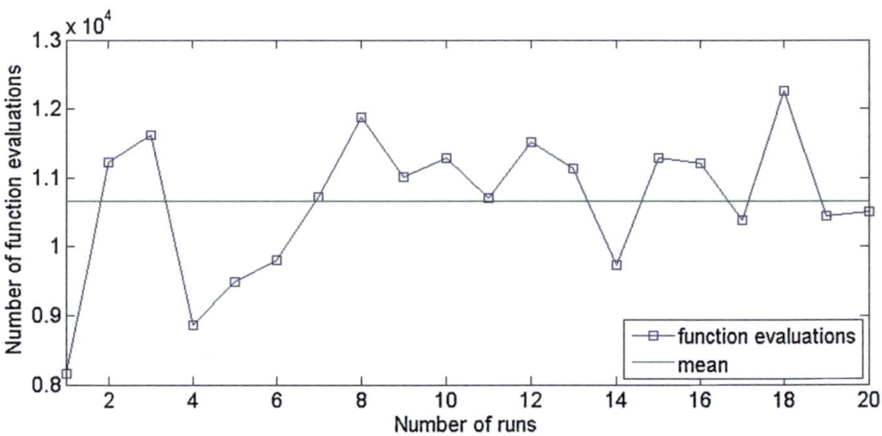

Fig. 7.8 Function evaluations varying with each run

SFLA is run for the Rosenbrock function for 20 iterations under the conditions that nm is 5, nf is 10, q is 50% of the frogs in memeplex, maximum step Smax is 1 and the number of evolutions Ne is 5 before shuffling is done again. The mean of the function evaluations to reach the optima is 10,660 evaluations. The graph in Fig. 7.8 shows the number of function evaluations with number of runs.

Figure 7.9 shows all the function values for 20 runs on one plot along with the mean (in thick red line) with the number of iterations in log-log scale.

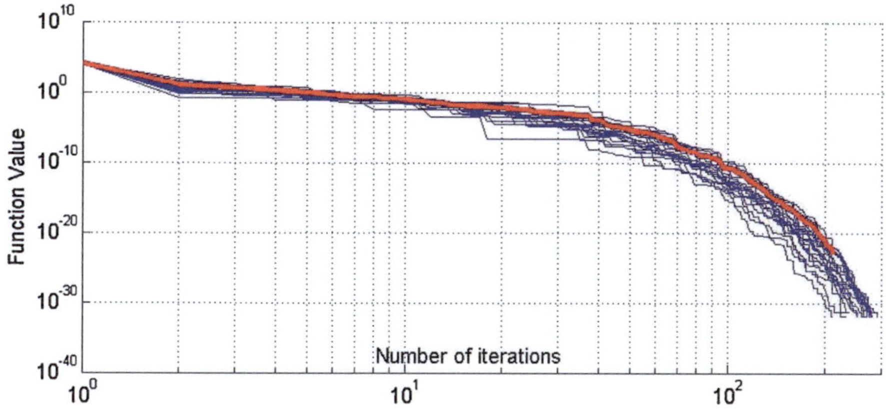

Fig. 7.9 Function value with number of iterations log-log

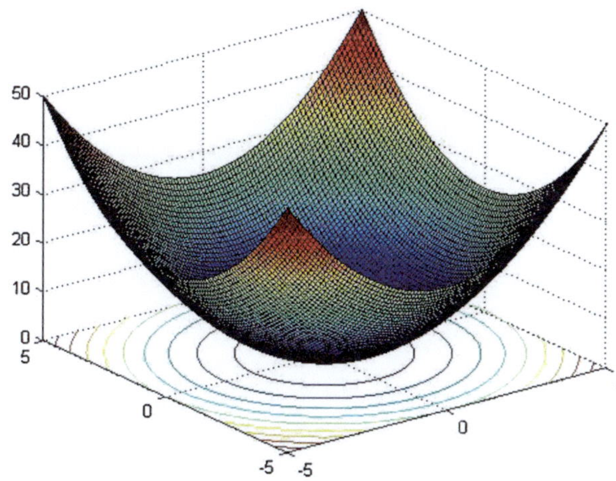

Fig. 7.10 Variation of Sphere function over the domain

7.3.3 Sphere Function

The Sphere function is defined in the Eq. 7.4 given below. It is a two variables function and the minimum value of of the function is 0 at (0,0). The variation of the function in the domain is shown in Fig. 7.10.

$$f(x, y) = x_1^2 + x_2^2 \tag{7.4}$$

SFLA is run for the Sphere function for 20 iterations under the conditions that n_m is 5, n_f is 10, q is 50% of the frogs in memeplex, maximum step S_{max} is 1 and the number of evolutions N_e is 5 before shuffling is done again. The mean of the

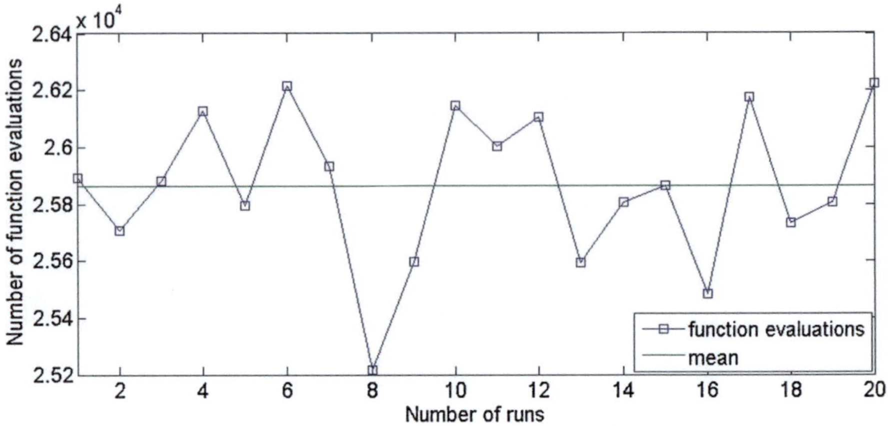

Fig. 7.11 Number of function evaluations with no of runs

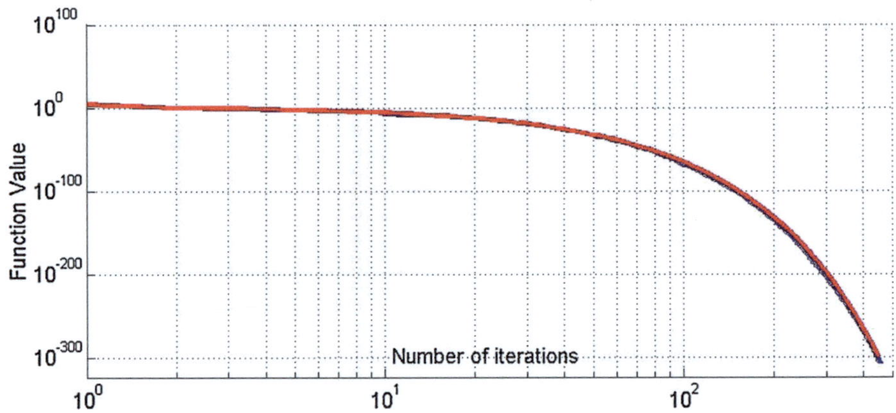

Fig. 7.12 Function value with number of iterations for all the 20 runs (log-log)

function evaluations to reach the optima is 25,865 evaluations. The graph in Fig. 7.11 shows the number of function evaluations with number of runs. Figure 7.12 shows all the function values for 20 runs on one plot along with the mean (in thick red line) with the number of iterations in log-log scale.

7.4 Conclusions

Shuffled Frog Leaping Algorithm (SFLA) is a memetic meta-heuristic algorithm to solve the optimization problems. The sample of virtual frogs which acts as memes evolve by leaping in the swamp. The algorithm is based on local and global searches for food and cultural evolution. The change in the position of frogs is obtained by the infection of best ideas to the worst frogs in each memeplex. The algorithm can be applied in diversified fields of engineering, science, and technology [4]. It is relatively very fast compared to the traditional evolutionary meta-heuristic genetic algorithm. The genetic algorithm is relatively slow because the evolution happens through generations or genetic transfer of ideas whereas the memetic algorithms are very fast due to the cultural transfer of ideas. The robustness of the algorithm lies in choosing the optimal parameters affecting the result. The parameters to be chosen and the effect of the parameters on the algorithm are discussed in this chapter. However, the selection of parameters highly depends on the nature of the function and the number of decision variables. As such utmost care must be taken for the selection of appropriate parameters based on the problem chosen for solving. The SFLA can be used for even mixed integer problems [7].

References

1. Dawkins R (1976) The selfish Gene. Oxford University Press, Oxford
2. Duan Q, Sorooshian S, Gupta V (1992) Effective and efficient global optimization for conceptual rainfall-runoff models. Water Resour Res 28(4):1015–1031
3. Eberhart RC, Kennedy J (1995) A new optimizer using particles swarm theory. In: Proceedings of the 6th international symposium on micro machine and human science, Nagoya, Japan, 1995, IEEE Service Center, Piscataway, pp 39–43
4. Eberhart RC, Dobbins RW, Simpson P (1996) Computational intelligence PC tools. Academic, Boston
5. Eusuff MM, Lansey KE (2003) Optimization of water distribution network design using the shuffled frog leaping algorithm (SFLA). J Water Resour Plan Manag Am Soc Civ Eng 129(3):210–225
6. Kennedy J (1997) The particle swarm: social adaptation of knowledge. In: Proceedings of the IEEE international conference on evolutionary computation, Indianapolis, IN, USA, IEEE Service Center, Piscataway, pp 303–308
7. Eusuff M, Lansey K, Pasha F (2006) Shuffled frog-leaping algorithm: a memetic meta-heuristic for discrete optimization. Eng Optim 38(2):129–154. https://doi.org/10.1080/03052150500384759

Chapter 8
Groundwater Management Using Coupled Analytic Element Based Transient Groundwater Flow and Optimization Model

Komal Kumari and Anirban Dhar

Abstract The groundwater management problems are broadly solved using the coupled simulation-optimization model. In the coupled simulation-optimization model, Analytic element method (AEM) based simulation model and particle swarm optimization (PSO) based optimization model is adopted to find out the best groundwater management strategy. Moreover, the coupled model is applied to a illustrative example to check its efficiency in solving the groundwater management problems involving large number of decision variables. Two-dimensional isotropic, homogeneous and confined aquifer is considered for the groundwater flow simulation. AEM is used to simulate the groundwater flow and to penalize the constraints, whereas PSO is used to evaluate the objective function (time varying operating cost) of the management problem. The optimization model calls the simulation model repeatedly until the stopping criterion is achieved. To evaluate the performance of the developed methodology optimization model using Nelder-Meade method is also considered. The obtained results are compared with PSO based optimization model. Obtained results demonstrate the potential applicability of the model to solve groundwater management problems.

Keywords Groundwater management · Analytic element method · Particle swarm optimization · Present value cost

8.1 Introduction

Management of groundwater systems still remains a challenging problem due to its complex nature. The groundwater management problems are generally

---

K. Kumari · A. Dhar (✉)
Department of Civil Engineering, Indian Institute of Technology Kharagpur, Kharagpur, West Bengal, India
e-mail: anirban@civil.iitkgp.ernet.in

F. Bennis, R. K. Bhattacharjya (eds.), *Nature-Inspired Methods for Metaheuristics Optimization*, Modeling and Optimization in Science and Technologies 16,
https://doi.org/10.1007/978-3-030-26458-1_8

solved using simulation-optimization technique. The fundamental capability of this technique is to simulate the groundwater flow, and simultaneously identify the best management strategy to achieve objectives under prescribed set of constraints. In the coupled simulation-optimization technique, the optimization model repeatedly calls the simulation model until the termination criteria is achieved to get the values of groundwater variables, such as groundwater velocity, concentration, head. This repetition enhances the computational complexity and in turn increases the computation time to achieve best solution. However, several researchers have used particle swarm optimization (PSO) as optimization tool and found it suitable for computationally demanding problems [20].

The Analytic Element Method (AEM) is one of the most popular methods to solve regional groundwater flow problem, developed by Strack [23–25]. AEM contributes in solving groundwater flow problems through superposition of elementary solutions of governing differential equation of flow. These solutions represent the variety of aquifer features such as well, line sink, area sink, line doublet etc. AEM is suitable for both finite and infinite domains and typically applied to the regional scale groundwater flow problems. Hunt [15] provided a review of various applications of AEM in groundwater modeling. A transient AEM based hydraulic management model is proposed. AEM based transient simulation model is linked with PSO based optimization model to ascertain the best groundwater management strategy. PSO is a heuristic, evolutionary and a population based stochastic computational technique proposed by Kennedy and Eberhart [16]. It is a computationally efficient approach for solving complex groundwater optimization problems and, in several cases, outperforms other evolutionary computation techniques. PSO has a broad range of applications in the field of engineering research and found to be suitable for computationally demanding problems with large number of variables. It has been applied in various domain of water resources management, such as parameter estimation of non-linear flood routing models [6, 21], groundwater remediation [19, 20], multi-purpose reservoir operation [2, 5, 17], groundwater management [1, 7, 9–12, 18, 20, 22]. However, limited applications of AEM-PSO based simulation-optimization models are available in which dynamic pumping rate and multiple management periods are considered.

Hsiao and Chang [14] developed groundwater management model incorporating genetic algorithm (GA) and Constrained Differential Dynamic Programming (CDDP) considering dynamic nature of groundwater resources system into account [4, 22]. Chang et al. [4] proposed an approach to optimize the total cost (fixed cost and time-varying pumping cost) in unconfined aquifers. They considered time-varying pumping rates, installation schedule, and well locations as the decision variables. Moreover, several researchers presented comparative study of PSO with other optimization techniques (ant colony optimization (ACO), genetic algorithm (GA), simulated annealing) and concluded that PSO is computationally more efficient than other optimization techniques [20, 22]. Although Gaur et al. [10] developed the AEM-PSO model to get the optimal solution of groundwater system and, investigated the benefits of AEM based simulation-optimization model over the FDM based model [11] in identifying the optimal location and determining the optimal cost. Their model does not consider time-varying pumping rate and hydraulic head in account while solving the optimization problem. In reality, the groundwater demand is time varying because of the rise in population and economic development. Therefore, dynamic simulation-optimization models need to be considered in account to get satisfactory solutions of management problems [4, 14].

It is evident that there is scope for development of Simulation-optimization model incorporating transient analytic element model and PSO. Therefore, the objective of the present study is to minimize the present value of the pumping cost while maintaining the minimum hydraulic head and satisfying the demand [14] with pumping within specified limits.

8.2 Formulation of AEM-PSO Model

The present model incorporates the AEM based flow model and PSO based optimization model to get the optimal pumping cost. AEM is used to simulate the flow and to assure minimum or zero constraint violations (groundwater head, demand and pumping rates) in the model. PSO is used to evaluate the objective function utilizing the simulation model to satisfy the constraints.

8.2.1 AEM Flow Model

The Analytic Element Method (AEM) is based on the superposition of fundamental solutions (elements) of governing differential equation of flow [13, 23]. These solutions represent the variety of aquifer features such as well, line sink, area sink, line doublet. The streams and their tributaries (line sinks) are broken up into many segments for accurate modeling. Each segment either receives groundwater (a gaining stream segment i.e., with positive sink density) or recharges the groundwater (a losing stream segment i.e., with negative sink density). The governing differential equation [26] is 2D groundwater flow in terms of discharge potential (Φ), aquifer diffusivity (α) and sink term (E).

$$\frac{\partial^2 \Phi}{\partial x^2} + \frac{\partial^2 \Phi}{\partial y^2} = E + \frac{1}{\alpha}\frac{\partial \Phi}{\partial t} \tag{8.1}$$

Where,

$$\Phi = kh\phi - \tfrac{1}{2}kh^2 \quad for\ confined\ aquifer$$
$$\Phi = \tfrac{1}{2}k\phi^2 \quad for\ unconfined\ aquifer$$

The transient analytic element model [26] incorporates many analytic elements including Constant, Wells, Head-line sinks, area-sinks etc. For two-dimensional flow problems, the components of discharge vector Q_x and Q_y can also be expressed in terms of the discharge potential as follows,

$$Q_x = -\frac{\partial \Phi}{\partial x}, \quad Q_y = -\frac{\partial \Phi}{\partial y} \tag{8.2}$$

8.2.2 Optimization Model

PSO is a heuristic, evolutionary and a population-based stochastic computational technique proposed by Kennedy and Eberhart [16]. It is a computationally efficient approach for solving complex groundwater optimization problems and, in several cases, outperforms other evolutionary computation techniques. The working steps of PSO method for solution of the current optimization problem are as follows:

- Initialize a population of particles randomly with positions in the range of [Qmin, Qmax] and velocities in the range of [$(Q_{min} - Q_{max})/2$, $(Q_{max} - Q_{min})/2$] in the $D = N \times T$-dimensional search space.
- Evaluate the fitness value of each particle using the fitness function.
- The solutions are compared for each evaluation, and the following criteria are always enforced [8].
  - Any feasible solution is preferred to any infeasible solution.
  - Among two feasible solutions, the one having better objective function value is preferred.
- Among two infeasible solutions, the one having smaller constraint violation is preferred.
- Compare particle's fitness function with the best solution (particle's best: pbest) so far. For each particle: analyze if the current fitness value is better than pbest, then set pbest = fitness value.
- Choose the particle with the best fitness value among all the particles (population's overall best: gbest).
- Update the velocity and position of the particle according to Eqs. (8.3) and (8.4) given below:

$$v_i^{t+1} = \chi \left[w v_i^t + c_1 r_1 \left(P_i^t - x_i^t \right) + c_2 r_2 \left(P_g^t - x_i^t \right) \right] \tag{8.3}$$

$$x_i^{t+1} = v_i^{t+1} + x_i^t \tag{8.4}$$

- Loop to step 2 until the stopping criteria is achieved, mostly a maximum number of iterations or a satisfactorily good fitness value is reached.

where c_1 and c_2 are acceleration constants; r_1 and r_2 random real numbers in range (0, 1); w is the inertia weight which is used to control the impact of the previous velocities on the current one; χ is the constriction coefficient, which is used to restrain velocity; P_i^t (pbest) is the best value achieved by individual particle i; P_g^t (gbest) is the global best value achieved by the population so far.

A conventional groundwater management model can be designed using the values of decision variables under consideration of the prescribed constraints. Optimization models are used to minimize or maximize an objective function without violating the constraints. In this study, the objective of the optimization model is to minimize the present value of well-operating cost. Mathematical formulations of the optimization model, including the objective function and required constraints, are as follows:

Minimize:

$$\min_{I \subset \Omega_S, K \subset \Omega_T} c \sum_{i \in I} \sum_{k \in K} \frac{Q_i^k \left(G_l - h_i^k\right)}{(1 + i_d)^{t_p^k}} + \beta_1 P_h + \beta_2 P_D \tag{8.5}$$

Subject to:

$$h = f(Q) \tag{8.6}$$

$$h_i^k \geq h_{\min}^k, \ \forall i, k \tag{8.7}$$

$$\sum_i Q_i^k \geq D^k, \ \forall k \tag{8.8}$$

$$Q_i^k|_L \leq Q_i^k \leq Q_i^k\Big|^U, \quad \forall i, k \tag{8.9}$$

$$P_h = \max\left(0, \ h_{\min}^k - h_i^k\right) \tag{8.10}$$

$$P_D = \max\left(0, \ D^k - \sum_i Q_i^k\right) \tag{8.11}$$

where Eq. (8.5) is the objective function; i = 1,2,...,N_w (number of wells); k = 1,2,...,T (number of management period); c is the is the pumping cost coefficient which is formulated as $c = \rho \times g \times c_p \times \Delta t$; coefficient; c_p = electric power cost per unit work; Δt = simulation time step length; ρ = water density; g = acceleration due to gravity; $Q_i^k$is the pumping rate of well i at operation time step k; G_l is the ground surface elevation from lower bottom; id is the interest rate; β_1 and β_2 are the penalty coefficients; P_h and P_D is the penalty terms of the constraint violation with respect to the minimum hydraulic head constraint and the demand constraint; D_k is the water demand at operation time step k; $Q_i^k|_L$ *and* $Q_i^k|^U$ is the lower and upper limit of pumping rate; $h_i^k$is the hydraulic head in well i at management period k.

8.2.3 Simulation-Optimization Model

The simulation and optimization models are formulated and validated with standard functions. After development and validation, both models are linked to solve groundwater management problems. Figure 8.1 shows the working steps of the

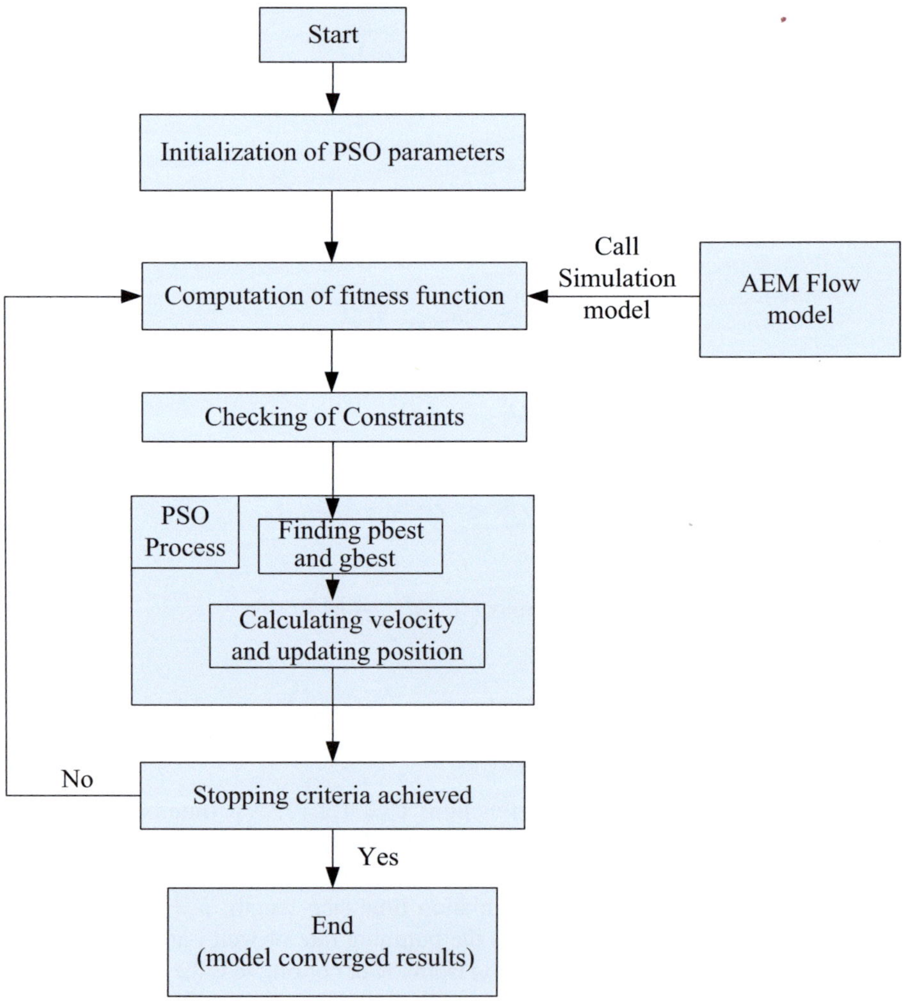

Fig. 8.1 Flow chart for coupled AEM-PSO model

linked AEM-PSO model, where the PSO model repeatedly calls the AEM model to find the global solution of the management problems.

8.3 Model Application and Discussions

Proposed methodology is applied on the illustrative system (5000 m × 3000 m) adapted from Hsiao and Chang [14] as shown in Fig. 8.2. The aquifer is having constant-head along left and right boundaries and no-flow boundaries along north and south direction. The aquifer in the flow domain is assumed to be isotropic, homogeneous and confined. The study area is described with 35-potential well locations as represented in Fig. 8.2. Before the commencement of pumping, hydraulic head distribution is assumed to be in steady state with the aquifer parameters listed in Table 8.1. The planning horizon is divided into 12 time steps over 3 years (1 time step = 90 days). The total pumping at each management period must satisfy the demand as depicted in Fig. 8.3, with 0.5 and 0.01m^3/s as maximum and minimum

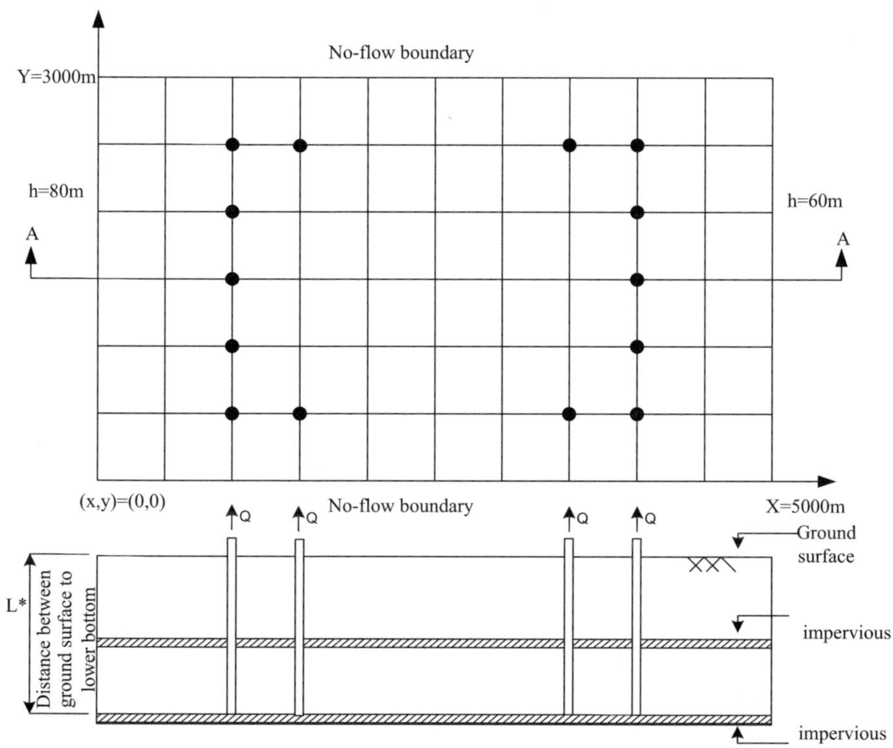

Fig. 8.2 Hypothetical aquifer system for water supply example

Table 8.1 Parameters for illustrative study area and coupled simulation-optimization model

Aquifer Properties	Value
Base elevation of aquifer [m]	0
Thickness of aquifer [m]	50
Hydraulic conductivity of aquifer [m/s]	0.000431
Porosity of aquifer [−]	0.200000
Storativity [−]	0.001000
Simulation parameters	
Simulation time step length [Days]	90
Planning horizon [Years]	3
Electric power cost (Dollars per kwh)	0.045
PSO parameters	
Inertia weight	0.8
Population size	20, 200
Acceleration constants	$c_1 = 1.3$, $c_2 = 2.7$
Constriction coefficient	0.5

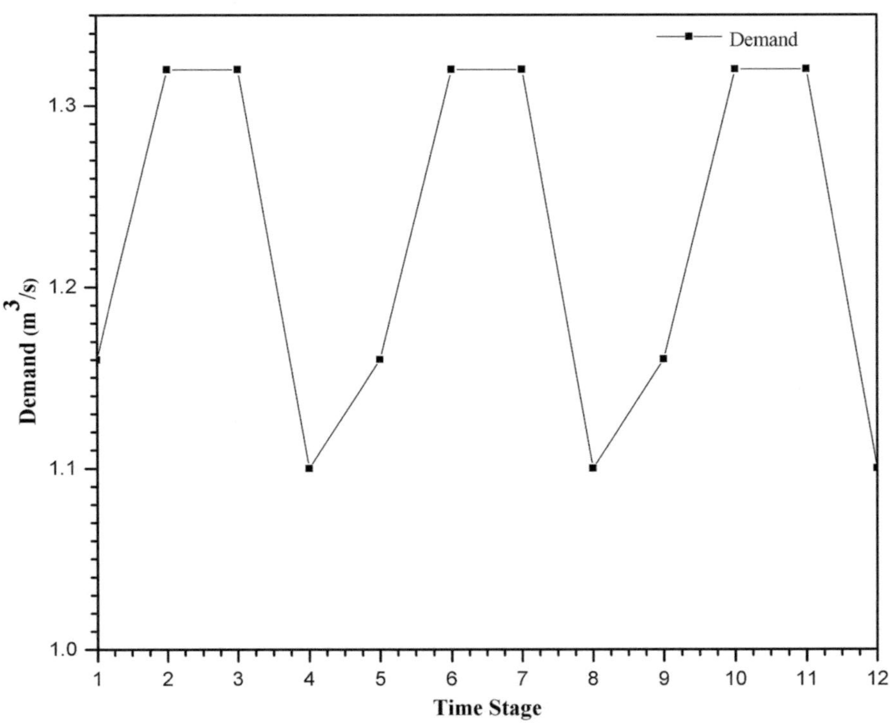

Fig. 8.3 Water demand for groundwater supply example

well capacities and 50 m as minimum hydraulic head. In all management periods 14 pumping wells are selected out of 35 potential well locations.

8.3.1 Sensitivity Analysis

To achieve the best management practice, the coupled simulation optimization model is executed for several times with different sets of PSO parameters and the best result among them is considered here. For the faster convergence of simulation-optimization model, it is important to select the appropriate values of PSO parameters. Among all of the PSO parameters, inertia weight is considered as the most critical parameter which is used for balancing the global and local search. Therefore, in this study the sensitivity analysis is performed with constant inertia weight (0.8) and randomly varying inertia weight between 0.4 and 0.8 (U(0.4,0.8)). It is observed that $w = U(0.4,0.8)$ helps the coupled model to converge faster than that with $w = 0.8$. However, the PSO model converged to a lower objective function value with constant inertia weight than the random one. Hence, the constant inertia weight as 0.8 is adopted in this study. Moreover, the model is run with 20 and 40 particles keeping other PSO parameters same and, the result shows that increasing the number of particles does not reduce the number of generations of convergence but increases the simulation time. The parameters adopted are $c_1 = 1.3$, $c_2 = 2.7$, $\chi = 0.5$ and $w = 0.8$, which give stable result. Reduced variable optimization model is solved based on symmetry of the well locations with respect to physical domain.

8.4 Results and Discussions

The developed AEM-PSO model is applied in two management scenarios. In Scenario-I, the pumping rate remains static with respect to different time stages. Therefore, the number of decision variables is equal to the number of wells considered in the problem. However, in the second scenario, dynamic pumping rates are considered where the number of variables is equal to the number of wells times the management period. The termination criterion is assumed to be satisfied if there is no considerable improvement in the solution over 100 successive iterations upto a convergence limit of four decimal places. Table 8.2 summarizes the optimal pumping rates, total pumpage at each time steps and pumping costs for Scenario-I and II. Table 8.2 indicates that the total pumpage at the end of each management period is satisfying the total demand (in some cases significantly higher than the total demand) for the illustrative system. The contour plots at the end of each time step for Scenario-II reveals that with increase in the demand, the total pumpage increases.

Table 8.2 Optimal Pumping rates of 14 wells for scenario-I and scenario-II

	Scenarios													
	Scenario I:		Scenario II:											
	Static pumping (m^3/s)		Dynamic pumping (m^3/s)											
Well	AEM-ND	AEM-PSO	1	2	3	4	5	6	7	8	9	10	11	12
1	0.1286	0.0553	0.1519	0.1411	0.2000	0.1434	0.1395	0.0101	0.0175	0.1091	0.1313	0.0819	0.1202	0.1461
2	0.0567	0.0797	0.0365	0.1206	0.1276	0.0764	0.0616	0.0316	0.0553	0.0100	0.1376	0.0910	0.0875	0.1133
3	0.0768	0.0812	0.0527	0.0245	0.0461	0.0388	0.0585	0.1057	0.0721	0.0671	0.0237	0.0555	0.0877	0.1143
4	0.0830	0.0716	0.1114	0.0988	0.0190	0.0512	0.1107	0.0933	0.1243	0.0660	0.0366	0.0982	0.0100	0.0100
5	0.0674	0.0902	0.1494	0.0176	0.1123	0.1224	0.1332	0.0869	0.0829	0.1226	0.0793	0.0845	0.1309	0.0903
6	0.1114	0.0901	0.0293	0.1037	0.0594	0.1139	0.0991	0.0606	0.1403	0.0912	0.1306	0.1181	0.0624	0.0100
7	0.1053	0.0923	0.1091	0.0864	0.1819	0.0753	0.0100	0.1004	0.1569	0.0272	0.0620	0.0701	0.1430	0.0957
8	0.1235	0.1653	0.1844	0.1982	0.2000	0.0385	0.1470	0.1961	0.1311	0.1422	0.1162	0.1471	0.1477	0.1000
9	0.1235	0.1653	0.1844	0.1982	0.2000	0.0385	0.1470	0.1961	0.1311	0.1422	0.1162	0.1471	0.1477	0.1000
10	0.1053	0.0923	0.1091	0.0864	0.1819	0.0753	0.0100	0.1004	0.1569	0.0272	0.0620	0.0701	0.1430	0.0957
11	0.1114	0.0901	0.0293	0.1037	0.0594	0.1139	0.0991	0.0606	0.1403	0.0912	0.1306	0.1181	0.0624	0.0100
12	0.0674	0.0920	0.1494	0.0176	0.1123	0.1224	0.1332	0.0869	0.0829	0.1226	0.0793	0.0845	0.1309	0.0903
13	0.0830	0.0716	0.1114	0.0988	0.019	0.0512	0.1107	0.0933	0.1243	0.0660	0.0366	0.0982	0.0100	0.0100
14	0.0768	0.0812	0.0527	0.0245	0.0461	0.0388	0.0585	0.1057	0.0721	0.0671	0.0237	0.0555	0.0877	0.1143
Total	1.3201	1.32	1.461	1.3201	1.565	1.1	1.3181	1.3277	1.488	1.1517	1.1657	1.3199	1.3711	1.1
Cost (Million)	$0.83322	$0.841172						$0.84758						

8.4.1 *Scenario-I (Static Pumping Rate)*

In this scenario, static pumpage is considered and thus the model runs with only 14 (equal to the number of wells) decision variables. The model runs with 20 and 40 particles keeping other PSO parameters same and, the result shows that increment in the number of particles from 20 to 40 does not decreases the number of iterations very much but inhances the computation time. Therefore, inertia weight, $w = 0.8$ and 20 number of particles show better results than other combinations for the problem considered here. In this case, the best PSO parameters found are $c_1 = 1.3$, $c_2 = 2.7$, $\chi = 0.5$ and $w = 0.8$, with 20 number of particles. The value of optimal operating cost, in this case, is found to be equal to \$0.841172 Million. Figure 8.4 summarizes the change in the value of objective function in each generation for AEM-PSO model of management scenario-I, in which an inset plot is added to the main plot with reduced dimension (up to 21 generations) so that the main graph remains partially visible. The hydraulic head distribution for the present scenario at the end of 1st, 4th, 8th and 12th time steps is shown in Fig. 8.5. The contour plots at the end of each time step for scenario-I reveals that with the change in time stage,

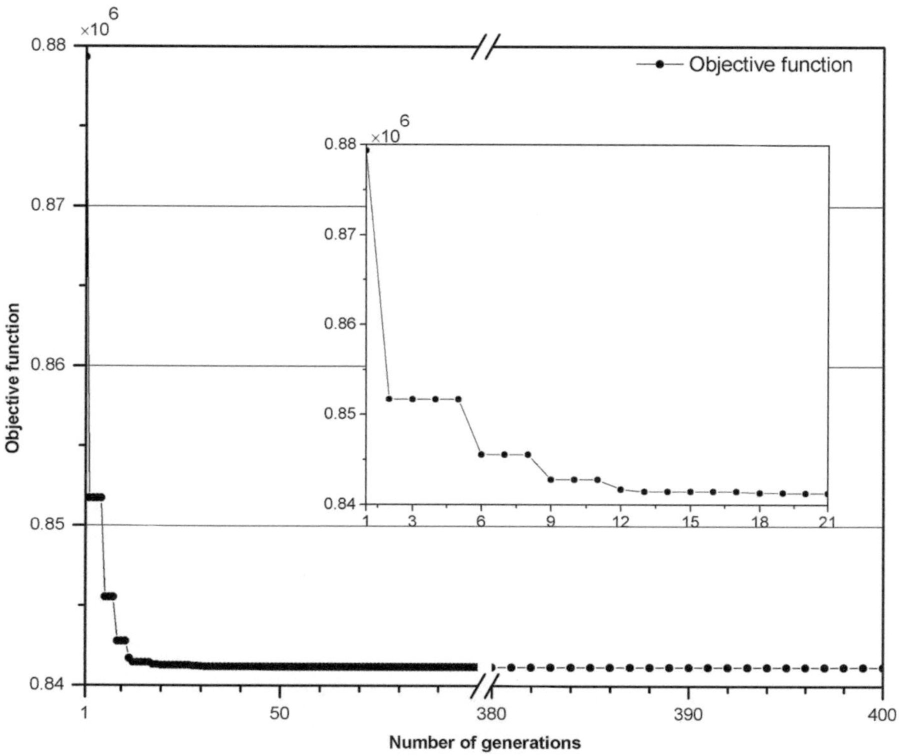

Fig. 8.4 Objective function versus number of iterations for AEM-PSO in scenario-I

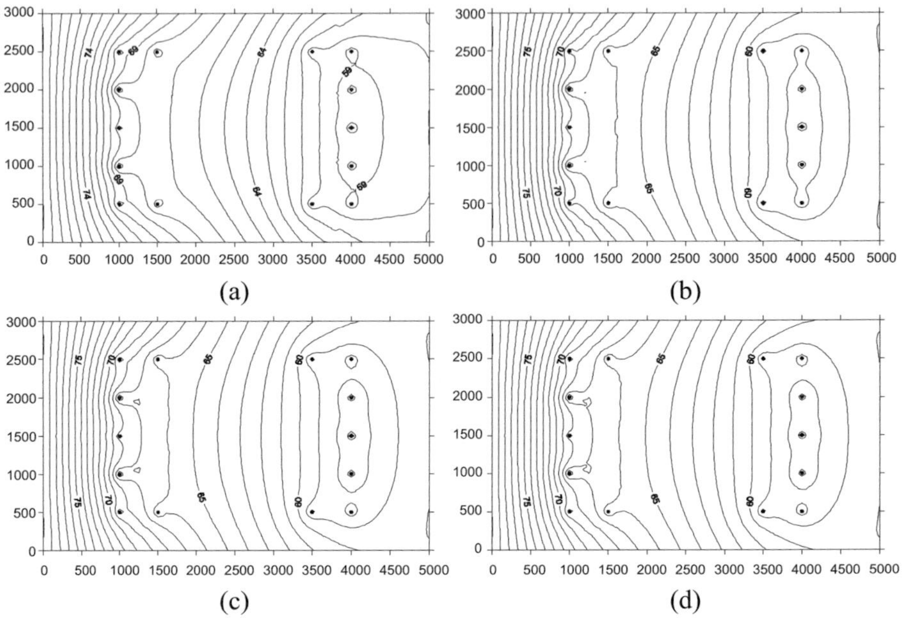

Fig. 8.5 Contour plot for scenario-I at four different time steps. (**a**) Contour plot at 1st (90 day) time step. (**b**) Contour plot at 4th (360 day) time step. (**c**) Contour plot at 8th (720 day) time step. (**d**) Contour plot at 12th (1080 day) time step

slight variation in hydraulic head is observed since the pumping rate remains static in this case.

To evaluate the performance of the present scenario, Nelder-Meade algorithm [3] based optimization model is also considered and its result is compared to that of PSO based optimization model. The optimal pumping values obtained in AEM-PSO model is used as the initial guess value in the Nelder-Meade method and linked with AEM (hereinafter referred as “AEM-ND”) to achieve the best solution. The variation of objective function and constraint violation versus number of generations for AEM-ND model of management scenario-I is shown in Fig. 8.6, in which an inset plot is added with reduced dimension (up to 50 function evaluations). It can be observed from Fig. 8.6 that constraint violation values are higher than that of the objective function at early function evaluations while the convergence starts nearby 1000th function evaluation with constant zero constraint violation and checked up to 1407 function evaluation. However, in the case of AEM-PSO model constraint violation is zero from the first iteration only while the convergence starts nearby 200th iteration and checked up to 400 iterations (Fig. 8.4).

The value of optimal operating cost for AEM-ND is found to be equal to \$0.83322 Million which is better than that of the result of AEM-PSO. Therefore, with a reasonable initial guess value, Nelder-Meade algorithm outperforms PSO for this hydraulic management problem.

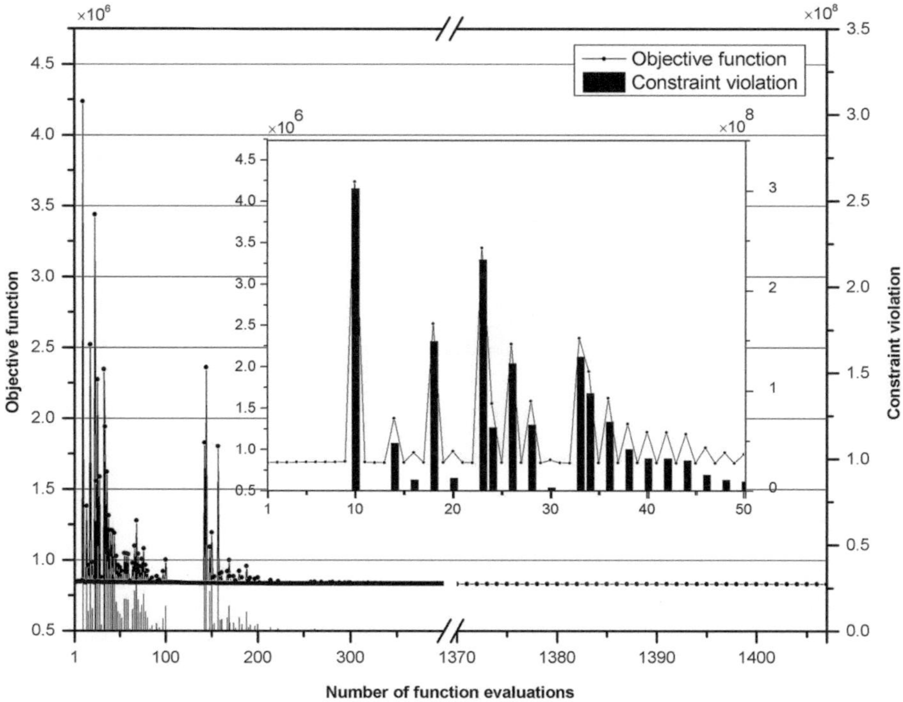

Fig. 8.6 Objective function and constraint violation values versus number of function evaluations for AEM-ND in scenario-I

8.4.2 Scenario-II (Dynamic Pumping Rate)

In this scenario, time varying pumpage is considered and thus the model runs with 14 × 12 (number of wells times the number of management periods) decision variables. The model runs with 200 and 400 particles keeping other PSO parameters same and, the result shows that increasing the number of particles does not reduce the number of generations very much but increases the computation time. Therefore, inertia weight, w = 0.8 and 200 particles show better results than other combinations for the problem considered here. In this case, the best PSO parameters found are $c_1 = 1.3$, $c_2 = 2.7$, $\chi = 0.5$, $w = 0.8$ and, 200 number of particles. The value of optimal operating cost in this case is found to be equal to $0.84758 Million.

The objective function and constraint violation value versus number of generations for AEM-PSO model of management scenario-II is shown in Fig. 8.7, in which an inset plot is added with reduced dimension (up to 21 generations). It can be observed from Fig. 8.7 that at early iterations minimal constraint violation are there while the solution starts converging nearby 100th iteration with constant zero constraint violation and checked upto 200 iterations. The hydraulic head distribution

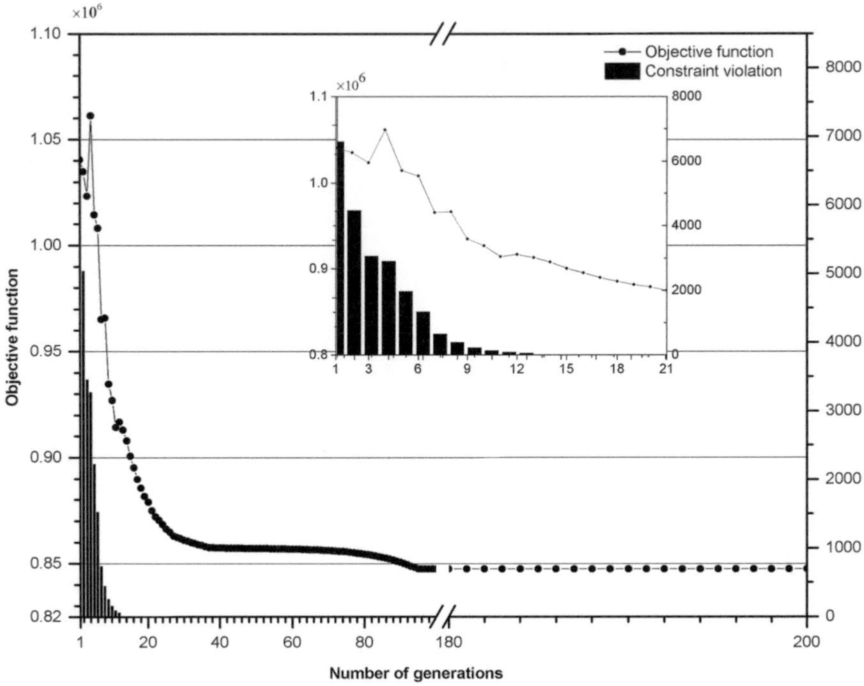

Fig. 8.7 Objective function and constraint violation values versus number of iterations for scenario-II

for the present scenario at the end of 1st, 4th, 8th and 12th time steps is shown in Fig. 8.8. The contour plots at the end of each time step for scenario-II reveals that with the increase in demand, the total pumpage increases. The optimal pumping rate for the wells located in the western region is higher than that of wells located nearby the right boundary of the groundwater supply system. Thus providing better solution of pumping cost, since specified hydraulic head in the left boundary is higher.

The present scenario can be further modified to couple the optimal pumping values obtained in the AEM-PSO model with the Nelder-Meade algorithm as done in the case of scenario-I.

8.5 Conclusions

A groundwater pumping management plan based on AEM and PSO model is developed. The proposed model links the PSO model with AEM to minimize present value of operating cost. The optimal solution of groundwater management problems are determined satisfying three constraints: maintaining minimum hydraulic head, satisfying total demand at each time steps with the optimal pumping lying in the

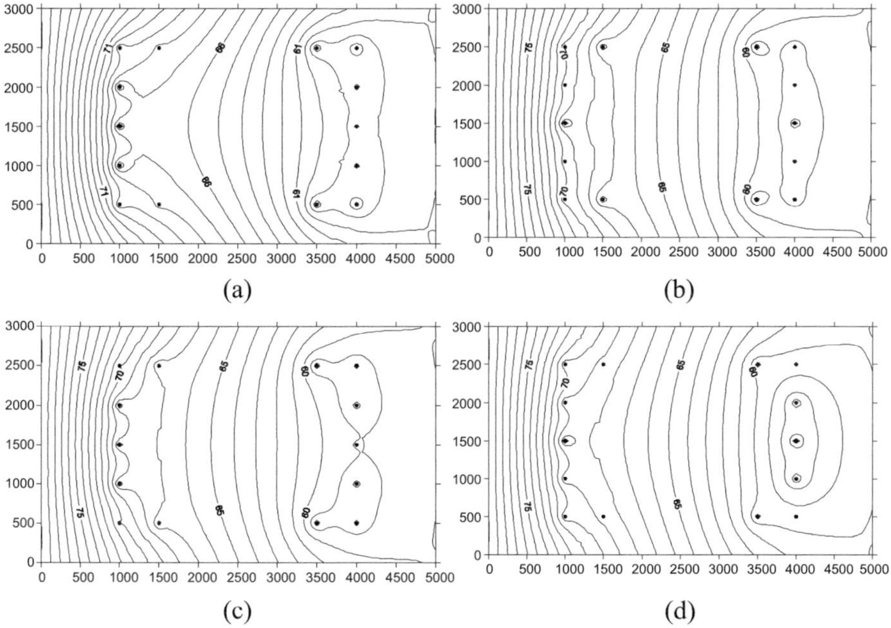

Fig. 8.8 Contour plot for scenario-II at four different time steps. (**a**) contour plot at 1st (90 day) time step. (**b**) contour plot at 4th (360 day) time step. (**c**) Contour plot at 8th (720 day) time step. (**d**) Contour plot at 12th (1080 day) time step

range of lower and upper bound of pumping values. It has been observed in the integrated AEM-ND model that the solutions are improved to that of AEM-PSO model in scenario-I. However, it alone can not suffice our management goal since it needs a proper initialization of decision variables, which is obtained using AEM-PSO model only. Although this study considers only confined aquifer, the developed AEM-PSO model can be further modified for the application of unconfined aquifers. In reality, the groundwater demand and hydraulic head are time varying components of groundwater supply system. Therefore, the developed model for Scenario-II can be efficiently applied to solve real field problems, since the time-varying pumping rates are considered. In future, the potential applicability of the developed model could be enhanced with the development of more efficient transient elements.

References

1. Ayvaz MT, Elci A (2013) A groundwater management tool for solving the pumping cost minimization problem for the Tahtali watershed (Izmir-Turkey) using hybrid HS-Solver optimization algorithm. J Hydrol 478:63–76
2. Baltar AM, Fontane DG (2008) Use of multiobjective particle swarm optimization in water resources management. J Water Resour Plan Manag 134(3):257–265

3. Belegundu AD, Chandrupatla TR (2002) Optimization concepts and applications in engineering. Pearson Education Pvt. Ltd, Delhi, India, p 432
4. Chang LC, Chen YW, Yeh MS (2009) Optimizing system capacity expansion schedules for groundwater supply. Water Resour Res 45(7):W07407
5. Chang J, Bai T, Huang Q, Yang D (2013) Optimization of water resources utilization by PSO-GA. Water Resour Manag 27(10):3525–3540
6. Chu HJ, Chang LC (2009) Applying particle swarm optimization to parameter estimation of the nonlinear Muskingum model. J Hydrol Eng 14(9):1024–1027
7. Dhar A, Datta B (2009) Saltwater intrusion Management of Coastal Aquifers. I: linked simulation-optimization. J Hydrol Eng 14(12):1263–1272
8. Deb K (2000) An efficient constraint handling method for genetic algorithms. Comput Methods Appl Mech Eng 186:311–338
9. El-Ghandour HA, Elsaid A (2013) Groundwater management using a new-coupled model of flow analytical solution and particle swarm optimization. Int J Water Resour Environ Eng 5(1):1–11
10. Gaur S, Chahar BR, Graillot D (2011a) Analytic element method and particle swarm optimization based simulation-optimization model for groundwater management. J Hydrol 402(3):217–227
11. Gaur S, Mimoum D, Graillot D (2011b) Advantages of the analytic element method for the solution of groundwater management problems. Hydrol Process 25(22):3426–3436
12. Gaur S, Sudheer C, Graillot D, Chahar BR, Kumar DN (2013) Application of artificial neural networks and particle swarm optimization for the management of groundwater resources. Water Resour Manag 27(3):927–941
13. Haitjema HM (1995) Analytic element modeling of groundwater flow. Academic, San Diego
14. Hsiao CT, Chang LC (2002) Dynamic optimal groundwater management with inclusion of fixed costs. J Water Resour Plann Manag 128(1):57–65
15. Hunt J (2006) Groundwater modeling applications using the analytic element method. Ground Water 44(1):5–15
16. Kennedy J, Eberhart RC (1995) Particle swarm optimization. In: Proceeding of the IEEE international conference on neural networks, IEEE service center, Piscataway, NJ, pp 1942–1948
17. Kumar D, Reddy J (2007) Multiple reservoir operation using particle swarm optimization. J Water Resour Plann Manag 133(3):192–202
18. Majumder P, Eldho TI (2016) A new groundwater management model by coupling analytic element method and reverse particle tracking with cat swarm optimization. Water Resour Manag 30(6):1953–1972
19. Mategaonkar M, Eldho TI (2012) Groundwater remediation optimization using a point collocation method and particle swarm optimization. J Environ Model Softw 32:37–48
20. Matott LS, Rabideau AJ, Craig JR (2006) Pump-and-treat optimization using analytic element method flow models. Adv Water Resour 29(5):760–775
21. Moghaddam A, Behmanesh J, Farsijani A (2016) Parameters estimation for the new four-parameter nonlinear Muskingum model using the Particle Swarm Optimization. Water Resour Manag 30(7):2143–2160
22. Sedki A, Ouazar D (2011) Swarm intelligence for groundwater management optimization. J Hydroinf 13(3):520–532
23. Strack ODL (1989) Groundwater mechanics. Prentice Hall, Englewood Cliffs
24. Strack ODL, Haitjema HM (1981a) Modeling double aquifer flow using a comprehensive potential and distributed singularities: 1. Solution for homogeneous permeability. Water Resour Res 17(5):1535–1549
25. Strack ODL, Haitjema HM (1981b) Modeling double aquifer flow using a comprehensive potential and distributed singularities: 2. Solution for inhomogeneous permeabilities. Water Resour Res 17(5):1551–1560
26. Zaadnoordijk WJ, Strack ODL (1993) Area sinks in the analytic element method for transient groundwater flow. Water Resour Res 29(12):4121–4129

Chapter 9
Investigation of Bacterial Foraging Algorithm Applied for PV Parameter Estimation, Selective Harmonic Elimination in Inverters and Optimal Power Flow for Stability

J. Prasanth Ram and N. Rajasekar

Abstract Inspired by the foraging behavior in bacteria (e-coli), Bacterial Foraging Algorithm (BFA) is designed for solving global optimization problems. Being unique in optimization, BFA has already received universal attention from researchers to apply for various engineering application. In order to investigate the BFA performance, mathematical equations for unimodal and multimodal functions are investigated for minimization problem and the optimal results are reported in this regard. Performance indices of BFA method are comprehensively gauged for all functions and a comparative study with standard algorithms like Genetic Algorithm (GA), Particle Swarm Optimization (PSO) and Geometric Search Optimization (GSO) Algorithm is reported. Finding more suitable to optimize non-linear problems, the modified BFA in fusion with nelder- meed method is also introduced. This chapter binds the collective knowledge on BFA that aids the researchers to apply for their research application. Taking BFA as their optimization tool, few authors have attained their research objectives in following areas: (i) PV parameter estimation (2) Selective Harmonic Elimination in inverters and (3) a modified BFA method for optimal power flow to maintain load stability. To have an interior understanding the applications, the simulation results of the above study are also presented more in detail. In all the cases, the due role of BFA to attain accurate results is always appreciated.

Keywords Parameter estimation · Optimization · PWM inverter · Optimal power flow

---

J. P. Ram
New Horizon College of Engineering (NHCE), Bengaluru, Karnataka

N. Rajasekar (✉)
Department of Energy and Power Electronics, School of Electrical Engineering, Vellore Institute of Technology (VIT)- Vellore, Vellore, India

F. Bennis, R. K. Bhattacharjya (eds.), *Nature-Inspired Methods for Metaheuristics Optimization*, Modeling and Optimization in Science and Technologies 16,
https://doi.org/10.1007/978-3-030-26458-1_9

9.1 Introduction

The nature and science are always a great source of education for the mankind. Moreover the interdependency in these sources is vital for the universe to explore and exploit engineering applications. To solve the complexity present in engineering applications, the nature inspired biological algorithms are investigated and applied over a large extent. Further the effectiveness in these algorithms is more suitable to solve non-linear problems to improve the accuracy in design applications [28].

In terms of biology, population defines to the compilation of inter-breeding organisms of a particular species. Further, population size in a one of the essential phenomenon's to play a major role in the development of Evolutionary Algorithms (EAs) [9]. So far, many EAs inspired by nature and biological process are proposed. However, the improper selection of population for a specified framework introduces redundant computation in the optimization process. The setback in computation is due to the factors such as: (1) the lack of knowledge in the relationship between population size and dimensionality; and (2) the undefined relationship between the size of population and complexity in optimization problem. To alleviate the drawback of fixing population size, few methods have investigated in selecting the optimal population size; however, the guarantee for their performance is found uncertain. Therefore, in most of the applications, the population size is determined based on trial and error procedure.

Various evolutionary algorithms found successful in recent years with optimal selection of population size are: Genetic Algorithm, Particle Swarm Optimization Ant colony optimization and Flower Pollination Algorithm [10]. Similarly, foraging behaviour of bacteria is formulated as Bacterial Foraging Algorithm by Passino [28, 35]. The search for nutrients via chemotaxis movement performed by bacteria ensures the search for global solutions in problem search space [7, 8]. Further the ability in BFA to handle non-linearity is widely received in various engineering applications. Interestingly multi dimensional problem are even handled with ease is a noteworthy feature in BFA optimization. Adapting BFA, Alanis et al. [3], have improved the performance of sliding mode controller. In another approach, Rajasekar et al. [29] have used BFA for PV parameter extraction. Witnessing the potential with the area, later in 2016, Awadallah [5] has performed an experimental work in estimating the performance of PV. Further the authors have managed to generate the PV characteristics or four different modules. Using BFOA, Babu et al. [6] have proposed a scholarly research for optimizing the switching angle to mitigate the voltage harmonics in an inverter. Understanding the potential of optimization in electric drives, Bhushan and Singh [11], has used BFA to find the unknown nonlinear dynamics of the motor and the load parameters.

Having played a vital role in drives, BFA have also been utilized in power electronics for its effective usage. In order to reduce peak over shoot as well as to improve the systems response, Arunkumar et al. [4] have tuned proportional and integral constants of a boost converter using BFA. With the of help of power electronics, the BFA has also made its survival to optimize the planning of passive power filters (PPFs) and distributed generations (DGs). The author Mohammadi et al. [22] has used the method intelligently to reduce Total Harmonic Distortion (THD) and power loss in DG. Understanding the load demand and power loss, Dhillon et al. [15] has optimized the load flow control using BFA. In a different approach, Ramyachitra and Veeralakshmi [31] have made an interesting approach of predicting protein structure using the properties of bacteria. Being different from the above, Turanoğlu and Akkaya [42] recently investigated the dynamic facility layout problem using hybrid heuristic algorithm.

In addition to the above, hybrid and modified works with BFA is also seen in the literature. Incorporating nelder-meed properties, Edward et al. [16] have attempted to reduce the generation cost by handling optimal power flow problem. Since BFA is a multi dimensional optimization, Daryabeigi et al. [14] proposed a novel smart BFA (SBFA) to select a suitable capacitor for parallel circuit. Similar to the earlier, the modifications are made in conventional to maintain the optimal power flow in hydro thermal power generation system [25]. In another approach, the author Naveen et al. [23] have proposed the modified version of BFA for network optimization problem in power systems. Owing to the flexibility in BFA, Elattar [17] fused GA method with BFA to solve economic dispatch problem. Identifying BFA as one of the accurate and low-invasive tool for solving non linearity, a 7.5 kW motor subjected to power quality issues are optimized. Further, the unbalance voltage condition and its efficiency are found improved in this regard.

Finding BFA a potential method to improve its conventional properties, an adaptive crossover bacterial foraging optimization algorithm (ACBFOA) to improve the chemotaxis behavior is proposed by Panda and Naik [26]. Considering three mathematical functions, the exploration made by the adaptive form of BFA outperforms conventional BFA method in terms o accuracy. Similarly, optimal foraging has proposed by Zhu and Zhang [45] has also verified its method for twenty mathematical expressions. Earlier to the above, Li et al. [20] has explored another version of BFA and the method was tested for various population sizes. Further, the decision on deciding the optimal population is comprehensively arrived and truthfully verified.

Thus the promise shown by BFA to solve the stochastic nature is well proven. From the above literature the dominance of BFA is seen in three major fields (1) PhotoVoltaics (2) Drives and (3) power flow in power system. Hence, In this book chapter the authors have investigated aforesaid three different fields in electrical engineering. The BFA applied for specific areas in above fields are displayed as follows (1) Renewables- Parameter estimation, (2) Power Electronics- Harmonic Elimination in inverter and (3) Power system – Optimal power flow. The remaining subsection is organized as follows.

Section 2 gives the outline of BFA method and describes the design procedure in BFA optimization. In Sect. 3, the selection of different PV parameters is analyzed, selective harmonic elimination using BFA is studied in Sect. 4, and the modified BFA using nelder-meed for an optimal power flow problem is investigated in Sect. 5. Conclusions are presented at last.

9.2 Bacterial Foraging Algorithm

Bacteria Foraging Algorithm (BFA) is one of the nature inspired algorithms proposed by Kevin passino in 2002. In BFA, foraging behavior refers to process involved in search for food. Based on the genes having successful foraging capability, evolution chain is formulated. It is important here to note here that E-Coli bacteria present in human intestine will also undergo similar strategy. Application of E-Coli bacterium foraging strategy is the hidden theory behind BFA to apply for multi dimensional optimization problems. Further BFA method search process involves four important steps to explore and exploit problem search space as follows:

1. Chemotaxis
2. Swarming
3. Reproduction
4. Elimination and dispersal.

9.2.1 *Chemotaxis*

In chemotaxis, the bacterium stimulates its movement towards search for food via two processes namely (i) swimming and (ii) tumbling. For the bacteria over entire lifetime, it follows only swimming or tumbling movement to locate food sources. But at the same it can perform both the action to move towards food location. The movement in bacteria is achieved by flagella. It is observed that when flagella move in clock wise direction, the bacterium is allowed to move independently on its own cell. Further, the bacteria tumble slowly to locate nutrients under harmful places. It is seen that when bacteria moves in anticlockwise direction, the swimming gets faster. The swim and tumble movement of a bacteria is presented in Fig. 9.1a. For illustration, if $\theta^i(j, k, l)$ is assumed to bacterium location in population the it represents i^{th} bacterium at j^{th} chemotactic movement, k^{th} reproductive action and l^{th} elimination-dispersal step; then the bacterium tumbles for an unit step size of C in different random direction $\varphi(i)$ so as to find a new direction of movement. The tumbling movement in bacterium followed by movement in new direction is represented by

$$\theta^i\ (j+1, k, l) = \theta^i\ (j, k, l) + C * \phi(i) \tag{9.1}$$

Where $\phi(i) = \frac{\Delta(i)}{\sqrt{\Delta^T(i)\ \Delta(i)}}$

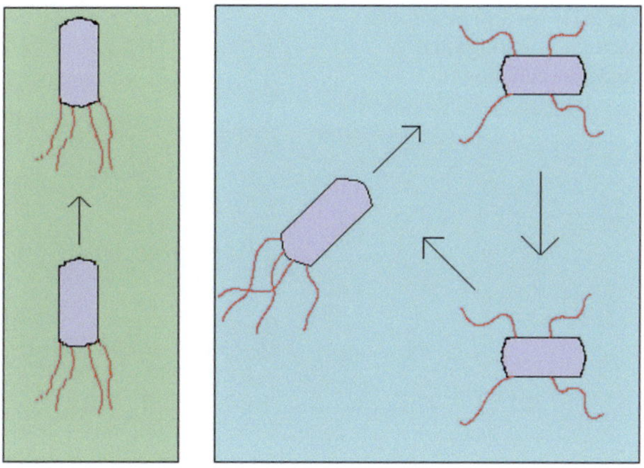

Fig 9.1a Swim and tumble of a bacterium

If the present direction is rich in nutrients then, the bacterium will continue to swim after the tumble, however limited up to a maximum number of steps N_s. Chemotaxis process is one of the vital steps in the BFA to create exploration. Chemotaxis in foraging strategy implements a local optimization in which the bacteria try to climb the nutrient concentration hence to avoid noxious substance.

9.2.2 Swarming

In swarming process, an interesting behavior is followed as bacteria already travelling in optimal path gains the attention of other bacterium so that, it attain desired location as a group. A group of bacterium *E.coli* cells arrange themselves in a traveling ring by moving up the nutrient gradient when placed amidst a semisolid matrix with a single nutrient chemo-effecter. The cells when stimulated by a high level of *succinate*, release an attractant *aspertate.* This helps the bacteria to move as group to particular location. This typical behavior in BFA is technically termed as swarming. The bacteria are formed in groups in concentric patterns will have high bacterial density. The mathematical representation of cell-to-cell signaling between bacteria can represented as

$$\sum_{i=1}^{S} J_{cc}\left(\theta, \theta^{i}\left(j, k, l\right)\right) = \sum_{i=1}^{S}\left[-d_{attract}\exp\left(-w_{attract}\sum_{m=1}^{p}\left(\theta_m-\theta_m^i\right)^2\right)\right] + \sum_{i=1}^{S}\left[-h_{repelent}\exp\left(-w_{repelent}\sum_{m=1}^{p}\left(\theta_m-\theta_m^i\right)^2\right)\right] \tag{9.2}$$

Fig 9.1b Swarm action of a bacteria in multimodal objective function

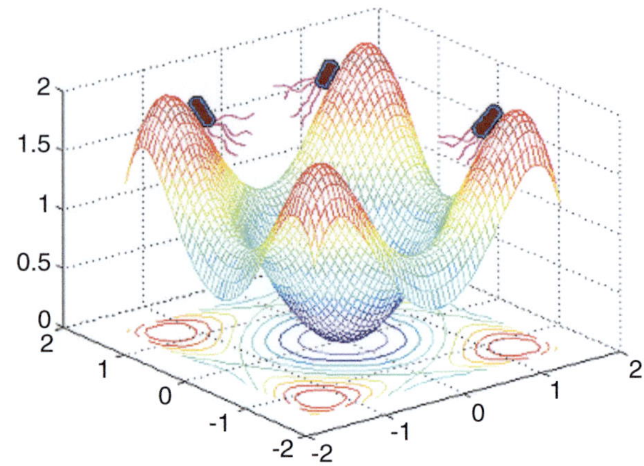

Where, J_{cc} denotes time varying objective function whose value relies on cell-to-cell signaling based on attractant-repellent profile. The swarm action of a bacteria for in a multimodal objective function is given in Fig. 9.1b.

9.2.3 Reproduction

In this step, every bacterium in the population will undergo health evaluation based on the mathematical equation defined as follows.

$$J_{health}^{i} = \sum_{j=1}^{N_c+1} J\,(i, j, k, l) \tag{9.3}$$

The attained health values of bacterium is arranged in ascending order while the least healthy bacteria in the population are removed; n the other hand healthy bacteria which can locate good food locations (Bacteria locating poorer locations/local solutions) are split into two, thus always helping the application to maintain constancy in population. This process will reserve healthier bacteria to stay in population and removes the awful ones. Further, it also helps the algorithm to speed up the searching process for pareto front.

9.2.4 Elimination and Dispersal

Due to numerous reasons, gradual or rapid changes in the local environment may cause either of occurrences (i) a group of bacteria is killed or (ii) the bacteria's will disperse into a new location. Occasionally this process will place the bacterium to near good nutrients/global solutions. Furthermore, this elimination and dispersion in BFA reduces the probabilities to converge for local solutions i.e., premature convergence. If the considered population is small then, this step in BFA has higher possibilities to explore global regions. Furthermore, this exclusive feature in the algorithm helps to avoid premature convergence problem and hence increases the exploration capability.

BFA processes towards locating global solutions involving chemotaxis, swarming, reproduction, elimination and dispersion for a non-linear problem is implemented by utilizing the following steps.

Step 1: Initialize the following parameters p, S, N_c, N_s, N_{re}, N_{ed}, P_{ed}, C, θ^i, $i = 1, 2, 3, \ldots S$

Step 2: Declare Elimination–dispersal loop as $l = l + 1$

Step 3: Declare Reproduction loop $k = k + 1$

Step 4: Declare Chemotaxis loop $j = j + 1$

(a) For, $i = 1, 2, 3, \ldots \ldots, S$, obtain chemotactic step for each bacterium.

(b) Determine the objective function $J(i, j, k, l)$.
Let $J(i, j, k, l) = J(i, j, k, l) + J_{cc}(\theta^i(j, k, l), P(j, k, l))$

(c) Let us assume $J_{last} = J(i, j, k, l)$, in order to save the fitness value. Since it will result us with better solution in further movements.

(d) Tumbling: Create a random vector as $\Delta(i)$, with each bacterium drawn from uniform distrubtion from $[-1, 1]$.

(e) Move: Let

$$\theta^i\ (j + 1, k, l) = \theta^i\ (j, k, l) + C * \phi(i)$$

This step size $C(i)$ results in the tumbling direction for bacterium$i$.
Compute fitness $J(i, j + 1, k, l)$ and let

$$J\,(i,\, j + 1, k, l) = J\,(i,\, j, k, l) + J_{cc}\left(\theta^i\,(j + 1, k, l)\,,\, P\,(j + 1, k, l)\right) \tag{9.4}$$

(f) Swim (another movement by bacterium):

(i) Let m = 0 (swim length counter)

(ii) If $m < N_S$

- Increment the counter $m = m + 1$
- If the current fitness $J(i, j + 1, k, l) < J_{last}$ (if doing better) then, let the last fitness $J_{last} = J(i, j + 1, k, l)$ and
$\theta^i\,(j + 1, k, l) = \theta^i\,(j, k, l) + C * \varphi(i)$, use above $\theta^i\,(j + 1, k, l)$ to calculate new $J(i, j + 1, k, l)$ (as performed in step f).
- Else, let the $m = N_S$. (End of if statement)

(g) Go for the next bacterium $(i + 1)$ if $i \neq S$.

Step5: If $j < N_c$, goto [step 4] and continue chemotaxis process, since the bacteria life is not over.

Step6: Reproduction:

(a) For given k, l, for every $i = 1, 2, 3, \ldots \ldots \ldots, S$, let us compute health of bacterium as $J^i_{health} = \sum_{j=1}^{N_c+1} J\ (i, j, k, l)$ for i^{th} bacterium and arrange the fitness J_{health} in ascending order

(b) The S_r bacteria having highest fitness (J_{health}) values die and left over S_r bacteria with the minimum fitness value gets split.

Step 7: If $k < N_{re}$, then switch to step 3. In this case, when reproduction steps specified in initialization are not reached, then start the next generation from the chemotactic loop.

Step 8: Elimination–dispersal step: For every bacterium $i = 1, 2, 3, \ldots \ldots \ldots \ldots, S$, an arbitrary number is created and if it is less than or equal to P_{ed}, then the corresponding bacterium is dispersed to new random location in search space else it remains at its current location.

***Step 9*:** If $l < N_{ed}$, then switch to step 2 else stop the process and print the result.

The flowchart for BFA implementation is shown in Fig. 9.2.

9.2.5 Movement of Bacteria in Search Space

Considering the movement of bacterium in a search space, an exclusive study for a minimization problem is considered. Taking θ_1, θ_2 in x-axis and y-axis, the convergence rate for a set of bacteria's is observed. The representation of bacterial movement is shown in Fig. 9.3. For the initial generation, the bacterial movement is seen in all over the search space. The chemotaxial movement made by bacterium has arrived them to reach near optimal fitness value. Once the bacterium finds its minimal fitness value, the convergence of all the bacterium to the minimal value below 10 is seen in 2nd generation. As intense progress made by bacterium in 1st and 2nd generation results the trajectories to surrender for minimal value in 3rd and 4th generation.

9.2.6 Verification of BFA with Mathematical Equations

To validate BFA for its success, various unimodal and multimodal benchmark functions are tested and its success rate is benchmarked with Genetic Algorithm (GA), Particle Swarm Optimization (PSO), Group Search Optimizer (GSO) and Fast

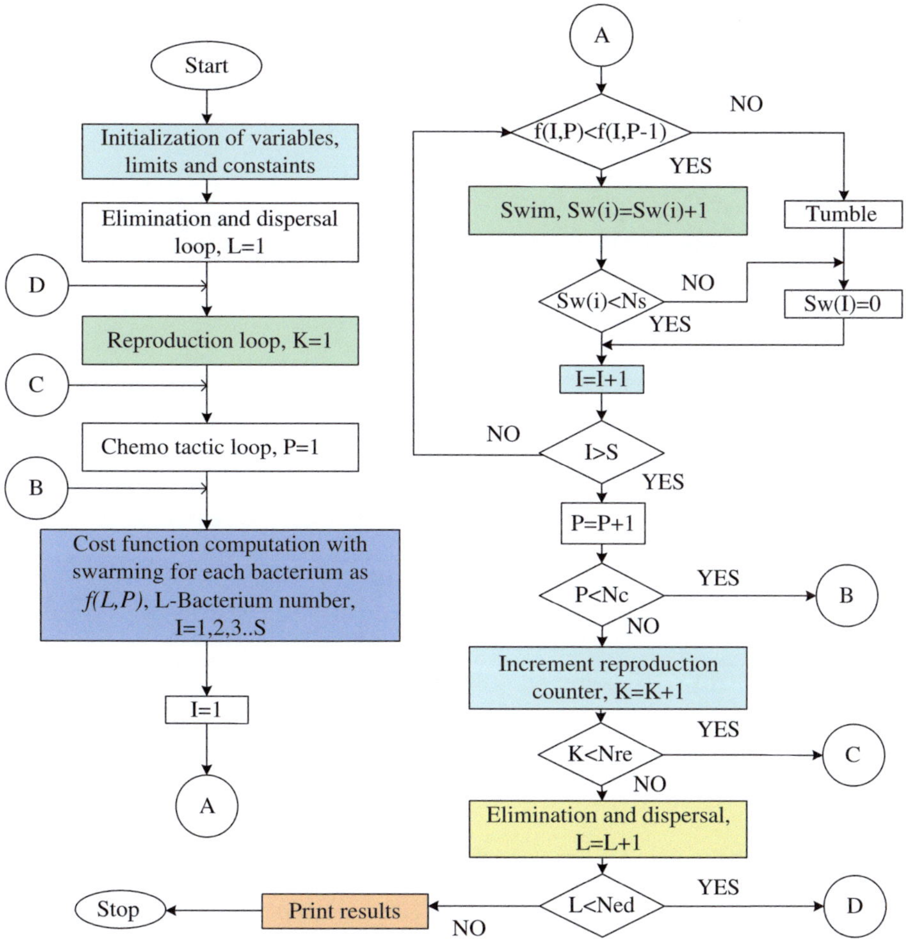

Fig. 9.2 Flowchart for Bacterial foraging algorithm

Evolutionary Programming (FEP) methods. All the aforesaid algorithm algorithms are coded and made to run for a trial of 50 runs. Based on the results obtained from various functions, the mean and standard deviation is tabulated in Table 9.1. For evaluations, five equations corresponding to unimodal function and five multidimensional bench mark functions are evaluated. The equations considered for evaluations are given below.

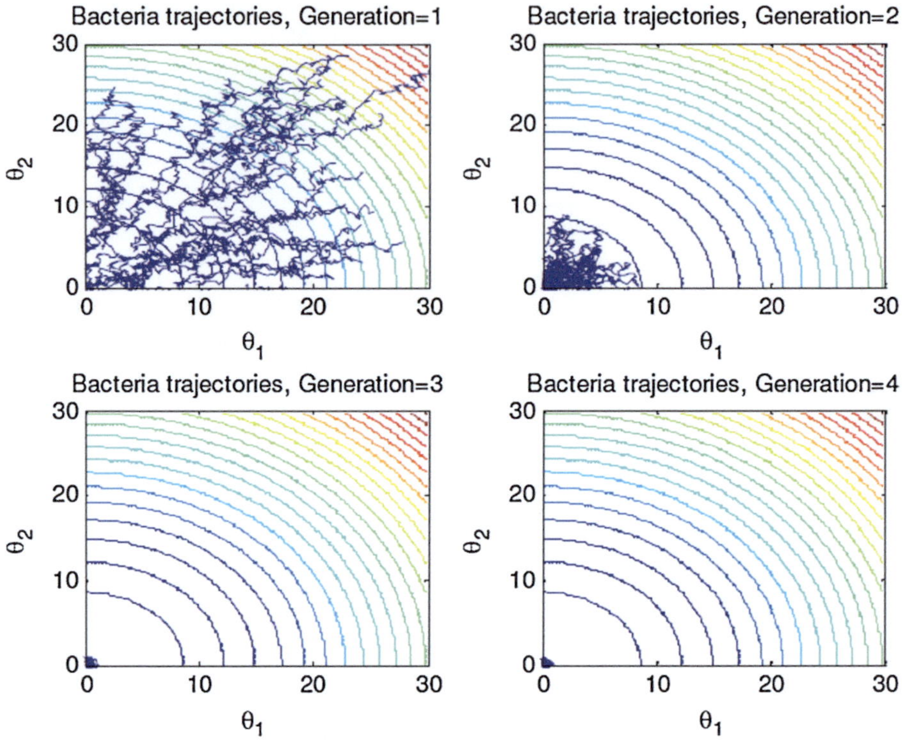

Fig. 9.3 Movement of Bacterium for 4 Generations

High Dimensional unimodal Benchmark Functions

1. Sphere model: $f_1(x) = \sum_{i=1}^{30} x_1^2 - 100 \quad -100 \le x_i \le 100$

   $\min(f_1) = f_1(0, \ldots\ldots, 0) = 0$

2. Schwefel's Problem 2.22: $f_2(x) = \sum_{i=1}^{30} |x_i| + \prod_{i=1}^{30} |x_i| \quad -10 \le x_i \le 10$

   $\min(f_2) = f_2(0, \ldots\ldots, 0) = 0$

3. Schwefel's Problem 2.21: $f_3(x) = \max_i \{|x_i|, 1 < i < 30\} \quad -100 \le x_i \le 100$

   $\min(f_3) = f_3(0, \ldots, 0) = 0$

4. Rosenbrock's function: $f_4(x) = \sum_{i=1}^{29} \left[100\left(x_{i+1} - x_i^2\right) + (x_i - 1)^2\right];$

   $-30 \le x_i \le 30$

   $\min(f_4) = f_4(1, \ldots\ldots, 1) = 0$

5. Step Function: $f_5(x) = \sum_{i=1}^{30} (|x_i + 0.5|)^2 \quad -100 \le x_i \le 100$

   $\min(f_5) = f_5(0, \ldots\ldots, 0) = 0$

High Dimensional Multimodal Benchmark Functions

6. Generalised Schwefel's Problem 2.26	$f_6(x) = \sum_{i=1}^{30} -x_i \sin\left(\sqrt{x}\right) \quad -500 \leq x_i \leq 500$ $\min(f_6) = f_6(420.9687, \ldots\ldots, 420.9687) = -12569.5$
7. Generalised Rastrigin's Function	$f_7(x) = \sum_{i=1}^{30} \left[x_i^2 - 10\cos(2\pi x_i) + 10\right]$ $-5.12 \leq x_i \leq 5.12$ $\min(f_7) = f_7(0, \ldots\ldots, 0) = 0$
8. Ackley's Function	$f_8(x) = -20 \ \exp - \left(0.2\sqrt{\frac{1}{n}\sum_{t=1}^{30} x_i^2}\right) - \exp\left(\frac{1}{30}\sum_{t=1}^{30} \cos(2\pi x_i)\right) + 20 + e$ $-32 \leq x_i \leq 32$ $\min(f_8) = f_8(0, \ldots\ldots, 0) = 0$
9. Generalised Girewank Function	$f_9(x) = \frac{1}{4000}\sum_{i=1}^{30} x_i^2 + \prod_{i=1}^{30} Cos\left(\frac{x_i}{\sqrt{i}}\right) + 1$ $-600 \leq x_i \leq 600$ $\min(f_9) = f_9(0, \ldots\ldots, 0) = 0$
10. Generalised Penalised Functions	$f_{10}(x) = \frac{\pi}{30}\left\{10\sin^2(\pi y_1) + \sum_{t=1}^{29} (y_t - 1)^2\left[1 + 10\sin^2(\pi y_{t+1})\right] + (y_{29})^2\right\} + \sum_{t=1}^{30} u(x_t, 5, 100, 4), \quad -32 \leq x_t \leq 32$ $\min(f_{10}) = f_{10}(1, \ldots\ldots, 1) = 0$ *where*, $y_i = 1 + \frac{1}{4}(x_i + 1)$ $u(x_i, a, k, m) = \begin{cases} k(x_i - 1)^m, x_i > a, \\ 0, -a < x_i < a, \\ k(-x_i - 1)^m, x_i < -a \end{cases}$

From Table 9.2, it is found that for the mathematical functions f_2–f_5, BFA method has showcased a far better performance in comparison to FEP, GSO and PSO. Nevertheless, PSO method on overall has regularly attained a better fitness to attain second position in order. More importantly, PSO has beaten BFA for sphere model by producing an accuracy of 3.6927 × 10–37. Overall, for the unimodal functions, BFA has rendered its excellence for all the functions except first sphere function. In case of multimodal functions, attaining convergence is quite challenging because, the controllability to have an edge over multi-dimensions is found critical. However, the BFA has yet again outplayed all the algorithms by producing a better fitness

Table 9.1 Mean and standard deviation for various unimodal and multimodal functions

Function	Parameter	BFOA	GA	PSO	FEP	GSO
f_1	μ, Mean	9.9008 × 10–9	3.1711	3.6927 × 10–37	5.7 × 10–4	1.9481 × 10–8
	σ, sigma	8.5307 × 10–9	1.6621	2.4598 × 10–36	1.3 × 10–4	1.9841 × 10–8
f_2	μ, Mean	5.2701 × 10–5	0.5771	2.9168 × 10–24	8.1 × 10–3	3.7039 × 10–5
	σ, sigma	3.3527 × 10–5	0.1306	1.1362 × 10–23	7.7 × 10–4	8.6185 × 10–5
f_3	μ, Mean	0.1204	7.961	0.4123	0.3	0.1078
	σ, sigma	0.1021	1.5063	0.25	0.5	3.9981 × 10–2
f_4	μ, Mean	7.7145 × 10–4	338.5616	37.3582	5.06	49.8359
	σ, sigma	2.1035 × 10–3	361.497	32.1436	5.87	30.1771
f_5	μ, Mean	0	3.697	0.146	0	1.6000 × 10–2
	σ, sigma	0	1.9517	0.4182	0	0.1333
f_6	μ, Mean	−12589.4882	−12566.0977	−9659.6993	−12554.5	−12569.4882
	σ, sigma	1.2582 × 10–2	2.1088	463.7825	52.6	2.2140 × 10–2
f_7	μ, Mean	0.3119	0.6509	20.7863	4.6 × 10–2	1.0179
	σ, sigma	0.3318	0.2805	5.94	1.2 × 10–2	0.9509
f_8	μ, Mean	1.6125 × 10–5	0.8678	1.3404 × 10–3	1.8 × 10–2	2.6548 × 10–5
	σ, sigma	2.9289 × 10–5	0.2805	4.2388 × 10–2	2.1 × 10–2	3.0820 × 10–5
f_9	μ, Mean	2.4549 × 10–2	1.0038	1.0633 × 10–2	2.6 × 10–2	3.1283 × 10–2
	σ, sigma	4.2353 × 10–2	6.7545 × 10–2	1.0895 × 10–2	2.2 × 10–2	2.8757 × 10–2
f_{10}	μ, Mean	1.9296 × 10–9	4.3572 × 10–2	3.9503 × 10–2	0.2 × 10–6	2.7648 × 10–11
	σ, sigma	2.0533 × 10–9	5.0539 × 10–2	9.1424 × 10–2	6.1395 × 10–6	9.1674 × 10–11

Table 9.2 Parameter values obtained using BFA for different irradiation conditions

Parameters	Mono crystalline S36		Thin Film ST40	
	GA	BFA	GA	BFA
G = 1000 W/m²				
Rs	1	1.6959	1	1.6489
Rp	0.4148	0.1281	1.654	1.01
a	200.391	478.4748	457.478	370.9784
G = 800 W/m²				
Rs	1	1.7931	1.8338	1.6234
Rp	0.6041	0.0414	1	1.01
a	461.3881	489.8589	372.434	370.9784
G = 600 W/m²				
Rs	1	1.7491	1.7243	1.6496
Rp	0.6178	0.0216	1.2199	1.0027
a	207.2336	475.3372	162.2678	299.2976
G = 400 W/m²				
Rs	1	1.6756	1.3607	1.6547
Rp	0.8074	0.0867	1.8495	1
a	441.8377	479.3708	372.434	309.5164
G = 200 W/m²				
Rs	1	1.6043	1.5797	1.6297
Rp	0.3891	0.02	1.1144	1.0187
a	451.6129	486.3113	193.5484	384.6959

value. To be specific, a qualitative comparison made in terms of mean and standard deviation, it is found that for functions f6–f9, the BFA was in high standard and in case of Generalised penalized function i.e., f10, GSO method comes out as a surprise to emerge better fitness.

9.2.7 Modified Bacterial Foraging Algorithm

Understanding the scope and visibility with BFA, the scope for improvisations with conventional BFA is found higher. Moreover, the possibilities of tuning parameters with BFA method are found very higher. Further, it is possible that good solutions arrived with BFA method can further be enhanced to achieve near optimal value. Thus, a nelder-meed property is introduced in BFA for improving the good solutions in BFA. Further on trial and error, the author Edward et al. [16] has proved that the nelder meed-BFA fusion can suit many of the engineering applications. The pseudo code with defined parameter values for attaining the optimal solutions is presented as a pseudocode below.

Pseudo Code for Nelder-Meed BFA Method

Pseudo code for nelder-meed BFA method:

```
Initialize the following parameters as follows (i) Number of bacteria S: 4, (ii) Number
of chemotactic steps Nc: 5 (iii) Swimming length Ns: 4, (iv) Number of reproduction
steps Nre: 4 (v)Number of elimination and dispersal events Ned: 2(vi) Probability of
elimination and dispersal Ped: 0.2. (vii) Depth of attractant: 0.01,(viii)Width of
attractant: 0.04, (xi) Height of repellent: 0.01, (x)Width of repellent: 10.0.

Obtain fitness for all the bacterium in population,
 For (l=0 to Ned)
    For(k=0 to Nre)
        For (j=0 to Nc)
                Chemotaxis and swim for given population.,
           For (bacterium € population)
                If (fitness(current bacterium<=global best)
                Global best= Current bacterium (location)
           End
        End
    End
    Sort the fitness in ascending order (population)
    Select the bacterium by health by using Bacterium (population, fitness sum/2)
    Select the best bacterium in population (Use nelder-meed algotithm to evaluate
    the best location)
    For (bacterium € population)
        If (rand() <=Ped)
            Create random location for FACTS devices
        End
    End
End
RETURN (BEST BACTERIUM IN POPULATION AND FITNESS) (I,E)FINAL
```

9.3 BFA for PV Parameter Estimation

In recent years, tremendous recognition gained by Solar PV among renewables has made it as a promising alternative for conventional power generation [30]. Although the PV has advantages like zero carbon emission and less maintenance, the accurate reproduction of PV characteristics is very much essential for a PV panel to deliver maximum power. Hence accurate modelling of solar PV is needed to be carried out by figuring out the suitable parameters. It is important here to mention that PV modelling parameters highly depend to the change in environmental characteristics.

The PV modelling in general demands the manufacturer's datasheet values like Voltage at MPP (V_{MPP}), Current at MPP (I_{MPP}), Short circuit current (I_{SC}) and Open circuit Voltage (V_{OC}). However, despite these data few more parameters like PV current (I_{PV}), series resistance (R_S), diode ideality factor (a) and Parallel resistance (R_P) also becomes a mandatory obligation that can be estimated using an optimization procedure [30]. Two important modelling procedures available in

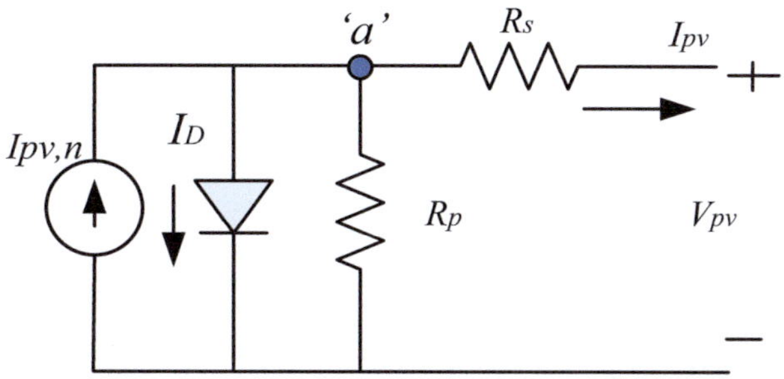

Fig. 9.4 Single diode model of solar PV

literature are (i) Analytical and (ii) Numerical. The procedure adopted in earlier approach is time consuming and non-accurate while, in case of the later approach use accurate mathematical procedure in consideration to individual points on the I–V curve [29].

Two common models that are prevalent in the field of PV modelling are one diode and two diode model. The two diode model has an excessive diode compared to one diode model which makes it to generate accurate PV characteristics even under low irradiation conditions. Various optimization procedures involved in extracting PV parameters are: Genetic Algorithm, Bird Mating Optimization, Simulate Annealing, Flower Pollination Algorithm and Bee pollinated Flower Pollination Algorithm. In 2013, Rajasekar et al. [29] has investigated PV modelling using BFA. The Excellency of BFA to explore and exploit search space has yielded appreciable results with fair convergence speed. The results and optimization and the problem formulation are explained in further subsections.

9.3.1 *PV Modelling*

The commonly used one diode electric model to represent the behavior of a PV module is shown in Fig. 9.4. Further, one diode model is given importance for its less complexity and simple accessibility. The schematic of one diode model comprises of a current source (I_{pv}), diode, series resistance (R_s) and parallel resistance (R_P). Series resistance (R_s) is formulated with an effect to cumulative resistance of semiconductor material and metallic contacts in panel. Parallel resistance (R_P) is formulated to measure the leakage losses.

Applying KCL to the Fig. 9.2., the PV module output current equation can be written as:

$$I_{PV} = I_{pv,n} - I_D - \frac{V_D}{R_p} \tag{9.5}$$

Where, V_D is the diode voltage and I_D is the diode current.

The diode current equation can be given as:

$$I_D = I_o \left(e^{V_D/\alpha V_t} - 1\right) \tag{9.6}$$

Where, 'a' is the diode ideality factor and thermal voltage 'V_t' is given as:

$$V_t = N_s kT/q \tag{9.7}$$

Where, k is the Boltzmann constant, T is the cell temperature in Kelvin, q is the electron charge, N_s is the number of cells in series. Using KVL, the output voltage of the module can be written as:

$$V = V_D - IR_s \tag{9.8}$$

Thus, from the above discussion it can inferred that for a single-diode model five parameters are needed to be computed (I_{pv}, I_o, R_s, R_p, and a).

9.3.2 Problem Formulation

Among the five parameters (I_{pv}, I_o, R_s, R_p, and a.) I_{pv} and I_o are calculated analytically to reduce computational complexity, and the remaining R_s, R_p, and a are arrived via BFA such that, the error between the computed and actual characteristics is minimized. The PV current can be measured using

$$I_{pv} = (I_{scn} + k_i dT)\frac{G}{G_n} \tag{9.9}$$

Where, I_{scn} is the nominal short circuit current at Standard Test Condition (STC), k_i is the short circuit current temperature coefficient taken from panel datasheet. G_n is the irradiation at STC (i.e.,) 1000 W/m^2 0and G is the specific irradiation for which the panel is exposed and change in temperature $dT = T - T_n$, Where, T_n is the standard temperature at STC condition i.e. $25^o C$ and T is the surface temperature of the PV panel.

The reverse saturation current in diode can be calculated is obtained using the equation formulated as below:

$$I_o = \frac{I_{pv}}{\exp\left[(V_{oc} + k_v dT)/a/V_t\right] - 1} \tag{9.10}$$

The Open circuit voltage (V_{oc}) and Voltage at maximum power (V_{mp}) values are determined using the following equation

$$V_{oc} = V_{ocn} + V_t \ln\left(\frac{G}{G_n}\right) + k_v dT \tag{9.11}$$

$$V_{mp} = V_{mpn} + V_t \ln\left(\frac{G}{G_n}\right) + k_v dT \tag{9.12}$$

Pseudo Code for PV Parameter Estimation Using BFA

Pseudo code for PV parameter estimation using BFA:

```
Initialize the following parameters as (i) Chemotaxis loop counter , Nc= 200, (ii)
Reproduction loop counter, Nre = 4, (iii) Elimination and dispersal loop counter,
Ned = 2, (iv) Population size = 50, (v)Elimination and dispersal probability =
0.25, 0<Rs<2, 50<Rp<500,1<a<2
Choose random locations of Rs, Rp, a in search space.
Obtain fitness for all the bacterium in population,
 For (l=0 to Ned)
    For(k=0 to Nre)
        For (j=0 to Nc)
                Chemotaxis and swim for given population.,
           For (bacterium € population)
                If (fitness(current bacterium<=global best)
                Global best= Current bacterium
           End
       End
    End
    Sort the fitness in ascending order (population)
     Select the bacterium by health by using Bacterium (population, fitness sum/2)
    Select the best bacterium in population
    For (bacterium € population)
        If (rand() <=Ped)
             Create random location for Rs,Rp,a
        End
    End
End
Return (best bacterium in population and fitness) (i,e)final Rs, Rp, a and Gbest
```

In order to improve accuracy the above equations are modified and represented as

$$V_{oc} = V_{ocn} + V_t \ln\left(\frac{G}{G_n}\right) + k_v dT - \alpha \log\left(\frac{G}{G_n}\right) \tag{9.13}$$

$$V_{mp} = V_{mpn} + V_t \ln\left(\frac{G}{G_n}\right) + k_v dT - \beta V_t \log\left(\frac{G}{G_n}\right) \tag{9.14}$$

Similarly, maximum power point current can be given as:

$$I_{mp} = I_{mpn} \ln\left(\frac{G}{G_n}\right)\{1 + k_i dT\} \tag{9.15}$$

The calculated V_{oc} and V_{mp} values using the above equations accurately match with the manufacturer's datasheet and has the ability to vary according to the solar panel. The values of α is taken as 0.9 and β is taken as 1.65 in the work to appreciate accuracy.

Based on the mathematical condition at Maximum Power Point (MPP), the derivative of the power with respect to voltage is equal to zero. i.e. $\frac{dP}{dV} = 0$.

The power equation in general can be written as $P = VI$. Applying the condition at MPP, the above equation becomes

$$\frac{dP}{dV} = V\frac{dI}{dV} + I \tag{9.16}$$

The objective function is a minimization function and it can be represented as,

$$J = \left|\frac{dI}{dV}\right|_{(V_{mp}, I_{mp})} + \frac{I_{mp}}{V_{mp}} \tag{9.17}$$

Where, dI/dV is obtained from single diode model basic current equation as

$$\left|\frac{dI}{dV}\right|_{(V_{mp}, I_{mp})} = \frac{I_o\,\Gamma\,\exp\left\{\Gamma\,\left(V_{mp} + I_{mp}R_s\right)\right\} - G_p}{1 + I_o\,\Gamma\,R_s\exp\left\{\Gamma\,\left(V_{mp} + I_{mp}R_s\right)\right\} - G_pR_s} \tag{9.18}$$

Where, $G_p = 1/R_p$ and $\Gamma = 1/aV_t$

The optimal solution is obtained in regard to a specific value for which the fitness value is calculated after each iteration.

9.3.3 Results and Discussion

The initial random guesses for the PV model parameters are randomly chosen within the boundary limit. It is general fact that, the value of series resistance is observed to have very small which is to reduce the loses and parallel resistance is chosen as high to restrict the recombination losses.

With above inputs, the MATLAB code for the BFA is programmed for BFA and Genetic Algorithm. Further two module characteristics (Shell S36, ST40) characteristics are reproduced for five irradiations respectively (200 W/m^2, 400 W/m^2, 600 W/m^2, 800 W/m^2 &1000W/m^2) and shown in Figs. 9.5 and 9.6. respectively. The corresponding parameters extracted using BFA and GA is tabulated in Table 9.2. From Table 9.2 it is understood that BFA provides accurate reproduction of manufacturers curves with a better computational efficiency. The search process in BFA to effectively use chemotaxis movement has improved the solution quality via swimming and tumbling hence convergence to global solutions via exploitation is assured via BFA method. Although solution quality with high accuracy is seen, the convergence time of BFA is inferior to GA as BFA converge at 200th iteration while GA converges at 80th iteration.

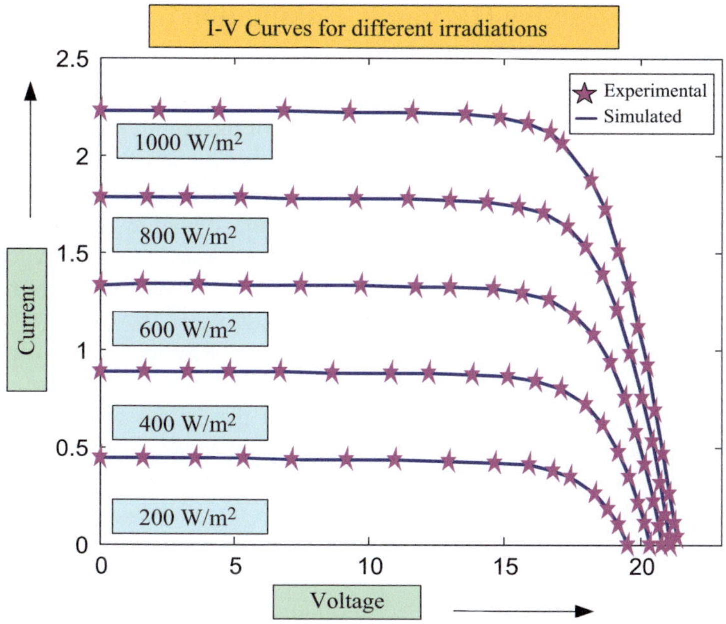

Fig. 9.5 I-V Curve of Shell S36 panel for different irradiance and temperature conditions

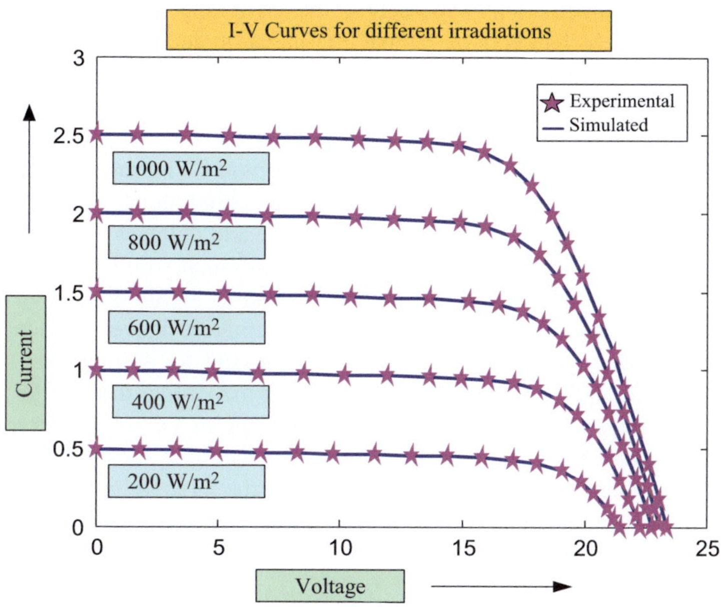

Fig. 9.6 I-V Curve of ST40 panel for different irradiance and temperature conditions

9.4 BFA for Selective Harmonic Elimination in PWM Inverter

Pulse Width Modulation (PWM) technique is one of the commonly used switching scheme to regulate the inverter output voltage. In general, the voltage and frequency of PWM inverter is regulated by controlling the switching sequences of power semiconductor devices [44]. Having witnessed the dynamic growth of digital processors and recent advancements in low price semiconductor switches, the inverters are built cost efficient. Further the efficiency in PWM inverters has made them inevitable in industrial and commercial applications [32].

Being benefitted with several advantages like (i) Elimination of triplen harmonics, (ii) The ability to deliver satisfactory performance even at low switching frequency and (iii) Controlled harmonic limits in the output waveform help the PWM inverter to stand tall among other traditional inverters [6]. After getting discovered in [43], the output voltage waveforms of SHE PWM inverter are analyzed in Fourier domain and developed as non-linear transcendental equation which can be solved using iterative Newton-Raphson method [38]. However the method is derivative dependent and it may converge to local convergence. In another work, Walsh functions are used to optimize the switching angles of power semiconductor devices [21, 40]. But then the method fails to optimally locate an angle which is continuously varying for the given interval.

Later to the mathematical approaches, the switching sequence optimization is quite alternatively handled by global optimization techniques. Further these methods are found to locate accurate global solutions irrespective of the non-linearity present in the system. Several algorithms like Genetic Algorithm (GA) [37] Particle Swarm Optimization (PSO) [33] and are Ant Colony Optimization (ACO) [39] and Artificial Bee Colony (ABC) [19] are proposed for harmonic elimination in PWM inverters. Inspired by genetics, GA based switching sequence generation was attempted in [37], however premature convergence and delayed computation adds to GA as a drawback. To solve the issues in GA, a much simpler and faster PSO is attempted is proposed in [33], but the convergence is highly dependent to initial exploration in control variables. In consequence, premature convergence may occur. In another approach ACO and ABC are also proposed for setting gate sequence to eliminate harmonics in PWM inverter. However the similar drawbacks found in GA and PSO is also found here and in addition parameter setting increases the computational burden. As a replacement to earlier methods, BFA based switching sequence is designed and discussed in [6]. Further the method has come up with a positive result in comparison to the earlier methods. The formulation ob objectives and the results attained via BFA are discussed below.

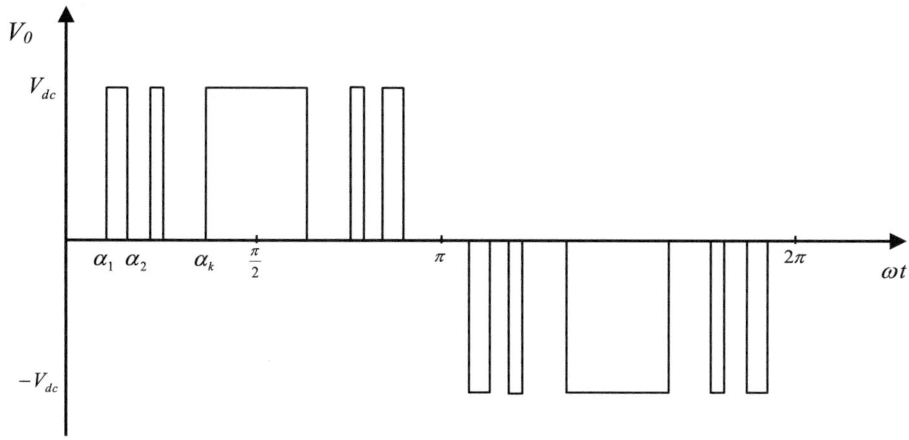

Fig. 9.7 Output Voltage waveform for a PWM inverter

9.4.1 Problem Formulation

The output voltage for a 1-Φ PWM inverter having '*k*' pulse in a half cycle is illustrated in Fig. 9.7. For realization, it is assumed that output voltage have '*k*' pulses in a half cycle where the switching angles are found symmetrical with respect to $\pi/2$. Further, '*k*' is an odd number and varies in the range from three, five, seven, nine etc. The output voltage of a PWM inverter using the Fourier series is represented as,

$$V_0 = a_0 + \sum_{n=1}^{\infty} A_n \mathrm{Cos}\,(n\omega t) + \sum_{n=1}^{\infty} B_n \sin\,(n\omega t) \tag{9.19}$$

It is well fact that for a symmetrical signal, even harmonics are absent. Hence, the Fourier coefficients A_n and a_0 in above equation becomes zero due to symmetry. Further the output voltage equation becomes,

$$V_0 = \sum_{n=1}^{\infty} B_n \sin\,(n\omega t)\,, \tag{9.20}$$

The Fourier coefficient B_n can be computes as

$$B_n = \frac{4V_{dc}}{n\pi}[\cos\,(n\omega t)]_{\alpha_2,\alpha_4,\ldots\ldots\pi/2}^{\alpha_1,\alpha_3,\ldots\ldots\alpha_k} \tag{9.21}$$

The value of fundamental component value is mathematically realized using

$$B_1 = \frac{4V_{dc}}{\pi}[\cos(n\omega t)]_{\alpha_2,\alpha_4,\ldots\ldots\pi/2}^{\alpha_1,\alpha_3,\ldots\ldots\alpha_k} \tag{9.22}$$

The basic objective behind the proposed method is to generate switching instance (α_1, α_2. ...α_k) for the output voltage waveform for which the harmonic components B_3, B_5, ...B_{2k-1} are made zero. It is also important o note here that the fundamental component (‘B_1’) equals the output voltage. Hence output voltage of inverter is regulated along with harmonic regulation. This problem is suitably addressed by using BFA as an iterative based optimization function. The minimization objective function can be mathematically presented as follows.

$$F(\alpha) = F(\alpha_1, \alpha_2, \ldots\ldots, \alpha_k) = e_r + h_c \tag{9.23}$$

Constraints for the Objective:

$$0 \le \alpha_1 \le \alpha_2 \ldots\cdots \le \alpha_{k-1} \le \alpha_k \le \pi/2$$

Where,

$$e_r = \left|V_0^* - B_1\right| \text{ and } h_c = |B_3| + |B_5| + |B_7| + |B_9| + \ldots\ldots\cdots + |B_{2k-1}| \tag{9.24}$$

The research objective is twofold (i) to maintain desired output voltage level of inverter. i.e., (er). (ii) selective harmonic elimination of harmonic content i.e., (h_c). Based on generation of pulses in a half cycle (odd number) harmonics (even number) can be eliminated. For explanation, if k-pulses exist in a output voltage for a half cycle, then $k - 1$ harmonics can be reduced.

9.4.2 Simulation Results and Discussion

To minimize the objective function defined in (9.23), the identification of switching sequence is mandatory. Further the BFA method is utilized here to generate the optimum switching instances ($\alpha_1, \alpha_2, \ldots \ldots, \alpha_k$). The ability in BFA to explore an exploit the search space has satisfactorily addressed two important concerns (i) Lower order harmonics elimination and (ii) Fundamental component improvement. A dedicated MATLAB program is developed for BFA to undergo performance evaluations. Further the PSO and GA method is also coded in MATLAB for comparison. To code the algorithm the parameters set in BFA are $d_{attract}$, $W_{attract}$, $h_{repellent}$, $W_{repellent}$, C, P_{ed} and S are set to 0.01, 0.04, 0.01, 10, 0.1, 0.25 and 50 respectively. In case of PSO, inertia weight factor w_I and social weight (c_I) and cognitive weight factor (c_2) are set to 0.4, 1.6 and 1.43 respectively.

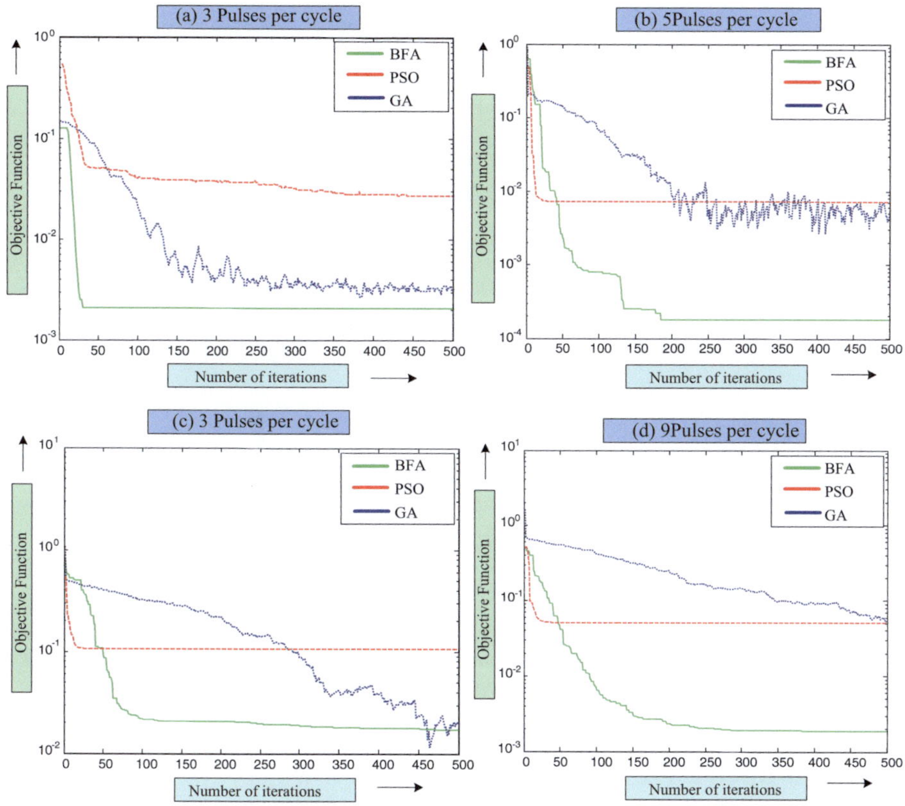

Fig. 9.8 Convergence characteristics of BFA,PSO and GA 3,5,7&9 pulse generation sequence

In order to determine the solution quality, the convergence characteristics for GA, PSO and BFA methods are plotted for 3, 5, 7, 9 pulses per half cycle. Further the graphs correspond to convergence can be seen in Fig. 9.8a–d. From the figure it is clear that, BFA method has very low fitness value for the switching sequence generation which indicates the accuracy of BFA to explore more search space to arrive global solutions. Although GA and PSO methods has performed better in 3 and 5 pulse generation sequence, due to the smaller value difference in firing angle for 7 and 9 pulse per half cycle, the methods got fallen for local optima. Hence it can be concluded that the BFA method investigated for harmonic elimination is superior to PSO and GA methods.

To further strengthen the harmonic spectra generated at 3, 5, 7 and 9 pulse per half cycle sequences are plotted and presented in Fig. 9.9a–d. The lesser harmonics present in BFA compared to GA and PSO illustrates the suitability of BFA to the research objective and emerges as an implementable solution to reduce voltage harmonics.

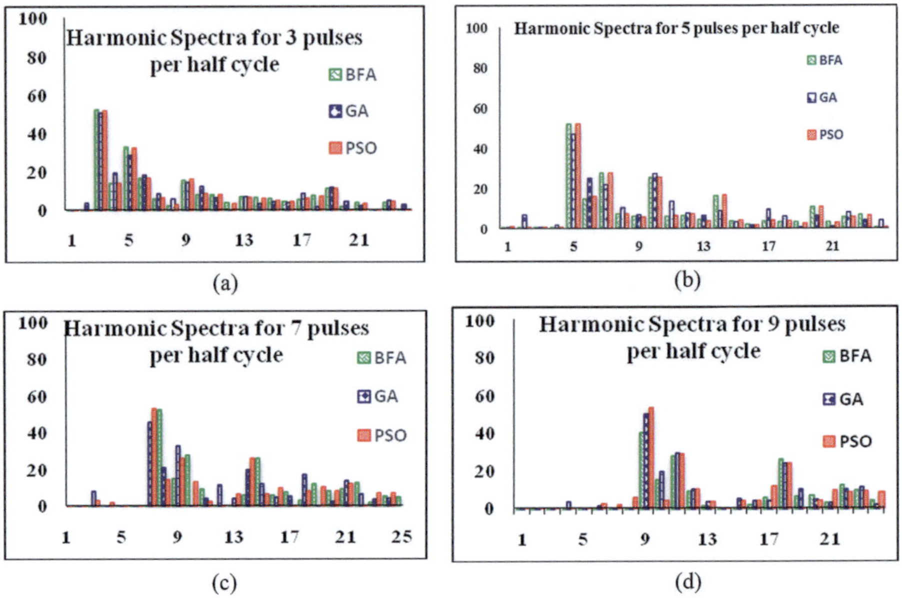

Fig. 9.9 Harmonic spectra Simulated for (a) $k = 3$ (b) $k = 5$ (c) $k = 7$ and (d) $k = 9$ pulses per half cycle. (**a**) 3 pulses per half cycle. (**b**) 5 pulses per half cycle. (**c**) 7 pulses per half cycle. (**d**) 9 pulses per half cycle

9.5 Modified BFA for Optimal Power Flow

J. Carpentier in 1962 has evolved a new non-linear optimal power flow (OPF) problem [12]. The problem formulation made via mathematical approach was large and took decades to arrive at solutions via efficient algorithms [18]. Numerous works have been carried out to solve OPF problem. Some of the notable and intelligent algorithms applied for OPF problem are Genetic Algorithm [41], Simulated Annealing [34], Differential Evolution [2] and Particle Swarm Optimization [1].

Recent development in Flexible AC Transmission Systems (FACTS) utilizing power electronic components has increased the flexibility of a power system to control power flow [16]. It is noteworthy to mention here that the stability and safety of the power system is increased drastically even under abnormal operating conditions. Some of the important FACTS devices used in transmission system are Unified Power Quality Controller (UPQC), Unified Power Flow Controller (UPFC), Thyristor, TCSC and Static VAR Compensator (SVC). These FACTS devices can easily regulate power flow parameters like (i) bus voltages, (ii) line impedances and (ii) Phase angle. In addition, incorporating FACTS devices thermal stress over increased transmission line can be limited. One can see the first paper proposed for OPF in [24]. The authors have used phase shifters and series capacitors to control flow. Later GA [13] and PSO [36] are utilized to solve non-linear OPF problem.

Witnessing the advantages of using BFA in an optimization, Edward et al. [16], introduced BFA for a optimal power flow an IEEE-30 bus system. Further the author commended that BFA is one of the powerful optimization tools in real world. However the selection of optimal solutions is found difficult in BFA hence, nelder-meed property is fused with BFA for an OPF. A similar method is used for an economic dispatch problem in [27].

9.5.1 Modelling of FACTS Devices

To obtain optimal power flow, two FACTS devices are constructed (i) Static VAR Compensator (SVC) for Reactive power compensation (ii) TCSC is used to change the reactance of the line.

9.5.1.1 Modeling of SVC

The SVC structure has a Thyristor Controlled Reactor (TCR) connected in parallel with capacitor bank as shown in Fig. 9.10. The operation of SVC resemble to the variable reactance connected in parallel. This SVC can absorb or generate reactive power to regulate voltage at PCC (Point of Common Coupling). Further the mathematical representation of controllable reactance 'X_V' in terms of reactive impedance 'X_L' is given as.

$$X_V = X_L \frac{\pi}{2\pi - 2\alpha + \sin(2\alpha)} \tag{9.25}$$

Where α denotes the firing angle of the thyristor.

The Fixed Capacitor-Thyristor Controlled Reactor (FC-TCR) effective susceptance 'B' in SVC is given by:

Fig. 9.10 Representation of VAR compensator

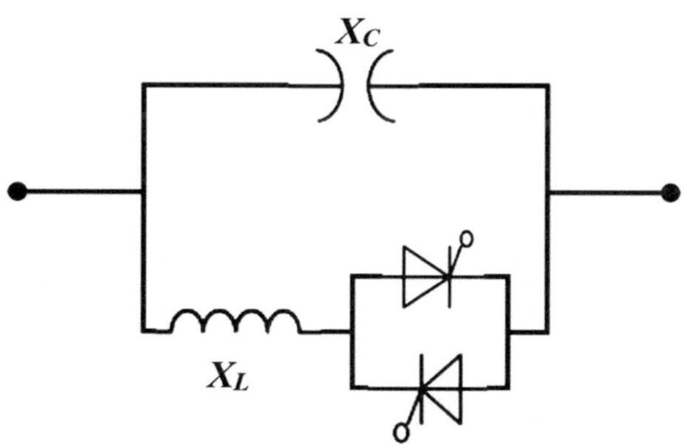

$$B = \frac{X_V + X_C}{X_V X_C} \tag{9.26}$$

Where 'X_C' denotes the capacitive reactance of the fixed capacitor.

The SVC is able to perform both inductive and capacitive compensation. Further it is modeled as an ideal reactive power injection at bus 'i'. The power injected at i^{th} bus is given as.

$$\Delta Q_{is} = Q_{SVC} \tag{9.27}$$

9.5.1.2 Modeling of TCSC

The purpose of attaining controllable reactance utilizing a TCSC inserted in related transmission line can be realized by schematic shown in Figs. 9.11 and 9.12. The mathematical equation for the power flow in transmission line is realized as follows:

$$P_{mn} = V_m^2 g_{mn} - V_m V_n g_{nm} \cos(\delta_m - \delta_n) + V_m V_n b_{nm} \sin(\delta_m - \delta_n) \tag{9.28}$$

$$Q_{mn} = V_m^2 g_{mn} - V_m V_n g_{nm} \sin(\delta_m - \delta_n) + V_m V_n b_{nm} \sin(\delta_m - \delta_n) \tag{9.29}$$

Where

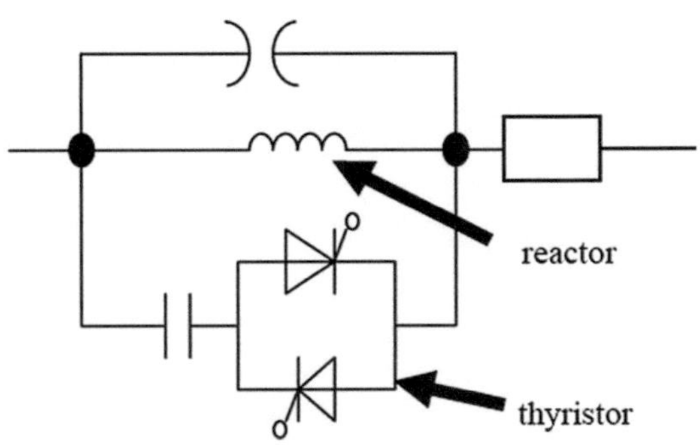

Fig. 9.11 Schematic of TCSC

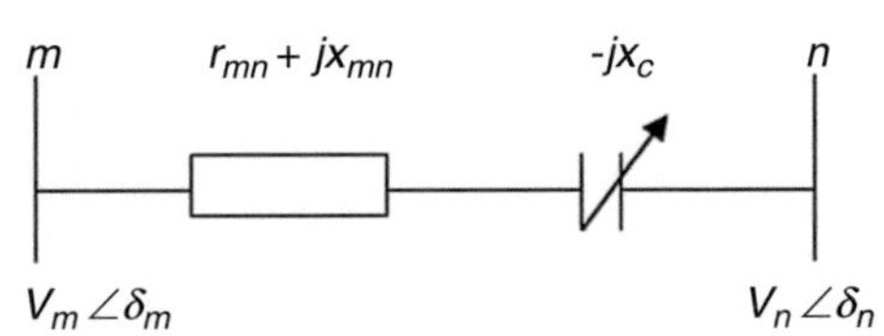

Fig. 9.12 Single line diagram for TCSC regulation

$$g_{mn} = \frac{r_{mn}}{r_{mn}^2 + (x_{mn} - x_c)^2}, \quad b_{mn} = -\frac{x_{mn} - x_c}{r_{mn}^2 + (x_{mn} - x_c)^2}$$

The conventional OPF objective function is been modified to apply for the FACTS devices. The objective defined for optimization is discussed in followed section.

9.5.2 *Formulation of Objective Function*

9.5.2.1 Cost Function

The primary objective of this work is locate FACTS devices in order to minimize the overall cost function. This cost function consists of generation cost and investment cost of FACTS devices.

(i) *Conventional Cost Function*

The conventional generation cost function can be approximated as a quadratic function in terms of US$/Hour as represented in (9.30):

$$C_x\left(P_g\right) = aP_g^2 + bP_g + c \tag{9.30}$$

Where 'P_g' denotes a vector of active power outputs for the generators in (MW), a, b and c are the cost coefficients of generator.

(ii) *FACTS Devices Cost Functions*

Based on Siemens AG Database, the cost functions of SVC and TCSC are developed and mathematically represented as follows:

(i). *The SVC cost function is:*

$$C_{SVC} = 0.0003S^2 - 0.315S + 127.38\,(US\$/kVar) \tag{9.31}$$

(ii). *The TCSC cost function is:*

$$C_{TCSC} = 0.0015S^2 - 0.7130S + 153.38\,(US\$/kVar) \tag{9.32}$$

Total cost investment function for FACTS devices can be given as

$$C_y(f) = C_{SVC} + C_{TCSC} \tag{9.33}$$

9.5.3 Optimal Cost Minimization Using BFA

9.5.3.1 Optimal FACTS allocation

The overall cost function C_T involves two functions (i) cost of generation '$C_x(P_g)$' and (ii) average investment costs of FACTS devices '$C_x(P_g)$'.

$$C_T = C_x\left(P_g\right) + C_x\left(P_g\right) \tag{9.34}$$

The objective function is to minimize the total cost C_T

$$\text{Min}\left(C_T\right) = \text{Min}\left(C_x\left(P_g\right) + C_y(f)\right) \tag{9.35}$$

Subjected to the following constraints

$$E\ \left(f, g\right) = 0 \tag{9.36}$$

$$B_1(f) < b_1,\ B_2(g) < b_2 \tag{9.37}$$

Where, $E(f,\ g)$:is the conventional power flow equations. $B_1(f)$ and $B_2(g)$ are the FACTS devices inequality constraints to maintain the conventional power flow.

The derived minimization cost function is further solved to find the best rated values of FACTS devices as well to locate devices. The implementation procedure for BFA-nelder meed (BFA-NM)method applied for optimal power flow represented above as pseudo code.

9.5.4 Results and Discussion

To demonstrate the effectiveness of BFA method for optimal power flow, IEEE 30 bus system is considered for testing. Further the method is tested for various operating conditions in order to find the suitable optimal location and its consequent FACTS controller rating to ensure optimal power flow. Four different cases along with optimized FACTS devices location corresponding to different levels of loading are analyzed for different buses and presented. The cases study performed are listed as follows (1) Regular Loading for IEEE 30 bus system. In order to analyze the effect of loading in the system, different levels of loading are done at different bus. (2): increased loading only at bus 2, (3): increased loading at only bus 15, and (4): increased loading at buses 2 and 4 concurrently.

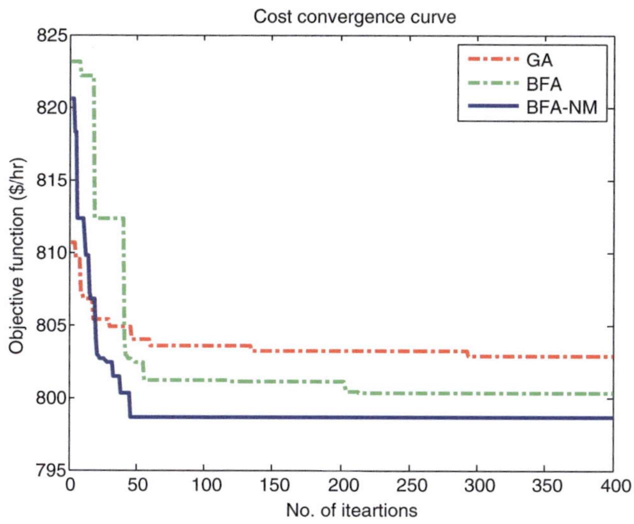

Fig. 9.13 Cost convergence for case1

Case 1: Regular Loading of IEEE 30 Bus Systems
In this case the hybrid BFA- NM method is utilized to solve the conventional OPF Problem without the inclusion of FACTS devices. To test the computation ability of the scheme, the BFA-NM convergence along with conventional BFA and GA is also coded and plotted in Fig. 9.13.

Case 2: With Increased Loading at Bus 2
With the additional load of 30 MW at bus no. 2, the FACTS device (SVC) is installed at line 36 of 73.38 MVAr rating. Further the location of FACTS device-SVC rating is found via trial and error method. On performing several runs using BFA-NM method, the generators outputs values by including SVC are found to be 201.46 MW, 49.76 MW, 22.27 MW, 13.97 MW, 11.05 MW and 17.42 MW respectively. Furthermore the ability of BFA-NM to locate SVC in the bus meets the requirement in the entire run.

Case 3: With Increased Loading at Bus 15
With an increased load of 20 MW at 15th bus is considered for this case and TCSC is selected at 10th line. In similar to case 1, the location of TCSC is arrived via the trial and error. The SVC rating is taken as 72.90MVAr. A dedicated MATLAB code is now developed and simulated for case 3. In consequence, the generators outputs are found are listed as follows: 199.73 MW, 50.16 MW, 19.71 MW, 11.48 MW, 12.97 MW and 12.94 MW respectively. The results in with case 2 have reduced the total cost slightly but at the same time the total loss in the system increases significantly.

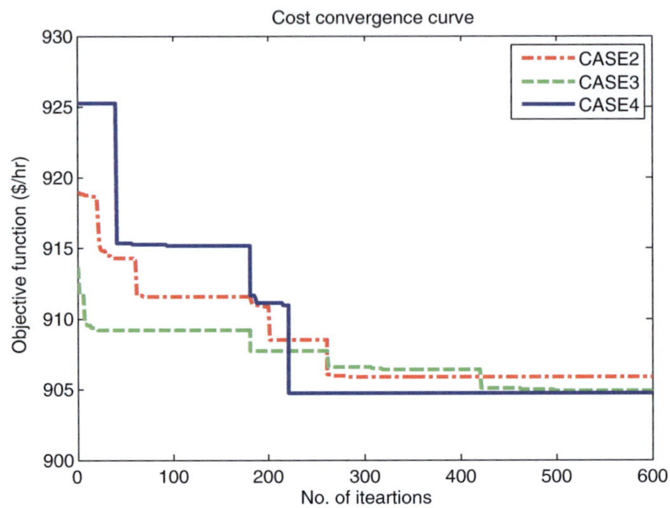

Fig. 9.14 Cost convceregnce cureve for Case 2, 3 & 4

Case 4: With Increased Loading at Buses 2 and 4 Simultaneously
Having simultaneous increase in load of 15 MW at 2nd and 4th bus, in order to meet the surmount loading, the SVC is placed at 34th line of 65.04 MVAr rating. The generators outputs corresponding to different trial are found to be 213.63 MW, 44.72 MW, 20.04 MW, 12.18 MW, 10.32 MW and 12.49 MW respectively. In the present case; since the addition of SVC has produced reduced total loss in above cases, in this case SVC is considered to meet the new load demand. The convergence characteristics for Case2, Case3& Case4 is shown in Fig. 9.14.

Thus from the above discussions, it is inferred that, the FACTS device maintains the power flow and decreased the total losses in system. However generation cost is increased. Therefore initial investment is needed in the beginning to increase generation but the cost can be met with the savings made via loss reduction.

9.6 Conclusion

In this book chapter the Bacterial Foraging Algorithm for an optimization problem is presented. It is noticed that chemotactic movement based on swimming and tumbling are more helpful for BFA method to find its success in optimization domain. An Enhanced exploration created in chemotactic movement is key reason for exploration in BFA. Based on various observations made with bacterial foraging optimization method, the following inferences are observed:

(i) The tumbling movement represented in mathematical form helps the bacteria to create specific exploitation in the search space.
(ii) On mathematical evaluations, the BFA method has performed its best in all the trial runs.
(iii) It is found that BFA has all scope to fuse with any of the soft computing methods to further improve its solutions.
(iv) Three various research problems investigated in this chapter clearly demonstrates that BFA is far superior to GA and PSO.
(v) Computational efficiency and convergence rate with global solutions are more appropriate for BFA to be an eventual choice for non-linear optimization problems.
(vi) A wide and huge scope for BFA to apply for modern engineering applications is seen at foresight and possibilities to replace existing methods with BFA as an optimization tool are its future scope.

Acknowledgements The authors would like to thank the Managements, Vellore Institute of Technology, Vellore and New Horizon College of Engineering, Bengaluru, India for providing the support to carry out research work. This work is carried out at Solar Energy Research Cell (SERC), school of Electrical Engineering, Vellore Institute of Technology, Vellore.

References

1. Abido MA (2002) Optimal power flow using particle swarm optimization. Int J Electr Power Energy Syst 24(7):563–571
2. Abou El Ela AA, Abido MA, Spea SR (2009) Optimal power flow using differential evolution algorithm. Elect Eng (Archiv fur Elektrotechnik) 91(2):69–78
3. Alanis AY, Arana-Daniel N, Lopez-Franco C (2015) Bacterial foraging optimization algorithm to improve a discrete-time neural second order sliding mode controller. Appl Math Comput 271:43–51
4. Arunkumar G, Gnanambal I, Karthik PC, Naresh S (2016) Proportional and integral constants optimization using bacterial foraging algorithm for boost inverter. Energy Procedia 90:535–539
5. Awadallah MA (2016) Variations of the bacterial foraging algorithm for the extraction of PV module parameters from nameplate data. Energy Convers Manag 113:312–320
6. Babu TS, Priya K, Maheswaran D, Kumar KS, Rajasekar N (2015) Selective voltage harmonic elimination in PWM inverter using bacterial foraging algorithm. Swarm Evol Comput 20:74–81
7. Back T (1996) Evolutionary algorithms in theory and practice: evolution strategies, evolutionary programming, genetic algorithms. Oxford University Press, Oxford
8. Bar Tana J, Howlett BJ, Koshland DE (1977) Flagellar formation in Escherichia coli electron transport mutants. J Bacteriol 130(May 2):787–792
9. Berg HC (2000) Motile behavior of bacteria. Phys Today 53(January 1):24–29
10. Bounds DG (1987) New optimization methods from physics and biology. Nature 329(September):215–219
11. Bhushan B, Singh M (2011) Adaptive control of DC motor using bacterial foraging algorithm. Appl Soft Comput 11(8):4913–4920
12. Carpentier J (1962) Contribution a l'etude du dispatching economique. Bulletin de la Societe Francaise des Electriciens 3(1):431–447

13. Chung TS, Li YZ (2000) A hybrid GA approach for OPF with consideration of FACTS devices. IEEE Power Eng Rev 20(8):54–57
14. Daryabeigi E, Zafari A, Shamshirband S, Anuar NB, Kiah MLM (2014) Calculation of optimal induction heater capacitance based on the smart bacterial foraging algorithm. Int J Electr Power Energy Syst 61:326–334
15. Dhillon SS, Lather JS, Marwaha S (2016) Multi objective load frequency control using hybrid bacterial foraging and particle swarm optimized PI controller. Int J Electr Power Energy Syst 79:196–209
16. Edward JB, Rajasekar N, Sathiyasekar K, Senthilnathan N, Sarjila R (2013) An enhanced bacterial foraging algorithm approach for optimal power flow problem including FACTS devices considering system loadability. ISA Trans 52(5):622–628
17. Elattar EE (2015) A hybrid genetic algorithm and bacterial foraging approach for dynamic economic dispatch problem. Int J Electr Power Energy Syst 69:18–26
18. Huneault M, Galiana FD (1991) A survey of the optimal power flow literature. IEEE Trans Power Syst 6(2):762–770
19. Kavousi A, Vahidi B, Salehi R, Bakhshizadeh MK, Farokhnia N, Fathi SH (2012) Application of the bee algorithm for selective harmonic elimination strategy in multilevel inverters. IEEE Trans Power Electron 27(4):1689–1696
20. Li MS, Ji TY, Tang WJ, Wu QH, Saunders JR (2010) Bacterial foraging algorithm with varying population. Biosystems 100(3):185–197
21. Liang TJ, O'Connell RM, Hoft RG (1997) Inverter harmonic reduction using Walsh function harmonic elimination method. IEEE Trans Power Electron 12(6):971–982
22. Mohammadi M, Rozbahani AM, Montazeri M (2016) Multi criteria simultaneous planning of passive filters and distributed generation simultaneously in distribution system considering nonlinear loads with adaptive bacterial foraging optimization approach. Int J Electr Power Energy Syst 79:253–262
23. Naveen S, Kumar KS, Rajalakshmi K (2015) Distribution system reconfiguration for loss minimization using modified bacterial foraging optimization algorithm. Int J Electr Power Energy Syst 69:90–97
24. Noroozian M, Andersson G (1993) Power flow control by use of controllable series components. IEEE Trans Power Deliv 8(3):1420–1429
25. Panda A, Tripathy M, Barisal AK, Prakash T (2017) A modified bacteria foraging based optimal power flow framework for hydro-thermal-wind generation system in the presence of STATCOM. Energy 124:720–740
26. Panda R, Naik MK (2015) A novel adaptive crossover bacterial foraging optimization algorithm for linear discriminant analysis based face recognition. Appl Soft Comput 30:722–736
27. Panigrahi BK, Pandi VR (2008) Bacterial foraging optimisation: Nelder–Mead hybrid algorithm for economic load dispatch. IET Gener Transm Distrib 2(4):556–565
28. Passino KM (2002) Biomimicry of bacterial foraging for distributed optimization and control. IEEE Control Syst 22(3):52–67
29. Rajasekar N, Kumar NK, Venugopalan R (2013) Bacterial foraging algorithm based solar PV parameter estimation. Sol Energy 97:255–265
30. Ram JP, Babu TS, Rajasekar N (2017) A comprehensive review on solar PV maximum power point tracking techniques. Renew Sust Energ Rev 67:826–847
31. Ramyachitra D, Veeralakshmi V (2017) Bacterial foraging optimization for protein structure prediction using FCC & HP energy model. Gene Rep 7:43–49
32. Ray RN, Chatterjee D, Goswamie SK (2008) A modified reference approach for harmonic elimination in pulse-width modulation inverter suitable for distributed generations. Electr Power Compon Syst 36(8):815–827
33. Ray RN, Chatterjee D, Goswami SK (2010) A PSO based optimal switching technique for voltage harmonic reduction of multilevel inverter. Expert Syst Appl 37(12):7796–7801
34. Roa-Sepulveda CA, Pavez-Lazo BJ (2003) A solution to the optimal power flow using simulated annealing. Int J Electr Power Energy Syst 25(1):47–57

35. Santos VS, Felipe PV, Sarduy JG (2013) Bacterial foraging algorithm application for induction motor field efficiency estimation under unbalanced voltages. Measurement 46(7):2232–2237
36. Saravanan M, Slochanal SMR, Venkatesh P, Abraham JPS (2007) Application of particle swarm optimization technique for optimal location of FACTS devices considering cost of installation and system loadability. Electr Power Syst Res 77(3):276–283
37. Shi KL, Li H (2005) Optimized PWM strategy based on genetic algorithms. IEEE Trans Ind Electron 52(5):1458–1461
38. Sundareswaran K, Kumar AP (2004) Voltage harmonic elimination in PWM AC chopper using genetic algorithm. IEE Proc Electr Power Appl 151(1):26–31
39. Sundareswaran K, Jayant K, Shanavas TN (2007) Inverter harmonic elimination through a colony of continuously exploring ants. IEEE Trans Ind Electron 54(5):2558–2565
40. Swift F, Kamberis A (1993) A new Walsh domain technique of harmonic elimination and voltage control in pulse-width modulated inverters. IEEE Trans Power Electron 8(2):170–185
41. Todorovski M, Rajicic D (2006) An initialization procedure in solving optimal power flow by genetic algorithm. IEEE Trans Power Syst 21(2):480–487
42. Turanoğlu B, Akkaya G (2018) A new hybrid heuristic algorithm based on bacterial foraging optimization for the dynamic facility layout problem. Expert Syst Appl 98:93–104
43. Turnbull FG (1964) Selected harmonic reduction in static DC—AC inverters. IEEE Trans Commun Electr 83(73):374–378
44. Tutkun N (2010) Improved power quality in a single-phase PWM inverter voltage with bipolar notches through the hybrid genetic algorithms. Expert Syst Appl 37(8):5614–5620
45. Zhu GY, Zhang WB (2017) Optimal foraging algorithm for global optimization. Appl Soft Comput 51:294–313

Chapter 10
Application of Artificial Immune System in Optimal Design of Irrigation Canal

Sirajul Islam and Bipul Talukdar

Abstract Artificial Immune System (AIS) is an emerging field of computational intelligence used in engineering and scientific applications. Like other biologically inspired algorithms, they are evolutionary in nature. Over the years, these algorithms have been applied in many problems of Engineering Optimization and they have proved to be superior search methods. An optimization algorithm based on artificial immune system is formulated in the present study and is applied in optimal design of an irrigation canal section. Results are compared with those obtained from optimization using Genetic Algorithm (GA). The results compare very keenly. Moreover, the due to the modifications incurred into the AIS algorithm, it consumes much less time in comparison to GA while converging to the optimal solution.

Keywords Optimization · Artificial immune system · Clonal selection algorithm · Affinity · Cloning · Mutation

10.1 Introduction

Artificial Immune System (AIS) is a comparatively new computational paradigm which is derived from the natural immune system. AIS is used in a wide range of applications for solving real world science and engineering problems such as machine learning, pattern recognition, data mining, intrusion detection and engineering system optimization etc.

The biological immune system is a complex, distributed, and adaptive system that defends the body from invasions by using a multilevel defense mechanism. This defense mechanism works by recognizing all cells within the body and then categorizing them as self (body's own cells) and non-self (foreign cells). The immune system learns through evolution to distinguish between body's own cells and foreign

S. Islam (✉) · B. Talukdar
Department of Civil Engineering, Assam Engineering College, Guwahati, India

F. Bennis, R. K. Bhattacharjya (eds.), *Nature-Inspired Methods for Metaheuristics Optimization*, Modeling and Optimization in Science and Technologies 16,
https://doi.org/10.1007/978-3-030-26458-1_10

cells, also called the antigens. The body's immune defense mechanism generally coexists with cells that carry distinctive "self" marker molecules. But when immune defenders encounter cells or organisms carrying "non-self" molecules, they quickly launch an attack. The immune system is characterized by its adaptability, learning and memory mechanisms that can be easily applied in many computational tasks. Several concepts have been derived from the specific immunological theories for development of search algorithms. Some of these concepts are- Immune Network Theory, Negative Selection Mechanism and Clonal Selection Principle. These theories or combinations of them have been used to develop algorithms that are applied for problem solving in science and engineering.

The concept of AIS emerged in the mid 1980s with the work of Farmer, Packard and Perelson [11] on theoretical immune network models proposed to describe the maintenance of immune memory. Ishida [23] proposed the first immune network algorithm which was subsequently re-defined and implemented by Timmis et al. [21]. A negative selection algorithm was proposed for the first time by Forrest et al. [12] to detect data manipulation caused by a virus in a computer system. De Castro and Von Zuben [10] developed a Clonal Selection Algorithm (CSA) named CLONALG for learning and optimization, which is by far the most popular and widely used AIS algorithm. Being one of the pioneers of research in the field of AIS, Dasgupta [5–7, 9] conducted extensive studies on various aspects of AIS. Dasgupta et al. [8] presented a detailed account of recent developments in AIS research and new computational techniques being developed by using various immunological theories. Attempts have also been made to develop hybrid algorithms by combining AIS with other evolutionary computation methods such as genetic algorithm (GA), particle swam optimization (PSO) etc. A PSO-AIS hybrid algorithm was proposed by Ge et al. [13] for job-shop scheduling and was found to be very effective. Zheng and Ponnambalam [25] used a combined GA-AIS algorithm for optimization of turning operations and reported that this hybrid technique performed better than other heuristics.

In the recent time, AIS has seen few applications in water resources management. Chu et al. [4] proposed an optimization procedure based on a CSA framework to optimize the designs of water distribution networks. In a comparative evaluation of different evolutionary algorithms for parameter estimation of a hydrologic model, Zhang et al. [24] found the performance of CLONALG to be at par with other evolutionary algorithms. Luo and Xie [18] used a CSA for parameter estimation of the non- linear Muskingum model for stream flow routing. Wang et al. [22] employed AIRS as a new approach of data mining to extract operating rules on a water-supply reservoir and reported that AIRS could effectively extract reservoir operating rules. In authors' earlier works, CSA was used for optimization of a single reservoir operation [14] as well as for conjunctive management of the irrigation system [15]. In the present study, CSA is used for optimal design of an irrigation canal system.

10.2 Overview of AIS Algorithms

Inspired by various theories about the function and behavior the adaptive immune system, the following computational techniques have been derived from specific immunological theories.

10.2.1 Clonal Selection Algorithm

A class of algorithms inspired by Burnet's clonal selection theory of acquired immunity [2] that explains the basic features of an immune response to antigenic stimulus. These algorithms focus on the Darwinian attributes of the theory that only those cells that recognize the antigens can proliferate and produce clones subject to a mutation process known as somatic hypermutation. Clonal selection algorithms are most commonly applied to optimization and pattern recognition domains.

10.2.2 Negative Selection Algorithm

These Algorithms are inspired by the mechanism of the immune system to train the T-Cells (thymus cells) to recognize the antigens and to prevent them from recognizing the body's own cells [8]. The T-cells undergo a process of Negative selection whereby the self-reacting cells are destroyed and the rest are allowed to proliferate. This class of algorithms is typically used for classification and pattern recognition problem domains where the problem space is modeled in the complement of available knowledge.

10.2.3 Immune Network Algorithms

These Algorithms are inspired by the idiotypic network theory proposed by Jerne [16] that describes the regulation of the immune system by an idiotypic network of interconnected B-cells. Two B-cells are interconnected when the affinities they share exceed a certain threshold, and the strength of the connection is directly proportional to the affinity they share. Immune network algorithms are used in clustering, data visualization, control, and optimization domains, and share properties with artificial neural networks.

Apart from the above classes of AIS algorithms, attempts have also been made to develop algorithms based on Danger Theory, Dendritic Cell and some hybrid techniques involving various immunological theories as described in Dasgupta et al. [8].

10.3 Formulation of AIS Algorithm

Among the different versions of AIS, CSAs have mostly been applied in numerical optimization. The feature of selection, cloning and mutation etc. make these algorithms suitable for finding global optima of engineering optimization problems. In the present study, a modified CSA is formulated for application in water resources system optimization.

The basic ingredients of a CSA as defined in CLONALG are -generation of antibody cells (self-cells), affinity computation, selection, cloning and mutation. A single decision variable of the optimization problem is known as a chromosome of an antibody cell. These chromosomes are represented in a string structure. Next step is affinity computation, which refers to the evaluation of the objective function. After measuring the affinity of the antibodies, a sizeable percentage of the best affinity population is selected and each of them is replicated by cloning. The clones are then subjected to random genetic changes called somatic hypermutation (or simply mutation). The process is repeated for specific number of generations or until some stopping criteria are fulfilled.

Two unique features of a CSA [1] that make this algorithm more suitable for finding global optima are-

1. Affinity proportionate cloning, which means that the number of clones to be produced for a particular antibody is proportional to its affinity.
2. Affinity inverse proportionate mutation, which means that the quantum of genetic change effected to a particular chromosome is inversely proportional to its affinity.

In the present work, a CSA is formulated with certain modifications to the traditional CSAs. The algorithmic steps of the formulation are presented in Fig. 10.1. In the traditional CSAs such as CLONALG, the lower affinity population of antibodies is replaced by new ones during transition to each new generation, so as to maintain population diversity. This necessitates computation of the affinity again at the beginning of the next generation for obtaining the best affinity population, resulting in huge computational burden. But in the present formulation, a chosen percentage of random population is added to the existing population just after mutation. The affinity is then computed and a selected best affinity population is sent to the next generation. This avoids the need to compute the affinity again at the beginning of the next generation. Thus, the number of functions to be evaluated at each generation in the new CSA formulation is equal to the number of clones produced. It is expressed as:

$$NC = \sum_{ab=1}^{Ab} ab \times RC \tag{10.1}$$

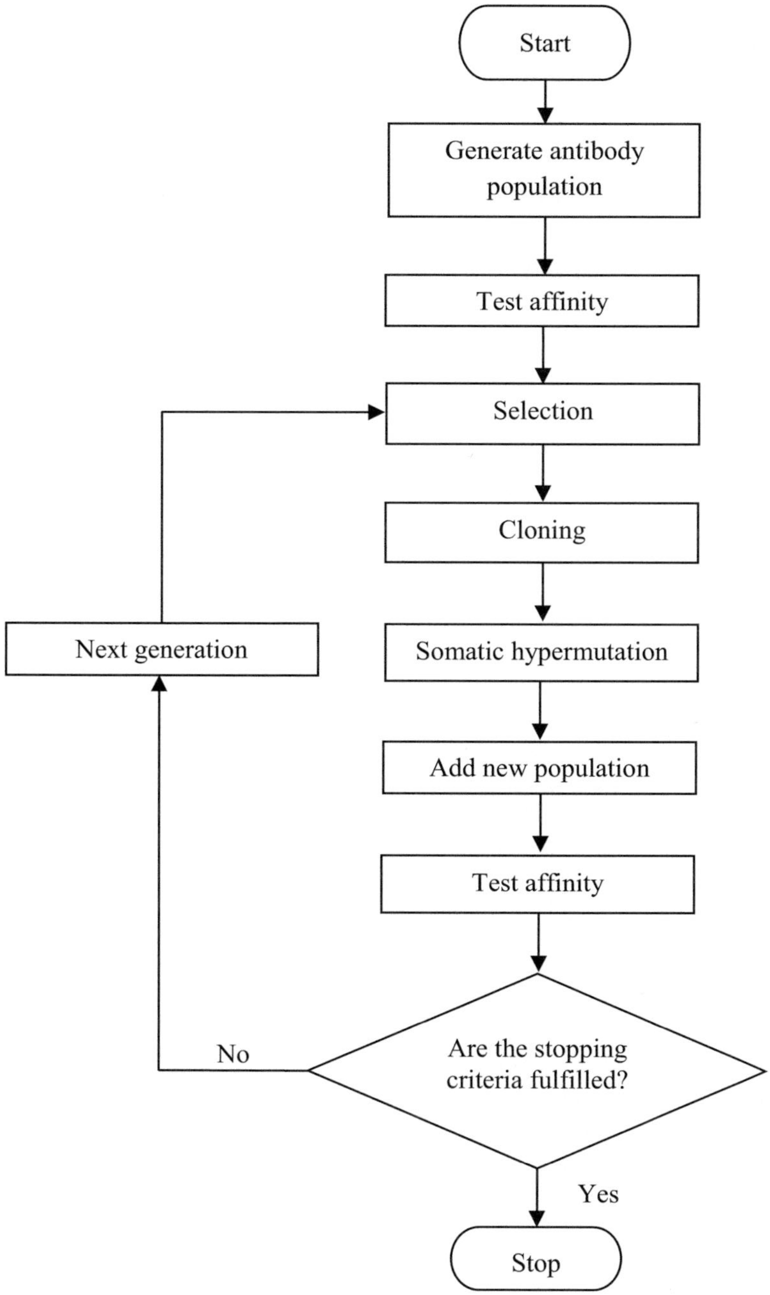

Fig. 10.1 Flowchart representing the formulated clonal selection algorithm

where, NC is the total number of clones produced; RC is the rate at which cloning occurs; ab is the antibody index in ascending order of affinity and Ab is the number of selected best affinity antibodies.

10.4 Model Application

The formulated AIS algorithm is applied on a problem of optimal design of an irrigation canal. The objective of the problem is to minimize water loss from the canal subject to fulfillment of design criteria. The total water loss from a canal section is given by

$$q_w = q_s + q_e \tag{10.2}$$

where, q_w is the total water loss from the channel section, which is the sum of seepage discharge per unit length of the canal, q_s (m^2/s) and evaporation discharge per unit length of the canal q_e (m^2/s). q_s and q_e are given by Swamee et al. [20]

$$q_s = ky_n F \tag{10.3}$$

$$q_e = ET \tag{10.4}$$

where, k is the hydraulic conductivity of the porous medium (m/s), y_n is the depth of water in the canal (m), F is a seepage function depending on channel geometry, E is the evaporation discharge per unit free surface area (m/s) and T is the width of free surface (m). Replacing q_s and q_e from Eq. 10.3 and Eq. 10.4 in Eq. 10.1, the optimization problem can be expressed as

$$\text{Minimize} : f = ky_n F + ET \tag{10.5}$$

Where, f is the objective function to be minimized. The minimization problem expressed by Eq. 10.5 is subject to the constraint of resistance of flow in canal. The most commonly used flow equation in design of uniform open canal is Manning's equation [3] which is applicable to rough turbulent flow and with a limited bandwidth of relative roughness. The present study uses the modified equation proposed by Swamee et al. [20] which relaxes the above mentioned restrictions. It is given by,

$$Q = -2.457A\sqrt{gRS_0}\ln\left(\frac{\varepsilon}{12R} + \frac{0.221\nu}{R\sqrt{gRS_0}}\right) \tag{10.6}$$

where, Q = canal discharge (m3/s); A = flow area (m2); g = gravitational acceleration (m/s2); R = hydraulic radius (m) defined as the ratio of the flow area to the wetted perimeter P (m); ε = average roughness height of the canal lining (m); and ν = kinematic viscosity of water (m2/s); S_0 = bed slope (dimensionless).

The above equation is inserted as equality constraint to the minimization problem. In addition to the above constraint, optimal design of canal section is also subject to limitation on the flow velocity. The average flow velocity V_a for the designed section is expressed by the continuity equation

$$V_a = \frac{Q}{A} \tag{10.7}$$

This average velocity V_a has to be less than the limiting velocity $V_{l.}$

10.4.1 Design Problem

The present study considers the design of a concrete lined canal with rectangular section. Figure 10.2 shows the dimensional notations for the canal section. For rectangular section, width of free surface T is same with that of bed width b. Swamee et al. [20] derived the expression for seepage function, F for rectangular section as

$$F = \left[\left(4\pi - \pi^2\right)^{0.77} + \left(\frac{b}{y_n}\right)^{0.77}\right]^{1.33} \tag{10.8}$$

Thus, Eq. 10.5 may be expressed as

$$f = ky_n\left[\left(4\pi - \pi^2\right)^{0.77} + \left(\frac{b}{y_n}\right)^{0.77}\right]^{1.33} + bE \tag{10.9}$$

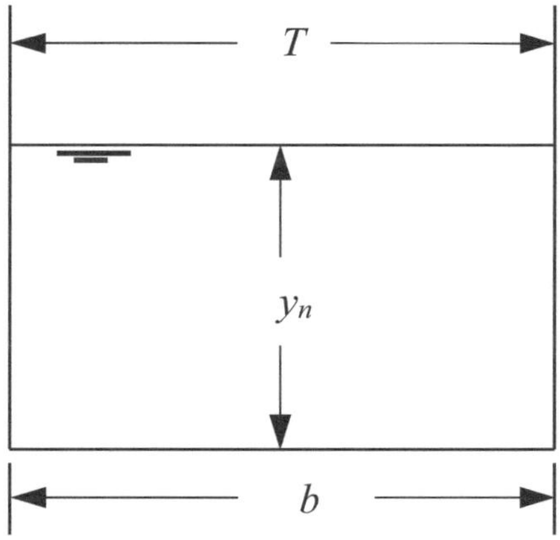

Fig. 10.2 Rectangular canal section with dimensional notations

In order to evaluate the performance of AIS in comparison to other evolutionary algorithms, the present study considers the input parameters from literature [17], where GA was used for optimal design of concrete lined canal with rectangular section. The different parameters are $Q = 10\ m^3/s$; $\varepsilon = 1$ mm; $k = 10^{-6}$ m/s; $E = 2.5 \times 10^{-6}$ m/s; $\nu = 1.1 \times 10^{-6}\ m^2/s$; $V_l = 2.5$ m/s; $S_0 = 0.001$; $g = 9.79\ m/s^2$.

10.4.2 Optimization Using CSA

The proposed CSA is implemented in MATLAB platform using real coding scheme. The objective function as well as initialization, cloning and mutation functions are stored in separate m-files, which are called from the main optimization module. The decision variables in this optimization problem are the bed width, b and the depth of water, y_n. It's necessary to impose upper and lower bounds on the decision variables in a CSA. Depending on the discharge and flow requirements, the upper and lower bounds for both the variables are chosen to be 3 m and 2 m respectively. The MATLAB code for the optimization algorithm is presented in Appendix.

Like most of the evolutionary search algorithms, Penalty function approach [19] is used to handle the constraints in a CSA. In penalty function method, the set of constraint functions is added to the objective function as a cost for violation of constraints, and as a result, the constrained optimization problem turns into an unconstrained one in the following manner.

$$f_p = f + \sum_{m=1}^{M} \beta\ f_{c_m} \tag{10.10}$$

where, f_p is the penalty function; f is the objective function; f_c is the constraint function; m is the constraint function index; M = the number of constraints; β is a multiplier such that, β = 1, if constraint is violated or β = 0, if constraint is satisfied.

The present problem consists of one equality constraint i.e. $Q = 10\ m^3/s$ and one inequality constraint i.e. $V \leq 2.5$ m/s.

10.5 Results and Discussion

A sensitivity analysis is carried out for the different CSA parameters used in the present algorithms. The most important user defined parameters of the formulated CSA that influence the convergence speed and computational time are Ab and RC. Sensitivity of the algorithm is tested using different values of these two parameters. The results are tabulated in Table 10.1. Ab = 20 and RC = 2 are the best values

Table 10.1 Sensitivity analysis of CSA parameters

ab	RC	Best objective function value (f) ($\times 10^{-5}$)	Constraint violation (value of β)
1	2	3	4
10	2	1.6215	0
15	2	1.6178	0
20	2	1.5997	0
20	3	1.5984	0
20	5	1.5980	0

Table 10.2 Optimal design parameters for rectangular canal section

Number of generations	Best objective function value (f) ($\times 10^{-5}$)	b	y_n	A	V
1	2	3	4	5	6
50	1.601	2.273	2.285	5.195	1.925
100	1.593	2.241	2.324	5.207	1.920
200	1.587	2.269	2.290	5.196	1.924

Table 10.3 Comparative results of optimization from CSA and GA

Search method	Best objective function value (f) ($\times 10^{-5}$)	b	y_n	A	V
1	2	3	4	5	6
AIS	1.587	2.269	2.290	5.196	1.924
GA	1.579	2.170	2.414	5.238	1.909

obtained. Using these values, the model is run three times each for 50,100 and 200 generations respectively. At each run, the algorithm yields separate results, but the objective function value remains the same for the same number of generations. The results are shown in Table 10.2. The best result is obtained from 200 generations of iterations with objective function value of 1.587E-05. The values of $Q = 9.999$ and $V = 1.924$ show that the constraints are not violated. The best results are also shown in comparison with those obtained by using GA [17], in Table 10.3.

The results compare very keenly. The object function value obtained from the present algorithm is nearly the same with that obtained from GA optimization. Moreover, due to the modifications incurred to the CSA, it involves lesser computational burden as compared to GA. Hence the present algorithm may be considered a good option to obtain accurate results with.

10.6 Summary and Conclusions

This study presented an overview of the algorithms based on AIS and their applications in real life problem solving. A CSA was formulated and applied in the design of a rectangular canal section using input data from literature. The results obtained from optimization using CSA were found to be satisfactory. The results were also shown to be comparable with those obtained from established evolutionary optimization algorithm like GA.

Appendix: MATLAB Code for Real Coded Clonal Selection Algorithm

Optimization Module

User defined parameters

```
1. clear all                                   % Remove data from memory
2. Objectfun = @ Canaldesign;                  % Objective function
3. Initializationfun = @initAB;                % Initialization function used
4. Cloningfun =@clonefun;                      % Cloning function used
5. Mutationfun =@mutfun;                       % Mutation function used
6. ABpop = 100;                                % Size of antibody population to be
                                                 generated
7. Bounds =[2 3; 2 3];                         % bounds on the variables ( antigens)
8. bestaffinABpop =20;                         % best population to select on the basis
                                               of affinity
9. rateclone =2;                               % rate of afinity proportionate cloning
10. ratemut =1;                                % rate of affinity inverse proportionate
                                               mutation
11. sizeclone = rateclone*bestaffinABpop*....
    (bestaffinABpop+1)/2;                      % cloning size
12. np = 0.2;                                  %  size of new antibodies for next gen.
13. Gen =100;                                  % No. of generations to be iterated

14. genAB = initAB(ABpop,bounds);              % generate random population with the
                                               help of user defined function "initAB"
15. options = [];                              % options on the input variables
16. for  G=1:Gen
17. for  k=1:ABpop
18. [x,aff] = Objectfun (genAB (k,:),options); % test the affinity
19. affinval(:, k) = aff;
20. affinx(:, k)=x;
21. end
22. affinmatrix (:,1)=affinval;                % matrix of affinity values
23. affinmatrix (:,2:size (bounds,1)+1)=affinx';
24. sortmatrix = (sortrows(affinmatrix,1));    % arrange the affinity values in descending
                                               order
25. bestaffin = sortmatrix(ABpop-bestaffinABpop...
    +1:ABpop,1:size(bounds,1)+1);              % select the best affinity antibodies
26. cloneABpop = clonefun (bounds,......
    bestaffinABpop, bestaffin, rateclone);     % cloning by using function "clonefun"
27. [mutposition,mutABpop]=mutationfun.....
    (bounds,cloneABpop,rateclone,ratemut,...
    bestaffinABpop);                           % Mutation using "mutationfun" function
28. for k =1:sizeclone
29. [x,aff] = Objectfun(mutABpop(k,:),options);  % test the affinity after mutatation
30. mutaffinity(:,k)=aff;
31. mutx (:,k) = x;
32. end
33. memoryABpop (:,1) = mutaffinity;
34. memoryABpop (:,2:size (bounds,1)+1)=mutx';
```

```
35. sortmemoryAB=flipud….
    (sortrows (memoryABpop,1));
36. bestsol (G,:)=max(sortmemoryAB(:,1));      % best value for affinity
37. xval = sortmemoryAB……
    (1,2:size (bounds,1)+1);
38. bestmutABpop=sortmemoryAB…
    (1:ABpop*(1-np),2:size(bounds,1)+1);       % keep the best population after mutation
39. addpop =(ones(ABpop*np,1)*…..
    (bounds(:,2)-bounds(:,1))').*….
40. (rand(ABpop*np,size(bounds,1))).....
    +(ones(ABpop*np,1)*bounds(:,1)');          % add new population of antibodies
41. genAB(1:ABpop*np,1:size(bounds,1))….
    = addpop;
42. genAB(ABpop*np+1:ABpop,…..
    1:size(bounds,1))=bestmutABpop;            % new generation of popuation
43. end
44. xval                                       % best values at the end of specified no. of
                                                 generations.
```

Initialization Function

```
1. function  genAB= initAB(ABpop,bounds)
   genAB = (ones(ABpop,1)*(bounds(:,2)…
   bounds(:,1))').*(rand(ABpop,size(bounds,1))).....
     +(ones(ABpop,1)*bounds(:,1)');            % Initialize population
```

Cloning Function

```
1. function [cloneABpop] =clonefun…
   (bounds,bestaffinABpop,bestaffin,rateclone);
2. for p=1:bestaffinABpop
3. clonematrix(1+rateclone*p*(p-1)/2:rateclone*p*(p+1)/2,...
   1:size(bounds,1)+1)= repmat….
   (bestaffin(p,:),p*rateclone,1);             % affinity proportionate cloning
4. cloneABpop=clonematrix…..
   (1:rateclone*p*(p+1)/2,2:size(bounds,1)+1);
5. end
```

Mutation Function

```
function [mutposition,mutABpop] =……..
    mutfun(bounds,cloneABpop,rateclone,ratemut,bestaffinABpop);
sizeclone = rateclone*bestaffinABpop*(bestaffinABpop+1)/2;
for p =1:bestaffinABpop
mutposition = round(rand * (size(bounds,1)-1))+1;
randmut(1+rateclone*p*(p-1)/2:rateclone*p*(p+1)/2-1).......
    = ratemut/p.^2*randn(rateclone*p-1,1)...
    +cloneABpop(rateclone*p-1, mutposition);  % random mutation of antibodies inversely
                                                proportional to affinity
end
mutABpop = cloneABpop;
mplr = bounds(mutposition,1);
mpur = bounds(mutposition,2);
for m =1:sizeclone-1
if randmut(m)<mplr
randmut(m)=mplr;
elseif randmut(m)>mpur
randmut(m) = mpur;
end
mutABpop(m,mutposition) = randmut(m);
end
```

References

1. Brownlee J (2007) Clonal selection algorithms. CIS technical report 070209A, Centre for Information Technology Research, Swinburne University of Technology, Melbourne, Australia
2. Burnet FM (1959) The clonal selection theory of acquired immunity. Cambridge University Press
3. Chow VT (1973) Open channel hydraulics. McGraw-Hill Book Co. Inc, New York
4. Chu CW, Lin MD, Liu GF, Sung YH (2008) Application of immune algorithms on solving minimum cost problem of water distribution network. Math Comput Model 48(11–12):1888–1900
5. Dasgupta D (1999) Artificial immune systems and their applications. Springer, Berlin
6. Dasgupta D, Ji Z, Gonzalez F (2003) Artificial immune system (AIS) research in the last five years. In: Congress on evolutionary computation (CEC'03), 1, pp 123–130
7. Dasgupta D, Ji Z (2007) Revisiting negative selection algorithms. Evol Comput 15(2):223–251
8. Dasgupta D, Yua S, Nino F (2011) Recent advances in artificial immune systems: models and applications. Appl Soft Comput 11:1574–1587
9. Dasgupta D, Silva GC (2014) Steps toward developing an artificial cell signaling model applied to distributed fault Detection. In: 13th international conference on unconventional computation and natural computation (UCNC), University of Western Ontario, Canada
10. De Castro LN, Von Zuben FJ (2002) Learning and optimization using the clonal selection principle. IEEE Trans Evol Comput Spec Issue Artif Immune Syst 6(3):239–251
11. Farmer JD, Packard N, Perelson A (1986) The immune system, adaptation and machine learning. Physica D 2:187–204

12. Forrest S, Perelson AS, Allen L, Cherukuri R (1994) Self-nonself discrimination in a computer. In: Proceedings of the IEEE symposium on research in security and privacy, Los Alamitos, CA, pp 202–212
13. Ge HW, Liang YC, Qian F (2008) An effective PSO and AIS based hybrid intelligent algorithm for job-shop scheduling. IEEE Trans Syst Man Cybern Syst Hum 38(2):358–368
14. Islam S, Talukdar B (2012) Application of artificial immune system in optimization of reservoir operation. Int J Water Resour Environ Manag 3(2):241–254
15. Islam S, Talukdar B (2016) A linked simulation–optimization (LSO) model for conjunctive irrigation management using clonal selection algorithm. Journ Inst Eng India Ser A 97(3):181–189
16. Jerne NK (1974) Towards a network theory of the immune system. Ann Immunol 125(C):373–389
17. Kentli A, Mercan O (2014) Application of different algorithms to optimal design of canal sections. J Appl Res Technol 12:762–768. 768
18. Luo J, Xie J (2010) Parameter estimation for nonlinear Muskingum model based on immune clonal selection algorithm. J Hydrol Eng 15(10):844–851
19. Smith AE, Coit DW (1996) Penalty functions. In: Handbook of evolutionary computation, Oxford University Press and Institute of Physics Publishing
20. Swamee PK, Mishra GC, Chahar BR (2000) Design of minimum seepage loss canal sections. J Irrig Drain Eng 126(1):28–32
21. Timmis J, Neal M, Hunt J (2000) An artificial immune system for data analysis. Biosystems 55:143–150
22. Wang XL, Cheng JH, Yin ZJ, Guo MJ (2011) A new approach of obtaining reservoir operation rules:artificial immune recognition system. Expert Syst Appl 38(9):11701–11707
23. Ishida Y (1990) Fully distributed diagnosis by PDP learning algorithm: towards immune network PDP model. IEEE International joint conference on neural networks, San Diego, USA
24. Zhang X, Srinivasan R, Zhao K, Liew VM (2008) Evaluation of global optimization algorithms for parameter calibration of a computationally intensive hydrologic model. Hydrol Process 23(3):430–441
25. Zheng LY, Ponnambalam SG (2010) In: Liu H et al (eds) A hybrid GA-AIS heuristic for optimization of multipass turning operations, ICIRA, Part-II, LNAI 6425. Springer, Berlin, pp 599–611

Chapter 11
Biogeography Based Optimization for Water Pump Switching Problem

Vimal Savsani, Vivek Patel, and Mohamed Tawhid

Abstract This chapter introduces the basic concepts of biogeography based optimization (BBO) algorithm and its application to a combinatorial water switching problem. Water switching optimization is a pump scheduling problem which considers minimization of total electrical energy requirement as an objective function. Pump status (switch on/switch off) of pumping stations are considered as a discrete (binary) decision variables for the optimization problem. Suction and discharge pressure are considered as constraints in the procedure. A case study with 10 pumping station and 40 pumps is presented for the experimentation. The performance of BBO is tested against other state-of-the art algorithms that includes genetic algorithm (GA), branch & bound method (B&B), harmony search (HS) algorithm, particle swarm optimization (PSO) and ant colony optimization (ACO) algorithms. Water pump switching problem is also investigated by using different constraint handling techniques and by considering different pump situations in a pumping station. The Computational results indicate that BBO is an appropriate algorithm to solve water pump switching problem and is effective over other optimization methods. Moreover, 20 alternative optimum solutions are presented to demonstrate water switching problem as a multi-modal problem with different optimum solutions and the search capability of BBO to find alternate optimum solutions.

Nomenclature

E pumping energy (hp)
Q flow rate (cfs)

V. Savsani (✉) · V. Patel
Department of Mechanical Engineering, Pandit Deendayal Petroleum University, Gandhinagar, Gujarat, India

M. Tawhid
Department of Mathematics and Statistics, Thompson Rivers University, Kamloops, BC, Canada

F. Bennis, R. K. Bhattacharjya (eds.), *Nature-Inspired Methods for Metaheuristics Optimization*, Modeling and Optimization in Science and Technologies 16,
https://doi.org/10.1007/978-3-030-26458-1_11

P pump pressure (psi)
x binary variable
n pumping station
m pump

Greek letters

η motor-pump efficiency
γ specific weight (lb/ft^3)

Subscript

L lower limit
U upper limit

Superscript

D discharge
L pressure loss
S suction

Biogeography-Based Optimization

Biogeography-based optimization (BBO) is an evolutionary algorithm (EA) that is used for the optimization of a mathematical function, especially discrete in nature, by using stochastic and iterative process. It was developed by Dan Simon in 2008 based on the mathematical modeling of immigration and emigration of species from one place to the other. The basic mathematical models for the biogeography of the species were reported in the early 1960s by Robert MacArthur and Edward Wilson. Their work was published in 1967 titled as 'The Theory of Island Biogeography'. The basic mathematical models of biogeography is based on how species migrate from one island (habitat) to another, how new species arise, and how species become extinct. Geographical areas that are well suited as residences for biological species are said to have a high habitat suitability index (HSI).The factors (natural factors) that characterize habitability are called suitability index variables (SIVs). Habitats with a high HSI tend to have a large number of species, while those with a low HSI have a small number of species. Habitats with a high HSI have many species that emigrate to nearby habitats, because of the large number of species that they host. Habitats with a high HSI have low species immigration rate because they are already nearly saturated with species.

'Biogeography Based Optimization' works on the concept of immigration and emigration of the species also written as BBO in abbreviate form. To solve an optimization problem by BBO, a good solution is comparable to an island with a high HSI, and a poor solution represents an island with a low HSI. High HSI solutions resist change more than low HSI solutions. By the same expression, high HSI solutions tend to share their characteristics with low HSI solutions. This approach is the key driving factor for the working of biogeography-based optimization (BBO).

BBO searches for the solution by using two natural phenomenon called, migration and mutation. Migration is a process in which different species transfer from one habitat to the other. Consider a population of candidate solution which is represented by design variables. These design variables are mimicked for suitability index variables (SIVs). The solutions that are good can be considered to be habitats with a high HSI, and those that are poor can be considered to be habitats with a low HSI. This goodness of solution is obtained from the value of objective function. It can be assumed that each solution (habitat) has an identical species curve shown in Fig. 11.1 (with E = I), but the S value represented by the solution depends on its HSI. As observed from the figure, the maximum possible immigration rate occurs when there are zero species on the island. As the number of species increases, the island becomes more crowded, fewer species are able to survive immigration, and the immigration rate decreases. The largest possible number of species that the habitat can support is S_{max}, at which the immigration rate is zero. If there are no species on the island, then the emigration rate is zero. As the number of species on the island increases, it becomes more crowded, more species are able

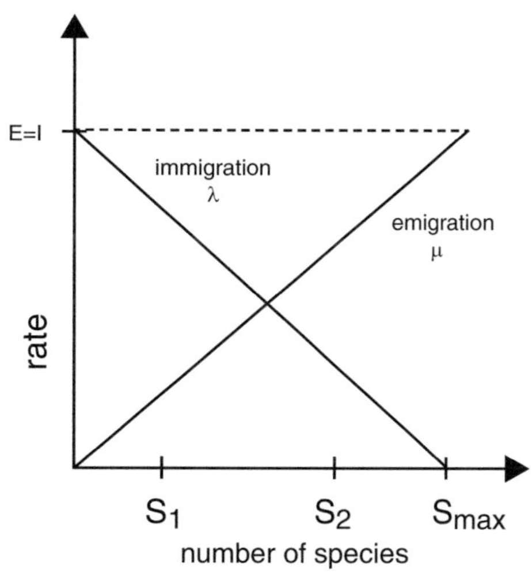

Fig. 11.1 Species model for single habitat showing two candidate solutions

to leave the island, and the emigration rate increases. When the island contains the largest number of possible species S_{max}, the emigration rate reaches its maximum possible.

Let S_1 and S_2 represent two solutions with different HSI. The emigration and immigration rates of each solution are used to probabilistically share information between solutions (habitats). If a given solution is selected to be modified, then its immigration rate λ is used to probabilistically decide whether or not to modify each design variable in that solution. If a given design variable in a given solution S_i is selected to be modified, then the emigration rates μ of the other solutions is used to probabilistically decide which of the solutions should migrate a randomly selected SIV to solution S_i. As with other population-based optimization algorithms, some sort of elitism is incorporated in order to retain the best solutions in the population. BBO updates the solution by migration by using Algorithm-1.

Algorithm-1: Migration in BBO

---

1. For each individual, map the fitness to the number of species
2. Calculate the immigration rate λ_i and the emigration rate μ_i for each individual *Solution*$_i$
3. For $i = 1$ to *N(N=number of habitats)*
4. Select *Solution*$_i$ with probability proportional to λ_i
5.   If rand(0, 1) < λ_i
6.   For $j = 1$ to N
7.   Select *Solution*$_j$ with probability proportional to μ_j
8.   If rand(0, 1) < μ_j
9.   Randomly select a variable x from *Solution*$_j$
10.   Replace the corresponding variable in *Solution*$_i$ with x
11.   Endif
12.   Endfor
13.   Endif
14. Endfor

In nature a habitat's HSI can change suddenly due to apparently random events (unusually large flotsam arriving from a neighbouring habitat, disease, natural catastrophes, etc.). This phenomenon is termed as SIV mutation, and species count probabilities are used to determine mutation rates. The SIV mutation can be understood easily from the Algorithm-2:

Algorithm-2: Mutation in BBO

1. For i = 1 to N
2. Compute the probability Pi
3. Select design variable $X_i(j)$ with probability (P_i)
4. If rand (0, 1) < m_i (m_i is user defined parameter for mutation)
5. Replace $X_i(j)$ with a randomly generated SIV
6. End if
7. End for

BBO improves the current solutions by only considering the objective function value. BBO is a metaheuristics technique since it implements several variations and it does not make any assumptions about the problem to be solved. The above advantages make BBO to be applied to a wide class of problems. BBO was specifically developed to solve discrete optimization problems that can be multidimensional real-valued functions and it does not use the gradient of the function. So, BBO do not require the function to be differentiable as required by many classic optimization methods. BBO can therefore be used on discontinuous functions as well. So, BBO optimizes a problem by maintaining a population of candidate solutions, and creating new candidate solutions by combining existing ones according to simple rules of migration and mutation. In this way the objective function is simply treated as a black box.

The basic working of BBO can be understood by following pseudo code:

1. Initialize the BBO parameters (maximum species count (N), migration rates (E *and* I), maximum mutation rate m_i)
2. Initialize a random set of solutions (habitats). (Migration)
3. For each solution, map the HSI to the number of species S and find immigration rate λ, and emigration rate μ for each solution.
4. Probabilistically use immigration and emigration rates to modify each solution. (Mutation)
5. For each solution, update the probability and perform SIV mutation.
6. Preserve elite solution and go to step (3) for the next iteration till the termination criteria is reached

The population size N is a tuning parameter. If N is too small or too large, then the performance of BBO will suffer on different problems. Typical implementations of BBO use a value of N somewhere between 20 and 200, but it can be noted that the value of N largely depends on the type of problem to be solved. The initial population of candidate solutions is usually generated randomly. However, it could be generated in a problem-dependent way based on some logical and realistic guesses or previously-known good solutions to the optimization problem required to be solved. The termination criterion is also other parameter which requires some attention. In most applications the termination criterion can be a generation count limit or function evaluations.

Many variants have been proposed to the basic BBO algorithm in the literature so far. Elitism is implemented in most EAs to make sure that the best candidate solution is not lost from one generation to the next. This can be implemented in a variety of ways, but one common way is to save the best candidate solutions at the beginning of each generation that can replace the worst solutions with the elite solution at the end of the generation, after migration and mutation have completed. The number of elite solution can also make a significant difference in the performance of BBO, and it is always required to experiment with different number of elite solutions. Duplicate replacement is often implemented in BBO. This is a procedure at the end of each generation that replaces duplicate individuals in the population. Scanning for duplicates can be computationally intensive because it is an $O(N^2)$ operation, so it is often performed only every few generations, rather than in every generation. Blending can be implemented in BBO to improve its performance. In blending, instead of replacing design variables in an immigrating solution with the design variable from the emigrating solution, the immigrating solution is updated based on the linear combination of immigrating and the emigrating solution. Blended BBO is based on blended crossover in genetic algorithms and has been shown to outperform standard BBO [14]. Other approaches for selecting the immigrating and emigrating candidate solutions have also been proposed [11, 23]. The migration curves in the basic BBO are linear, but nonlinear migration curves often give better performance [13]. Hybridization of BBO is also reported with several other EAs, such as particle swarm optimization [11, 30], differential evolution [3], evolution strategy [6], opposition-based computing [7], case-based reasoning, artificial bee colony algorithm [1], bacterial foraging optimization [12], harmony search [28], ant colony algorithm and artificial immune algorithm [20] and the simplex algorithm [26]. BBO had been combined with local search to develop a memetic algorithm that performs much better than basic BBO. BBO has been mathematically analyzed using Markov models [24] and dynamic system models [22]. BBO has been implemented to noisy functions, constrained functions [17, 18], combinatorial functions [25], engineering design problems [16, 19] and multi-objective functions [5, 17, 18].

This chapter extends the use of standard BBO for the challenging combinatorial water switching problem. The description of the problem along with its mathematical formulation is explained in the subsequent section.

Water Pump Switching Problem

Rising awareness of environmental issues, sustainable living, and limited energy supply has prompted interest in energy saving research in various industrial and commercial applications. In the present chapter energy saving of a pump system in water pump switching problem is investigated. Water pump switching system is to deliver water from water sources to demanding place. Water pump switching system consists of number of pumping station and each pumping station operates with number of pumps in series which adds pressure to deliver the water.

In general, savings in pump energy can be achieved in two ways. One is by designing more efficient pumps [4, 8]. Another is improving pump performance with effective scheduling strategies [2, 27, 29]. The objective of the scheduling model is to minimize energy consumption of the pump system while maintaining the performance of the hydraulic load in an expected range. An optimal schedule determines the most preferred configuration of pumps.

Traditional techniques for optimization like iterative optimization, linear & non-linear programming, geometric programming etc. are point search methods which may require large number of function evaluation and so, such methods are computationally not suitable to solve challenging water switching problem. Also, because of point search method the traditional methods are likely to get trapped at local optima and it do not guarantee global optimum solution. The water pump switching problem is a combinatorial multi-modal problem that involves binary variables which increases with the increase in number of pumps. Hence, it is difficult to solve water pump switching problem with traditionally optimization techniques.

The main purpose of the present chapter is to introduce Bio-geographic based optimization [21], algorithms for water pump switching problem. The objectives of this study are (i) to optimize the operating status of the pumps for minimum energy requirement by using PSO, ACO, and BBO algorithms (ii) to demonstrate the effectiveness of BBO algorithm to solve water pump switching problem. (iii) To compare the results of BBO with the results obtained by using PSO and ACO and also with the published results of GA [10] and HS [9].

Mathematical Formulation for Water Pump Switching Optimization Problem

Typically, for any water pump switching system the water flow rate (i.e. requirement of water at demanding place) as well as the number of pumping station (n) and pumps (m) within each pumping stations are obtainable. The problem is then to decide that how many pumps should be operated at each pumping stations and in which sequence so that the total energy requirements are minimum. While deciding the operating status of the pumps for minimum energy requirement the suction and discharge pressure requirement at each pumping stations should also be fulfill. So, there are $m \times n$ numbers of decision variable in water pump switching problem. Figure 11.2 shows the pumping system considered in the present work.

Water can be delivered from water source into demanding place by the pressure which is added by each pump and consumed along each pipeline due to the friction between water and pipe. For example, in any pumping station i, the pump j can add pressure P_{ij} using energy E_{ij} if pump is turn on. The energy E_{ij} (in hp) required by

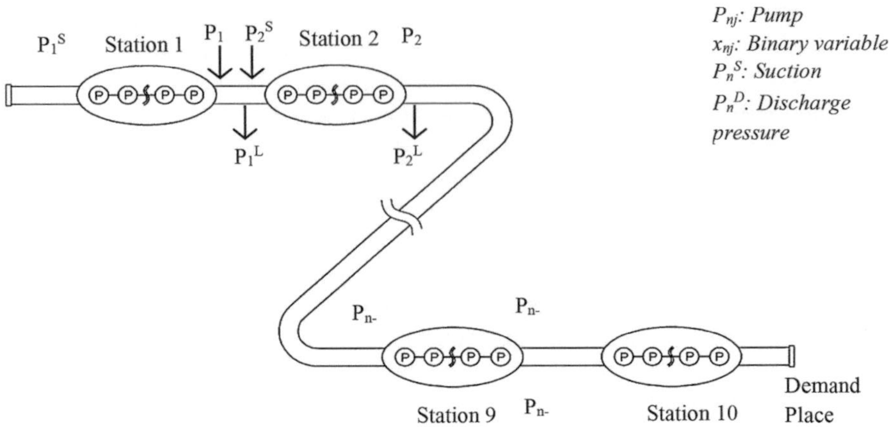

Fig. 11.2 Layout of water pump switching problem

the pump is a function of specific weight of water (γ in lb/ft^3), flow rate Q (in cfs), Pressure rise P_{ij} (in psi) across pump and motor pump efficiency η_{ij} and given by [9].

$$E_{ij} = \frac{\gamma\ Q\ P_{ij}}{238.3\eta_{ij}} \tag{11.1}$$

The on/off scheduling of the pump is denoted by assigning a binary variable x_{ij} $\in\{0,1\}$ to each pump at each station.

Objective Function

The objective function of the problem is to minimize the total electrical energy requirement of the pump station. Moreover, to take in to account the effect of constraints violation during the optimization process an arbitrarily large value (known as penalty function) is also added to the objective function. So, finally the objective function for the present problem is represented as,

$$Minimize\quad f(x) = \sum_{i=1}^{n}\sum_{j=1}^{m} E_{ij}x_{ij} + \sum_{j=1}^{m} R1\left(g_j(X)\right)^2 \tag{11.2}$$

Where, $R1$ is the penalty parameter having a static value. $g_j(X)$ indicates violation of jth constraint. The term $\sum_{j=1}^{m} R1\left(g_j(X)\right)^2$ takes into account the effect of constraints violation. The above mention objective function for water pump switching problem is subjected to following constraints.

Discharge Pressure Constraints

A discharge pressure P_i^D in pumping station i should be equal to the summation of suction P_i^S and all operating pressures $\sum_{j=1}^{m} P_{ij}x_{ij}$

$$P_i^D = P_i^S + \sum_{j=1}^{m} P_{ij}x_{ij}, \quad i = 1, \ldots\ldots, n \tag{11.3}$$

Discharge Pressure Bound Constraints

Discharge pressure in any station should be placed less than upper limit discharge pressure ${}_U P_i^D$

$$P_i^D \leq {}_U P_i^D, \quad i = 1, \ldots\ldots, n \tag{11.4}$$

Suction Pressure Constraints

A suction pressure in pumping station $i + 1$ should be calculated by subtracting pressure loss P_i^L, which is occurred along the pipe line i, from discharge pressure in pumping station i,

$$P_{i+1}^S = P_i^D - P_i^L, \quad i = 1, \ldots\ldots, n-1 \tag{11.5}$$

Suction Pressure Bound Constraints

Any suction pressure should be placed between lower suction pressure ${}_L P_i^S$ and upper suction pressure ${}_U P_i^S$.

$${}_L P_i^S \leq P_i^S \leq {}_U P_i^S, \quad i = 1, \ldots\ldots, n \tag{11.6}$$

Initial Suction Pressure Constraints

Initial suction pressure (i.e. suction pressure of first pumping station) is assumed to be zero

$$P_i^S = 0 \tag{11.7}$$

Binary Decision Variable Constraints

The binary value (0 or 1) is assigned to decision variable x_{ij}.

$$x_{ij} \in \{0, 1\}, \quad i = 1, \ldots, n \quad j = 1, \ldots, m \tag{11.8}$$

Present problem is solved by using the steps outlined below:

Step1: Assuming values of a set of decision variables (i.e. operating status of the pumps) and estimating the suction and discharge pressure at each pumping stations energy requirements of the pumps.

Step2: Evaluation of the energy requirements of the pumps at each pumping stations and formulation of the objective function.

Step3: Verifying suction and discharge pressure constraint at each pumping station and evaluation of the objective function

Step4: Utilization of the PSO, ACO, and BBO, algorithms to select a new set of values for the decision variables.

Step5: Iteration of the previous steps until a minimum of the objective function is found.

Results and Discussion

In the present case study the water pump switching system consists of 10 pipes ($n = 10$) and 10 pump stations ($n = 10$) with 4 pumps ($m = 4$) in series within each station. Thus, the number of decision variables (pump status variables) is 40 (*i.e.* $n \times m$). Each decision variable x_{ij} ($i = 1, \ldots, n$, $j = 1, .., m$) has a binary value (0 or 1), representing pump status: if x_{ij} equals to 1, pump is on; and if x_{ij} equals to 0, pump is off. Pumping pressure P_{ij} across each pump is tabulated in Table 11.1, and corresponding energy E_{ij} can be calculated using Eq. 11.1, where motor pump efficiencies η_i for ten stations are tabulated in Table 11.2.

The pressure loss P_i^L along the pipeline i, which can be calculated using Darcy-Weisbach and Colebrook-White equations is tabulated in Table 11.3. The lower suction pressure bound ${}_L P_i^S$, upper suction pressure bound ${}_U P_i^S$, and upper discharge pressure bound ${}_U P_i^D$ in pumping station i are also shown in Table 11.3.

Table 11.1 Pumping pressure data

Pump number (i, j)	Pumping pressure P_{ij} (psi)	Pump number (i, j)	Pumping pressure P_{ij} (psi)
(1, 1)	173.22	(6, 1)	209.82
(1, 2)	173.22	(6, 2)	209.82
(1, 3)	173.22	(6, 3)	209.82
(1, 4)	86.62	(6, 4)	104.91
(2, 1)	191.78	(7, 1)	229.13
(2, 2)	191.78	(7, 2)	229.13
(2, 3)	191.78	(7, 3)	229.13
(2, 4)	95.89	(7, 4)	114.56
(3, 1)	191.78	(8, 1)	169.51
(3, 2)	191.78	(8, 2)	169.51
(3, 3)	191.78	(8, 3)	169.51
(3, 4)	95.89	(8, 4)	84.75
(4, 1)	100.07	(9, 1)	211.32
(4, 2)	100.07	(9, 2)	211.32
(4, 3)	100.07	(9, 3)	211.32
(4, 4)	50.04	(9, 4)	105.66
(5, 1)	110.35	(10, 1)	192.31
(5, 2)	110.35	(10, 2)	192.31
(5, 3)	110.35	(10, 3)	192.31
(5, 4)	55.18	(10, 4)	96.15

Table 11.2 Motor-pump efficiency data

Pump station (n)	Motor-pump efficiency (η)
1	0.9789
2	0.9810
3	0.9810
4	0.9630
5	0.9660
6	0.9830
7	0.9840
8	0.9700
9	0.9800
10	0.9600

Experiments are carried out to find out the proper value of common controlling parameters (i.e. population size(N) and maximum number of generation) and algorithmic specific parameters of different algorithms considered in the present work. For the experiments 10–200 population size and 50–1000 numbers of generations is investigated initially to identify the best combination. Finally, after experimentations, the common controlling parameters as well as algorithmic specific parameters used in the present work are listed in Table 11.4.

The performance of all the algorithms is sensitive to the tuning of control parameters. The algorithms experimented with arbitrary values of control parameters

Table 11.3 Pressure loss and pressure limit data

Pump stations (n)	P_i^L (psi)	$_LP_i^S$ (psi)	$_UP_i^S$ (psi)	$_UP_i^D$ (psi)
1	309.85	00	200	900
2	154.98	25	200	900
3	258.19	25	200	800
4	309.85	25	400	900
5	154.98	25	250	900
6	309.85	25	350	900
7	309.85	25	450	1100
8	284.07	25	550	1100
9	82.675	25	400	1100
10	51.658	25	400	1100

Table 11.4 Control parameters of the different algorithms used in the present work

PSO algorithm	
Population size	25
Number of generations	200
Inertia weight (w)	0.4
Cognitive parameter (c_1)	1.6
Social parameter (c_2)	1.8
ACO algorithm	
Population size	25
Number of generations	200
Initial pheromone value	0.02
Pheromones update constant	50
Global pheromone decay rate	0.5
Local pheromone decay rate	0.5
BBO algorithm	
Population size	25
Number of generations	200
Maximum immigration rate	1
Maximum emigration rate	1
Mutation coefficient	0.1

required more computational effort (i.e. more number of function evaluations) or it can be possible for the algorithm to trap in the local optima. Table 11.5 shows the optimized pump status variable of the case study obtained by using PSO, ACO, and BBO algorithms and its comparison with the published optimized pump status variables obtained by using GA and B&B method [10] and HS method [9].

As seen from Table 11.5, solution provided by Goldberg and Kuo [10] by using GA results with 15 pumps in operation requiring 11263.19 HP power. Goldberg and Kuo [10] provided two more solution of the same problem by branch & bound method (B&B, B&B modified) using two different branch & bound method code. B&B[a] results in 12 pumps in operation with 11,187 HP requirements while B&B modified results in 13 pumps in operation with 11181.37 HP requirements. The

Table 11.5 Comparison of the solutions from different algorithms

Method	Energy (HP)	Pump status									
GA	11263.19	1100	1001	1001	1101	1001	1100	1100	0000	0000	0000
B&B	11187.00	1100	1000	1110	1000	0000	1100	1101	0000	0000	0000
B&B modified	11181.37	1011	0001	1110	0000	1010	0110	0110	0000	0000	0000
HS	11169.43	1100	0010	1011	0110	0000	1011	1010	0000	0000	0000
PSO	11169.43	1100	0010	1011	0110	0000	1011	1010	0000	0000	0000
ACO	11174.22	1100	0010	1101	0100	0100	0101	1011	0001	0000	0000
BBO	11165.95	1010	1001	0111	0001	0001	1011	1010	0000	0000	0000

solution provided by Geem [9] by using HS results in 13 pumps in operation with 11169.43 HP requirements. The solution obtained using PSO algorithm also results in 13 pumps in operation with 11169.43 HP requirements. The solution obtained using ACO algorithm result in 11174.22 HP requirements with 14 pumps in operation. While, the solution obtained using the BBO algorithm results in 14 pumps in operation with 11165.95 HP requirements. The operating sequence of the pumps obtained by using BBO algorithm is different compared to rest of five solutions.

Results shows that by using BBO algorithm, horsepower requirements of the present problem reduced by 3.48HP compared to HS approach given by Geem [9]. The solution obtained using PSO algorithm is alike to the solution obtained by HS approach. The solution obtained by using the ACO approach requires additional 4.79 HP compared to the solution obtained by using HS, and PSO algorithms. The present approaches using BBO results in one additional pump in operation compared to HS, and PSO approaches which reduced the energy requirement of the other pumps to increase the pressure of the fluid, and this reduction is more than the extra energy requirement for one additional pump in operation, so the overall energy requirement is reduced in the BBO approach.

The comparative results of the suction and discharge pressure solutions of GA, B&B modified, HS, PSO, and BBO are reported in Tables 11.6 and 11.7 respectively. It is observed from the results that all the solutions satisfy the upper and lower pressure bounds constraints. The suction and discharge pressure solutions of GA, B&B modified, HS, PSO, ACO and BBO are profiled in Figs. 11.3 and 11.4 respectively.

For the performance comparison of PSO, ACO, and BBO algorithms, two hundred trial runs of all the algorithms are performed on the considered case study. The performances of the algorithms are compared based on three different criteria: best value, mean value and standard deviation obtained through the two hundred independent run. In this comparison, best value gives the global optima, and mean value serves the purpose for the average performance to search optimum result. Standard deviation gives the deviation of best result from the mean result. Table 11.8 shows the performance comparison of considered algorithms for the present case study. Table 11.8 also shows the performance comparison of the considered approaches with GA and HS approaches where both the algorithms

Table 11.6 Suction pressure solution obtained by different algorithms

	B&B modified	B&B	GA	HS	PSO	ACO	BBO
Pumping station	Suction pressure (Psi)						
1	0	0	0	0	0	0	0
2	123.21	36.59	36.59	36.59	36.59	36.59	36.59
3	64.12	73.39	169.28	73.39	73.39	73.39	169.28
4	381.27	390.54	198.76	294.65	294.65	294.65	390.54
5	71.42	180.76	139.09	184.94	184.94	84.87	130.73
6	137.14	25.78	149.64	29.96	29.96	40.24	30.93
7	246.93	135.57	259.43	244.66	244.66	45.12	245.63
8	395.34	398.54	407.84	393.07	393.07	308.09	394.04
9	111.27	114.47	123.77	109	109	108.77	109.97
10	28.595	31.795	41.095	26.325	26.325	26.09	27.295

Table 11.7 Discharge pressure solution obtained by different algorithms

	B&B modified	B&B	GA	HS	PSO	ACO	BBO
Pumping station	Discharge pressure (Psi)						
1	346.44	433.06	346.44	346.44	346.44	346.44	346.44
2	228.37	219.1	324.26	228.37	228.37	228.37	324.26
3	648.73	639.46	456.95	552.84	552.84	552.84	648.73
4	490.61	381.27	448.94	494.79	494.79	394.71	440.58
5	180.76	292.12	304.62	184.94	184.94	195.22	185.91
6	445.42	556.78	569.28	554.51	554.51	354.97	555.48
7	708.39	705.19	717.69	702.92	702.92	617.94	703.89
8	398.54	395.34	407.84	393.07	393.07	392.84	394.04
9	114.47	111.27	123.77	109	109	108.77	109.97
10	31.795	28.595	41.095	26.325	26.325	26.09	27.295

were implemented with the same criteria on the considered problem by previous researchers [9, 10]. It is observed from the result that BBO algorithm performs better compared to other algorithms.

In order to identify the effect of different constrained handling methods on the performance PSO, ACO and BBO, four different constrained handling methods namely Superiority of Feasible Solutions (SF), Adaptive Penalty (AP), ε-Constraint (EC) and Stochastic Ranking (SR) [15] are experimented with the considered algorithms. The considered algorithms are experimented with different constrained handling methods with 5000 maximum function evaluations. Two hundred trial runs of each algorithm are performed for each constrained handling method. The comparative results of the algorithms are reported in Table 11.9. Here, the comparison is carried out based on three different criteria: best value (B), mean value(M) and standard deviation (SD) obtained through two hundred independent runs. It is observed from the results that Stochastic Ranking method of constraint handling perform better than rest of the methods with considered algorithms.

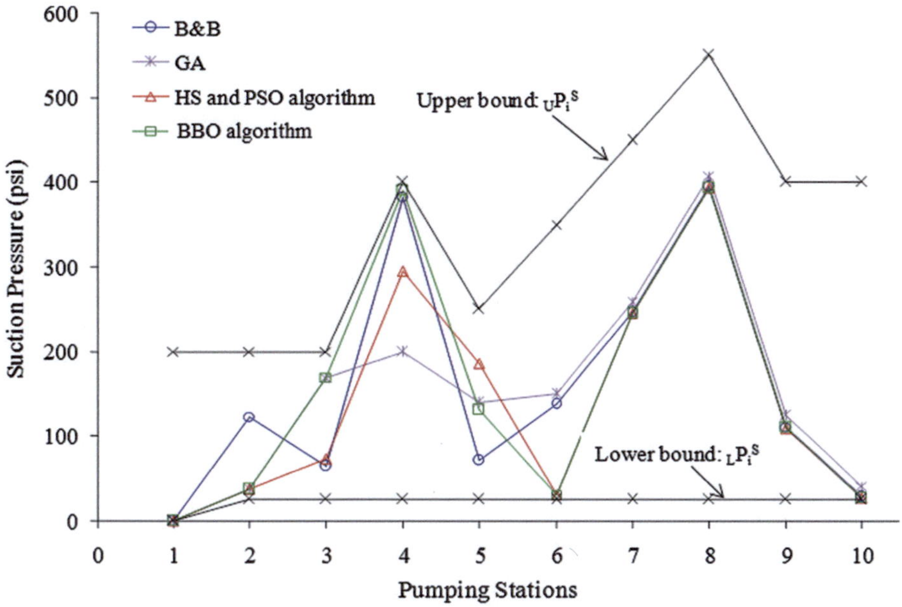

Fig. 11.3 Suction pressures solution of different algorithm

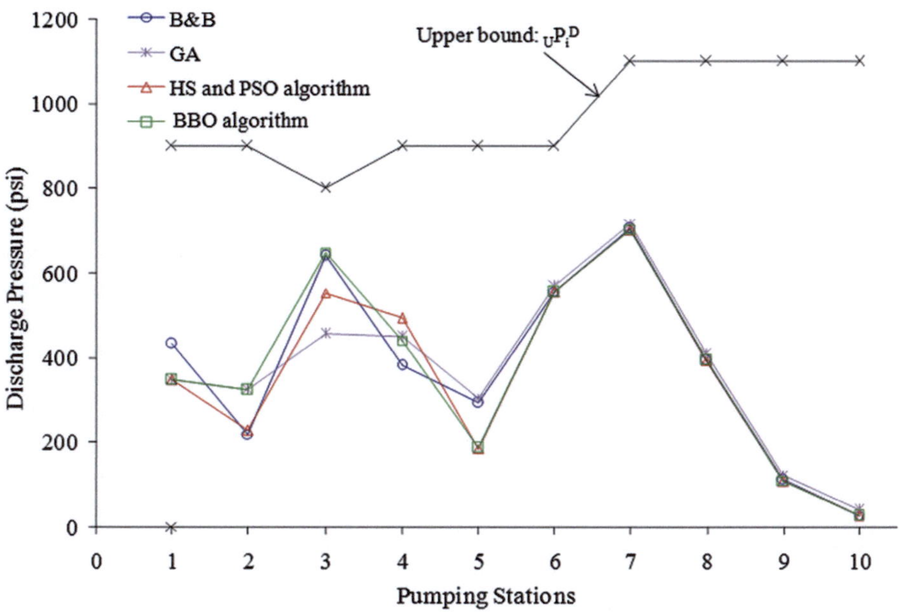

Fig. 11.4 Discharge pressures solution of different algorithm

Table 11.8 Performance comparison of GA, HS, PSO, ACO, and BBO algorithms

	Best solution	Mean solution	Standard deviation	Computational time (s)
GA	11263.19	11320.19	196.61	20.00
HS	11169.43	11215.11	48.13	15.00
PSO	11169.43	11211.73	46.57	16.00
ACO	11174.22	11219.73	52.12	24.00
BBO	11165.95	11194.14	17.91	21.00

Table 11.9 Comparative results of different constraint handling techniques with considered algorithms

		SP	SF	AP	EC	SR
PSO	B	11169.43	11169.43	11169.43	11169.43	11169.43
	M	11211.73	11210.62	11208.81	11209.56	11204.33
	SD	46.57	42.23	41.47	42.54	39.3
ACO	B	11174.22	11174.22	11174.22	11174.22	11174.22
	M	11219.73	11216.64	11213.66	11216.21	11209.71
	SD	52.12	49.63	43.75	48.94	41.13
BBO	B	11165.95	11165.95	11165.95	11165.95	11165.95
	M	11194.14	11190.21	11186.64	11188.73	11182.27
	SD	17.91	15.58	14.47	14.93	12.43

SP Static Penalty, *SF* Superiority of Feasible Solutions, *AP* Adaptive Penalty, *EC* ε-Constraint, *SR* Stochastic Ranking
B Best value, *M* Mean value, *SD* Standard deviation

The considered case study of water pump switching problem is multimodal problem. So, the problem has more than one solution for minimum energy requirement. Each solution has different operating status of pump; hence in case of any pump failure we can switch on to the alternate solution for same energy requirement. Table 11.10 shows the alternative solutions for minimum energy requirements with pump status obtained using the BBO algorithm. Similarly, Table 11.11 shows the energy requirement and pump status with different numbers of pump in operation obtained by using the BBO algorithm. Finally, the convergence comparison of the proposed approaches for minimum energy requirement consideration is shown in Fig. 11.5. It is observed from the Fig. 11.5 that objective function converges within about 4000 function evaluations using PSO and ACO algorithms, and 2000 function evaluations using BBO algorithm. Geem [9] described that using HS algorithm, convergence of the objective function was obtained within about 3500 function evaluations. Thus, BBO algorithm converges to the optimum value of objective function quite rapidly with better function value compared to other approaches.

Table 11.10 Alternative solutions for minimum energy requirements with pump status using BBO algorithm

Energy (HP)	Pump status
11165.95	1 1 0 0 1 0 0 1 0 1 1 1 0 0 0 1 0 0 0 1 1 0 1 1 1 0 1 0 0 0 0 0 0 0 0 0 0 0 0 0
11165.95	0 1 1 0 1 0 0 1 0 1 1 1 0 0 0 1 0 0 0 1 1 0 1 1 1 0 1 0 0 0 0 0 0 0 0 0 0 0 0 0
11165.95	1 0 1 0 0 1 0 1 1 1 0 1 0 0 0 1 0 0 0 1 1 0 1 1 0 1 1 0 0 0 0 0 0 0 0 0 0 0 0 0
11165.95	1 1 0 0 0 1 0 1 0 1 1 1 0 0 0 1 0 0 0 1 1 0 1 1 1 0 1 0 0 0 0 0 0 0 0 0 0 0 0 0
11165.95	0 1 1 0 0 1 0 1 1 1 0 1 0 0 0 1 0 0 0 1 1 0 1 1 0 1 1 0 0 0 0 0 0 0 0 0 0 0 0 0
11165.95	1 0 1 0 0 1 0 1 1 1 0 1 0 0 0 1 0 0 0 1 1 0 1 1 0 1 1 0 0 0 0 0 0 0 0 0 0 0 0 0
11165.95	1 1 0 0 0 1 0 1 1 0 1 1 0 0 0 1 0 0 0 1 1 0 1 1 1 0 1 0 0 0 0 0 0 0 0 0 0 0 0 0
11165.95	1 0 1 0 0 1 0 1 1 0 1 1 0 0 0 1 0 0 0 1 1 0 1 1 0 1 1 0 0 0 0 0 0 0 0 0 0 0 0 0
11165.95	1 1 0 0 0 1 0 1 1 1 0 1 0 0 0 1 0 0 0 1 1 0 1 1 1 0 1 0 0 0 0 0 0 0 0 0 0 0 0 0
11165.95	1 0 1 0 0 1 0 1 1 0 1 1 0 0 0 1 0 0 0 1 0 1 1 1 0 1 1 0 0 0 0 0 0 0 0 0 0 0 0 0
11165.95	1 1 0 0 0 1 0 1 1 1 0 1 0 0 0 1 0 0 0 1 0 1 1 1 1 0 1 0 0 0 0 0 0 0 0 0 0 0 0 0
11165.95	1 0 1 0 0 1 0 1 1 0 1 1 0 0 0 1 0 0 0 1 1 1 0 1 0 1 1 0 0 0 0 0 0 0 0 0 0 0 0 0
11165.95	1 1 0 0 0 1 0 1 1 1 0 1 0 0 0 1 0 0 0 1 1 0 1 1 0 1 1 0 0 0 0 0 0 0 0 0 0 0 0 0
11165.95	1 0 1 0 0 1 0 1 1 0 1 1 0 0 0 1 0 0 0 1 1 0 1 1 0 1 1 0 0 0 0 0 0 0 0 0 0 0 0 0
11165.95	1 1 0 0 0 1 0 1 1 1 0 1 0 0 0 1 0 0 0 1 1 0 1 1 0 1 1 0 0 0 0 0 0 0 0 0 0 0 0 0
11165.95	0 1 1 0 0 1 0 1 1 0 1 1 0 0 0 1 0 0 0 1 1 0 1 1 0 1 1 0 0 0 0 0 0 0 0 0 0 0 0 0
11165.95	1 0 1 0 0 1 0 1 1 0 1 1 0 0 0 1 0 0 0 1 1 0 1 1 1 0 1 0 0 0 0 0 0 0 0 0 0 0 0 0
11165.95	0 1 1 0 0 1 0 1 0 1 1 1 0 0 0 1 0 0 0 1 1 0 1 1 0 1 1 0 0 0 0 0 0 0 0 0 0 0 0 0
11165.95	1 0 1 0 1 0 0 1 0 1 1 1 0 0 0 1 0 0 0 1 1 0 1 1 1 0 1 0 0 0 0 0 0 0 0 0 0 0 0 0

Table 11.11 Energy requirement and pump status with different numbers of pump in operation using BBO algorithm

No. of pumps	Energy (HP)	Pump status
12	11177.59	0 1 1 0 1 0 0 0 1 1 1 0 1 0 0 0 0 1 0 0 1 1 0 0 1 1 0 0 0 0 0 0 0 0 0 0 0 0 0 0
13	11170.91	1 1 0 0 0 1 0 0 1 0 1 1 1 1 0 0 0 0 0 0 1 1 0 1 1 1 0 0 0 0 0 0 0 0 0 0 0 0 0 0
14	11165.95	1 0 1 0 1 0 0 1 0 1 1 1 0 0 0 1 0 0 0 1 1 0 1 1 1 0 1 0 0 0 0 0 0 0 0 0 0 0 0 0
15	11179.28	0 1 1 1 0 0 1 0 0 0 1 1 1 1 0 1 0 0 0 1 1 0 1 1 1 1 0 0 0 0 0 0 0 0 0 0 0 0 0 0
16	11181.25	0 1 1 0 0 1 0 1 0 0 1 1 1 0 1 1 0 0 0 1 0 0 1 1 0 1 1 1 0 0 0 1 0 0 0 0 0 0 0 0
17	11204.35	1 1 0 1 0 0 0 1 0 0 1 1 1 0 1 1 1 1 0 1 1 0 0 1 1 0 1 0 0 0 0 1 0 0 0 0 0 0 0 0
18	11308.62	0 1 1 1 0 0 0 1 0 1 1 1 1 1 0 1 0 1 1 1 0 0 1 1 0 0 0 1 1 0 0 1 0 0 0 0 0 0 0 0
19	11440.74	1 1 0 1 0 0 0 1 0 0 1 1 1 0 1 1 1 1 1 1 0 0 1 0 0 0 1 0 0 1 1 0 0 0 0 1 0 0 0 0
20	11887.70	1 1 0 1 0 0 0 1 1 0 0 1 1 1 1 1 0 1 1 1 0 0 1 1 0 0 0 1 0 0 1 1 0 0 0 1 0 0 0 1
21	12757.14	1 1 0 1 0 0 0 1 0 0 1 1 1 1 1 1 1 0 1 1 0 0 1 1 0 0 0 1 1 1 0 1 0 0 0 1 0 0 0 1
22	13753.80	0 1 1 1 0 0 0 1 0 0 1 1 1 1 1 1 1 1 0 1 0 1 0 1 0 0 0 1 0 1 1 1 0 0 0 1 1 0 0 1
23	14623.23	0 1 1 1 0 0 0 1 1 0 0 1 1 1 1 1 0 1 1 1 1 0 0 1 0 0 0 1 1 1 1 1 0 0 0 1 0 0 1 1
24	15619.89	1 0 1 1 0 0 0 1 1 0 0 1 1 1 1 1 1 0 1 1 0 1 0 1 0 0 0 1 1 1 1 1 0 0 0 1 0 1 1 1
25	16616.54	1 0 1 1 0 0 0 1 1 0 0 1 1 1 1 1 0 1 1 1 0 0 1 1 0 0 0 1 1 1 1 1 0 0 0 1 1 1 1 1

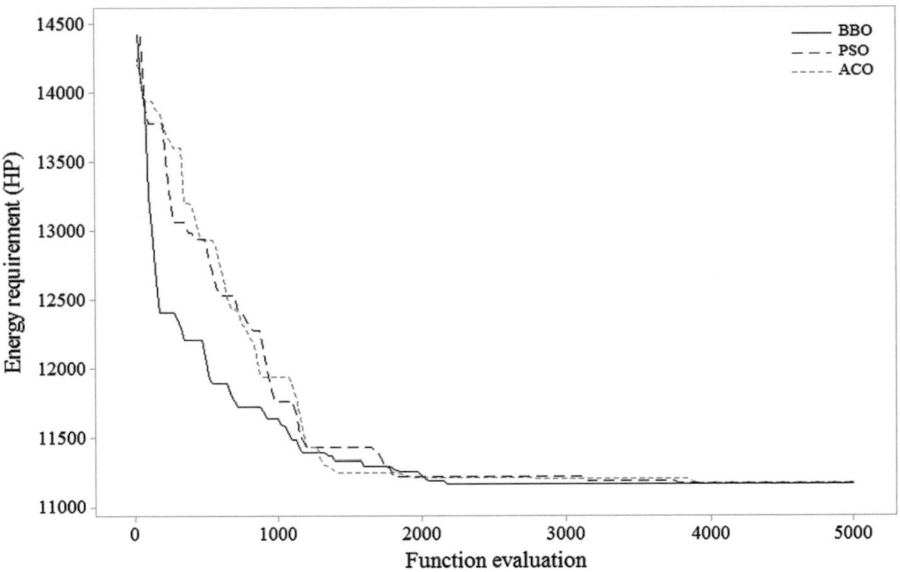

Fig. 11.5 Convergence comparison of PSO, ACO, BBO, ABC and TLBO algorithms

Summary

The basic concepts of biogeography based optimization algorithm are explained to understand the use of BBO for the optimization of a problem. This chapter also demonstrates the application of PSO, ACO, and BBO for the combinatorial water pump switching problem to find minimum energy consumption by appropriately scheduling the pumps in a pumping station. Case study with 40 pumps is optimized in order to identify the minimum energy requirement of the water pump switching system. The performance of BBO is judged against other methods such as GA, B&B, HS, PSO and ACO. The results indicate that BBO is effective over other methods. Effect of constrained handling methods on the performance of the proposed algorithms is also explored. Moreover, the alternate solutions for minimum energy require view point in case of pump failure as well as the energy requirement for different numbers of pumps in operation are also presented.

References

1. Arora P, Kundra H, Panchal VK (2012) Fusion of biogeography based optimization and artificial bee colony for identification of natural terrain features. Int J Adv Comput Sci Appl 3(10):107–111
2. Barán B, von Lücken C, Sotelo A (2005) Multi-objective pump scheduling optimisation using evolutionary strategies. Adv Eng Softw 36(1):39–47

3. Bhattacharya A, Chattopadhyay PK (2010) Hybrid differential evolution with biogeography-based optimization for solution of economic load dispatch. IEEE Trans Power Syst 25(4):1955–1964
4. Bonaiuti D, Zangeneh M, Aartojarvi R, Eriksson J (2010) Parametric design of a waterjet pump by means of inverse design, CFD calculations and experimental analyses. J Fluids Eng 132(3):031104
5. Di Barba P, Dughiero F, Mognaschi ME, Savini A, Wiak S (2016) Biogeography-inspired multiobjective optimization and MEMS design. IEEE Trans Magn 52(3):1–4
6. Du D, Simon D, Ergezer M (2009) Biogeography-based optimization combined with evolutionary strategy and immigration refusal. In: Systems, man and cybernetics, 2009, October, SMC 2009, IEEE international conference on. IEEE, pp 997–1002
7. Ergezer M, Simon D, Du D (2009) Oppositional biogeography-based optimization. In: Systems, man and cybernetics, 2009, October, SMC 2009, IEEE international conference on. IEEE, pp 1009–1014
8. Fan J, Eves J, Thompson HM, Toropov VV, Kapur N, Copley D, Mincher A (2011) Computational fluid dynamic analysis and design optimization of jet pumps. Comput Fluids 46(1):212–217
9. Geem ZW (2005) Harmony search in water pump switching problem. In: International conference on natural computation. Springer, Berlin/Heidelberg, pp 751–760
10. Goldberg DE, Kuo CH (1987) Genetic algorithms in pipeline optimization. J Comput Civ Eng 1(2):128–141
11. Kundra H, Sood M (2010) Cross-country path finding using hybrid approach of PSO and BBO. Int J Comput Appl 7(6):15–19
12. Lohokare MR, Pattnaik SS, Devi S, Panigrahi BK, Das S, Bakwad KM (2009) Intelligent biogeography-based optimization for discrete variables. In: Nature & biologically inspired computing, 2009, December, NaBIC 2009, World Congress on. IEEE, pp 1088–1093
13. Ma H (2010) An analysis of the equilibrium of migration models for biogeography-based optimization. Inf Sci 180(18):3444–3464
14. Ma H, Simon D (2011) Blended biogeography-based optimization for constrained optimization. Eng Appl Artif Intell 24(3):517–525
15. Mallipeddi R, Suganthan PN (2010) Ensemble of constraint handling techniques. IEEE Trans Evol Comput 14(4):561–579
16. Rao RV, Savsani VJ (2012) Mechanical design optimization using advanced optimization techniques. Springer Science & Business Media, London
17. Roy PK, Ghoshal SP, Thakur SS (2010a) Biogeography based optimization for multi-constraint optimal power flow with emission and non-smooth cost function. Expert Syst Appl 37(12):8221–8228
18. Roy PK, Ghoshal SP, Thakur SS (2010b) Multi-objective optimal power flow using biogeography-based optimization. Electr Power Compon Syst 38(12):1406–1426
19. Savsani VJ, Rao RV, Vakharia DP (2009) Discrete optimisation of a gear train using biogeography based optimisation technique. Int J Des Eng 2(2):205–223
20. Savsani P, Jhala RL, Savsani V (2014) Effect of hybridizing biogeography-based optimization (BBO) technique with artificial immune algorithm (AIA) and ant colony optimization (ACO). Appl Soft Comput 21:542–553
21. Simon D (2008) Biogeography-based optimization. IEEE Trans Evol Comput 12(6):702–713
22. Simon D (2011) A dynamic system model of biogeography-based optimization. Appl Soft Comput 11(8):5652–5661
23. Simon D (2013) Evolutionary optimization algorithms. Wiley, Chichester
24. Simon D, Ergezer M, Du D, Rarick R (2011) Markov models for biogeography-based optimization. IEEE Trans Syst Man Cybern B Cybern 41(1):299–306
25. Song Y, Liu M, Wang Z (2010) Biogeography-based optimization for the traveling salesman problems. In: Computational science and optimization (CSO), 2010, May, Third international joint conference on. IEEE, vol 1, pp 295–299

26. Wang L, Xu Y (2011) An effective hybrid biogeography-based optimization algorithm for parameter estimation of chaotic systems. Expert Syst Appl 38(12):15103–15109
27. Wang JY, Chang TP, Chen JS (2009) An enhanced genetic algorithm for bi-objective pump scheduling in water supply. Expert Syst Appl 36(7):10249–10258
28. Wang G, Guo L, Duan H, Wang H, Liu L, Shao M (2013) Hybridizing harmony search with biogeography based optimization for global numerical optimization. J Comput Theor Nanosci 10(10):2312–2322
29. Zhang Z, Zeng Y, Kusiak A (2012) Minimizing pump energy in a wastewater processing plant. Energy 47(1):505–514
30. Zhang Y, Wang S, Dong Z, Phillip P, Ji G, Yang J (2015) Pathological brain detection in magnetic resonance imaging scanning by wavelet entropy and hybridization of biogeography-based optimization and particle swarm optimization. Prog Electromagn Res 152:41–58

Chapter 12
Introduction to Invasive Weed Optimization Method

Dilip Kumar, B. G. Rajeev Gandhi, and Rajib Kumar Bhattacharjya

Abstract The weeds are generally defined as the unwanted plants growing in an agricultural field. The weeds are not very useful and occupy the space in the field to successfully outnumber the plants that are cultivated for regular use. Thus, a popular agronomical belief is that "The Weeds Always Win". Weeds typically generate large numbers of seeds, supporting their spread by wind or some other natural factors. They can also grow in adverse conditions and are very adaptable. These unique properties of weed growth shows the way for the development of optimization techniques. One of the algorithms motivated by this common phenomenon in agriculture field is based on the expansion of invasive weeds. The algorithm is known as Invasive Weed Optimization (IWO). In this chapter, we have described the IWO algorithm and its use in obtaining the optimal solution of common popular functions.

Keywords Invasive weeds · Seeds · Optimization

Introduction

Weeds occupy an agricultural field by means of spreading of seeds from the plants and take control of places between the crops. Weeds generally occupy the unused space of the field and further grow to a flowering weed. This flower provides new weeds and the cycle repeats. The production capacity of the flowering plant to produce weed depends on the adaptation of the flowering weeds in the field [6].

D. Kumar (✉)
Department of Civil Engineering, G B Pant Engineering College, Pauri, Uttarakhand, India

B. G. R. Gandhi · R. K. Bhattacharjya
Department of Civil Engineering, Indian Institute of Technology Guwahati, Guwahati, Assam, India
e-mail: b.rajeev@iitg.ac.in; rkbc@iitg.ac.in

F. Bennis, R. K. Bhattacharjya (eds.), *Nature-Inspired Methods for Metaheuristics Optimization*, Modeling and Optimization in Science and Technologies 16,
https://doi.org/10.1007/978-3-030-26458-1_12

The weeds with better adaptation to the environment will be able to occupy more unused space in the field. They will grow faster and will generate more seeds. These newly produced weeds are again scattered over the field and grow as new flowering weeds [5]. This cycle lasts until the number of weeds reaches its maximum capacity. Due to the competition between different weed plants, only the weeds with better adaptation capability can survive and produces new weeds i.e. they follow Darwin's principle of "survival of the fittest" [7, 8]. In brief, we can simply define the concept behind the invasive weed theory as:

(a) A measurable number of seeds are scattered over the field, known as the initial population.
(b) Each spread seed turns to a flowering plant and produces new seeds based on their adaptability with the surroundings, known as reproduction.
(c) The produced seeds are being scattered by some natural agency over the rest of the field and turn to new plants known as the spatial distribution of weeds.
(d) This cycle continues until the number of plants reaches a maximum. Once the number of plants reaches the maximum, the plants with the best fitness and adaptability survives and produce new seeds. Rest of the weak plants are eliminated which is called as competitive elimination or survival of the fittest.
(e) At last, the weed plant with the best fitness and adaptability are only present in the field.

These processes are used to design the optimization algorithm called the Invasive Weed growth Algorithm (IWO). The initialization of the population, reproduction, spatial distribution, competitive exclusion and the optimal solution are the steps involved in the design of the IWO algorithm [14, 15]. Each of these steps is described in detail in the following section.

Working Procedure of Invasive Weed Optimization Algorithm (IWO)

Initialize a Population

In general terms, the population is defined as the total number of people or residents living in a country, state or any specified region. If the region is specified to be the whole agricultural field, the initial weeds that emerge randomly over the field are the initial population. So, to replicate that in the IWO, we can define the population as initial number of seeds scattered over the field [1]. Mathematically we can define the population as the bunch of initial solutions which being expanded over the D dimensional problem space with random or blind positions.

Consider a set of population as a matrix X, with elements that are referred as plants or agents. Each plant represents possible solution in defined population size, max plant. For a D-dimensional problem and a population of size N, the matrix 'X' representing the population is given as.

$$X = \begin{bmatrix} x_{11} & x_{12} & \dots & x_{1D} \\ x_{21} & x_{22} & \dots & x_{2D} \\ \dots & \dots & \dots & \dots \\ x_{N1} & x_{N2} & \dots & x_{ND} \end{bmatrix} \tag{12.1}$$

The ith population consists of D variables as the dimension of the problem is 'D'. Therefore the coordinates of the elements x_{i1} is limited between the boundaries, U1 and L1, *i.e.*, ($L1 < xi_1 < U1$) and x_{iD} is limited between two other boundaries, UD and LD i.e., ($LD < x_{iD} < UD$). Let the lower and upper bounds be the vectors of length 'D' represented by L and U as given in Eq. 12.2 as follows.

$$\begin{aligned} L &= [L1 \;\; L2 \;\; \dots \;\; LD] \\ U &= [U1 \; U2 \dots UD] \end{aligned} \tag{12.2}$$

Pseudo Code for Initialization of Population

Define Vectors L and U and dimension D
for dim = 1:D
for pop = 1:N
    *X(pop,dim) = L(1,dim)+rand()*x*{U(1,dim)-L(1,dim)}*
end
end

Using this similar code in MATLAB, the population can be initialized in the same way as the initial weeds are produced in the field.

Reproduction

The weeds present in the filed have different strengths, *i.e.* the some plants are highly adaptable and can produce a large number of offsprings whereas some of the plants die off quickly. So, the fitness of the plant decides the number of offsprings it will produce. To reproduce the same process in the IWO algorithm, the fitness of the initial population is calculated. The plant in the population is then allowed to produce seeds depending on its own fitness and the adaptability as well as the lowest and the highest fitness value of the plants in the colony [7, 8]. The number of seeds produced by each plant will increases linearly (Fig. 12.1) from the minimum possible seed (S_{min}) to its maximum level (S_{max}). The fitness of any plant 'p' can be represented as 'F_p'. The Eq. (12.3) can be used to calculate the number of seeds produced by a plant according to its fitness.

$$S_p = S_{\min} + (F_p - F_W)\left({}^{(F_B - F_w)}\big/_{(S_{\max} - S_{\min})}\right) \tag{12.3}$$

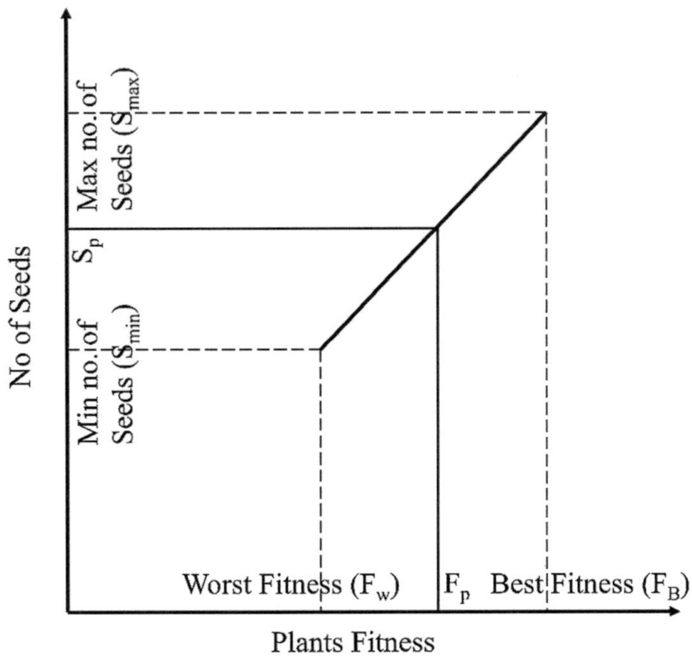

Fig. 12.1 Procedure of reproduction

Pseudo Code for Reproduction

Calculate fitness of 'N' Population as F
Scale the fitness from worst to best
Assign F_W as worst fitness and F_B as best fitness
Define minimum and maximum no of seeds as S_{min} and S_{max}
for pop = 1:N
    $S(pop) = S_{min}+(F(pop)-F_W)\text{x}\{(S_{max}-S_{min})/(F_B-F_W)\}$
end

This reproduction part is coded in MATLAB to get the number of seeds to be produced by each weed. The reproduction part decides only the number of offsprings (seeds) that are to be produced by each weed. The spread of the seeds of each of the weeds is decided in the next subsection.

Spatial Distribution of Seeds

The number of seeds that are generated according to the fitness values are to be dispersed around the parent weed. The dispersion happens by means of mostly wind. Initially, when there is enough space in the filed for the weeds to grow, the dispersion of seeds will be over a greater radius. But, as the empty space in the field gets occupied by the new weeds, the dispersion has to shrink to a minimum area close to the parent weed [3]. This process is replicated in the IWO algorithm as follows.

The seeds generated are distributed randomly over the search space. The random distribution is followed by a normal distribution with a zero mean but a variable variance. This ensures that the seeds will be randomly distributed such that they stand near the parent plant. However, standard deviation (σ) is generally reduced once you are reaching the optimal solution. As such, the standard deviation is varying from an initial value of σ_{initial} to a final value of σ_{final} in every step (generation). The following equation (Eq. 12.4) can be used to change the standard deviation in each iteration.

$$\sigma_{iter} = \frac{(iter_{\max} - iter)^n}{iter_{\max}^n}\left(\sigma_{initial} - \sigma_{final}\right) + \sigma_{final} \tag{12.4}$$

Where, $iter_{max}$ is the maximum number of iterations, σ_{iter} is the standard deviation at the present step and n is the nonlinear modulation index. The variation of the standard deviation with each iteration from maximum of 0.5 to minimum of 0.001with and 'n' value of 2 is given in Fig. 12.2.

Pseudo Code for Spatial Distribution

Define σ_{final}, $\sigma_{initial}$, *and exponent n.*
for iter = 1:Maxiter
  σ_{iter} = *{(Maxiter-iter)/Maxiter}*^{n}x *(*$\sigma_{initial}$ - σ_{final}*)*+ σ_{final}
for pop = 1:N
for dim = 1:D
for seedpop = 1:S(pop)
  X_{seed}*(pop,seedpop,dim) = X(pop,dim)+*σ_{iter}x*randNormal()*
end
end
end
end

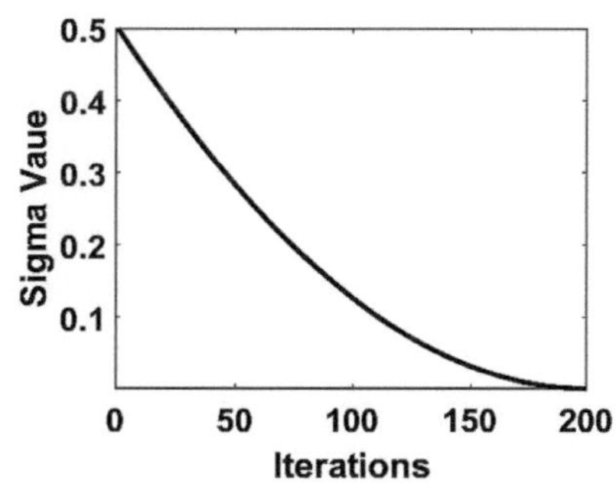

Fig. 12.2 Standard deviation varying with the iterations

This gives the population of seeds for each of the weeds distributed spatially over the area according to their respective normal distribution and the standard deviation. The spatial distribution part is also coded in MATLAB according to the pseudo code.

Competitive Elimination

By reproduction and the spatial dispersal of the seeds over a number of generations, the plants will occupy the whole field. The growth of the weeds and the seeds combined have to be limited by some principle so that the maximum number of plants and weeds in each generation is minimized to a specified value [2, 4]. Based on the minimum number of seeds (when set to zero), if a plant leaves no offspring then it would go extinct. Also, the plants producing a lesser number of offsprings would also be outnumbered by the seeds produced by the fitter weeds [5, 14]. These two processes can be incorporated into IWO by combining the whole population of weeds and seeds and then excluding the population which is not the fittest based on the maximum number of allowable population (P_{max}). This ensures that in each generation only a maximum of P_{max} population exists throughout the algorithm.

Pseudo Code for Competitive Elimination

Define the maximum population allowed P_{max}
Combine X *and* X_{seed} *to* X_{temp} *and calculate the fitness* F
Scale the fitness from best to worst and sort X_{temp}
If size(F) $< P_{max}$
        $X_{new} = X_{temp}$
else
        $X_{new} = X_{temp}(1{:}P_{max})$
end

This eliminates the inferior solutions of the population and the better solutions are allowed to go to the next generation. The process is then continued from reproduction to competitive elimination until the termination criteria is reached. The flow chart of the algorithm is shown in Fig. 12.3.

We have considered an area of 20 × 20 units to demonstrate the initialization of the population, reproduction and competitive exclusion in the following example. The function is taken as Ackley function with optimal solution at (0,0) and the initial population is 50, maximum population is 100. Figure 12.4a gives the scatter plot of the initial population, Fig. 12.4b gives the scatter plot of the reproduced population and the Fig. 12.4c gives the scatter plot of the population reduced to maximum after competitive exclusion.

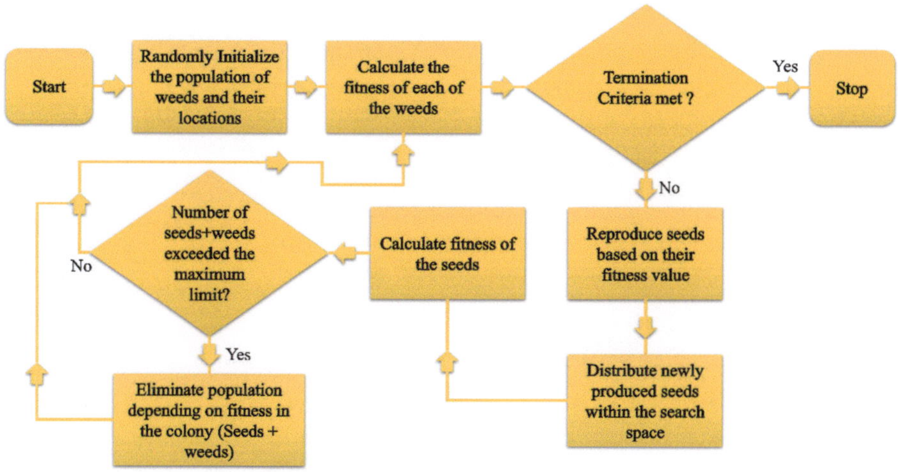

Fig. 12.3 Flowchart showing the IWO algorithm

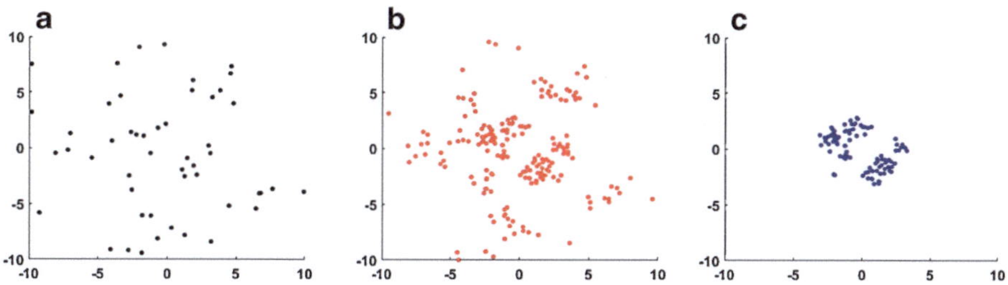

Fig. 12.4 Scatter plots of the population at different stages of the IWO algorithm. (**a**) Initial population. (**b**) Population after reproduction. (**c**) Population after competitive elimination

Standard Examples

The IWO algorithm can be used for solving the non-linear non convex problems. However, the robustness of the algorithm can be observed by solving some standard mathematical functions [11]. These mathematical functions are of different kinds such as one global optima (sphere function) and multiple global optima (Himmelblau function). There are many other functions containing of many local optima and only one global optima (Ackley function). All these functions represent different types of problems that arise in engineering optimization. The efficiency of an algorithm can be measured by obtaining the solution of these functions. The solutions for all the three functions considering two variables are discussed in the next sections. The parameters used for the optimization are as given below in Table 12.1.

Table 12.1 Parameters used for solving the problems

Parameter	Value used
Maximum iterations	200
Initial population size	10
Maximum population size	25
Minimum no. of seeds	0
Maximum no. of seeds	5
Exponent 'n'	2
Sigma-initial	0.5
Sigma-final	0.001

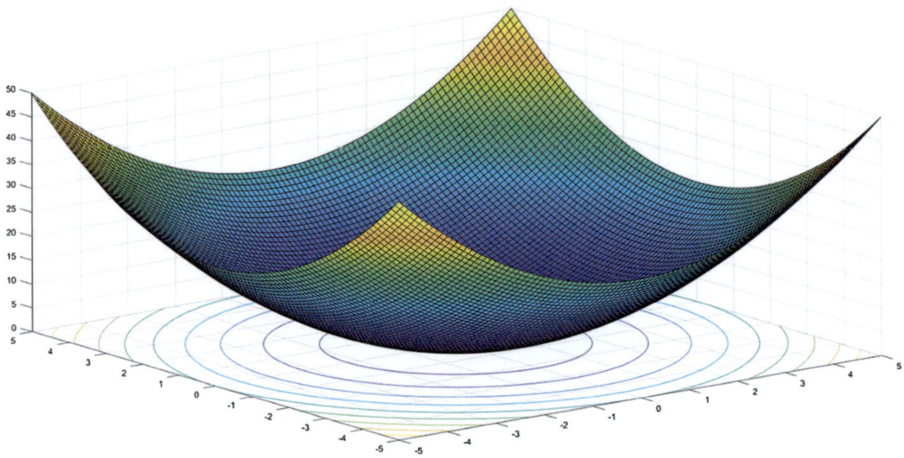

Fig. 12.5 Sphere function contour and the surface plot

Sphere Function

The sphere function is a simple part of the sphere in between the bounds. The minimum value of the function is at (0,0). The variation of the function in its bounds is given in Fig. 12.5. The equation of the sphere function is given in Eq. 12.5.

$$f(x, y) = x^2 + y^2 \tag{12.5}$$

This function is solved 20 number of times with the same parameters and the results of the best fitness for each iteration in each of the solution are presented in Fig. 12.6. It can be observed that, every time, the optimal value is reached over the total number of iterations.

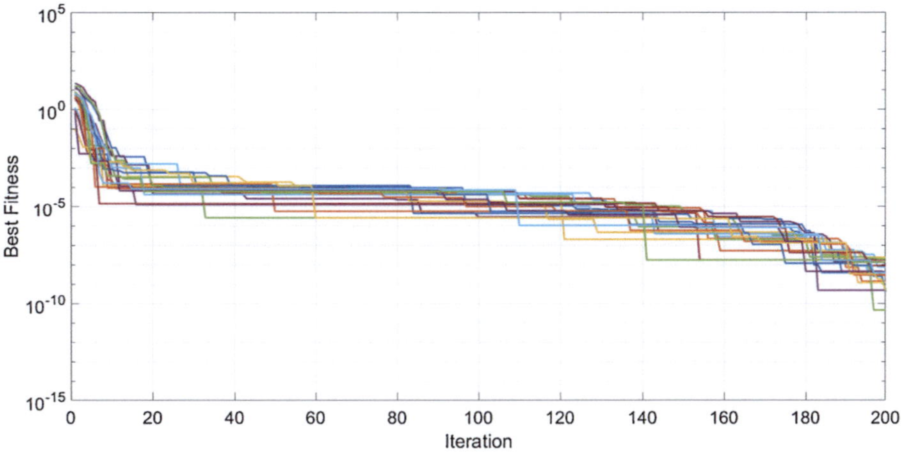

Fig. 12.6 Best fitness with number of iterations over 20 runs

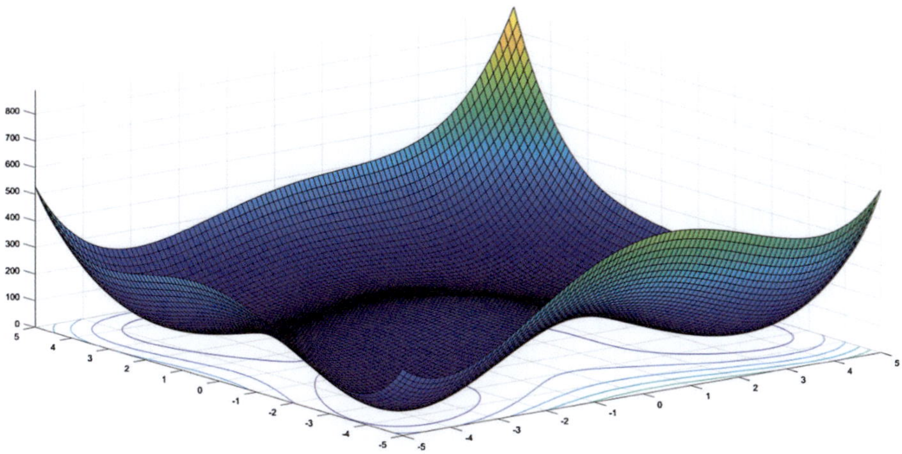

Fig. 12.7 Contour and surface plots of the Himmelblau function

Himmelblau Function

This function is defined by the equation below which has four identical minimum at (3,2), (−2.805118,3.131312), (−3.779310,−3.283186) and (3.584428,−1.848126) as (x,y) pairs where the function value is zero. The function is given in Eq. 12.6. The variation of the function can be seen in Fig. 12.7.

$$f(x, y) = \left(x^2 + y - 11\right)^2 + \left(x + y^2 - 7\right)^2 \tag{12.6}$$

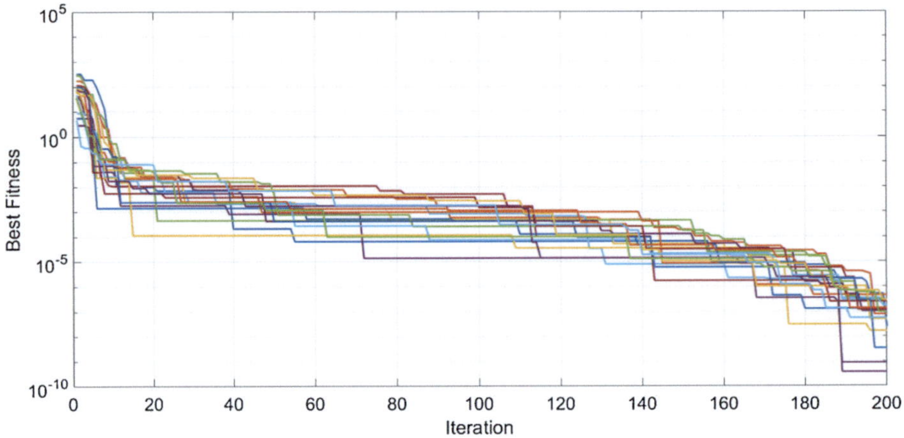

Fig. 12.8 Best fitness with number of iterations over 20 runs

The function with four identical optima is also solved 20 times with the same parameters as mentioned in the Table 12.1. The results of the best fitness with the number of iterations over 20 runs can be seen in the Fig. 12.8.

Ackley Function

This function resembles most of the functions in engineering optimization. This contains many local optima and only one global optima at (0,0). The function is given in Eq. 12.7. The figure for the variation of the function along the interval is given in Fig. 12.9.

$$f(x, y) = 20\left(1 - e^{-0.2\sqrt{\frac{x^2+y^2}{2}}}\right) + e - e^{\frac{\cos(2\pi x)+\cos(2\pi y)}{2}} \tag{12.7}$$

The solution of the Ackley function for 20 runs is carried out. The best fitness with the iterations for all the 20 runs are given in Fig. 12.10.

For all the three functions, the behavior of the algorithm for the best fitness did not change much. This is because of the typical behavior of the invasive weed growth optimization. The fitness initially decreases rapidly, then decreases slowly for most number of iterations and then again decreases rapidly over the end of the iterations. This can be observed in the Figs. 12.6, 12.8 and 12.10. This assures that IWO is a pure metaheuristic optimization approach. The nature of the problem does not have an effect on the solution. Thus this algorithm is efficient in solving different optimization problems in engineering and many other fields.

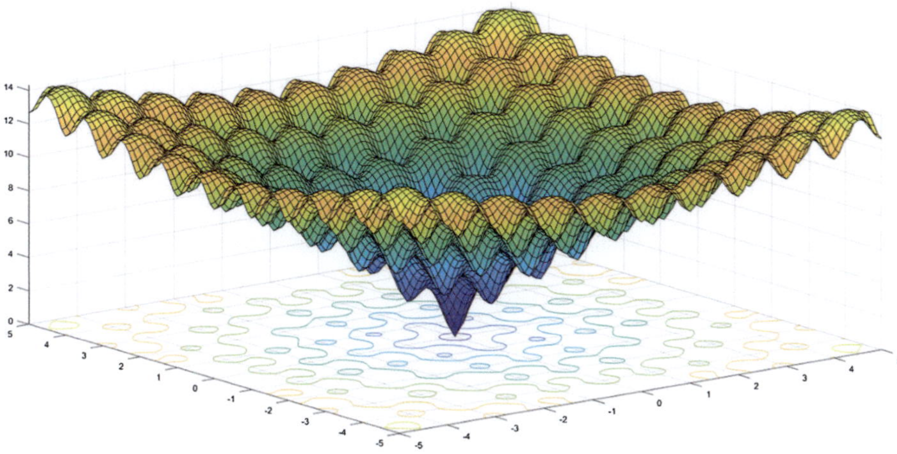

Fig. 12.9 Ackley function contour and surface plots

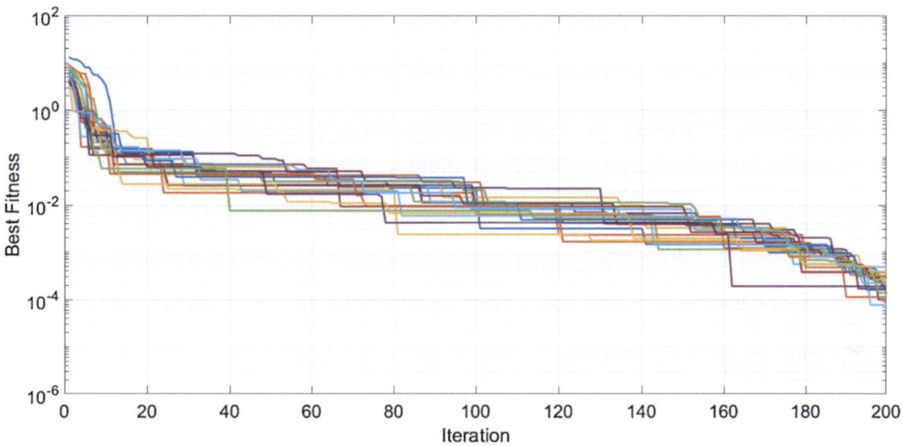

Fig. 12.10 Best fitness along with iterations for 20 runs

Conclusions

This chapter provides an overview of an IWO algorithm described from an evolutionary comparison or natural phenomena. The applications and growth of natural computing in the last decade has increased vastly. Numerous optimization problems in computer networking [13], bioinformatics [15], data mining, game theory, power systems [10], image processing, industry and engineering, robotics, applications involving the security of information systems etc [9, 12]. have been using such nature inspired optimization methods. The simulation results obtained by using IWO indicate the effectiveness and robustness of the algorithm in solving nonlinear non-convex problems.

References

1. Bevelacqua PJ, Balanis CA (2007) Minimum sidelobe levels for linear arrays, antennas and propagation. IEEE Trans Antennas Propag 55:12
2. Chen TB, Chen YB, Jiao YC, Zhang ES (2005) Synthesis of antenna array using particle swarm optimization. In: Asia-Pacific microwave conference proceedings 2005 (APMC 2005), vol 3, December 4{7, 2005}
3. Kacem I, Hammadi S, Borne P (2002) Approach by localization and multiobjective evolutionary optimization for flexible job shop scheduling problems. IEEE Trans Syst Man Cybern 32:245–276
4. Khodier MM, Christodoulou CG (2005) Linear array geometry synthesis with minimum sidelobe level and null control using particle swarm optimization. IEEE Trans Antennas Propag 53(8):2674–2679
5. Mallahzadeh AR (2008) Application of the invasive weed optimization technique for antenna configurations. Prog Electromagn Res PIER 79:137–150
6. Mallahzadeh AR, Es'haghi S, Alipour A (2009) Design of an E shaped MIMO antenna using IWO algorithm for wireless application at 5.8 Ghz. Prog Electromagn Res PIER 90:187–203
7. Mehrabian AR, Lucas C (2006a) A novel numerical optimization algorithm inspired from weed colonization. Eco Inform 1:355–366
8. Mehrabian AR, Lucas C (2006b) A novel numerical optimization algorithm inspired from weed colonization. Eco Inform 1:355–366
9. Mekni S, Châar Fayéch B, Ksouri M (2010) TRIBES application to the flexible job shop scheduling problem. IMS 2010 10th IFAC workshop on intelligent manufacturing systems, Lisbon, Portugal, July 1st -2nd 2010
10. Rattan M, Patterh MS, Sohi BS (2008) Design of a linear array of half wave parallel dipoles using particle swarm optimization. Prog Electromagn Res M 2:131–139
11. Sakarovitch M (1984) Optimisation combiantoire. Méthodes mathématiques et algorithmiques. Hermann, Editeurs des sciences et des arts, Paris
12. Xia W, Wu Z (2005) An effective hybrid optimization approach for multiobjective flexible job shop scheduling problems. J Comput Ind Eng 48:409–425
13. Zaharis Z, Kampitaki D, Papastergiou A Hatzigaidas A, Lazaridis P, Spasos M (2006) Optimal design of a linear antenna array using particle swarm optimization. In: Proceedings of the 5th WSEAS international conference on data networks, communications and computers, 69{74, Bucharest, Romania, October 16}
14. Zaharis ZD, Skeberis C, Xenos TD (2012) Improved antenna array adaptive beamforming with low side lobe level using a novel adaptive invasive weed optimization method. Prog Electromagn Res 124:137–150
15. Zhang X, Wang Y, Cui G, Niu Y, Xu J (2009) Application of a novel IWO to the design of encoding sequence for DNA computing. Comput Math Appl 57:2001–2008

Chapter 13
Single-Level Production Planning in Petrochemical Industries Using Novel Computational Intelligence Algorithms

Sandeep Singh Chauhan and Prakash Kotecha

Abstract Optimal production planning requires the solution of combinatorial optimization problems that need to be efficiently modelled so as to be solved with computational intelligence (CI) techniques. In this work, we report the computational performance of five recently proposed CI techniques, namely the sanitized-teaching-learning-based optimization (s-TLBO), moth flame optimization (MFO), flower pollination optimization, water cycle optimization, and adaptive wind driven optimization on the single level production planning problem which involves complex domain-hole constraints, nonlinearities in the production and investment costs, resource and unique process constraints. In this work, the domain-hole constraints are handled using a hard-penalty approach. The performance is evaluated on the case study of the Saudi Arabia petrochemical industry to determine optimal portfolio from 54 processes for producing 24 products with three production levels. Based on 2040 (8 Cases $\times$ 51 runs $\times$ 5 algorithms) unique instances of the problem, it was observed that s-TLBO and MFO are consistently able to determine efficient solutions and that s-TLBO was able to quickly discover feasible solutions and had relatively low variance.

Keywords Production planning · Evolutionary computation · Optimization · Combinatorial optimization · Sanitized-teaching-learning-based optimization algorithm · Moth flame optimization algorithm · Flower pollination algorithm · Water cycle optimization algorithm · Adaptive wind driven optimization algorithm

S. S. Chauhan · P. Kotecha (✉)
Department of Chemical Engineering, Indian Institute of Technology Guwahati, Guwahati, Assam, India
e-mail: sandeep.chauhan@iitg.ac.in; pkotecha@iitg.ac.in

F. Bennis, R. K. Bhattacharjya (eds.), *Nature-Inspired Methods for Metaheuristics Optimization*, Modeling and Optimization in Science and Technologies 16,
https://doi.org/10.1007/978-3-030-26458-1_13

Nomenclature

a_{sr}	amount of raw material r required in process s for producing per ton of product
k	production capacity level, K denotes the total number of production capacity level of a process
n_t	number of active processes employed to produce product t
r	index for raw material, $r \in (1, .., R)$; R denotes the total number of raw materials
s	index for process, $s \in (1, .., S)$; S denotes the total number of processes
t	index for product, $t \in (1, .., T)$; T denotes the total number of products
x_s	total amount of product produced from process s
y_s^k	binary variable to indicate the production level from process s; $y_s^k = 1$, if $\left(L_s^k \leq x_s \leq L_s^{k+1}\right)$, else 0
z_s	binary parameter to indicate the production in domain hole; $z_s = 0$ if $(x_s = 0)$ or $(L^s \leq x_s \leq U^s)$, else 1
B	total available monetary resources ($)
C_s	total production cost of process s
E_s	selling price per ton of the product produced from process s
I_s	total investment cost of process s
L_s^k	production capacity level of level k for process s
$P_r^{rawmaterial}$	penalty incurred in fitness function due to insufficient raw material of type r
P^{Budget}	penalty incurred in fitness function due to insufficient investment cost
P_t^{uni}	penalty incurred in fitness function due to violation of unique process constraint for product t
Q_s^k	production cost for production capacity level k of process s
S_t	set of processes that can produce product t
U_s	maximum production from process s
V_s^k	investment cost for production capacity level k of process s
λ, γ	static penalty factor

Introduction

Petrochemical industries use petroleum and natural gas as feedstock to produce a variety of petrochemicals that are subsequently transformed into everyday consumer goods. The products of these industries act as raw materials for a large number of industries in the manufacturing sector and thus potentially represent the financial backbone of a country. The global market value of petrochemical industries has been reported to be $514.5 billion in 2014 and it is estimated that the petrochemical industries would be worth $758.3 billion by 2022 (available at https://bit.ly/2kdX8UA). In addition to the variety of products that can be produced from a particular raw material, a large number of processes are also available which can convert a particular raw material into a specified product. These industries usually operate on large scale, which makes them capital as well as energy–intensive [5], and is encountering stiff competition from several environments friendly alternatives including synthetic biology and synthetic DNA [13]. It has also been proposed that petrochemical industries can utilize unconventional raw materials such as coal and biofuels [26]. Moreover, with an increasing emphasis on biorefineries, integration of biorefineries into petrochemical plants has also received increased attention. The abundance of choices from the selection of raw materials, processes, and products gives rise to combinatorial optimization problems. In order to sustain its economic viability, it has become essential for the petrochemical industries to solve these complex problems and to implement optimal production plans so as to enable the best use of its limited resources [1, 19]. To this end, artificial intelligence and soft computing tools [2, 27] have been increasingly employed for challenging problems in the oil industry and a concerted effort is being made to realize the strategic importance of operations research [16, 22].

In this work, we use the term Computational Intelligence techniques to refer to computational intelligence techniques designed for solving optimization problems and thus includes evolutionary, swarm and other stochastic optimization techniques. CI techniques have been increasingly employed for solving optimization problems in the oil sector. In view of its popularity, genetic algorithm has been predominantly used for wide ranging applications including reservoir characterization, gas storage, seismic inversion, and oil field development [29]. Other works include the use of simulated annealing to analyze and predict the estimations of petroleum exergy production and consumption [24], the use of ant colony optimization to estimate the energy demand of Turkey [28] using factors such as population, gross domestic product, import and export, and the use of Particle Swam Optimization (PSO) to estimate the oil demand in Iran based on socio–economic indicators [3]. The intelligent design of sensor networks [6] and design optimization of LPG thermal cracker [23] with respect to conflicting objectives using multi–objective evolutionary algorithms have also been reported.

CI techniques have also been used to solve combinatorial optimization problems. Despite their benefits over gradient based mathematical programming techniques, the use of CI techniques for solving optimization problems in the petrochemical sector has been relatively limited, particularly in the case of combinatorial optimization problems. The success of CI techniques depends on a large number of factors such as efficient encoding of the solution, modification of existing operators or incorporation of problem–specific operators [15]. CI techniques are able to solve black–box optimization problems and do not require the postulation of the optimization problem in terms of conventional equalities and inequalities. This benefit has to be harnessed to efficiently model the problem and the use of artificial decision variables to postulate the problem in the conventional form needs to be avoided. Some of the works which have successfully used CI techniques to solve combinatorial optimization problem include the solution of refinery crude oil scheduling problem by hybridizing genetic algorithm with finite state method [12] as well as integration with graph representation as structure adapted genetic algorithm [25]. The refinery production scheduling problem involving operational transition has been solved using an efficient hybridization of discrete PSO and linear programming which was otherwise intractable by Mixed Integer Linear Programming (MILP) approaches [31]. The use of moth flame optimization and teaching–learning–based optimization for solving the combinatorial production planning have also been reported [7, 17]. A comparison of CI techniques has also reported for multi-unit production planning with only continuous variables [8] for nonlinear cost function and as a mixed integer problem for piece-wise linear cost functions [9].

The use of recent optimization techniques, which have shown promising performance on benchmark optimization problems, remains relatively unexplored in the petrochemical sector, primarily as these have not been incorporated in the popular mathematical software. Moreover in recent years, a plethora of CI techniques have been proposed for solving optimization problems and conflicting results have been reported on the performance of these algorithms. In a majority of the cases, the algorithms are demonstrated on different sets of problems thereby making it difficult to compare their performance. In this article, we study the performance of five recently proposed CI techniques *viz.,* (i) Adaptive wind driven optimization algorithm (AWDO), (ii) Water cycle optimization algorithm (WCO), (iii) Flower pollination algorithm (FP), (iv) Moth flame optimization algorithm (MFO) and (v) Sanitized-teaching–learning–based optimization algorithms (s-TLBO) to solve the combinatorial production-planning problem. The performance is evaluated based on 2040 unique instances arising from the eight cases of the Saudi Arabian petrochemical production–planning problem reported in the literature. Though demonstrated in the context of petrochemical industry, the results can be extended to production planning problems in other industries. The article is structured as follows: In Sect. 13.2, we provide the problem description and follow with a discussion of the solution strategy. In Sect. 13.4, we provide a brief description of the five CI techniques. Section 13.5 discusses the performance of the algorithm from various perspectives. The article is subsequently concluded by summarizing the developments in this work and discussing potential future work.

Problem Description

In the production–planning problem, a set of processes (*S*) are available which can be used to produce a set of products *(T)* from a set of raw materials *(R)*. A particular product can be produced by more than one process. Tables 13.1, 13.2, 13.3 and 13.4 is an example of production planning problem that has been widely studied in literature for the Saudi Arabian petrochemical industry [1] and consists

Table 13.1 Data for propylene products and processes [1]

Product name	Product *ID*	Sale price ($/ton)	Process *ID*	Process name	Propylene used/ton a_{s2}	Capacity (10^3 ton/yr)			Production cost ($\$10^6$/yr)			Investment cost ($\$10^6$)		
						L_s^1	L_s^2	L_s^3	Q_s^1	Q_s^2	Q_s^3	V_s^1	V_s^2	V_s^3
Polypropylene copolymer	T1	975	S1	Amoco/Chisso	0.9480	70	135	270	50.7	90.1	170.7	55	81.1	131.6
			S2	BASF	0.9432	75	150	300	56.8	103.8	196.2	58	85.1	132.4
			S3	Himont	0.9490	77.5	155	310	56.9	103.7	195.7	60.2	86.8	134.1
Polypropylene block copolymer	T2	975	S4	Sumitomo (gas phase)	0.9546	70	145	290	51.7	97.6	184.8	55.1	83.1	132
			S5	UCC/Shell	0.9550	47.5	95	190	38.2	69.8	130.4	43.3	66.8	104.3
Polypropylene homopolymer	T3	780	S6	Borealis	1.0450	40	80	160	38.5	65.2	120.7	66.2	92.8	153.2
			S7	UCC/Shell	1.0500	40	80	160	31.8	57.1	105.5	40	61.4	95.1
Phenol	T4	735	S8	From C_6H_6/C_3H_8 via cumene	0.5103	45	90	180	37.8	57.7	94.9	106.6	151.7	231.5
Acrylic acid (ester grade)	T5	1450	S9	Two-stage oxidation	0.6289	40	80	160	38.5	65.6	119.1	82.8	125.4	207

(continued)

Table 13.1 (continued)

Product name	Product *ID*	Sale price ($/ton)	Process *ID*	Process name	Propylene used/ton	Capacity (10^3 ton/yr)			Production cost ($10^6/yr)			Investment cost ($10^6)		
					a_{s2}	L_s^1	L_s^2	L_s^3	Q_s^1	Q_s^2	Q_s^3	V_s^1	V_s^2	V_s^3
Propylene oxide	T6	1130	S10	Arco process (styrene product)	0.8648	90	180	360	92.2	159.2	290.9	233.5	390.7	698.7
			S11	Texaco (T butanol byproduct)	0.9546	90	180	360	86.7	154.1	287.7	185.8	304.5	537.1
			S12	Chlorohydrin	0.8265	90	180	360	95.8	175	330.9	119	179.4	289.2
			S13	Acro process (T butanol byproduct)	0.7875	90	180	360	87.5	157.2	294.9	212.3	362.7	657.7
			S14	Cell liquor neutralization	0.8101	90	180	360	105.9	196.6	375.2	109.8	164.3	263.1
			S15	Shell process (styrene byproduct)	0.8782	90	180	360	93.1	131.1	239.4	221.7	376.1	672.7
N-butanol	T7	830	S16	Via cobalt hydrocarbonyl catalyst	0.8150	50	100	200	41.4	68.7	117.2	115.5	180.4	287.4
			S17	Via N butryaldehyde Rh catalyst	0.6994	50	100	200	34.9	62	111.6	63.7	100.2	156.3
Cumene	T8	450	S18	From C_6H_6 and propylene	0.3784	60	120	240	36.6	62.1	120.8	23.1	33.2	50.7

Table 13.2 Data for ethylene products and processes [1]

Product name	Product *ID*	Sale price ($/ton)	Process *ID*	Process name	Ethylene used/ton a_{s1}	Capacity (10^3 ton/yr) L_s^1	L_s^2	L_s^3	Production cost ($\$10^6$/yr) Q_s^1	Q_s^2	Q_s^3	Investment cost ($\$10^6$) V_s^1	V_s^2	V_s^3
Poly vinyl Chloride	T9	740	S19	Suspension polymerization		100	200	400	67.6	125.2	237.2	117.6	186	307.5
			S20	Bulk polymerization		50	100	300	33	63.1	163.8	62.5	114	209.6
Poly vinyl Chloride (dispersion)	T10	1250	S21	Batch emulsion polymerization		25	50	100	28.7	48.3	86	73.1	101.1	148
			S22	Continuous emulsion polymerization		25	50	100	24	43.1	79.5	46.5	70.7	110.1
Vinyl chloride	T11	430	S23	TOSOH technology		125	250	500	63.8	123.5	241	49.2	74.4	112.8
			S24	Pyrolysis		125	250	500	68.5	134.5	264	79.1	144.2	258.1
			S25	Chlorination/ Oxychlorination	0.4678	250	500	1000	101.5	195	377	134	229.9	392.2
Ethylene glycol	T12	600	S26	Hydration of EO all EO for EG	0.7267	90	180	360	50.3	90	165.6	142.6	234.8	397.5

(continued)

Table 13.2 (continued)

Vinyl acetate	T13	690	S27	From ethylene and acetic acid	0.3930	67.5	135	200	53.9	101.2	146.4	82.7	133.6	181.3
High density polyethylene	T14	860	S28	UCC process	1.0200	70	135	270	42.1	75.1	141.8	56.9	84.5	131.5
			S29	Du Pont process	1.0200	70	135	270	44.6	77.5	147.7	63.4	84.5	136.9
			S30	Philips process	1.0200	70	135	270	44.6	78.8	148	66.5	96.2	147.7
Linear low density polyethylene	T15	900	S31	Dry mode gas phase univation process	0.9461	100	200	400	55.7	106.8	208.4	51.4	83.0	144.5
			S32	Bimodal grade by mixed mettal-locene/Ziegler catalyst	0.9387	75	150	300	48.3	90.2	172.8	46.9	66	98.6
			S33	Bimodal grade by unipol process	0.9430	122.5	245	490	92	174	336.2	82.4	116.6	175.6
Low density polyethylene	T16	870	S34	High pressure tubular reactor	1.0600	50	100	200	34.9	63.9	120.4	72	117.7	199.7

Table 13.3 Data for synthesis gas products and processes [1]

Product name	Product *ID*	Sale price ($/ton)	Process *ID*	Process name	Methane used/ton a_{s3}	Capacity (10^3 ton/yr)			Production cost ($\$10^6$/yr)			Investment cost ($\$10^6$)		
						L_s^1	L_s^2	L_s^3	Q_s^1	Q_s^2	Q_s^3	V_s^1	V_s^2	V_s^3
Acetic acid	T17	480	S35	Low pressure carbonylation (Rh.Catalyst solution)		182.5	365	540	63.2	111.4	156.6	125.6	195.9	259.6
			S36	Low pressure carbonylation supported Rh. catalyst		182.5	365	540	60.3	103	142.6	116.4	168.2	213.5
			S37	Low pressure carbonylation Rh. Halide catalyst		180	360	550	64.7	110.2	154.6	133.2	196.3	248.9
Ammonia	T18	160	S38	ICI AMV process	6.3500	300	430	590	48.3	65	85	210.9	278.2	356
			S39	MW Kellog process	5.9280	300	430	590	52.8	71.4	92.7	243.5	322.4	412.6
			S40	ICI LCA process	6.6780	105	170	340	19.4	27.4	47.3	87	119.5	196.7
Formaldehyde 1	T19	500	S41	From methanol using silver catalyst	6.6780	15	25	50	6.6	9.7	17.7	15.3	20.2	32.7
			S42	From methanol using Fe Mo catalyst	6.6780	15	25	50	6.9	10.6	19.4	17.9	26.2	44.9
Methanol	T20	150	S43	Lurgi process	7.8670	415	830	1660	55.2	96.3	184.3	224.6	365.5	682.1
			S44	ICI process copper catalyst	7.7780	415	830	1660	56.5	100.5	194.3	228.5	384.6	727.6
			S45	ICI LCM process	7.6610	415	830	1660	51.9	98	187.6	199.1	371.5	702.9

Table 13.4 Data for aromatics (BTX) products and processes [1]

Product name	Product *ID*	Sale price ($/ton)	Process *ID*	Process name	Ethylene used/ton a_{s1}	Capacity (10^3 ton/yr) L_s^1	L_s^2	L_s^3	Production cost ($\$10^6$/yr) Q_s^1	Q_s^2	Q_s^3	Investment cost ($\$10^6$) V_s^1	V_s^2	V_s^3
Styrene	T21	760	S46	Liquid phase alkyl/adiabatic dehydrogenation	0.2891	225	450	680	105.8	204.8	306	116.9	190	265.1
			S47	Liquid phase alkyl oxidative reheating	0.2878	225	450	680	108	209.7	313.5	115.6	191.8	266.8
			S48	Vapor phase alkyl/adiabatic dehydrogenation	0.2843	225	450	680	105.6	202.5	302.6	125.2	192.7	269
			S49	Vapor phase alkyl/isothermal dehydrogenation	0.2874	225	450	680	106.7	206.1	308.1	125.2	202	285.5
Pthalic anhydride	T22	700	S50	Attochem/Nippon		12.5	25	50	9.4	16.4	28.4	26	40.8	63.9
			S51	From O-Xylene by Alsuisse Italia process		12.5	25	50	9	15.4	27.3	27.7	39.6	56.9
Phenol	T23	735	S52	Liquid phase oxidation of toluene		45	90	180	36.8	64	118.7	108.8	157.2	251.6
PTA	T24	680	S53	Hydrolysis of dimethyl terephthalate		125	250	500	81.4	145.8	275.5	208.1	308.6	515.5
			S54	From P-xylene by bromine promoted air oxidation		125	250	500	78.4	145	277	170.5	267.3	452.7

of 24 different products (represented by T1 to T24) which can be produced by 54 processes (represented by S1–S54). For example, high–density polyethene (T14 in Table 13.2) can be produced by UCC process (S28 in Table 13.2), Du point process (S29 in Table 13.2) and Philips process (S30 in Table 13.2). The investment cost (V) and the production costs (Q) of each process are dependent on the production capacity and is provided for K capacity levels. In the current example, these details are available at three different levels (low, medium and high denoted as L_s^1, L_s^2, and L_s^3 respectively). For example, Table 13.1 shows that the production cost of process S3 for producing 77.5×10^3 tons/year of product T1 is \$ 56.9×10^6. Similarly, the production cost is \$ 103.7×10^6 per year for producing 155×10^3 tons/year, and it is \$ 195.7×10^6 for producing 310×10^3 tons/year. The minimum production from a process is 0 whereas the maximum production of a single unit is restricted by the production capacity of the highest level i.e., L_s^K (in this case L_s^3). It should be noted that the amount of production from a process has to be either 0 or greater than or equal to L_s^1, i.e. if a process is selected, it has to produce a minimum specified amount of product for operational reasons. The production cost and investment cost for any production between the lowest and the highest level capacity are determined using a piece–wise linear function between the two successive levels as shown in Fig. 13.1. In this figure Q_s^1, Q_s^2, and Q_s^3 denote the production cost of the three levels whereas V_s^1, V_s^2 and V_s^3 denotes the investment cost. It should be noted that the production cost is assumed to constitute all the various costs such as raw material

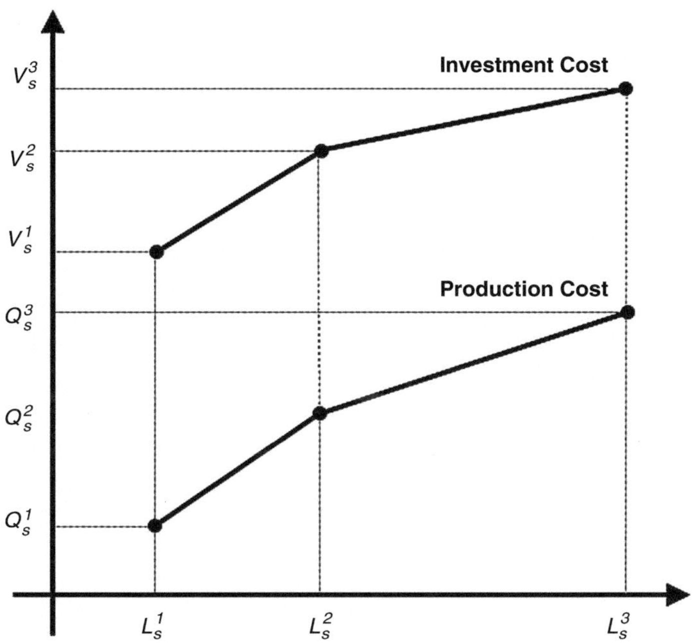

Fig. 13.1 Production cost and investment cost as a function of production capacity. [1]

cost, labor cost, and utility cost [1]. The amount of raw material r that is required to produce the product using the process s is also known and indicated by a_{rs}. The maximum amount of raw material r that is available is given by R_r. Additionally, the maximum amount of monetary resource that is available (B) and the selling price (E_s) of the product from process s are known.

The objective is to determine the optimal production plan in order to maximize the profit, which is the difference between the revenue realized by selling the products and the production cost of the products subject to the constraints on the raw material and investment cost. In particular, the optimal production plan requires the identification of (i) the products to be produced, (ii) the processes to be employed to produce the selected products, and (iii) the amount of the product that has to be produced from each of the selected processes. Another special constraint is the unique process constraint, which requires that an identical product should not be produced from more than one processes. The incorporation of the unique process constraint can lead to a reduction in the profitability but may be required to potentially diversify the product portfolio and reduce the complexity of the plant by avoiding the use of different type of processes to produce an identical product. The solution of the optimization problem with unique process constraint can be relatively challenging for CI techniques and hence we have included it to evaluate their performance.

Solution Strategy

The appropriate choice of the decision variables is crucial for the successful application of CI techniques to combinatorial optimization problems. In this strategy [7], the use of binary variables is avoided and the problem is formulated with the help of continuous variables without compromising on the rigor of the formulation. Moreover, unlike mathematical programming techniques, CI techniques do not rely on the gradient information and are designed to even solve black–box optimization problems. This benefit of CI technique is exploited in this strategy to efficiently model this problem without using artificial decision variables. In this strategy, a single decision variable is employed to represent the production from each process. If there are S processes, an equal number of continuous decision variables are employed and the production from a particular process s is denoted by x_s. The lower bound of the decision variables are set to zero to accommodate no production from a process.

The upper bound of the decision variable for the process s is denoted by U_s and can be determined by using Eq. (13.1)

$$U_s = \max_{\forall k=1,2,\ldots K} \left(L_s^k \right) \tag{13.1}$$

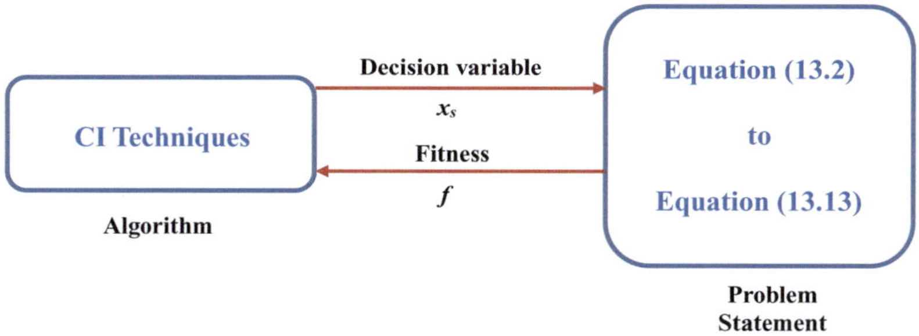

Fig. 13.2 Information flow between CI technique and the optimization problem

In the above equation, L_s^k indicates the production capacity of level k for process s. Thus, the decision variables and its bounds are given by Eq. (13.2).

$$0 \le x_s \le U_s \qquad \forall s = 1, 2, \ldots, S \tag{13.2}$$

If there is a process with three production levels (say low, medium and high) of 50 units, 70 units and 80 units, then $L_s^1 = 50$, $L_s^2 = 70$, $L_s^3 = 80$ and the value of U_s would be 80 (= max $(50, 70, 80)$). It can be observed that the upper bound can differ for each process depending on the maximum amount that can be produced from the various levels available for it. Figure 13.2 depicts the typical relationship between a CI technique and the optimization problem that is being solved. In every iteration, the CI techniques provide candidate solutions (the set of decision variables) whose fitness is to be determined using the problem statement. The fitness value of the candidate solution is to be returned to the CI technique, which utilizes this information to potentially discover better solutions. In the rest of the discussion, we describe the determination of the fitness based on the decision variables provided by a CI technique. Equation (13.3) helps in the determination of the processes, which violate the domain–hole constraint. Based on the value of the decision variable x_s, the binary parameter z_s is set to a value of 1 if the production in process s violates the domain–hole constraint and is set to 0 if it does not violate the domain–hole constraint.

$$z_s = \begin{cases} 0 & x_s = 0 \ or \\ 0 & (\exists x_s : L_s \le x_s \le U_s) \\ 1 & \text{otherwise} \end{cases} \qquad \forall s = 1, .., \mathrm{S} \tag{13.3}$$

Equation (13.4) assigns a value of 1 to the binary parameter y_s^k if the amount of production lies in between the level k and $k + 1$, and zero otherwise for the process s. The notation V_s^k and Q_s^k denote the investment and the production cost of production capacity level k respectively for process s.

$$y_s^k = \begin{cases} 1 & L_s^k \leq x_s \leq L_s^{k+1} \\ 0 & \text{otherwise} \end{cases} \quad \forall k = 1, 2, .., K-1; \forall s = 1, 2, .., S \tag{13.4}$$

Equations (13.5) and (13.6) determine the investment and production cost of the process s respectively. It should be noted that Eqs. (13.5) and (13.6) ensure that the production and investment cost for any production between two successive levels (say k and $k+1$) is calculated based on the linear interpolation between the two successive costs (refer to Fig. 13.1).

$$I_s = \sum_{k=1}^{K-1} y_s^k \left(V_s^k + \left(\frac{V_s^{k+1} - V_s^k}{L_s^{k+1} - L_s^k} \right) \left(x_s - L_s^k \right) \right) \quad \forall s = 1, 2, .., S \tag{13.5}$$

$$C_s = \sum_{k=1}^{K-1} y_s^k \left(Q_s^k + \left(\frac{Q_s^{k+1} - Q_s^k}{L_s^{k+1} - L_s^k} \right) \left(x_s - L_s^k \right) \right) \quad \forall s = 1, 2, .., S \tag{13.6}$$

Equation (13.7) determines the number of processes, which are employed to produce product t with S_t indicating the set of processes that can produce the product t.

$$n_t = \sum_{s=1}^{S_t} \sum_{k=1}^{K-1} y_s^k \quad \forall t = 1, 2, .., T \tag{13.7}$$

Equation (13.8) determines the penalty to be incorporated in the objective function for the violation of the raw material constraint where R_r indicates the amount of raw material r that is available and a_{sr} indicates the amount of raw material r that is required in process s. The equation ensures that the penalty is zero if the amount of raw material required is lower than or equal to the amount of raw material available.

$$P_r^{\text{rawmaterial}} = \left[\min \left\{ \left(R_r - \sum_{s=1}^{S} a_{sr} x_s \right), 0 \right\} \right]^2 \tag{13.8}$$

Similarly, Eq. (13.9) determines the penalty to be assigned if the budget constraint is violated where B indicates the monetary resource that is available for investment.

$$P^{\text{Budget}} = \left[\min \left\{ \left(B - \sum_{s=1}^{S} I_s \right), 0 \right\} \right]^2 \tag{13.9}$$

The unique process constraint is implemented using Eq. (13.10) wherein a penalty is assigned only if the product t is being produced by more than one process.

$$P_t^{\text{uni}} = \begin{cases} 1000^{n_t} & n_t > 1 \\ 0 & \text{otherwise} \end{cases} \qquad t = 1, 2, .., T \tag{13.10}$$

In Eq. (13.11), E_s denotes the selling price per unit of the product produced from the process s, which is used to calculate the total profit from the production plan.

$$\text{Profit} = \sum_{s=1}^{S} (E_s x_s - C_s) \tag{13.11}$$

Equation (13.12) determines the sum of all the penalties associated with domain–hole constraints, raw material constraints, investment constraints and the unique process constraints.

$$P = \left\{ \left(\gamma \sum_{s=1}^{S} z_s \right) + \left(\sum_{r=1}^{R} P_r^{\text{rawmaterial}} \right) + \left(P^{\text{Budget}} \right) + \left(\sum_{t=1}^{T} P_t^{\text{uni}} \right) \right\} \tag{13.12}$$

The objective is to maximize the profit but since most CI algorithms are designed for minimization problems, the fitness function involves minimizing the negative of profit. The fitness function that can be used with CI techniques to solve this combinatorial optimization problem is given by Eq. (13.13).

$$\text{Minimize} \qquad f = -\text{Profit} + \lambda(P) \tag{13.13}$$

Primary purpose of the fitness function in CI techniques is to discriminate the quality of the solutions. The fitness function ensures that a feasible solution has better fitness than an infeasible solution as the value of P would be zero. Moreover, between two feasible solutions, the solution with a larger profit would have a better fitness value. In the case of two solutions, which are infeasible, the solution with lower penalty would be selected as per the Eq. (13.13).

It should be noted that the decision variables to be used in the CI techniques for this production planning problem are only $x_s \ \forall \ s = 1, 2, .., S$. All the other variables used in Eq. (13.3) to Eq. (13.13) are uniquely fixed, if the values of the set of variables denoted by x_s are known, and should not be used as decision variables in the CI techniques.

Brief Description of CI Techniques

In this section, we briefly describe five algorithms that have been proposed in the recent past. One of the reasons for selecting these algorithms is that the authors have provided their codes and it is possible to do a critical analysis of the code and the algorithm. In view of the nature of the computational intelligence algorithm, it is often observed that there have been conflicting results [10, 20] which primarily arise due to the incomplete description of the algorithm. However, with the availability of the codes, the claims can be independently verified by researchers. Moreover, these algorithms have claimed superiority over many of the conventional algorithms such as Genetic Algorithm, Differential Evolution and Particle Swarm Optimization. It should be noted that a detailed explanation of these algorithms can be obtained from the literature and is not provided here for the sake of brevity.

Sanitized–Teaching–Learning–Based Optimization Algorithm

Teaching–learning–based optimization algorithm is a stochastic population–based algorithm that has been proposed multiple times and is inspired by the teaching–learning process of the classroom. The algorithm consists of two phases, *viz.*, the teacher phase, which is based on a student learning from the teacher, and the student's phase in which a student learns from other classmates. The decision variables correspond to the marks secured by a student and the student with the best fitness function is considered as the teacher of the class.

TLBO algorithm requires only two parameters in terms of the population size and the number of generations to be used for terminating the algorithm. There have been several issues in the implementation of TLBO, which have been described in literature [11]. The implementation used for this work does not employ any duplication removal step as it increases the computational complexity, requires additional evaluation of the objective function and does not significantly aid in the discovery of better results. Moreover, the rest of the four algorithms do not employ any duplication removal step and in order to provide a fair comparison, the duplicate removal step has not been included. In every generation, each member of s-TLBO helps in the discovery of two potentially new solutions (one each in the student and the teacher's phase) whose objective function needs to be evaluated. Thus the total number of fitness evaluations is given by {(2 × population size × number of generations) + population size}. Recent variants of TLBO include FTLBO Kommadath and Kotecha [18]. We have used the online available s-TLBO codes (available at https://goo.gl/vk8gLb).

Moth Flame Optimization Algorithm

Moth flame optimization algorithm is a stochastic population–based approach [21] and its performance has been demonstrated on benchmark real parameter optimization problems and computationally expensive optimization problems. It is inspired by the natural transverse orientation employed by moths to navigate at nights for travelling long distances in straight paths by maintaining a fixed angle with respect to distant celestial objects such as the moon. This mechanism becomes ineffective if the light source is nearby and often misleads the moths to spiral around in useless paths. In MFO, the decision variables of the optimization problem corresponding to the positions of the moth and a flame population are employed to correspond to the best positions of the moths. The moth positions are iteratively updated using a spiral function around the flame to ensure exploration and exploitation of the search space. Similar to s-TLBO, this algorithm also requires only two user–defined parameters in terms of the number of moths and the number of iterations. However, it requires only a single evaluation of the objective function per member of the population. Thus the total number of fitness evaluations is given by {(moth size $\times$ number of generations) + moth size. We have used the online available MFO code (available at https://goo.gl/usMXG4).

Flower Pollination Optimization Algorithm

Flower pollination algorithm is population–based algorithm [30] and is inspired by the flower pollination process associated with the transfer of pollen by biotic and abiotic pollinators. The solutions in the search space are termed as pollens. In every generation, each population member has to probabilistically either undergo global or local pollination. The global pollination is designed to model the transfer of pollen to the large distance by pollinators such as insects and utilizes the global best solution and Levy distribution. On the contrary, local pollination is designed to model the transfer in the neighborhood and relies on randomly selected solutions from the population. FP requires three user–defined parameters, *viz.*, population size (flowers), the termination criteria in terms of the maximum number of generations and switch probability, which is used to select the type of pollination for every member. The newly generated solution is accepted if it is better than the solution undergoing the pollination. In every generation, FP evaluates one solution per population member. The number of total functional evaluations in FP is given by {(population size $\times$ number of generations) + population size}. We have used the online available FP algorithm code (available at https://goo.gl/vqfftk).

Water Cycle Optimization Algorithm

Water cycle optimization algorithm is a population–based metaheuristic optimization algorithm [14] that is conceived from the water cycle process and the flow of rivers and streams into the sea. The population is divided into sea, river and streams based on the value of the fitness function. The solution representing the sea has the best fitness function value which is followed by rivers and streams. The streams are updated on the basis of a river whereas the rivers are updated based on the sea. In both the cases, a better-updated solution would be used to replace the corresponding river or sea. In order to avoid being trapped at the local optima, an evaporation condition is incorporated to model the merging of a river to the sea. These are replaced with new streams generated randomly to mimic the raining process. In addition to the population size and the number of iterations, WCA requires three user–defined parameters (i) the number of rivers, (ii) a tolerance value to indicate the merging of the river into the sea and (iii) a parameter to update the streams. The number of total functional evaluations in WCA is given by {(population size × number of iteration) + population size}. We have used the online available WCA optimization algorithm code (available at https://goo.gl/HCgVg7).

Adaptive Wind Driven Optimization Algorithm

Adaptive wind driven optimization algorithm [4] is the extension of the classical Wind driven optimization. It is based on the concept of atmospheric motion and reportedly captures the movement of wind from regions of high pressure to low pressure. The population members are considered as air parcels and are scaled in the domain of $[-1, 1]$. The pressure is analogous to the cost function and is utilized to rank the air parcels. A good solution is characterized by low pressure whereas a high pressure indicates an inferior solution. The velocity of the parcels used to determine new positions are modelled using the pressure gradient, friction, gravitational and the coriolis force. Thus it requires parameters such as friction coefficient, gravitational constant, the universal gas constant, temperature and the rotation of the Earth to be set by the user. AWDO employs the popular Covariance Matrix Adaptation Evaluation Strategy to adaptively tune these parameters and hence requires only two user–defined parameters, *viz.,* (i) population size and (ii) number of generations to be used for terminating the algorithm. The number of total functional evaluations in AWDO algorithm is given by {(population size × number of generations) + population size}. We have used the online available AWDO optimization algorithm code (available at http://www.thewdo.com/).

Results and Discussion

In this section, we evaluate the performance of the five computational intelligence algorithms to solve the combinatorial problem of single–level production planning problem of the Saudi Arabian petroleum industry. In this case study, 24 products can be produced by using 54 different processes. It should be noted that not all the products are produced by using every process and a product can be produced by more than one process. There are 18 processes available to produce 8 propylene derivative by–products, 16 processes available to produce 8 ethylene derivative by–products, 11 processes available to produce 4 synthesis gas derivative by–products and 9 processes available to produce 4 aromatic based derivative by-products. The production and investment costs for three different production capacities are also provided. Two different raw materials propylene (denoted as R1) and ethylene (denoted as R2), required for the production of various products, are available in limited quantities. Details about the production costs, raw material requirements and investment costs for each process to produce their respective product, along with selling price of the product is provided in Tables 13.1, 13.2, 13.3 and 13.4.

Two categories of problem are reported in the literature on the basis of the requirement of unique process constraint. Each category has four distinct cases. Case 1 – Case 4 form Category – I and require that an identical product should not be produced by more than a process. For example, there are three different processes (Lurgi process (S43 in Table 13.3), ICI process copper catalyst (S44 in Table 13.3) and ICI LCM process (S45 in Table 13.3) available to produce methanol (T20 in Table 13.3). For the four cases in Category – I, due to the requirement of the unique process, only one of these three process can be used to produce methanol. Category – II comprises of Case 5 – Case 8 that permit the production of an identical product from more than one process. The investment cost available for Case 1 and Case 2 is $1,000 million whereas $2,000 million is available for Case 3 and Case 4. The amount of raw material (R1 and R2) are available for Case 1 and Case 3 is 500,000 tons/year whereas 100,000 tons/year of raw materials are available for Case 2 and Case 4. The resource availability for Case 5 – Case 8 is identical to Case 1 – Case 4 respectively (as shown in Table 13.5). The current strategy leads to 54 decision variables corresponding to the 54 processes. The number of constraints in Category I and Category II are 81 and 57 respectively. The difference in the number of constraints is due to the presence of 24 unique process constraints (corresponding to the 24 products) in Category I. This is in addition to the bound constraints on the 54 decision variables.

All computations are performed using MATLAB R2015a on an Intel(R) Core(TM) i7–4790 CPU@3.60GHz processor with 16 GB memory running Windows 7. In view of the stochastic nature of the algorithms, all the five algorithms are run for 51 times with different random seeds of the random number generator. The seed of the random number generator (‘v5normal’) in MATLAB R2015a is varied from 1 to 51 in order to realize the independent runs thereby creating 2040 (= 5 algorithm $\times$ 51 runs $\times$ 8 cases) unique instances. This ensures that

Table 13.5 Resource availability and utilization

Case	Raw Material 1 (10^3 tons/year)		Raw Material 2 (10^3 tons/year)		Investment Budget (B) ($ 10^6)	
	Available	Utilized	Available	Utilized	Available	Utilized
Case1	500.00	435.44	500.00	496.53	1000.00	989.06
Case2	1000.00	571.02	1000.00	829.73	1000.00	999.95
Case3	500.00	496.13	500.00	499.30	2000.00	1903.81
Case4	1000.00	979.52	1000.00	1000.00	2000.00	1844.79
Case5	500.00	499.20	500.00	389.02	1000.00	998.61
Case6	1000.00	577.15	1000.00	998.06	1000.00	1000.00
Case7	500.00	490.58	500.00	493.57	2000.00	1986.70
Case8	1000.00	999.10	1000.00	1000.00	2000.00	1973.16

the results reported in this work are reproducible which we believe is necessary for scientific progress. The population size for all the five algorithms is set to 200 and the stopping criteria for each of the five algorithms (in every single run) are set to 100,200 functional evaluations. This translates into 500 iterations for all the algorithms except for s-TLBO. In case of s-TLBO, as it requires two functional evaluations per member in an iteration, 250 iterations are required to account for 100,200 functional evaluations. The parcel velocity in AWDO is set to 0.3 whereas the switching probability is set to 0.8 for FP. In WCA, the number of rivers is set to 3, the distance between stream and river is set to 10^{-16} and the constant C of the algorithm is set to 2. These parameters have been set based on the recommendations in their original works. The value of γ and λ in Eqs. (13.12) and (13.13) are 10^5 and 10^{15} respectively.

The best fitness function value obtained by the five algorithms at the end of 100,200 function evaluations for each run of the eight cases are shown in Fig. 13.3. If an algorithm is unable to obtain a feasible solution, the value of the fitness function is very high due to the penalties. Thus for better representation, only the values obtained in the feasible runs are shown in the figure. It can be observed that s-TLBO and MFO are able to determine feasible solutions in all the instances. However, FP was not able to determine even a single feasible solution for any of the cases. Of the 408 (8 Cases $\times$ 51 runs) instances, AWDO was unable to determine a feasible solution in 332 instances. In particular, AWDO fails to determine a feasible solution for 41 instances in Case 1, 42 instances in Case 2, 43 instances in Case 3, 42 instances in Case 4, 45 instances in Case 5, 42 instances in Case 6, 43 instances in Case 7 and for 34 instances in Case 8. On the other hand, WCA was not even able to determine even a single feasible solution in 132 instances. In particular, it fails in 23 instances for Case 1, 17 instances in Case 2, 21 instances in Case 3, 22 instances in Case 4, 17 instances in Case 5, 16 instances in Case 6, 10 instances in Case 7 and 6 instances in Case 8.

The statistical results of all the 2040 instances (8 Cases $\times$ 51 runs $\times$ 5 algorithms) are consolidated in Table 13.6, which shows the best, the worst, the mean, the median and the standard deviation of the best fitness values obtained in all the runs. It can be seen from the "best" value of Table 13.6 that AWDO, WCA, MFO and

Table 13.6 Statistical performance of 51 runs

Case		AWDO	WCA	FP	MFO	TLBO
Case 1	Best	−597.01	−545.27	5.7E+21	−607.60	**−636.43**
	Worst	1.2E+22	1.1E+21	1.2E+22	−268.80	−309.75
	Mean	2.7E+21	6.0E+19	8.6E+21	−488.69	−499.54
	Median	1.6E+21	−195.35	8.4E+21	−512.35	−506.94
	St.dev	3.2E+21	1.8E+20	1.4E+21	81.87	71.24
Case 2	Best	−607.35	−635.20	4.8E+21	−709.10	**−745.14**
	Worst	8.7E+21	1.8E+21	1.2E+22	−304.00	−389.75
	Mean	1.7E+21	8.2E+19	8.1E+21	−570.06	−571.59
	Median	6.2E+20	−225.90	8.0E+21	−596.44	−572.72
	St.dev	2.3E+21	2.8E+20	1.6E+21	94.71	71.93
Case 3	Best	−494.35	−703.64	3.1E+21	−813.14	**−834.73**
	Worst	7.4E+21	1.1E+21	7.4E+21	−513.29	−554.25
	Mean	1.0E+21	5.7E+19	5.2E+21	−684.67	−695.18
	Median	6.3E+20	−360.33	5.1E+21	−683.19	−699.51
	St.dev	1.3E+21	2.0E+20	1.0E+21	70.90	60.86
Case 4	Best	−796.87	−912.28	1.6E+21	**−1047.69**	−1032.20
	Worst	6.2E+21	5.4E+20	6.7E+21	−625.32	−740.91
	Mean	1.0E+21	1.2E+19	4.4E+21	−849.62	−888.35
	Median	4.9E+20	−347.76	4.3E+21	−860.15	−881.85
	St.dev	1.3E+21	7.6E+19	1.1E+21	109.46	71.81
Case 5	Best	−409.80	−549.66	4.3E+21	−688.49	**−712.72**
	Worst	4.4E+21	1.6E+21	8.3E+21	−366.50	−417.08
	Mean	1.3E+21	7.9E+19	6.0E+21	−513.73	−545.46
	Median	1.3E+21	−223.70	6.0E+21	−512.60	−545.95
	St.dev	1.1E+21	2.4E+20	7.8E+20	81.90	64.14
Case 6	Best	−532.15	−662.35	3.7E+21	**−818.66**	−784.65
	Worst	3.0E+21	1.6E+21	7.3E+21	−322.95	−388.17
	Mean	1.0E+21	5.6E+19	5.6E+21	−626.29	−601.71
	Median	7.7E+20	−309.70	5.6E+21	−618.89	−593.01
	St.dev	9.6E+20	2.4E+20	8.1E+20	78.78	96.61
Case 7	Best	−701.00	−867.11	2.4E+21	−979.56	**−986.29**
	Worst	1.8E+21	4.7E+20	4.3E+21	−501.71	−612.05
	Mean	3.3E+20	2.7E+19	3.3E+21	−773.40	−790.84
	Median	1.3E+20	−480.29	3.2E+21	−795.30	−798.09
	St.dev	4.0E+20	9.8E+19	4.8E+20	118.52	97.35
Case 8	Best	−1079.95	−1022.26	1.7E+21	**−1249.41**	−1233.26
	Worst	8.6E+20	1.7E+20	3.7E+21	−624.50	−794.05
	Mean	1.8E+20	8.3E+18	2.8E+21	−1030.88	−1025.76
	Median	7.9E+19	−636.73	2.8E+21	−1038.50	−1026.05
	St.dev	2.3E+20	3.3E+19	4.9E+20	126.61	106.41

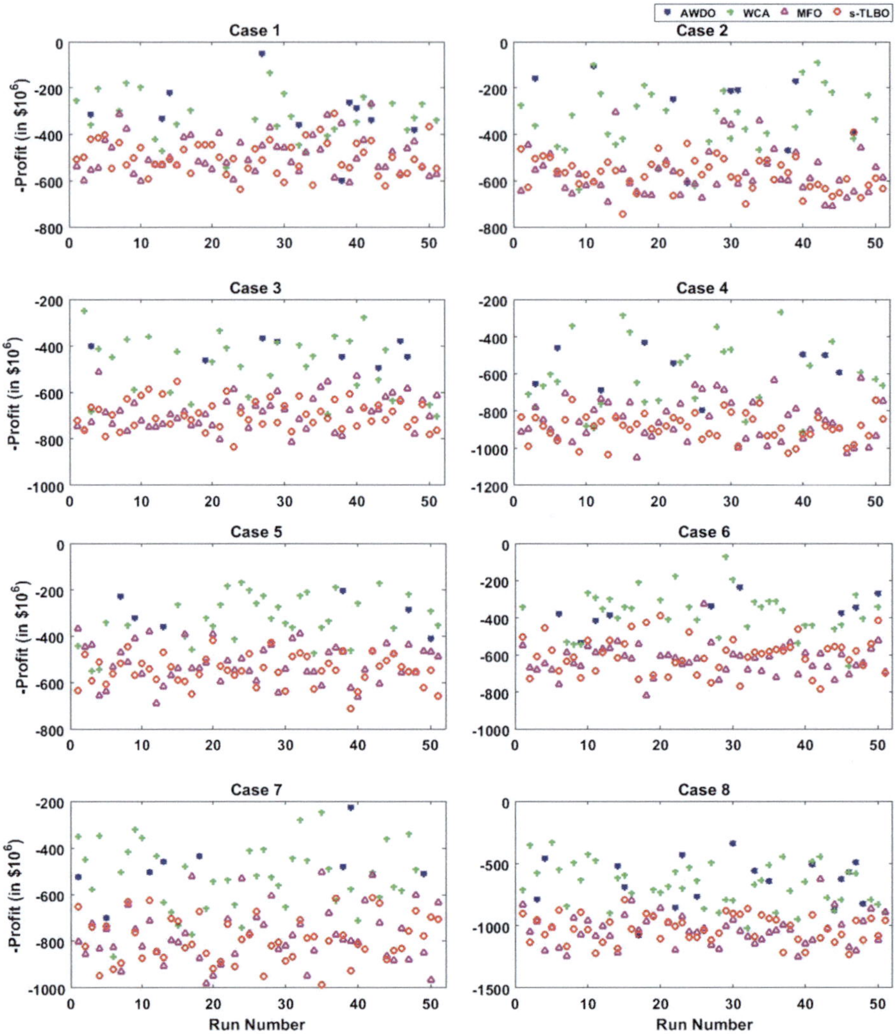

Fig. 13.3 Best objective function obtained in all the runs of the five algorithms

s-TLBO are able to identify the feasible solutions for all the eight cases whereas FP is not able to determine even a single feasible solution for any of the eight cases. It should be noted that this table provides a consolidated view of Fig. 13.3. Among the 2040 instances, it can be observed that for all the eight cases, the best value discovered by s-TLBO is better in five cases (Case 1, Case 2, Case 3, Case 5, and Case 7) than all other algorithms. While MFO determined the best value for three cases, (Case 4, Case 6 and Case 8) than other algorithms (AWDO, WCA, FP and s-TLBO). In addition to the best function value obtained, standard deviation obtained by s-TLBO in seven of the eight cases are lower than that of MFO. Interestingly, in all the eight cases, the worst solution reported by s-TLBO is better than the worst solution reported by MFO. The median determined using MFO is better than s-TLBO in four cases.

The convergence curve depicts the performance of the algorithm with respect to each functional evaluation. The best value determined by the algorithm (till that functional evaluation) is plotted against the number of times the objective function has been evaluated (as a percentage of the maximum functional evaluation). The convergence curve is plotted for the run that was able to determine the best objective function value. Due to the high penalty values, the curve is plotted only after the determination of the first feasible solution by an algorithm. From Fig. 13.4, it can

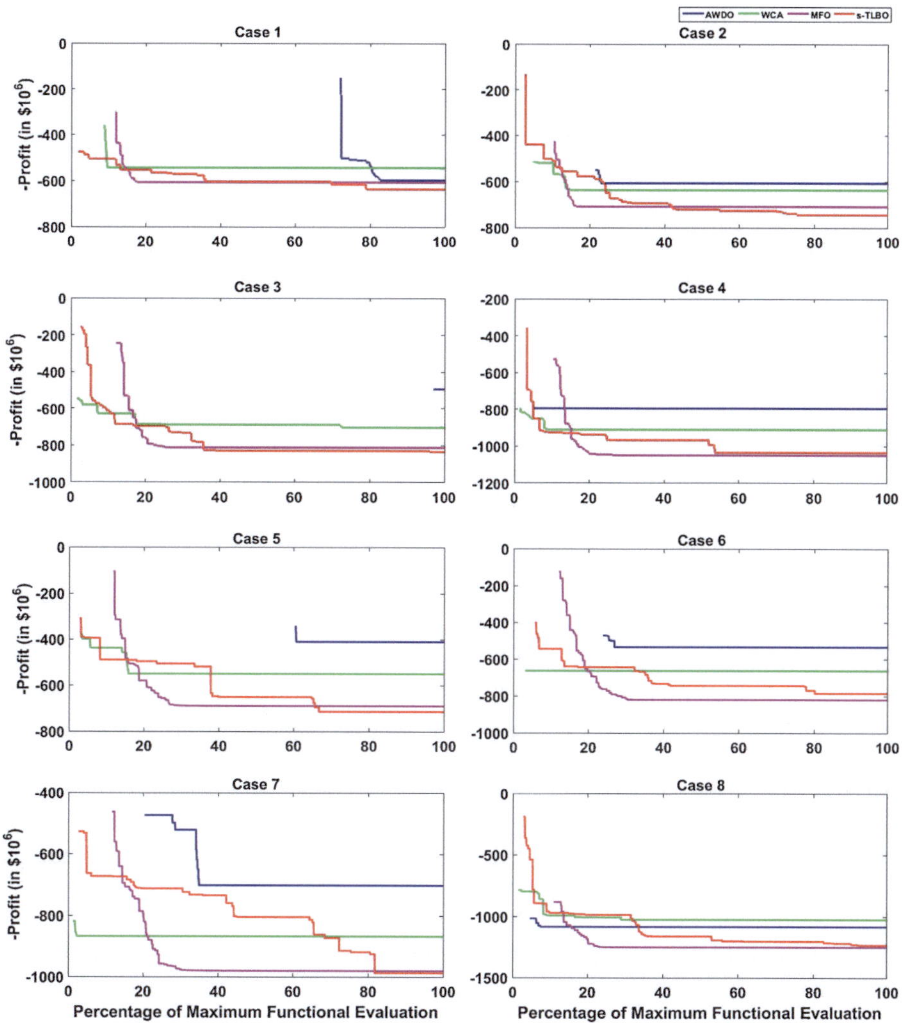

Fig. 13.4 Convergence curves of the best run

be observed that initially there is no feasible solution obtained by any of the five algorithms. The algorithms are able to determine a feasible solution after certain function evaluations and subsequently, the value of the objective function starts improving. Both s-TLBO and WCA are able to quickly determine a feasible solution in comparison to the other algorithms. In some of the cases, WCA determines a better feasible solution than s-TLBO. However WCA is not able to improve on this whereas s-TLBO determines a consistently better solution than WCA in all the eight cases. On the other hand, MFO requires a larger number of function evaluations to determine the first feasible solution. However, it quickly determines a better solution than s-TLBO but stagnates whereas s-TLBO is able to slowly determine a solution better than MFO (line corresponding to s-TLBO crossing the line corresponding to MFO) in many of the cases. It can be observed that in all the cases, the four algorithms (except FP) have converged to a final value.

Figure 13.5 shows the percentage of functional evaluations utilized for the algorithm to determine the first feasible solution by each algorithm in eight cases for all the feasible runs. From Fig. 13.5, it can be observed that s-TLBO is able to determine a feasible solution with very few function evaluations (in all the 51 runs for all eight cases) than the other four algorithms. AWDO was unable to determine feasible solutions in many instances and in other instances, it required a significantly larger number of function evaluations to determine the feasible solutions. MFO required a larger number of function evaluations in comparison to s-TLBO and WCA to determine the first feasible solution but lower than AWDO. As discussed earlier, FP was unable to determine even a single feasible solution.

The decision variables, i.e., the production plan corresponding to the best objective function value for each of the eight cases are given in Table 13.7. In Case 1 – Case 4, it can be observed that none of the products is produced by more than a single process thereby demonstrating the satisfaction of the unique process constraint. In contrast, multiple processes are used to produce profitable products in Case 5 – Case 8. For example, T1 and T21 are produced by multiple processes in Case 5. The product T1 is produced in six of the eight cases (except Case 3 and Case 7) whereas product T16 and T21 are produced in all the eight cases. In view of the high availability of resources and the unique process constraint, Case 4 has the most diverse product portfolio. The resource utilization is given under the column "Utilized" in Table 13.5 and it can be observed that the production plans utilize most of the available resources. Figure 13.6 depicts the profit distribution among the various processes for all the eight cases. For the cases in Category – I, the largest contribution to profit is from process P47 whereas in Category II, product T21 contributes the maximum to the profit. Thus these processes can be considered as critical and need to be intensively maintained. For the sake of brevity, details are not provided but several insights can be obtained from post–optimality analysis of the production plan. From the results and discussions, it can be observed that s-TLBO and MFO are better than the rest of the three algorithms for this problem.

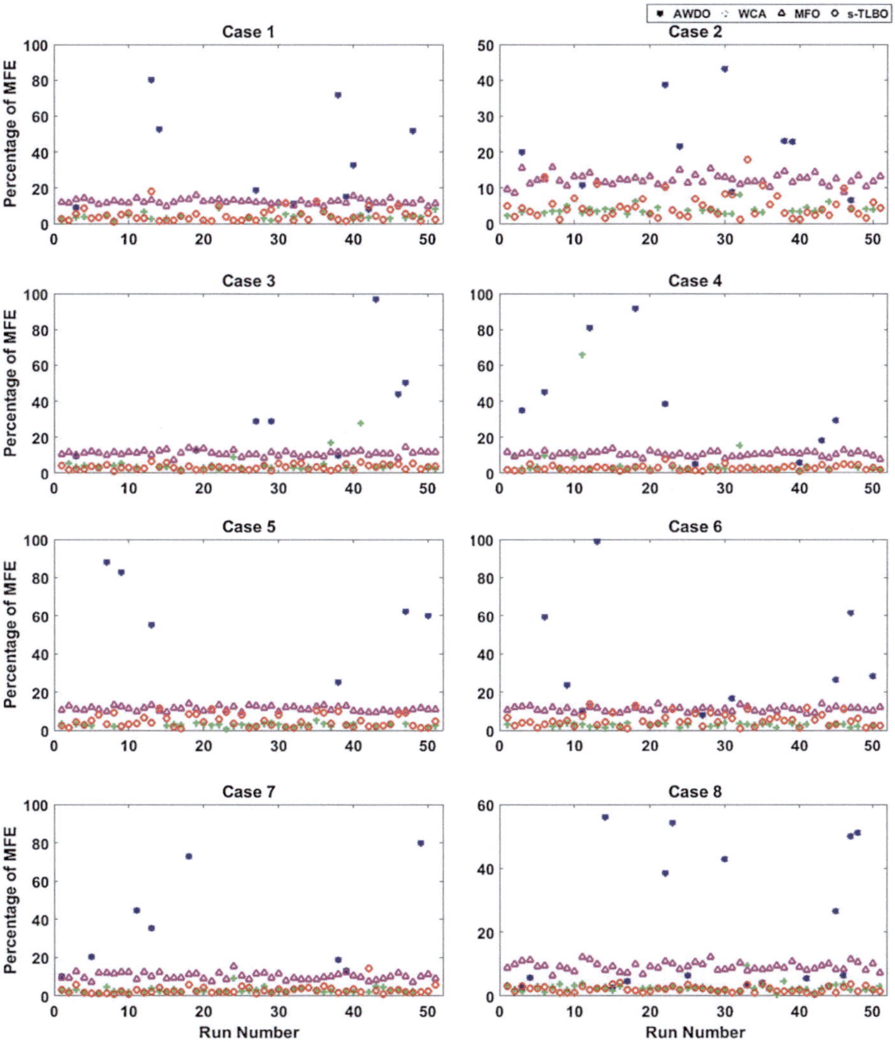

Fig. 13.5 Onset of feasibility in all the runs of the five algorithms

Table 13.7 Optimal production plan

Unique process constraint				Without unique process constraint			
Case	Product number	Process number	Net amount produced (10^3 ton/year)	Case	Product number	Process number	Net amount produced (10^3 ton/year)
Case 1	T1	P1	267.92	Case 5	T1	P1	270.00
	T2	P5	190.00			P3	256.31
	T10	P22	100.00		T17	P36	540.00
	T15	P31	317.96		T21	P47	680.00
	T17	P36	537.55			P48	680.00
	T19	P42	50.00				
	T21	P47	680.00				
Case 2	T1	P3	310.00	Case 6	T1	P2	300.00
	T2	P4	290.00			P3	310.00
	T14	P28	270.00		T15	P31	400.00
	T15	P31	400.00			P32	300.00
	T17	P36	540.00		T21	P48	680.00
	T21	P47	611.15			P49	503.43
Case 3	T3	P7	75.58	Case 7	T2	P4	218.43
	T5	P9	160.00			P5	190.00
	T6	P15	360.00		T5	P9	160.00
	T9	P20	300.00		T10	P21	100.00
	T10	P22	100.00			P22	42.54
	T15	P31	320.89		T14	P29	101.83
	T17	P36	540.00		T17	P35	540.00
	T19	P42	50.00			P36	540.00
	T21	P47	680.00		T19	P41	50.00
					T20	P45	519.86
					T21	P46	679.26
						P48	680.00
Case 4	T1	P3	310.00	Case 8	T1	P2	300.00
	T2	P4	290.00			P3	310.00
	T3	P7	160.00		T2	P5	190.00
	T5	P9	160.00		T5	P9	160.00
	T7	P17	200.00		T7	P17	200.00
	T10	P22	100.00		T9	P20	300.00
	T14	P28	270.00		T14	P28	228.23
	T15	P31	400.00		T15	P31	400.00
	T16	P34	141.93		T17	P36	540.00
	T17	P36	540.00		T21	P48	680.00
	T19	P42	49.99			P49	680.00
	T21	P47	680.00				
	T22	P51	50.00				

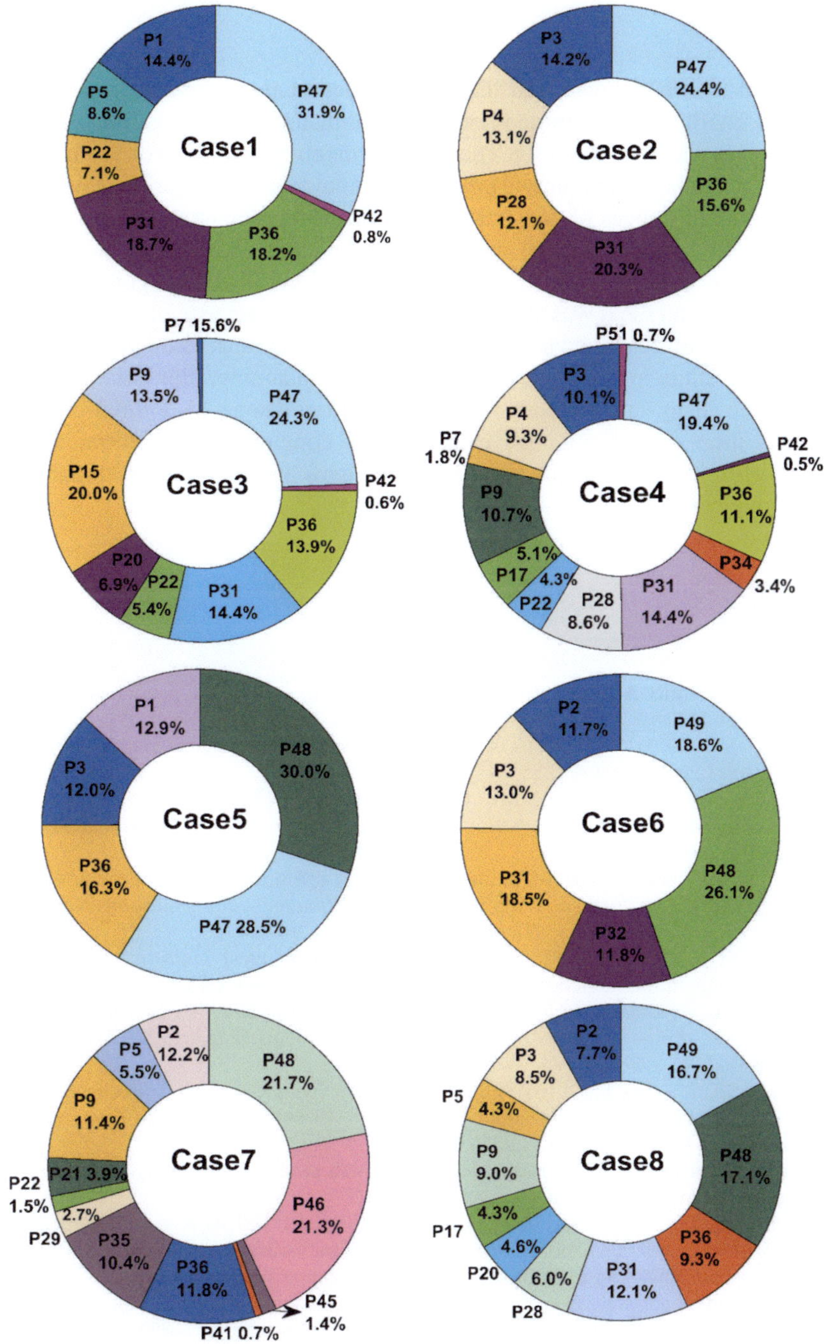

Fig. 13.6 Individual contribution of processes to profit

Conclusions

In this work, we have evaluated the performance of five recently proposed computational intelligence techniques on the combinatorial optimization problem of single–level production planning. The problem involves complex constraints arising from the domain holes and unique process requirements. The algorithms were tested for 408 unique instances of constrained optimization problem arising from 51 runs and 8 cases. The results of 2040 instances showed that s-TLBO was able to provide superior results as compared to other computational intelligence algorithms such as AWDO, WCA, and FP. It was also observed that the performance of MFO is competitive but had relatively high variance across the multiple runs. Though demonstrated for the case study of petrochemical production planning, this strategy can be applicable to many other industries that require the design of optimal production portfolio. Future work can include evaluating the computational performance in the presence of efficient constraint handling techniques and problems with conflicting multiple objectives.

References

1. Alfares H, Al Amer A (2002) An optimization model for guiding the petrochemical industry development in Saudi Arabia. Eng Optim 34(6):671–687
2. Aminzadeh F (2005) Applications of AI and soft computing for challenging problems in the oil industry. J Pet Sci Eng 47(2):5–14
3. Assareh E, Behrang MA, Assari MR, Ghanbarzadeh A (2010) Application of PSO (particle swarm optimization) and GA (genetic algorithm) techniques on demand estimation of oil in Iran. Energy 35(12):5223–5229
4. Bayraktar Z, Komurcu M (2016) Adaptive wind driven optimization. In: Proceedings of the 9th EAI international conference on bio-inspired information and communications technologies, New York, USA. https://doi.org/10.4108/eai.3-12-2015.2262424
5. Toledo CEE, Aranda CG, Mareschal B (2010) Petrochemical industry: assessment and planning using multicriteria decision aid methods. Technol Invest 1(2):118–134
6. Cecchini RL, Ponzoni I, Carballido JA (2012) Multi-objective evolutionary approaches for intelligent design of sensor networks in the petrochemical industry. Expert Syst Appl 39(3):2643–2649
7. Chauhan SS, Kotecha P (2016) Single level production planning in petrochemical industries using Moth-flame optimization. IEEE region 10 conference (TENCON), Singapore. https://doi.org/10.1109/TENCON.2016.7848003
8. Chauhan SS, Kotecha P (2018) An efficient multi-unit production planning strategy based on continuous variables. Appl Soft Comput 68:458–477
9. Chauhan SS, Sivadurgaprasad C, Kadambur R, Kotecha P (2018) A novel strategy for the combinatorial production planning problem using integer variables and performance evaluation of recent optimization algorithm, Swarm and Evolutionary Computation, 43, 225–243
10. Chinta S, Kommadath R, Kotecha P (2016) A note on multi-objective improved teaching–learning–based optimization algorithm (MO–ITLBO). Inf Sci 373:337–350
11. Črepinšek M, Liu SH, Mernik L (2012) A note on teaching–learning–based optimization algorithm. Inf Sci 212:79–93

12. Duan QQ, Yang GK, Pan CC (2014) A novel algorithm combining finite state method and genetic algorithm for solving crude oil scheduling problem. Sci World J 2014:1–11
13. Editorial (2015) Synthesizing tomorrow. Nat Plants 1:15047
14. Eskandar H, Sadollah A, Bahreininejad A, Hamdi M (2012) Water cycle algorithm – a novel metaheuristic optimization method for solving constrained engineering optimization problems. Comput Struct 110–111:151–166
15. Hasda RK, Bhattacharjya RK, Bennis F (2016) Modified genetic algorithms for solving facility layout problems. Int J Interact Des Manuf 11(3):713–725
16. Joly M, Rocha R, Sousa LCF, Takahashi MT, Mendonça PN, Moraes LAM, Quelhas AD (2015) The strategic importance of teaching operations research for achieving high performance in the petroleum refining business. Educ Chem Eng 10:1–19
17. Kadambur R, Kotecha P (2015) Multi-level production planning in a petrochemical industry using elitist teaching–learning-based-optimization. Expert Syst Appl 42(1):628–641
18. Kommadath R, Kotecha P (2017) Teaching learning based optimization with focused learning and its performance on CEC2017 functions, 2017 IEEE congress on evolutionary computation (CEC), San Sebastian, Spain. https://doi.org/10.1109/CEC.2017.7969595
19. Kadambur R, Kotecha P (2016) Optimal production planning in a petrochemical industry using multiple levels. Comput Ind Eng 100:133–143
20. Mernik M, Liu S-H, Karaboga D, Črepinšek M (2015) On clarifying misconceptions when comparing variants of the artificial bee Colony algorithm by offering a new implementation. Inf Sci 291:115–127
21. Mirjalili S (2015) Moth–flame optimization algorithm: a novel nature-inspired heuristic paradigm. Knowl-Based Syst 89:228–249
22. Moro LFL (2009) Optimization in the petroleum refining industry – I the virtual refinery. Comput Aided Chem Eng 27:41–46
23. Nabavi R, Rangaiah GP, Niaei A, Salari D (2011) Design optimization of an LPG thermal cracker for multiple objectives. Int J Chem React Eng 9(A80):1–4634
24. Ozcelik Y, Hepbasli A (2006) Estimating petroleum exergy production and consumption using a simulated annealing approach. Energy Sources 1(3):255–265
25. Ramteke M, Srinivasan R (2012) Large-scale refinery crude oil scheduling by integrating graph representation and genetic algorithm. Ind Eng Chem Res 51(14):5256–5272
26. Ren T, Daniëls B, Patel MK, Blok K (2009) Petrochemicals from oil, natural gas, coal and biomass: production costs in 2030–2050. Resour Conserv Recycl 53(12):653–663
27. Sheremetov L, Bañares-Alcántara MAR, Aminzadeh F, Mansoori GA (2005) Intelligent computing in petroleum engineering. J Pet Sci Eng 47(1–2):1–3
28. Toksarı MD (2007) Ant colony optimization approach to estimate energy demand of Turkey. Energy Policy 35(8):3984–3990
29. Velez-Langs, Oswaldo, (2005). Genetic algorithms in oil industry: An overview. Journal of Petroleum Science and Engineering 47:15–22
30. Yang XS (2012) Flower pollination algorithm for global optimization. In: Proceedings of the 11th international conference on unconventional computation and natural computation, Berlin, Heidelberg. https://doi.org/10.1007/978-3-642-32894-7_27
31. Zhang L, Jiang Y, Gao X, Huang D, Wang L (2016) Efficient two–level hybrid algorithm for the refinery production scheduling problem involving operational transitions. Ind Eng Chem Res 55(28):7768–7781

Chapter 14
A Multi-Agent Platform to Support Knowledge Based Modelling in Engineering Design

Ricardo Mejía-Gutiérrez and Xavier Fischer

Abstract Nowadays engineering design process requires the involvement of multiple partners from multiple disciplines throughout the Product Life Cycle (PLC). Consequently, the construction of numerical models became a difficult task due to the distribution of experts. This article proposes an agent based approach to support a coherent know-how elicitation, to enrich design problem analysis, based on the re-use of experiences and their storage in a standardized knowledge base. A set of Tutor-Agents (TAs) aid experts in the knowledge modelling process focusing on Variables, Domains and Constraints as a key component of engineering knowledge. A shared and coherent knowledge base is the main purpose of the proposed Multi-Agent System (MAS). The interaction among agents enables to highlight potential incoherencies during the modelling process to avoid inconsistent information. The Multi-Agent approach is implemented in a software prototype and a knowledge base can then be constructed, providing standardized Product Life Cycle (PLC) constraints (based on the product related knowledge) for creating models to be analyzed by traditional inference engines such as Optimization solvers, Constraint Satisfaction programming, etc.

Keywords Distributed knowledge modelling · Multi-agent system · Experts' knowledge reuse · Tutor agent · Engineering design

---

R. Mejía-Gutiérrez (✉)
Design Engineering Research Group (GRID), Universidad EAFIT, Medellín, Colombia
e-mail: rmejiag@eafit.edu.co

X. Fischer
ESTIA Engineering School, Technopôle Izarbel, Bidart, France

IMC-I2M, UMR CNRS 5295, Université de Bordeaux, Bordeaux, France
e-mail: x.fischer@estia.fr

F. Bennis, R. K. Bhattacharjya (eds.), *Nature-Inspired Methods for Metaheuristics Optimization*, Modeling and Optimization in Science and Technologies 16,
https://doi.org/10.1007/978-3-030-26458-1_14

Introduction

It can be assumed that particular classes of design problems can be formulated as optimisation, constraint satisfaction, or algebraic equation-solving problems [6]. Distributed knowledge may be formalised and used in early stages of product design as constraints in numerical models. The creation of such a model is not a simple task, especially in networked enterprises where multiple disciplines are involved in a common distributed project. Nowadays product design teams, working in a distributed environment, require new working methods and techniques in order to improve their design activities. The process of analysing problems in product development is a key task that requires the integration of experts' know-How and experience in early stages of product development. Engineering knowledge can be then considered as "a set of relations among a set of variables and their domains". That is why Constraint Satisfaction Problems fit to the Engineering Design needs as design problems can be modelled with this approach and then they can be solved by traditional inference engines such as Optimization or Constraint Satisfaction solvers, etc. There is evidence in the literature that CSP proofed to be a suitable numerical tool to help engineers in solving design problems [14, 28, 35]. Experiences are formalized trough Variables and the set of values that they can adopt, as well as the set of constraints (relations among variables) issued from experts' experience. The combination of Variables, Domains and Constraints enable the numerical modelling of design problems.

The ideas proposed in this article are located into the preliminary design phases. At this stage, several concepts are found and a set of variables start to emerge. The design process is a continuous decision making and that is why any help in Decision Support becomes critical. However, in early stages of the engineering process, design is still in a conceptual stage and values that those Variables can take are still fuzzy [11] converting the decision making a critical aspect. The modelling approach enable to consider several possibilities by using combinatorial analysis. The results comes from an inference engine that may produce an output corresponding to a set of possible solutions that meets a set of constraints. Any design can be defined by a vector in multidimensional space where each design variable represents a different dimension. Usually, if an optimisation is needed, an objective function expresses cost, weight, range, efficiency, return-on-investment, or any combination of parameters. It is subject to functional constraints in accordance to given relationships between variables and parameters and to variables' domains. These functional constraints will be considered from now on in this article as domain specific knowledge. These constraints define the permissible space where the optimum value has to be found, e.g. limits due to minimum sheet thickness, maximum stress, limit speed, etc. An agreement on those variables that compose functional constraints is a critical step in distributed environments. This issue will be addressed in this article in order to achieve an optimal distributed design process. Due to the importance of including experts' know-How in early stages of product development, the use of the company's knowledge is a new issue that arises.

Background

There are previous works on implementing constraint-based methods to model Engineering Design problems [1, 30, 32, 36] or to aid in the formalization of Decision Support Systems [21]. There are also generic implementations not necessarily for design problems, but including agents [33], and some other researchers work on novel approaches that integrates the collaborative design aspect, as it is the particular case of CoCSP (Cooperative Constraint Satisfaction Problem) [38]. However, some of them focused on the information gathering, others in the computational analysis and others are focused on the solution. The proposed work presented in this article, intends to go further by support the distributed approach with agents and characterizing the different environmental entities of the Multi-Agent System. This enable agents to have more information available to perform analysis and to communicate (ask or request other agent's information) in order to have a standardized knowledge base, that can be employed to re-use/restore experiences to build a consolidated model. Some previous experiences have shown the importance of Knowledge Based Systems (KBS), particularly from the Knowledge Management domain in organizational applications [2], knowledge intensive collaborative environments [8] and applications to the decision-making process in design projects [12], among others. An interesting survey was published around methods to automatically manage Knowledge Bases [23], being the key to support that standardized knowledge bases may help in further computational processing as optimization methods or combinatorial exploration, in order to be applied in engineering design.

The aim of this research is to help the design team through a tutored process to cooperatively create a model for analysis including Product Life Cycle (PLC) information. To achieve this, the use of Multi-agents technology has been considered, as those systems have proven to properly support a very large domain of applications, rich in knowledge and strong in the collaborative aspect [19]. Multi-Agent System (MAS) supports applications in areas such as collaborative building design [31], computer integrated manufacturing (CIM) and networked organizations [5], as well as virtual enterprises [22]. Multi-Agents systems (MAS) represents one of the most promising technological paradigms for the development of open, distributed, cooperative, and intelligent software systems [16]. Several applications in product development process have been tested in research and industrial applications. Some of these agent based systems have proved a great potential in creation and automation of collaborative and distributed environments for product design [15, 18, 26, 29, 34]; however, they are specific applications on a given domain with specific knowledge. There are also other Multi-Agent System (MAS) with application in Optimization [37], but they are often applied in the model solving, rather than helping experts to define constraints, based on their domain–specific knowledge. That is why a Tutor Agent (TA) is proposed to identify actors' know-how, focusing on the method to elicitate knowledge from different technical domains in order to be able to re-use that experts' knowledge to model a general formulation of an engineering design problem.

A well accepted aspect about agents is the notion of autonomous and interactive entities that exist as part of an environment shared with other agents [9]. However, agents should play an active role to facilitate the process of creating structured models by distributed teams. Therefore each agent is considered to be a *tutor*, able to guide a user through a reflection process according to the current design problem. The tutoring method and the concept of "actors" from the literature [10] are interesting to the guiding purposes that Tutor Agent (TA) s intend to offer. The use of standard Multi-Agent frameworks has been widely used for MAS development [20]. Additionally, some ontology based approaches for optimization have been presented in the literature [27] for automating the creation of mathematical models based on operations research principles. The prototype system of the Multi–Agents approach addressed in this research is implemented using Java Agent DEvelopment framework (JADE™) [4]. It provides FIPA compliant communication protocols as well as Ontology based messaging [3].

The contribution of this research is the combination of multi-agents technology to support engineering problems description by knowledge elicitation in early stages of product design. A methodology is proposed and implemented in a software prototype to support distributed design teams to (i) Identify relevant knowledge throughout the Product Life Cycle (PLC); (ii) Formalize and standardize variables and knowledge, and (iii) Construct numerical models for analysis.

Modeling the Knowledge

Technical Knowledge in Engineering Design may be considered as *"a relation among a set of variables and its domains"*. Therefore, a structured process of elicitation should be set up in terms of $P(\mathcal{V}, \mathcal{D}, \mathcal{C})$[1] where $\mathcal{V}$ is the set of n design variables $\{V_1, V_2, \ldots, V_n\}$, $\mathcal{D}$ is the set of n domains[2] of each variable $\{D_1, D_2, \ldots, D_n\}$ and $\mathcal{C}$ is the set of p relations among variables, called constraints $\{C_1, C_2, \ldots, C_p\}$.

This knowledge, issued from expert's experience, is desired to be elicitated properly in order to be reused to enrich future product developments. The idea is to help experts (involved as partners) in distributed engineering teams to work together in the definition of coherent models for analysis, by using their experiences to avoid problems related to potential design problems. By including this experiences since the beginning, downstream problems are minimized. The issue starts by coordinating efforts in distributed and heterogeneous knowledge modelling.

[1]From the triple $\langle X, D, C \rangle$ definition of Constraint Satisfaction Problem (CSP) theory that defines CSP as mathematical problems composed by as a set of objects whose state must satisfy a number of constraints

[2]Set of possible values that a variable V_i can take

As the design problems analysis will require a structured modelling process, a way to describe a variable in relation to its properties is proposed. Under an object oriented approach, variables are considered as objects. As such, a set of properties may be defined (e.g. Name, Representation, type, units, etc.), being the key to identify potential redundancies after their definition by experts. This section will describe the way a variable should me described. Accordingly, the properties are stored on a Knowledge Base (KB) for further instantiation in a specific problem modelling, aided by the modelling interface of the Tutor Agent (TA) prototype.

Variable Model

After previous identification of relevant knowledge, the first task for experts is to define the set of variables from the design problem V in order to be able to define relations among them (knowledge as constraints). Therefore, each variable should be characterized in order to facilitate the Tutor Agent (TA) task of analysis. Variables are composed of a set of properties that should make each variable "unique". However as humans think in a different way, multiple variables may be defined by different experts to measure the same parameter. The variables' properties enable a structured definition of knowledge by describing a variable in terms of its characteristics.

The reference model for a variable is composed by a tuple of sets where properties are instantiated from a set of values in order to be able to compare and analyze variables in succeeding analysis [24]. Properties are stored with the variable in the KB for further instantiation in a specific problem modelling. An instance of a variable from the set V, can be expressed in terms of its properties in the form of:

$$\begin{aligned} &Prop(v): \\ &V \rightarrow \mathcal{N} \times \mathcal{R} \times \mathcal{P} \times \Re^7 \times \mathcal{K} \times \mathcal{T} \times \mathcal{L} \times \mathcal{X} \times \wp(\mathcal{X}) \times \mathcal{S} \\ &V \mapsto Prop(v) \equiv (n_v, r_v, p_v, u_v, k_v, Tc_v, Lc_v, r_v, Us_v, Sc_v) \end{aligned} \tag{14.1}$$

From those properties, n_v and r_v correspond, respectively, to the "Name" and "Representation" of the variable. These two variables are assigned with a `string` and are freely filled by the user (according to suggestions from Tutor Agent (TA)). These are usually defined in regular modelling, being r_v the definition of the variable into the numerical model. The other properties are proposed to be included as Meta-Data as well, being important information for the standardization process performed by Tutor Agents (TAs). Among those properties the Measured parameter (p_v) belongs to a set of common parameters extracted from a review of different domains from physics and technical sciences and compiled in the $\mathcal{P} = \{speed, distance, force, \ldots, work\}$. The Units' dimensional exponents (u_v) is a seven tuple specifying the values of the dimensional exponents $\dim Q = u_v = \langle \alpha, \beta, \gamma, \delta, \varepsilon, \zeta, \eta \rangle$ from the unit of a quantity Q expressed in terms of the "The International System of Units (SI)" as $Q = l^{\alpha} m^{\beta} t^{\gamma} I^{\delta} T^{\varepsilon} n^{\zeta} I_l^{\eta}$ corresponding to the base quantities: length (l), mass (m), time (t), electric current (I), thermodynamic temperature (T), amount of substance (n), and luminous intensity (I_l). The Variable type (k_v) can be Critical (V_{Cr}), Design variable (V_D) or intermediary variable (V_I). The Technical classification (Tc_v) comes from a list of technical domains $\mathcal{T} = \{electronic, mechanical, fluids, \ldots, manufacture\}$. The stage from the Product Life Cycle (PLC) (Lc_v) belongs to the set $\mathcal{L} = \{design, manufacturing, dispatching, \ldots, recycling\}$ that contains the different Life Cycle stages. The variable's Responsible (r_v) is the user that creates a variable for the first time and will be responsible for further negotiations over his variable, if some other partner will use it. $r_v \in \mathcal{X}$, where $\mathcal{X} = \{partner_1, partner_2, \ldots, partner_i\}$ is the set of experts (also called "partners") involved in the modelling process. The variable's Users (Us_v) represent all users that exploit/share/use the same variable, $Us_v \in \wp(\mathcal{X})$. Finally, the source of the information Sc_v belongs to the set $\mathcal{S} = \{Table, Interview, \ldots, Empirical\}$.

Modelling Process

Traditional knowledge elicitation processes provide a link between experts and the storage of his know-how (ready-to-use experiences). In order to achieve this, a methodology is proposed (See Fig. 14.1) to support the distributed modeling process for describing problems in engineering design, by allowing knowledge creation or re-use. The main stages of the methodology are:

Knowledge determination This stage intends to locate throughout the Product Life Cycle (PLC), all the relevant information needed to create a coherent model for analyzing a specific design problem. The aim is to guide experts trough a structured process of reflection to inquire into their experiences or available information, in order to extract relevant knowledge from the corresponding discipline aided by a tutoring process [24]. Four basic ways to determine variables are: (i) Optimization objectives definition ($S_m = 1$), (ii) Morphologic analysis of physical interaction ($S_m = 2$), (iii) Physical behavior analysis ($S_m = 3$) and (iv) Domain knowledge analysis ($S_m = 4$) which are basically all relevant expert's experience relevant to a specific design problem. Agents provide a dynamic and a static help, where partners can follow a questionnaire in order to be guided to a set of defined knowledge. Additionally a set of information can be presented to users in order to help them in a new constraints definition based on traditional physics or domain specific knowledge (e.g. manufacturing, costing, etc).

Knowledge formalization At this stage, experts have to formalize their corresponding variables according to their discipline. Under an object oriented approach, variables are considered as objects with a set of properties: Name, Representation, type, units, etc. Experts can explore the knowledge base to search any equation

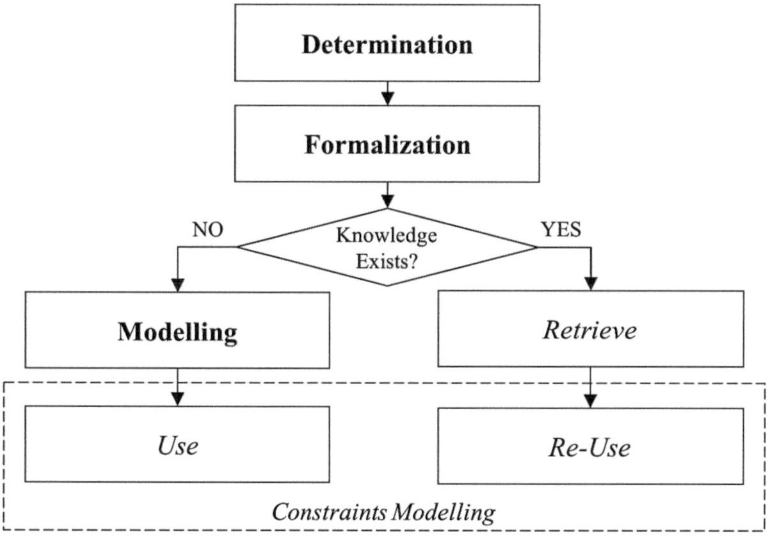

Fig. 14.1 Modelling process

that represents their interests among those already stored. If they find it, they just extract the variables from the base and standardize them to be coherent to the current problem. Otherwise, experts have to define a new constraint and its variables.

Knowledge modelling The description of the variable is supported by a tuple of sets where properties are instantiated from a set of values. Properties are stored with the variable in the knowledge base for further instantiation in a specific problem modelling. Those properties will be useful for redundancies analysis and they will state a basis for negotiation and decision making. Once variables are already stored in the knowledge base, experts can proceed to write their rules based on they knowledge and according to design problem. It is necessary to have all variables defined before, as knowledge will be constructed by retrieving variables from the base.

Multi Agent System (MAS)

The MAS environment can be formally represented by its structural parts. The formalism used to describe the system is based on the basic concepts, used to represent the structural parts of a multi-agent environment [17]. This representation helped to the characterisation of the proposed Multi-Agent System (MAS) in a distributed engineering design context. The different components are described below and Fig. 14.2 depicts the multi-level distribution of the different entities involved in a distributed knowledge modelling approach.

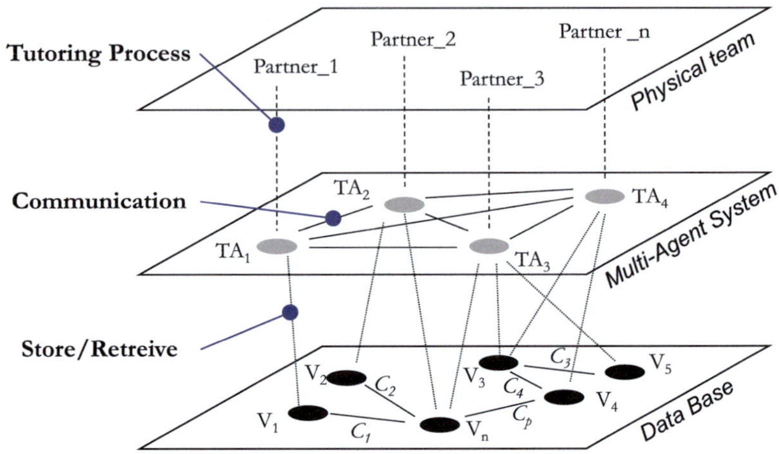

Fig. 14.2 Layers of the multi-agent environment. Adapted from [25]

Environmental Entities and Properties

Evolved from [25], the set of "Environmental Entities" is defined $E = \{e_1, e_2, \ldots, e_n\}$, and they can be partitioned into a set of disjoint subsets $\text{Part}_E = \{E_1, E_2, \ldots, E_k\}$, with each subset grouping entities of the same kind and $E_i \cap E_j = \emptyset, \forall i \neq j$. Under a distributed knowledge modelling approach, E is composed by several sets of partitioned environmental entities, and expressed $Part_E = \{\mathcal{T_A}, \mathcal{V}, \mathcal{D}, \mathcal{C}\}$, with:

$$\begin{array}{l|l} \mathcal{T_A} = \{TA_1, TA_2, \ldots, TA_i\} & \text{The set of } i \text{ tutor agents} \\ \mathcal{V} = \{V_1, V_2, \ldots, V_n\} & \text{The set of } n \text{ variables} \\ \mathcal{D} = \{D_1, D_2, \ldots, D_n\} & \text{The set of } n \text{ domains} \\ \mathcal{C} = \left\{C_1, C_2, \ldots, C_p\right\} & \text{The set of } p \text{ constraints} \end{array} \tag{14.2}$$

The set of "Environmental Properties" is defined $P = \{p_1, p_2, \ldots, p_m\}$, as well as the set of partitions of those properties in different kinds $Part_P = \{P_1, P_2, \ldots, P_l\}$. Then, for the distributed knowledge modelling environment it is possible to have an optimisation objective (that will compose the objective function to maximize or minimize). These design objectives can be defined as the partitioned environmental property $Part_P = \{\mathcal{O}\}$, where $\mathcal{O}$ is the set of j optimisation objectives $\mathcal{O} = \left\{O_1, O_2, \ldots, O_j\right\}$.

The State of the Constituents

The set of all constituents is defined by $C = E \cup P$. The formalism says that constituents can be partitioned according to their kind $Part_C = Part_E \cup Part_P = \{C_1, C_2, \ldots, C_{k+l}\}$. The "state of the constituents" S_{C_i} is the set of all possible states of constituents of kind C_i. The agent's communication analysis is based on all possible states for tutor agents $S_{\mathcal{T}_\mathcal{A}}$, for variables $S_\mathcal{V}$, for domains $S_\mathcal{D}$ and for constraints $S_\mathcal{C}$, and they are:

$$\begin{aligned} S_{\mathcal{T}_\mathcal{A}} &= \{Awaiting,\ Validating,\ Negotiating,\ Deciding\} \\ S_\mathcal{V} &= \{In\text{-}definition,\ Owned,\ Shared,\ Questioned\} \\ S_\mathcal{D} &= \{In\text{-}definition,\ Open,\ Flexible,\ Closed\} \\ S_\mathcal{C} &= \{In\text{-}definition,\ Relevant,\ Irrelevant\} \end{aligned} \tag{14.3}$$

Agents Embodied as Environmental Entities

Embodiment of agents: As described in Sect. 14, let $\mathcal{X}$ be the set of experts involved in the modelling process, that corresponds to the list of partners from the top level in Fig. 14.2. The set of partners will be instantiated into a set of agents (Tutor-Agents) to interact in the distributed environment. The function of instantiating external partners into agents is called "embodiment". Partners are embodied as environmental entities by $Embody : \mathcal{X} \rightarrow E$. It maps an agent to an environmental entity in the form of:

$$\begin{aligned} Embody : \mathcal{X} &\rightarrow \mathcal{T}_\mathcal{A} \\ partner_i &\mapsto TA_i \end{aligned} \tag{14.4}$$

Embodiment describes the relation between partners and the Multi-Agent environment. From the context of distributed product design, the elements from the set $\mathcal{X}$ will be the partners involved in the product development, which will be assigned to a TA. The environmental entity represents the tangible part [7], by means of which an agent exists in a particular environment. The embodiment of agents as environmental entities is defined as the Embody function that maps a partner to the environmental entity TA that embodies the partner:

$$\begin{aligned} Embody\,(Partner_1) &= TA_1 \\ Embody\,(Partner_2) &= TA_2 \\ Embody\,(Partner_i) &= TA_i \end{aligned} \tag{14.5}$$

Agents Analysis

Inter-Agent Analysis: Communication Among Agents

To clarify the communication aspect, consider the multi-agent system at a time t'. At this time, each component of the Multi-Agent System and the model has a predefined state (from Eq. 14.3). For example, the case of states for an Agent and a Variable, can be expressed:

$$\begin{aligned} Init(TA) &= S_{\mathcal{T}_A}^{t_0} = \langle Awaiting \rangle \\ &\longmapsto S_{\mathcal{T}_A}^{(t=t')} = \langle Negotiating \rangle \\ Init(V) &= S_{Variable}^{t_0} = \langle In\text{-}definition \rangle \\ &\longmapsto S_{Variable}^{(t=t')} = \langle Questioned \rangle \end{aligned} \tag{14.6}$$

The case of the TA's states (as a constituent from the environment) are very important in the communication among agents. The variation from one state to the other depends on communication status with other agents. Agents are by default in mode $S_{\mathcal{T}_A} = Awaiting$ and can be interacting with the Expert. After a validation of an input, when $S_{\mathcal{T}_A} = Validating$, the agent initiates a communication in order to validate information with other agents from the MAS. The result from standardisation is an input to Tutor Agent (TA) that can be positive if information is coherent (not redundant variables) or requested for validation if a redundancy is discovered or suggested. Then a new state begins where $S_{\mathcal{T}_A} = Negotiating$. At this state the Tutor Agent (TA) propose redundancies to the user to be analyzed and make a decision $S_{\mathcal{T}_A} = Decision$ in collaboration with the other Tutor Agent (TA) involved. Those states can be better identified from a sequence diagram from UML notation as depicted in Fig. 14.3.

Intra-Agent Analysis: Standardization

When a Tutor Agent (TA) receives information to be analysed during $S_{\mathcal{T}_A} = Validating$, it manipulates variables as objects and exchange corresponding properties (those explained in Sect. 14) in order to perform a redundancy analysis based on Graph Theory. In the context of this project, a "graph" is a collection of nodes (variables) and a collection of edges (relation between variables) that connect pairs of nodes. The relations are taken from constraints issued from different partners. To detect potential incoherencies, a new set is constructed, by the union of the variables set V and two sets of properties from $Prop(v)$ (See Eq. 14.1). This can be represented by a graph corresponding to graph $G = (U, A)$ where some properties and algorithms from graph theory, such as cycles' analysis, can highlight

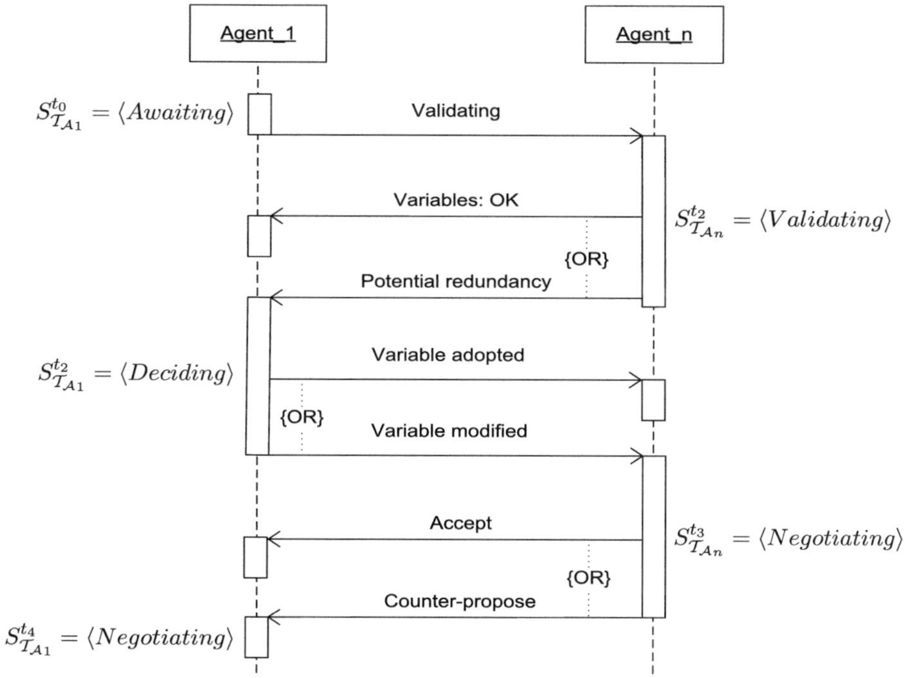

Fig. 14.3 States of Agents and communication principles

potential redundancies and partners may remark variables already defined by other Tutor Agent (TA). For example, the case of $U = V \cup \mathcal{P} \cup \mathcal{T}$ means that different technical domains $Tc_v \in \mathcal{T}$ are trying to measure a similar parameter $p_v \in \mathcal{P}$ if a cycle is found. This analysis can also be performed with other properties as we obtain a graph with multi-properties representation. If a closed path (cycle) is found, there is a potential incoherence caused by variables linked in such a way that a starting node will be the ending point of a path. Under a product design view, this may be an incoherence that is suggested to the expert by the Tutor Agent (TA) but it is up to him to react on this (negotiate or accept the stored variable if that is the case).

Implementation

A prototype system is proposed. It was developed with java™ and MySQL™ technologies. The system is composed of agents that are instantiated according to the number of partners involved (See Sect. 14) and the tutoring process is assigned to each expert. The Multi-Agents approach is supported by JADE™ framework and communication protocols are FIPA compliant. The system launches a Tutor

Agent (TA) for each partner involved in the project by instantiating a `TutorAgent` general class created by extending the `jade.core.Agent` class and redefining the `setup()` method in the form of:

```
public class TutorAgent extends Agent {
    protected void setup() {
        // Register the tutoring service in the yellow pages
        ...
        // Update the list of neighbor agents
        ...
        sendMessage();
        addBehaviour(new ReceiverBehaviour(this));
  }}
```

Each Tutor Agent (TA) has associated a main graphical user interface (GUI) in order to interact with the corresponding Partners from the physical world. Agents will interact in the virtual world (Multi-Agent System). This main GUI is the type of "desktop" (see Fig. 14.4) and in that space called Desktop, each phase of the methodology is processed and displayed independently. We therefore find three sub-windows according to the process explained in Sect. 14:

- Identification: containing sub-modules for the initial analysis, it is "Quality", "Morphology", "Behavioral" or "Technical".

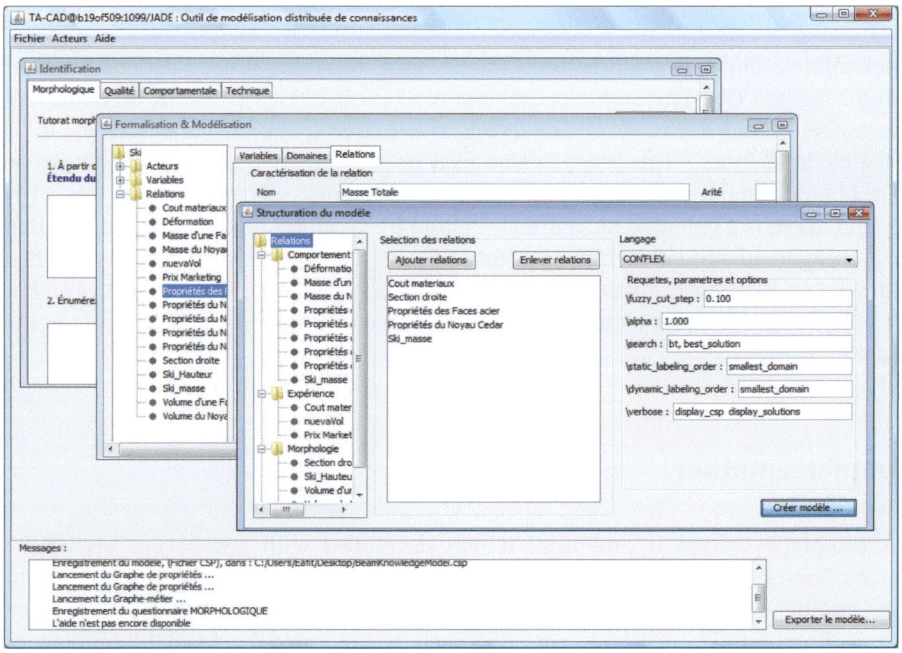

Fig. 14.4 TA's GUI main desktop

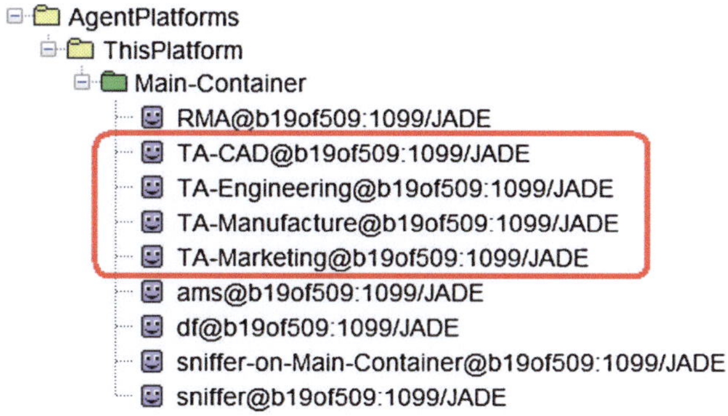

Fig. 14.5 Tutor agents instantiated into the JADE main container

- Formalisation & Modeling: containing the CSP basics to model the knowledge. Three tabs are proposed for "Variables", "Domains" and "Constraints" definition.
- Structure: the construction of the numerical model. This window is accessible from the button "Export model" and enables de sub-model integration in order to export it to the solver.

Agents Definition and Communication

The benefit of using agents, instead of a regular application, is the flexibility of creating and destroying agents without affecting the environment. Under a product development approach, the identification of experts is not fixed. It is a continuous process where agents can be created as new expertises are needed to be included in the modelling process. Each Tutor Agent (TA) instance is identified by an AID (`jade.core.AID`) which is composed of a unique name plus some addresses `<local-name>@<platform-name>`, and it is the name that TAs will receive from the agents' manager. The tutoring process of TAs is defined as `Behaviour` methods by extending the `jade.core.behaviours.Behaviour` class (as shown in Fig. 14.5).

Agents interacts among them through communication, as explained in Sect. 14. To send a message to other agents, Tutor Agent (TA) have to fill the fields of an `ACLMessage` object and then use the method `sendMessage` of the agent class. Under JADE framework, messages are FIPA compliant. Ontologies are used as a base to communicate and JADE propose the use of ontologies for messages' structure. An ontology is an explicit specification of a conceptualization, which is a body of formally represented knowledge [13]. The information is then represented as objects into the agents (ease to manipulate) and the use of Ontologies enable to convert those objects into a group of characters (to be more easy to transfer in the message). The Ontology helps to interpret those messages to convert them again in objects into the receiver agent (See Fig. 14.6).

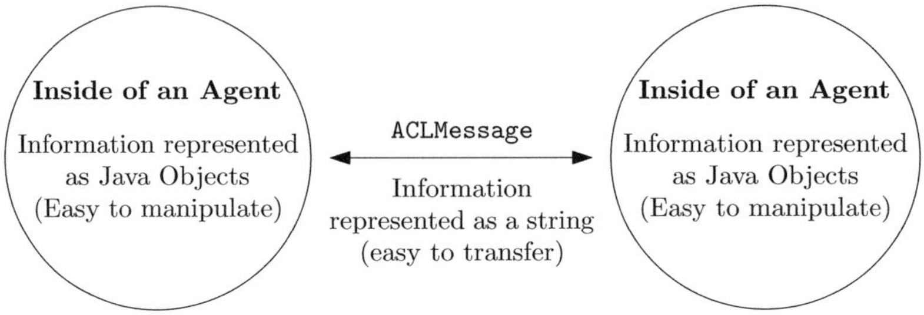

Fig. 14.6 Content language and ontologies with JADE

This is an advantage from the proposed MAS as it is easy to handle information within agents and easy to send messages, being an asset in the distributed environment. Messages are then exchanged easily under the `ACLMessage` context from the JADE framework enabling an organized and referenced communication standard. According to this, the implementation method for the communication protocols are described in the following extract of code:

```
private void sendMessage() {
  try {
    Variable variable = new Variable();
    ACLMessage msg = new ACLMessage(ACLMessage.INFORM);
    for (int i = 0; i < receiverAgents.length; ++i) {
          msg.addReceiver(receiverAgents[i]);
        }
    msg.setLanguage(codec.getName());
    msg.setOntology(ontology.getName());
    manager.fillContent(msg,variable);
    send(msg);
 }}
```

Once the system is running, the agents interact with their corresponding experts using the tutoring process. As the information that is relevant is identified, it is formalized and the Tutor Agent (TA) is in charge of sharing the variable's information to other Tutor Agent (TA) s, in order to be validated by the set of agents and minimize the risk of having duplicated information. Figure 14.7 depicts the FIPA compliant communication into the MAS.

The exchange of messages is carried out continuously and fosters collaborative and distributed work among partners involved. Similarly, the identification of redundancies allow partners to establish negotiations or conciliation to define the shared variables. The goal is that the information stored in the DataBase should be homogeneous in order to be able to construct a coherent knowledge model that supports the decision making process in conceptual engineering design.

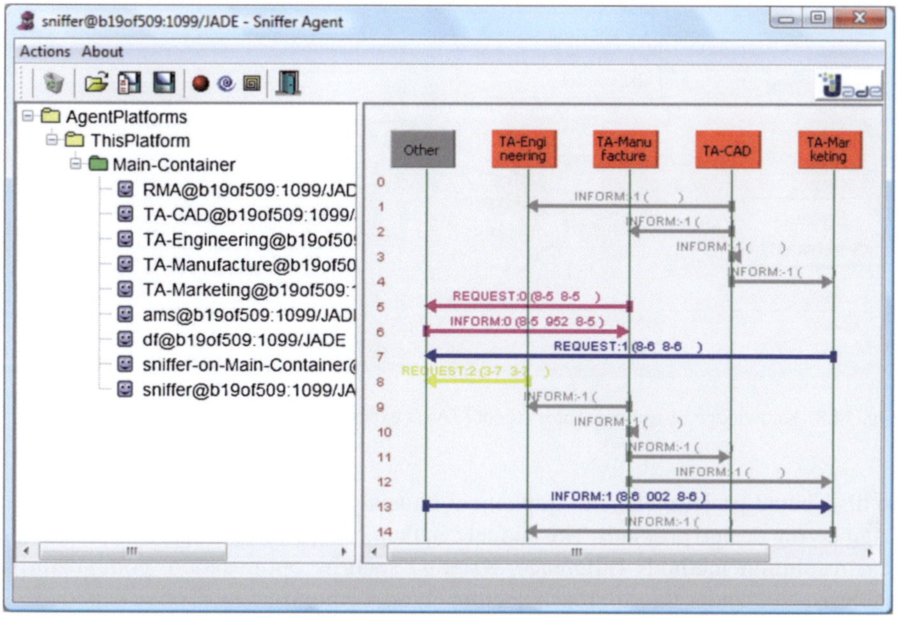

Fig. 14.7 FIPA compliant communication

Experiences Re-Use: Model Construction

Experts invited to participate in the engineering design team are able to collaborate through the aid of their corresponding Tutor Agent (TA) s in order to contribute, exploit and re-use the knowledge base for the construction of an experience based model. This model is built with variables, domains and constraints, related to a specific design problem and supported by expert's experiences. In the proposed software prototype, each Tutor Agent (TA) has a function able to "structure a model" while the GUI displays all constraints from the database. The expert can extract a subset of constraints that will be considered in the numerical evaluation of the knowledge model. The Tutor Agent (TA) is then able to "export" the model in an external file, according to the solver language. Tutor Agent (TA) evaluates the constraints taken into account and it automatically includes the variables and the corresponding domains. Figure 14.8 shows an example of exporting a partial model (sub-set of constraints) as a way to re-use expert's knowledge. Subsequently, the CSP file is exported with a specific programming language (the example shows a language model adapted to CON'FLEX software). In this way, Tutor Agent (TA) leads partners from the design team to construct a "granular", "standardized" and "Homogeneous" model of a engineering design problem. After having a coherent model, experts are able to build the knowledge model by converting the experiences stored in the Data-Base into a formal CSP model. The proposed approach enable

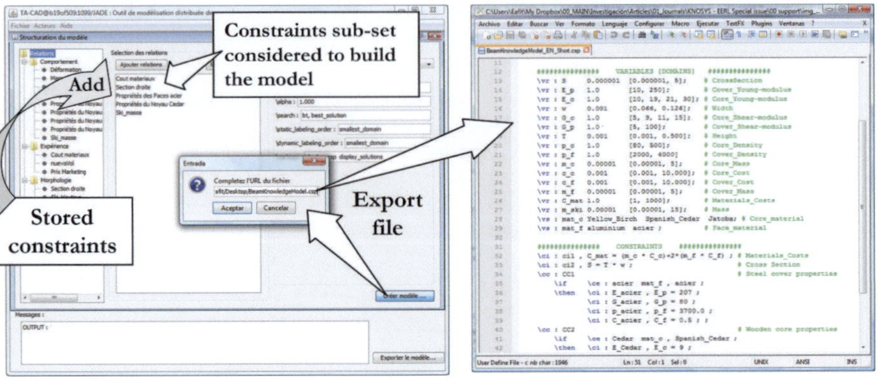

Fig. 14.8 Knowledge re-use and Tutor Agent (TA) 's export functionality

to distributed project teams to implement an analytical mathematical model of the engineering design problem. This model can then be treated with appropriate numerical resolution methods (inference systems) such as optimisation, combinatorial analysis, etc. Under this approach, a global model composed by diverse experiences from different technical areas are able to build local models that describe a specific design problem. Each local model includes the knowledge of each area of expertise needed to help the decision making. Once the model is structured, it can be exported to an inference engine, being able to execute the solution process in the solver and having as a result the possible solutions for the set of variables, respecting the included constraints (or one solution if it is an optimisation process).

Conclusion

Collaborative and distributed engineering approaches are leading nowadays engineering processes. The design process under these conditions demands more and more for tools and methods to help experts involved in the product development to facilitate their activities. The fact of having multiple partners implies a potential level of repetitive information, creating a problem as these issues may be tackled since early stages of the design process. The contribution of this research is the proposal of a tutoring methodology supported by multi-agents technology that supports distributed engineering teams to identify, formalize and model relevant experiences in a standardized way in order to construct models for analyzing problems in engineering design. There are several possibilities for inference engines, but they are limited to the numerical solution and they do not tackle the modelling issue leading to a complete absence of modelling techniques. The proposed approach gives an insight on how to aid engineering teams to collaboratively model a design problem by means of a tutor agent into a multi-agent system. A new interactive way to navigate the models and knowledge space is available. The software may serve as input for the current offer of numerical solvers for solutions searching. It is important to remark that this approach is a semiautomatic decision aid system, as no determination/decision is automatically set. Contrarily, it highlight possible inconvenient to the user and it is up to them to better decide. It enhance communication among partners (instead of avoiding it), as this approach foster focused discussion on specific information, as well as the exchange among experts in order to avoid downstream problems in the full product development process. The scope of this research does not include communication tools; hence, traditional communication methods are still necessarily. The proposed approach also serves as a way to capitalize and store expert's knowledge in order to consolidate them into a centralized database and able to be re-used in future projects. As further research, there are current sub-projects oriented to complete the prototype development focusing on: agents Behaviour improvement, database access optimisation, constraints and dimensional analysis, fuzzy domains analysis, Multi-language availability and a long-term project that intends to integrate a built-in solver to the system.

References

1. Aldanondo M, Vareilles E, Hadj-Hamou K, Gaborit P (2008) Aiding design with constraints: an extension of quad trees in order to deal with piecewise functions. Int J Comput Integr Manuf 21(4):353–365
2. Barao A, de Vasconcelos JB, Rocha A, Pereira R (2017) A knowledge management approach to capture organizational learning networks. Int J Inf Manage 37(6):735–740
3. Bellifemine F, Poggi A, Rimassa G (1999) JADE – A FIPA-compliant agent framework. In: Proceedings of PAAM, vol 99. pp 97–108
4. Bellifemine F, Poggi A, Rimassa G (2001) Developing multi-agent systems with JADE. In: Intelligent agents VII. Agent theories architectures and languages: 7th international workshop, ATAL 2000. Springer, Berlin/Heidelberg, pp 42–47
5. Camarinha-Matos L, Afsarmanesh H, Marík V (1999) Multi-agent systems applications. Rob Auton Syst 27:1–2
6. Chandrasekaran B (1990) Design problem solving: a task analysis. AI Mag 11(4):59
7. Ferber J, Michel F, Baez J (2005) Agre: integrating environments with organizations. In: Weyns D, Dyke Parunak H, Michel F (eds) Environments for multi-agent systems. Lecture notes in computer science, vol 3374. Springer, Berlin/Heidelberg, pp 48–56
8. Ferreira F, Faria J, Azevedo A, Marques AL (2017) Product lifecycle management in knowledge intensive collaborative environments: an application to automotive industry. Int J Inf Manage 37(1, Part A):1474–1487
9. Flores-Mendez R (1999) Standardization of multi-agent system frameworks. ACM Crossroads 5(4):18–24
10. Frasson C, Mengelle T, Aïmeur E, Gouardéres G (1996) An actor-based architecture for intelligent tutoring systems. In: Proceedings of ITS'96 conference. Lecture notes in computer science, vol 1086. Springer, Berlin/Heidelberg, pp 57–65
11. Giachetti R, Young R, Roggatz A, Eversheim W, Perrone G (1997) A methodology for the reduction of imprecision in the engineering process. Eur J Oper Res 100(2):277–292

12. Girodon J, Monticolo D, Bonjour E, Perrier M (2015) An organizational approach to designing an intelligent knowledge-based system: application to the decision-making process in design projects. Adv Eng Inform 29(3):696–713
13. Gruber T (1995) Toward principles for the design of ontologies used for knowledge sharing. Int J Hum Comput Stud 43(5/6):907–928
14. Guan Q, Friedrich G (1992) Extending constraint satisfaction problem solving in structural design. In: Belli F, Radermacher F (eds) Industrial and engineering applications of artificial intelligence and expert systems. Lecture notes in computer science, vol 604. Springer, Berlin Heidelberg, pp 341–350
15. Hao Q, Shen W, Zhang Z (2005) An autonomous agent development environment for engineering applications. Adv Eng Inform 19:123–134
16. Hao Q, Shen W, Zhang Z, Park S-W, J-K L (2006) Agent-based collaborative product design engineering: an industrial case study. Comput Ind 57:26–38
17. Helleboogh A, Vizzari G, Uhrmacher A, Michel F (2007) Modeling dynamic environments in multi-agent simulation. Auton Agent Multi Agent Syst 14(1):87–116
18. Jin Y, Zhou W (1999) Agent-based knowledge management for collaborative engineering. In: Proceedings of DETC'99 – 1999 ASME design engineering technical conferences, Las Vegas
19. Karacapilidis N (2002) Modeling discourse in collaborative work support systems: a knowledge representation and configuration perspective. Knowl Based Syst 15(7):413–422
20. Kravari K, Bassiliades N (2015) A survey of agent platforms. J Artif Soc Soc Simul 18(1):11
21. Liao S, Wang H, Liao L (2002) An extended formalism to constraint logic programming for decision analysis. Knowl Based Syst 15(3):189–202
22. Liu P, Raahemi B, Benyoucef M (2011) Knowledge sharing in dynamic virtual enterprises: a socio-technological perspective. Knowl Based Syst 24(3):427–443
23. Martinez-Gil J (2015) Automated knowledge base management: a survey. Comput Sci Rev 18(Supplement C):1–9
24. Mejia-Gutierrez R, Fischer X, Bennis F (2008) A tutor agent for supporting distributed knowledge modelling in interactive product design. International Journal of Intelligent Systems Technologies and Applications 4(3):399–420
25. Mejia-Gutierrez R, Fischer X, Bennis F (2008) Virtual knowledge modelling for distributed teams: towards an interactive design approach. Int J Networking Virtual Organ 5(2):166–189
26. Moon S-K, Kumara S, Simpson T (2006) A multi agent system for modular platform design in a dynamic electronic market environment. In: Proceedings of DETC/CIE 2006–ASME 2006 international design engineering technical conferences & computers and information in engineering conference
27. Munoz E, Capon-Garcia E, Lainez-Aguirre JM, Espuna A, Puigjaner L (2014) Integration of methods for optimization in a knowledge management framework. In: Klemes JJ, Varbanov PS, Liew PY (eds) Proceedings of 24th European symposium on computer aided process engineering. Computer aided chemical engineering, vol 33. Elsevier, pp 859–864
28. Nadel BA, Lin J 7 (1991) Automobile transmission design as a constraint satisfaction problem: modelling the kinematic level. Artif Intell Eng Des Anal Manuf 5:137–171
29. Ostrosi E, Fougères A-J, Ferney M (2012) Fuzzy agents for product configuration in collaborative and distributed design process. Appl Soft Comput 12(8):2091–2105
30. Qureshi AJ, Dantan J-Y, Bruyere J, Bigot R (2010) Set based robust design of mechanical systems using the quantifier constraint satisfaction algorithm. Eng Appl Artif Intell 23(7):1173–1186
31. Ren Z, Yang F, Bouchlaghem N, Anumba C (2011) Multi-disciplinary collaborative building design – a comparative study between multi-agent systems and multi-disciplinary optimisation approaches. Autom Constr 20(5):537–549
32. Scaravetti D, Nadeau J-P, Pailhes J, Sebastian P (2005) Structuring of embodiment design problem based on the product lifecycle. Int J Prod Dev 2(1/2):47–70
33. Wang H, Liao L (1997) A framework of constraint-based modeling for cooperative decision systems. Knowl Based Syst 10(2):111–120

34. Wang J, Tang M (2006) An agent based approach to collaborative product design. In: Proceedings of DETC/CIE 2006–ASME 2006 international design engineering technical conferences & computers and information in engineering conference
35. Yan-hong Q, Guang-xing W (2009) Product configuration based on CBR and CSP. In: International conference on measuring technology and mechatronics automation, ICMTMA'09, vol 3. IEEE, pp 681–684
36. Yang D, Dong M (2012) A constraint satisfaction approach to resolving product configuration conflicts. Adv Eng Inform 26(3):592–602
37. Yang S, Liu Q, Wang J (2017) A multi-agent system with a proportional-integral protocol for distributed constrained optimization. IEEE Trans Automat Contr 62(7):3461–3467
38. Yvars P-A (2009) A CSP approach for the network of product lifecycle constraints consistency in a collaborative design context. Eng Appl Artif Intell 22(6):961–970

Part II
Applications

Chapter 15
Synthesis of Reference Trajectories for Humanoid Robot Supported by Genetic Algorithm

Teresa Zielinska

Abstract This work presents biologically inspired method of gait generation. It uses the reference to the periodic signals generated by biological Central Pattern Generator (CPG). The coupled oscillators with correction functions are used to produce leg joint trajectories. The human gait is used as the reference pattern. The features of generated gait are compared to the human walk. The example illustrates well the profit offered by the optimization using genetic algorithm. The problem would be impossible to solve using traditional approach.

Keywords Robotics · Gait generation · Genetic algorithms

Introduction

Genetic algorithms belong to stochastic search methods inspired by biology, they use evolution as a strategy for optimizing complex and usually non-linear problems with a large number of variables.

The origins of genetic algorithms (GA) date back to fifties of XX c. and are associated with the beginnings of computer simulations of biological evolution processes. Computer simulation of evolution processes were initiated by Nils Aall Barricelli but his publications dated 1954 [1, 2] were not widely noticed. The roots of genetic algorithms are also seen in the works of Alan Turing who in 1950s issued the concept of learning machine [16]. From 1957 [7] the geneticist Alex Fraser published a series of papers on simulation of artificial selection process. The computer simulation of biological evolution became more common in the early 1960s Several methods were described in the books by Fraser and Burnell [8] and by Crosby [5]. Fraser's simulations included all key elements of modern genetic algorithms. Hans-Joachim Bremermann in his papers published in 1960s described

T. Zielinska (✉)
Faculty of Power and Aerospace Engineering, Warsaw University of Technology, Warsaw, Poland
e-mail: teresaz@meil.pw.edu.pl; http://www.meil.pw.edu.pl

F. Bennis, R. K. Bhattacharjya (eds.), *Nature-Inspired Methods for Metaheuristics Optimization*, Modeling and Optimization in Science and Technologies 16,
https://doi.org/10.1007/978-3-030-26458-1_15

solution to optimization problems with recombination, mutation, and selection. His research created a fundamentals for modern genetic algorithms. In 1960s and early 1970s Ingo Rechenberg's group was able to solve complex engineering problems through evolution process. Other relevant works were by Richard Friedberg, George Friedman, and Michael Conrad, many of those papers were later reprinted by Fogel [6].

In seventies the book by John Holland [10] marked a significant milestone. He had elaborated a formalized method for predicting the quality of next generation, it is now known as the Holland's schema indicating how to solve optimization problem. J.Holland underscored that the fittest individual in each biological generation represents the best solution reached so far, therefore evolution allows to obtain a set of improving solutions with the evolving generations. Finally it leads with approaching the optimum. Another approach for solving the optimization problems was the evolutionary programming method by Lawrence J. Fogel. Evolutionary programming originally used finite state machines for predicting environments, and used variation and selection for optimization. The concept of cellular automata was used by J. Holland in his book.

Until the end of eighties research on GA remain mainly theoretical. Starting from 1990 John Koza, the computer scientists published a series of papers on *genetic programming* describing the set of practical optimization examples. In his approach the computer program was initiated using some pre-solutions (represented by set of chromosomes) called population. The better solutions from one population were preserved and used for forming a next population. Idea was motivated by expectation that the next population will be better than the previous. Solutions which were preserved (offsprings) were selected considering their fitness – the more suitable had more chances being reproduced. Such scheme was repeated until reaching some condition – for example limit number of populations or defined fitness condition.

In 1992 J. Koza presented the genetic programming method using LISP.

The First International Conference on Genetic Algorithms was held in Pittsburgh, Pennsylvania in mid eighties. Afterwards the use of generic algorithms for solving practical problems become very popular.

Fundamentals of Genetic Algorithms

In genetic algorithms the analyzed quantities contributing to solution are called chromosomes. The most common way of encoding it are a binary strings. The examples of chromosomes are:

chromosome 1: **0101100100110110**
chromosome 2: **1101111000011111**

Gene is ether the single bit on chromosome or the set of adjacent bits. That encode an element in candidate solution. Population is the set of genes available to test. The crossover and mutation are the key parts of genetic algorithm. The method performance is influenced mainly by these two operators. The first step in algorithm is crossover which selects genes from parent chromosomes and creates a new offspring. Reproduction methods are mainly use two parents what is biologically inspired, but some researchers [2, 9] suggest that using more than two parents can produce higher quality chromosomes.

The simplest way of cross-over is to choose randomly some crossover point in parent chromosomes and take a part before this point from one parent and a part after crossover point from a second parent.

Example:

chromosome 1:	**01011 ‖ 00100110110**
chromosome 2:	**11011 ‖ 11000011111**
offspring 1:	**01011 ‖ 11000011111**
offspring 2:	**11011 ‖ 00100110110**

Then mutation take place. This prevents of falling all population into a local optimum. During mutation some bits in the offsprings are changed randomly. In binary encoding it mean that a few randomly chosen bits are changed from 1 to 0 or from 0 to 1.

For example:

original offspring 1:	**01011 ‖ 110000111**
original offspring 2:	**10011 ‖ 001001101**
mutated offspring 1:	**10001 ‖ 110000111**
mutated offspring 2:	**10011 ‖ 011001101**

The population size depends on the nature of the problem, and typically contains several hundreds or thousands of possible solutions (encoded in chromosomes) because genetic algorithms are often used for optimizing large-scale problems.

The optimization process starts from a population of randomly generated chromosomes or from pre-defined set if some (acceptable) solution to the problem is known. In consecutive iterations the new populations called also a generations are produced. For each generation, the fitness of every individual is evaluated;. The fitness is evaluated using the objective function of the optimized problem. The more fitting chromosomes are selected from actual population, and are used for forming next generation using cross-over and mutation operations. The algorithm terminates when either a pre-specified number of generations has been produced, or an objective function reached satisfactory fitness.

A fundamental components of genetic: algorithm are:

– a genetic representation of the solution domain,
– a fitness function evaluating the solution domain.

The fitness function is defined for genetic representation and measures the quality of generated solution. A representation of a solution might be the value of objective function or can be composed from an array of bits or integer numbers. In the array each position represents a fitness for different object or factor, or a penalty measure. In the case of penalty the value is set from the predefined range depends on how much a penalty condition is exceeded. The final fitness o is the sum of all partial fitness and the penalty components. However in some problems, it is hard or impossible to define the fitness formula; in these cases, a simulation may be used for obtaining the fitness value.

In genetic algorithms the crossover and mutation are the most important parts. The performance of the method is significantly influenced by these two operations.

During successive generations, a part of the population is selected to produce a new generation. Chromosomes are selected taking into account their fitness. Solutions (chromosomes) obtaining better fitness are more likely to be selected. Some selection methods rate the fitness for each chromosome from population. Other methods evaluate fitness only for random sample of the population, because evaluating the fitness foe each chromosome may take a lot of time.

It must be also noted that there are some limitations of genetic algorithms comparing to the classic optimization methods:

- fitness evaluation for complex problems may require several hours to several days, typical optimization methods can not deal with such types of problems, and it may be necessary to use a simplified fitness and simplified models of the problem,
- in genetic algorithms is observed an exponential increase of search space size with the size of problem, therefore the complex problems must be decomposed into a set of simpler representations,
- in some problems genetic algorithms may have tendency for converging in local optima or even in some random points, a common technique of overcoming it is to introduce a niche penalty, wherein for group of individuals of specified similarity (a niche radius) a penalty is considered, another technique is to replace that part of population with randomly generated individuals, when most of the population is too similar to each other,
- another issue concerns protecting good solutions from further destructive mutations what can be done by many ways,
- the improvement of a solution is judged only by comparison to other ones, it means. That he stop criterion can be not so obvious.

Gait Generation Using Coupled Oscillators

In recent years there are many attempts for obtaining the two-legged robots mowing as a humans [9, 12, 14]. Unfortunately achieving human like robot movement is still a problem due to the mechanical construction limits and due to the weaknesses of motion generation methods [13]. The biologically inspired methods of gait generation often refers to the neural generators called Central Pattern Generators (CPGs). CPG is the system of biological neural networks which produces the living rhythms, in that the rhythmic pattern of locomotion without the sensory feedback.

CPG was found in almost all vertebrates as the neural structure located in the vertebral column. During the organism development such structure learns by trials and errors how to generate the locomotion rhythm. Once gained, the rhythm is stored, it is be modified when needed by higher levels of neural system taking into account the sensory feedbacks.

A broad discussion of profits from using the CGP inspired motion generation methods in robotics can be found in [11]. The recent works are focusing on CPG based motion generation methods which allows the self-adjustment of gait dynamics [15]. The oscillators formula are commonly used for imitating the work of biological CPG's. Coupled oscillators are storing in the compact way the fundamental rhythmic pattern and they allow to generate relatively easy the motion transitions. An interesting overview of the different oscillators utilized for motion generation can be found in [4].

Considered by us method applies a special type of coupled oscillators as the leg joints trajectory generators – they are van der Pol oscillators. Equations describing the dynamical properties of those oscillators have the following general form:

$$\ddot{x}_{osc} - \mu \cdot (p^2 - x_{osc}^2) \cdot \dot{x}_{osc} + g^2 \cdot x_{osc} = q \tag{15.1}$$

The variables μ, p^2, g^2, q influence the properties of oscillators.

In our work cyclic solutions of four coupled oscillators formula was considered as the legs joint trajectories for a hip and knee [3]:

$$\begin{aligned}
\ddot{\alpha}_1 - \mu_1 \cdot (p_1^2 - x_a^2) \cdot \dot{\alpha}_1 + g_1^2 \cdot x_a &= q_1 \\
\ddot{\alpha}_2 - \mu_2 \cdot (p_2^2 - x_b^2) \cdot \dot{\alpha}_2 + g_2^2 \cdot x_b &= q_2 \\
\ddot{\alpha}_3 - \mu_3 \cdot (p_3^2 - x_c^2) \cdot \dot{\alpha}_3 + g_3^2 \cdot x_c &= q_3 \\
\ddot{\alpha}_4 - \mu_4 \cdot (p_4^2 - x_d^2) \cdot \dot{\alpha}_4 + g_4^2 \cdot x_d &= q_4
\end{aligned} \tag{15.2}$$

where

$$\begin{aligned}
x_a &= \alpha_1 - \lambda_{21} \cdot \alpha_2 - \lambda_{31} \cdot \alpha_3 \\
x_b &= \alpha_2 - \lambda_{12} \cdot \alpha_1 - \lambda_{42} \cdot \alpha_4 \\
x_c &= \alpha_3 - \lambda_{13} \cdot \alpha_1 - \lambda_{43} \cdot \alpha_4 \\
x_d &= \alpha_4 - \lambda_{24} \cdot \alpha_2 - \lambda_{34} \cdot \alpha_3
\end{aligned}$$

These equations have 24 parameters: μ_1, μ_2, μ_3, μ_4, p_1^2, p_2^2, p_3^2, p_4^2, g_1^2, g_2^2, g_3^2, g_4^2, q_1, q_2, q_3, q_4, λ_{13}, λ_{31}, λ_{12}, λ_{21}, λ_{24}, λ_{42}, λ_{43}, λ_{34}. The relationship between these parameters and oscillations is very complex. The angles α_1^s, α_2^s, α_3^s, α_4^s correspond to the adequately scaled α_1, α_2, α_3, α_4. The angles α_i^s are positive if the thigh or shank are in front of the vertical line and negative when behind (Fig. 15.1a). The variables α_i and α_i^s represent the values expressed in degrees. The oscillators have many stable cycles. With 24 parameters there are is a large space of possible solutions, therefore is a serious problem how to select the properly all the parameters for obtaining the trajectories similar to those in human gait assuring proper frequency of the limit cycle. On the other hand once the proper parameters are selected by modifying them accordingly the gait transitions can be obtained. This is main advantage of the method, because generation of motion transitions is still a problem in human like robots.

In this work we describe utilization of genetic algorithm for selecting the oscillator parameters. The advantage of using it is obvious, the manual search of unknown parameters for imitating accurately the human trajectories is almost impossible or it will take a lot of time.

Genetic Algorithm Applied for Parameters Search

Exact analytical solution of considered coupled oscillators can not be obtained. As it was already mentioned, coupled oscillators are characterized by many locally stable cycles. In the other works dedicated to coupled oscillators the approximate solutions are often used (e.g. for coupled chemical oscillators), where the analytical form of approximation is assumed first and later – through iterative search, the parameters of the assumed form are found. Unfortunately such method does not picture sufficiently the synergy of coupling and not obviously all the possible stable cycles.

Due to above difficulties and pretending to imitate the biological learning process a genetic algorithm for oscillators parameters evaluation was applied. The search started from the previously identified (by simple search) set of parameters [21, 22]: $\mu_1 = 1$, $\mu_2 = 2$, $\mu_3 = 1$, $\mu_4 = 2$, $p_1^2 = 1$, $p_2^2 = 1$, $p_3^2 = 1$, $p_4^2 = 1$, $g_1^2 = 17$, $g_2^2 = 20$, $g_3^2 = 17$, $g_4^2 = 20$, $q_1 = 12$, $q_2 = -20$, $q_3 = 12$, $q_4 = -20$, $\lambda_{13} = 0.2$, $\lambda_{31} = 0.2$, $\lambda_{12} = -0.2$, $\lambda_{21} = -0.2$, $\lambda_{24} = 0.2$, $\lambda_{42} = 0.2$, $\lambda_{43} = -0.2$, $\lambda_{34} = -0.2$.

The human gait trajectories recorded, processed and evaluated using the specialized software (combined with the VICON motion recording system) were used as the reference. The motion was sampled every 0.021 s. It was confirmed that considered person (healthy young man) was walking with statistically verified, representative gait (Fig. 15.1b). The gait period was 1.04 s, walking speed was 1.53 m/s, the support phase was about 60% of the gait period, in this the double support phase took 20% of the gait period. Those values match the norms established for the human gait [17, 20].

The biped robot consisting of the lower extremities and the torso was meant to serve as a research platform for presented method of motion synthesis.

Searched oscillator parameters were stored forming a chromosome. The search for parameters started with 100 chromosomes (for each identical at this stage) holding the previously identified set of parameters [21]. The next 200 chromosomes were created in following way: 50 chromosomes were produced using simple

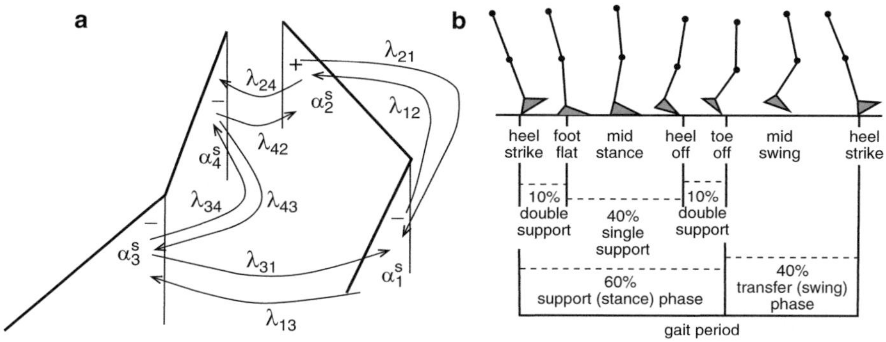

Fig. 15.1 Joints and the coupling between them considered in coupled oscillators (**a**), gait phases (**b**)

cross-over, next 50 chromosomes were obtained by the arithmetic cross-over, and another 100 was produced by simple mutations. In simple cross-over the a pair of chromosomes was draw, next the cross-over point was stochastically selected. First part of new chromosome was identical with the first parent chromosome till the cross-over point, the part after cross-over was taken from the next parent. In arithmetic cross-over $C_{offsping} = rC_{parent\ 1} + (1 - r)C_{parent}$ where $r \leq 1$ was the arbitrary selected parameter. In simple mutations only one gene was draw for modification. The oscillators formula was solved numerically for each obtained set of chromosomes (oscillator parameters). Only chromosomes resulting in periodic oscillations were accessed using the the primary fitness function and penalties. The fitness function was an square error between trajectories obtained by genetic algorithm parameters and those for a human:

$$f_j = \frac{\sum_{i=1}^{n} (\alpha_j^h(i) - \alpha_j^s(i))^2}{n} \tag{15.3}$$

where $\alpha_j^h(i)$ is the value of j-th joint angle during human walk in the i-th data instant, and $\alpha_j^s(i)$ is the angle obtained from the generator, n is the number of data registered for one walking step.

The six applied penalties had experimentally chosen ranges (over those ranges oscillators were becoming unstable). The penalties were as following: s_1 – penalty weight considering inter-limb coordination, this penalty was maximal ($s_1 = 5$) when both legs were moving together, s_2 – penalty weight for unstable oscillations (maximal $s_2 = 10$), s_3 – penalty weight for backward movement of torso (maximal $s_3 = 3$), s_4 – penalty weight for backward movement of the foot (maximal $s_4 = 3$), s_5 – penalty weight for lack of foot forward motion (maximal $s_5 = 3$), s_6 – penalty weight for abnormal knee bend (maximal $s_6 = 3$). If any of the listed above conditions was not fulfilled the appropriate penalty assumed its minimum – $s_i = 1$.

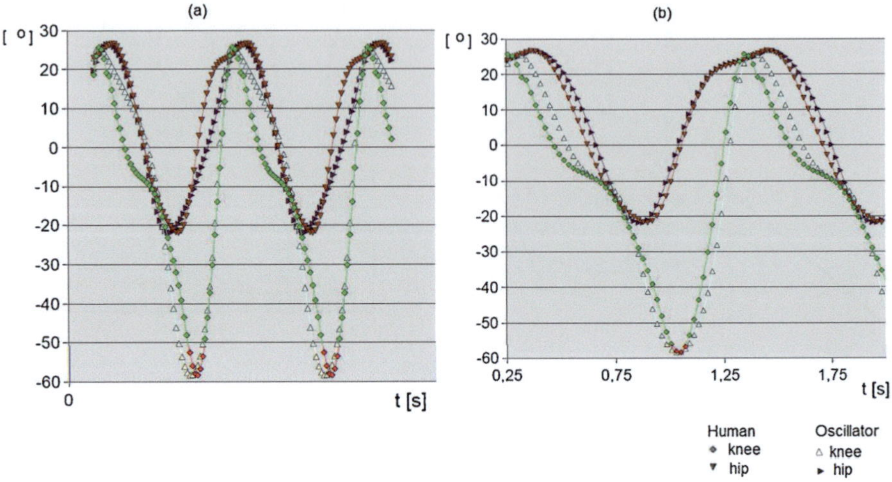

Fig. 15.2 Generated joint trajectories related to human trajectories: (**a**) before correction, after correction (**b**)

The penalties applied in our example can be referred as an equivalent of qualitative feedback experienced by a baby when learning the locomotion rhythm.

The overall fitness function was expressed by:

$$FF_G = -(f_1 + f_2 + f_3 + f_4) \cdot \prod_{k=1,\dots,6} s_k \tag{15.4}$$

Chromosomes were sorted taking into account the obtained fitness function. Next they are were ordered taking into account their fitness. The first ones had the smallest fitness and they were selected as the new parents without draw. The probability of selecting the chromosomes along in the sequence was decreasing (chromosomes with worse fitness were considered less often in the draw). By that way a 100 new parents were produced and the scheme was repeated.

In repeated trials it was found that the considerable convergence to human motion pattern takes place only for early mutations, the farther chromosome mutations were not much effective. Figure 15.2a illustrates obtained joint trajectories related to the human gait. Oscillator parameter are as following: $\mu_1 = \mu_3 = 3.59375$, $\mu_2 = \mu_4 = 2$, $p_1^2 = p_3^2 = 2$, $p_2^2 = p_4^2 = 1$, $g_1^2 = g_3^2 = 28.0039$, $g_2^2 = g_4^2 = 17.7031$, $q_1 = q_3 = 15.8516$, $q_2 = q_4 = -7.04492$, $\lambda_{12} = \lambda_{21} = \lambda_{34} = \lambda_{43} = -0.451172$, $\lambda_{24} = \lambda_{42} = \lambda_{31} = \lambda_{13} = 0.417969$.

In that stage the gait rhythm was acquired however the differences between generated trajectories and human trajectories sill remained. It was not possible to minimize the error using only genetic algorithm.

Fine Tuning of Gait Generator

It was observed that the difference (error) between the human and generated trajectories in each joint occurred only for some small parts of the gait. Moreover such error smoothly drifts from zero to some level (depends on the joint – positive or negative) and then returns to zero (Figs. 15.2a and 15.3). This brought the idea that proper addition (or substraction) of some smooth function will made the needed correction. The function should have its derivative equal to zero at moth ends of the domain. This will provide the needed correction for joint trajectories. Basis on the above we selected the following correcting function:

$$F_{corr} = A\ cos^2(\phi) \quad \phi = < -\Pi/2, \Pi/2 > \tag{15.5}$$

where amplitude A can be positive or negative. Negative correction is needed for a knee. It is activated at the moment when the knee angle is maximal (leg is

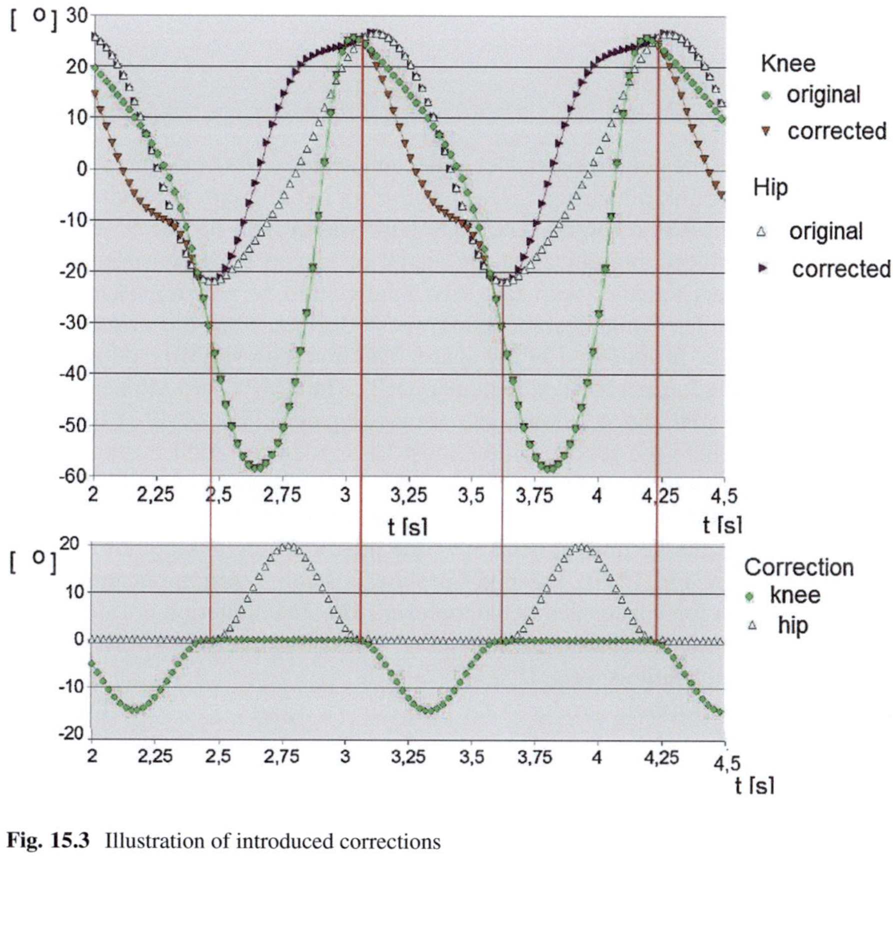

Fig. 15.3 Illustration of introduced corrections

in maximum forward position just before ground touch down) and lasts until the middle of the gait period (in our tests it was 48% of the gait period). Positive correction function is for a hip joint. It starts for minimum hip angle which is just after mid-stance. In the same moment the knee correction ends. Hip trajectory correction takes about half of the gait period (in our tests it was 52% of the gait period – Fig. 15.3). With introduced corrections the generated and real gait trajectories were almost overlapping (Fig. 15.2b).

The absolute cumulative errors expressing the discrepancy between human and generated gait in one gait period were calculated using the formula $|E_j| = \sum_i |\alpha_j^h(i) - \alpha_j^s(i)|/n$. The errors were as following: $|E_1| = 3.1037°$, $|E2| = 2.8389°$, $|E_3| = 3.7892°$, $|E_4| = 2.0064°$. The error are small taking into account that the hip trajectories are varying in range $-8°, +15°$ and the knee trajectories in range $-20°, +25°$.

The biggest observed error was for hip trajectory and it was less than 12° (Fig. 15.4). Such difference are not critical, in the normal walk of the healthy person the joint trajectories from step to step can differ by 8° or even more. In that way we concluded that important features of gait were achieved.

Final Proof

For the final proof the point mass model of the human body was considered. Masses of body parts and link dimensions were taken from anthropomorphic data for the person whose gait was recorded. The body built was typical for the 95% centile level. With introduced simplification the model of human body consisted of five parts – two shanks together with feet, two thighs and one part representing the whole upper part. The dynamical modeling was performed using our own software with Newton-Euler formulation in 3D space. The time courses of reaction forces were matching the forces recorded in human walk. The ZMP (Zero Moment Point) positions for the real and generated gait were compared. (The ZMP is the point on the ground where the reaction force must be applied to produce the moment compensating the tipping moments (M_x, M_y) due to gravity and inertial forces. The condition for equilibrating the moments is expressed with reference to the contact point P which is the vertical projection of the middle of the ankle joint onto the support plane (see Fig. 15.6). The body posture is stable when application point F_{ZMP} of reaction force is located in the footprint. The ZMP criterion is a simplified formulation of torques equilibrium condition. The details regarding ZMP can be found in many publications – e.g. [18, 19].

The following formula produces the coordinates x_{ZMP}, y_{ZMP} of the point F_{ZMP}:

$$x_{ZMP} = \frac{\sum_i^n m_i(\ddot{z}_i - g)x_i - \sum_i^n m_i\, \ddot{x}_i\, z_i - \sum_i^n I_i^y \ddot{\alpha}_i^y}{\sum_i^n m_i(\ddot{z}_i - g)} \tag{15.6}$$

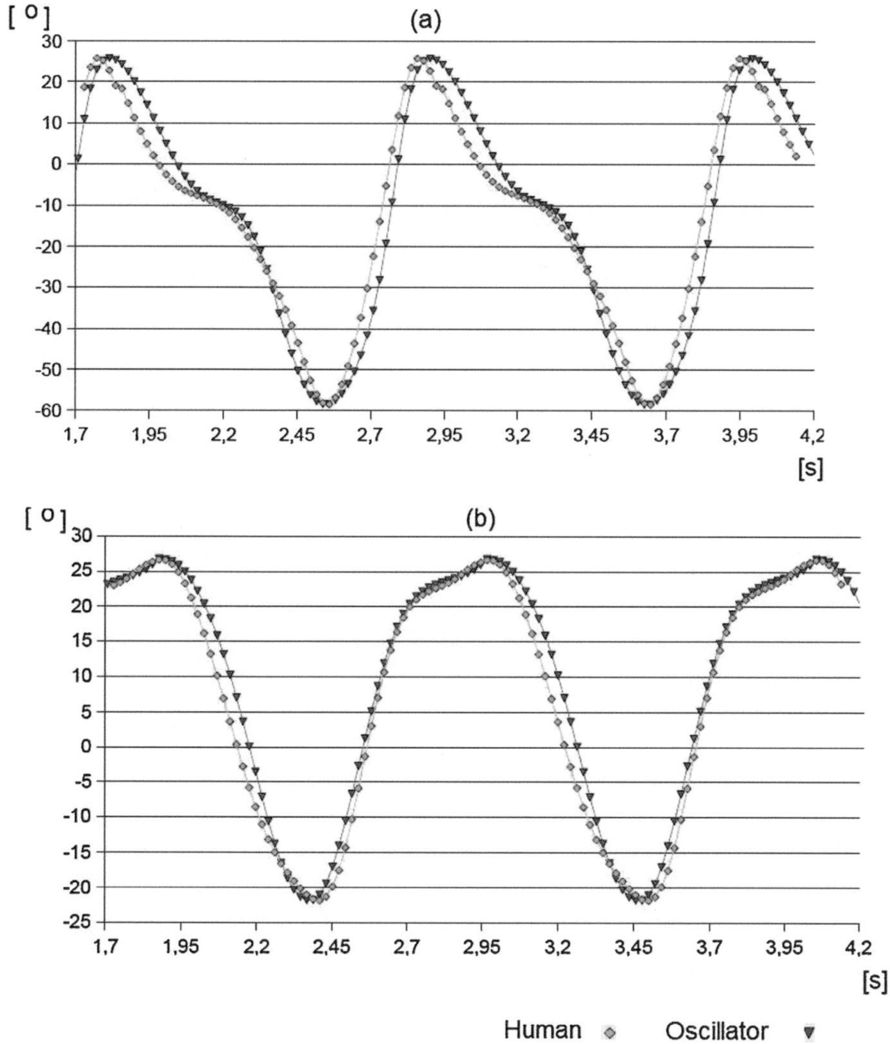

Fig. 15.4 Human and generated corrected trajectories for one leg: (**a**) knee joint, (**b**) hip joint

$$y_{ZMP} = \frac{\sum_i^n m_i(\ddot{z}_i - g)y_i - \sum_i^n m_i\ \ddot{y}_i\ z_i - \sum_i^n I_i^x \ddot{\alpha}_i^x}{\sum_i^n m_i(\ddot{z}_i - g)} \tag{15.7}$$

m_i is the mass of the i-th body part, x_i, y_i, z_i are the coordinates of mass center of i-th part expressed in the frame attached to the trunk (Fig. 15.5), $\ddot{x}_i$, $\ddot{y}_i$ $\ddot{z}_i$ are accelerations of those points with respect to this frame, I_i^x, I_i^y are the main moments of inertia for i-th body part about X and Y axis, and $\ddot{\alpha}_i^x$, $\ddot{\alpha}_i^y$ are the angular accelerations about those axes, g is the gravity constant. We assumed that feet are

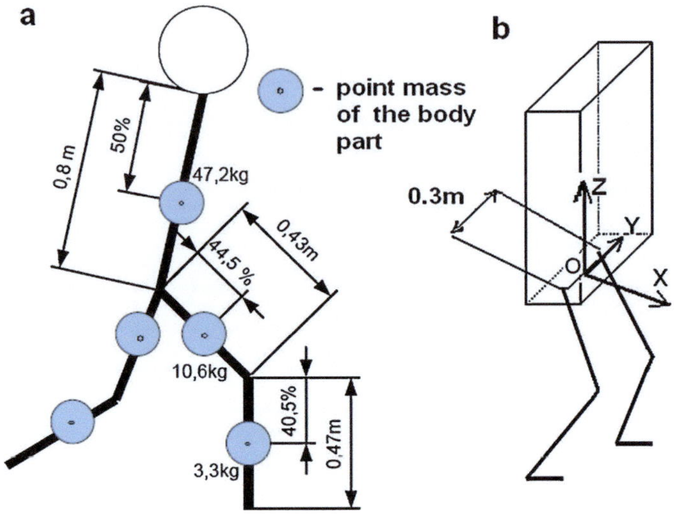

Fig. 15.5 Model of the human body (the distance between hip joints, and distance between foot, along y axis is 0.3 m)

Fig. 15.6 Model of the foot; reference point *P* is the vertical projection onto the support plane of the ankle joint center

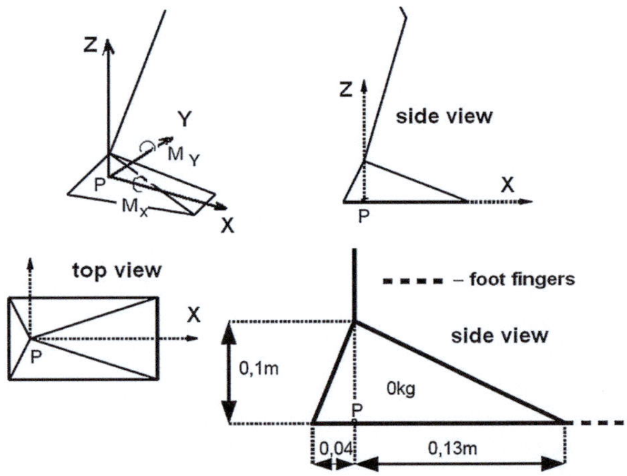

massless (their masses were taken into account when calculating the point masses of shanks and the positions of those point masses in relation to the shank segments). The proportions of body parts match those of the considered human (Fig. 15.6).

For both, the generated and the human gait the ZMP point varies in a narrow range. Figure 15.7 illustrates the ZMP trajectory during single support phase. In both gaits the ZMP trajectory shows similar trend, the ZMP moves from the inner side of the foot in the begging of single support to the middle and next to the outer side what indicates a side sway of the body. The ZMP stays inside the foot area what means that the posture is dynamically stable.

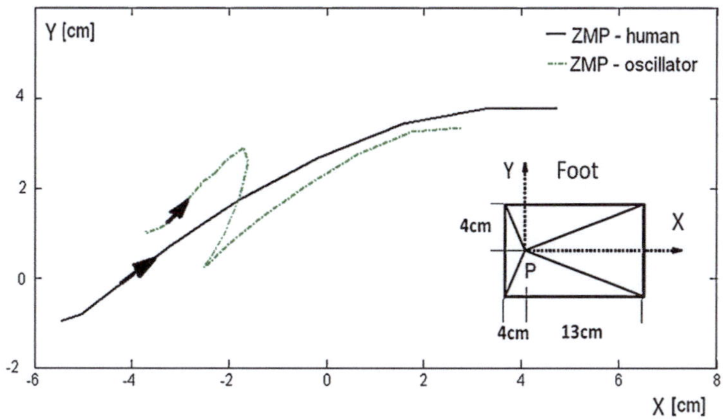

Fig. 15.7 ZMP trajectory of human and generated gait for the left leg

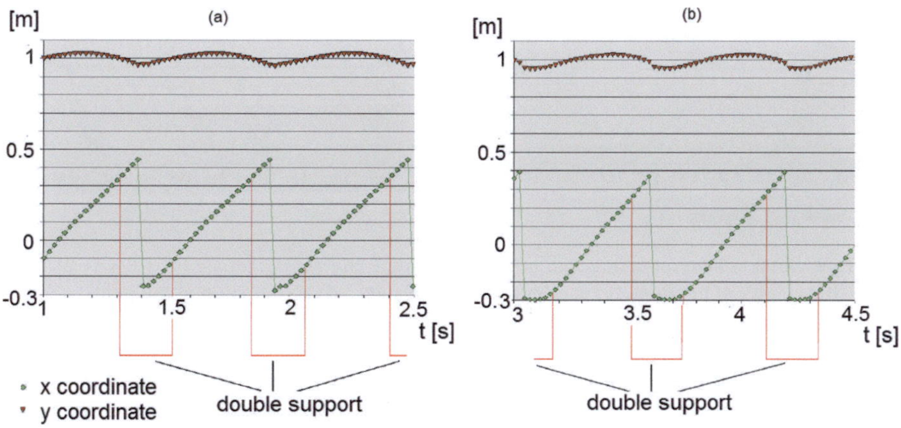

Fig. 15.8 Position of body mass center: (**a**) human gait, (**b**) generated gait

For final comparison, the trajectories of center of mass, angular velocities and accelerations of the hip and knee joints were calculated. Looking to Fig. 15.8 it can be noticed that coordinates x, z of center of mass (measured in the reference frame attached to the supporting foot – Fig. 15.6a, b) are changing similarly in human and generated gait. The small time shift in the time axes of those two figures was introduced intentionally following the start time of our simulations.

Figures 15.9a, b illustrate angular velocities and Figs. 15.9c, d show angular accelerations. For velocities the greatest difference between real and generated gait occurs for the knee joint during the double support phase and during the touch-down of next supporting leg (i.e., at time 2.4 to 2.6 s in the Fig. 15.9a). For the hip joint the differences are greatest in fully loaded single support phase (i.e., at time

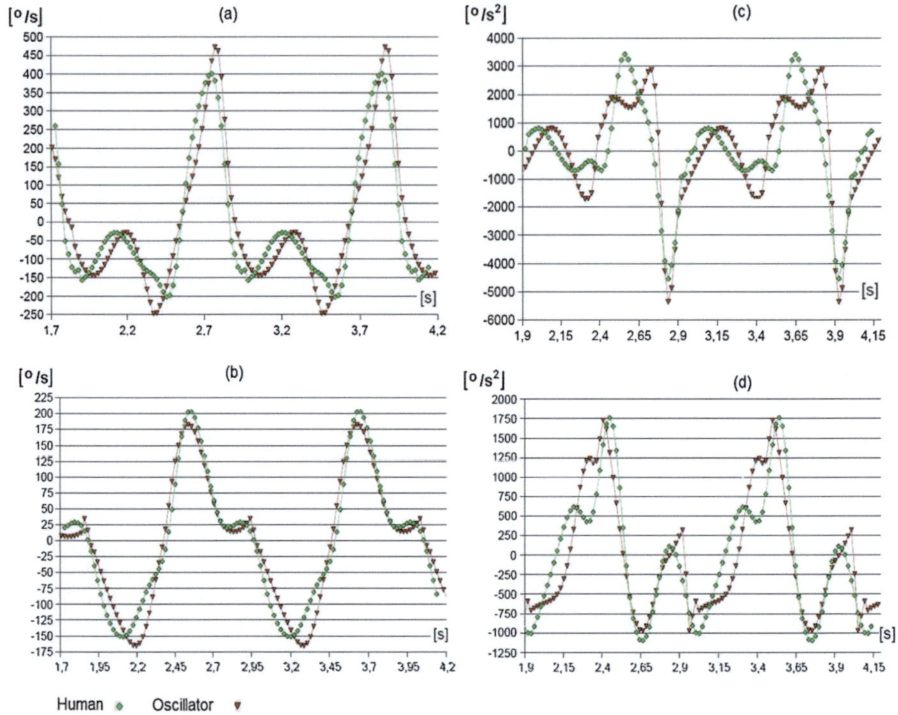

Fig. 15.9 Final validation taking into account velocities and accelerations: (**a**) velocity of knee joint, (**b**) velocity of hip joint, (**c**) acceleration of knee joint, (**d**) acceleration of hip joint

2.1 to 2.3 s in the Fig. 15.9b). The greatest difference in angular acceleration in the knee joint occurs during single support phase when the leg is almost straightened. This is during smallest angular transition in the knee joint which occurs near to $-10°$ (i.e., close to the time 2.2 s in the Fig. 15.9c). Another larger discrepancy is near the middle of the double support phase (i.e., at time 2.45 s). Considering the acceleration of the hip joint, the greatest difference is visible in the support phase when the thigh, moving backward towards the trunk, passes the vertical line (the moment when the hip angle drops bellow $0°$), and lasts until the end of the support phase which is close to minimal value of this angle (i.e., at time 2.1 to 2.4 in the Fig. 15.9d). Figure 15.10 shows the stick diagram of human and generated gaits. The differences obtained by analyzing ZMP position, velocities and accelerations are not noticeable in this diagram.

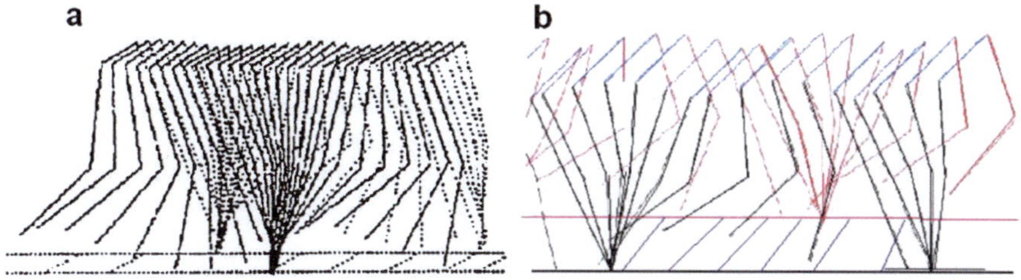

Fig. 15.10 Stick diagram: (**a**) – human gait, (**b**) – generated gait

Conclusions

The generic algorithm allowed final tuning of the oscillators parameters. This would be not possible by trial an errors due to large amount of parameters and factors which must be considered when obtaining the proper gait pattern. Adding the corrective function delivered finally the gait pattern very similar to the real gait. Discussed example illustrates well usability of genetic algorithm as un-typical optimization method and proposes biologically inspired method for humanoid robot gait generation.

References

1. Barricelli NA (1954) Esempi numerici di processi di evoluzione. Methods 6:45–68
2. Barricelli NA (1957) Symbiogenetic evolution processes realized by artificial methods. Methods 9:143–182
3. Bay JS, Hemami H (1987) Modeling of a neural pattern generator with coupled nonlinear oscillators. IEEE Trans Biomed Eng BME-34(4):297–306
4. Buchli J, Righetti L, Ijspeert AJ (2006) Engineering entrainment and adaptation in limit circle systems. From bilogical inspiration to application in robotics. Biol Cybern 95:645–664
5. Crosby JL (1973) Computer simulation in genetics. Wiley, London
6. Fogel DB (ed) (1998) Evolutionary computation: the fossil record. IEEE, New York
7. Fraser A (1957) Simulation of genetic systems by automatic digital computers. I. Introduction. Aust J Biol Sci 10:492–499
8. Fraser A, Burnell D (1970) Computer models in genetics. McGraw-Hill, New York
9. Harada K, Yoshida E, Yokoi K (eds) (2010) Motion planning for humanoid robots. Springer
10. Holland JH (1992) Adaptation in natural and artificial systems: an introductory analysis with applications to biology, control and artificial intelligence. MIT (reprint)
11. Ijspeert AJ (2008) Central pattern generators for locomotion control in animals and robots; a review. Preprint of Neural Networks 21/4:642–653
12. Kulic D, Venture G, Yamane K, Demircan E, Mizuuchi I, Mombaur K (2016) Anthropomorphic movement analysis and synthesis: a survey of methods and applications. IEEE Trans Robot 32(4):776–795
13. Nakanishi J, Morimoto J, Endo G, Cheng G, Schaal S, Kawato M (2004) Learning from demonstration and adaptation of biped locomotion. Robot Auton Syst 47:79–91

14. Ouezdou FB, Konno A et al (2002) ROBIAN biped project – a tool for the analysis of the human-being locomotion system. In: Proceedings of the 5th international conference on climbing and walking robots
15. Santos CP, Juan NA, Moreno C (2017) Biped locomotion control through a biomimetic CPG-based controller. J Intell Robotic Syst 85(1):47–70
16. Turing AM (1950) Computing machinery and intelligence. Mind LIX(238):433–460
17. Vaughan ChL, Davis BL, O'Connor JC (1992) Dynamics of human gait. Champaign: Human Kinetics Publishers
18. Vukobratovic M, Borovac B (2004) Zero-moment point – thirty five years of its life. Int J HR 1(1):157–173
19. Vukobratovic M, Stepanenko Y (1972) On the stability of anthropomorphic systems. Math Biosci 15:1–37
20. Winter DA (1991) Biomechanics and motor control of human gait: normal, elderly and pathological. University of Waterloo, Ontario
21. Zielinska T (1996) Coupled oscillators utilized as gait rhythm generators of a two-legged walking machine. Biol Cybern 74:263–273
22. Zielinska T, Chew ChM, Kryczka P, Jargilo T (2009) Robot gait synthesis using the scheme of human motion skills development. Mech Mach Theory 44(3):541–558

Chapter 16
Linked Simulation Optimization Model for Evaluation of Optimal Bank Protection Measures

Hriday Mani Kalita, Rajib Kumar Bhattacharjya, and Arup Kumar Sarma

Abstract This work proposes a new methodology for determination of cost effective river training work using groynes. The proposed methodology links the two dimensional (2D) hydrodynamic model with genetic algorithm based optimization model. The hydrodynamic model uses Beam and Warming implicit finite difference scheme for solution of the governing equations of unsteady free surface flows in general coordinate system. The optimization model minimizes the total construction cost of the groyne system required for achieving desirable training of the river. Binary coded GA is used for solving the proposed optimization problem. The efficiency and field applicability of the developed model is evaluated using two different test problems. The first problem considers a hypothetical meandering channel and the second problem deals with a vulnerable reach of river Brahmaputra in India. The performance evaluation of the model shows potential of the developed model for field application.

Keywords River training · Groyne · Beam and warming scheme · Genetic algorithm · Simulation- optimization · Hydrodynamic model

Introduction

River training is a very important activity for management of water resources of a region. Groynes or spur dikes are commonly used river training structures, which are constructed from the bank extending towards the main flow. Groynes are constructed

H. M. Kalita (✉)
Department of Civil Engineering, National Institute of Technology Meghalaya, Shillong, Meghalaya, India

R. K. Bhattacharjya · A. K. Sarma
Department of Civil Engineering, Indian Institute of Technology Guwahati, Guwahati, Assam, India
e-mail: rkbc@iitg.ernet.in; aks@iitg.ernet.in

F. Bennis, R. K. Bhattacharjya (eds.), *Nature-Inspired Methods for Metaheuristics Optimization*, Modeling and Optimization in Science and Technologies 16,
https://doi.org/10.1007/978-3-030-26458-1_16

for various purposes; one of which is the reduction of water speed to promote sedimentation and to prevent the bank erosion. The construction cost of a groyne is very expensive, as it requires huge quantity of high quality materials to carry out the construction. Therefore, an optimal combination of groynes in terms of their number, position and length is necessary to minimize the total construction cost of the project.

For obtaining the optimal combination of groynes, the 2D hydrodynamic flow simulation model is required to be incorporated with the optimization model. Extensive studies by numerous investigators [9, 18, 19, 23, 24, 33, 37, 39, 43] have been carried out for developing hydrodynamic flow simulation model using different numerical techniques such as, finite difference, finite element and finite volume methods. Among them, Beam and Warming implicit finite difference method is one of the excellent methods that can be used for solution of shallow water equations. Beam & Warming [7] developed this method for solution of the two dimensional Euler equations. The major advantage of the Beam & Warming scheme is that it is a non-iterative in nature method. It has the capability for quickly converge and has the ability of remain stable for a larger time step than that of explicit scheme. Many researchers [20, 23, 25, 28, 30, 34] have applied this scheme for solution of the unsteady free surface flow equations to simulate wide varieties of hydrodynamic problems. Beam and Warming scheme is used in this study for developing the 2D hydrodynamic model.

Numerous experimental as well as mathematical model studies [1, 17, 27, 28, 35, 41] have been carried out for simulating the flow process around groyne. However, only few studies [3, 11, 15] have focused to find the optimal combination of groynes. The studies so far conducted considers some pre-decided combination of groynes and the best one is found out from the chosen alternatives. Recently, Kalita et al. [23] have proposed a linked simulation-optimization model for finding true optimal combination of groynes by utilizing optimization model. The restriction of their model is that it is applicable to straight channel only. In this present study, a more general model is proposed, which is capable to handle any kind of river shape.

In order to obtain an optimal combination, the flow model needs to incorporate with the optimization model. The flow model can be incorporated with the optimization model using linked simulation-optimization approach [4, 6, 10, 23, 31, 38, 40]. The present study also uses the linked simulation-optimization approach where the hydrodynamic model which solves the governing equations of unsteady free surface flow is linked with an optimization model. This approach can automatically evaluate the optimal combination of groynes leading to minimum construction cost.

An optimization technique is necessary to solve the simulation-optimization model. Many investigators [2, 12, 14, 16, 22, 32, 42] applied traditional gradient based techniques and global search methods such as the genetic algorithms for solving various problems related to water resources engineering. Genetic algorithms are generally suitable for solving non-linear non-convex problems as they have the mechanism to overcome the local optimal solutions. Genetic algorithm is an iterative based robust search technique which is based on the evolution process of natural species. One of the major differences between genetic algorithms and the other classical optimization search techniques is that the genetic algorithm works with a population of possible solutions, whereas the classical optimization techniques work with a single initial solution. Further, binary coded genetic algorithm is also efficient in handling integer variable problem. As such, this study uses genetic algorithm for solving the proposed linked simulation-optimization model for finding optimal combination of groynes.

Hydrodynamic Model

Governing Equations and Solution Technique

The equations describing 2D unsteady free surface flow in a boundary fitted coordinate system (ξ, η) are [8]:

$$\frac{\partial L}{\partial t}+\frac{\partial M}{\partial \xi}+\frac{\partial N}{\partial \eta}=\mathrm{P} \tag{16.1}$$

Where the variables L, M, N, and O are defined in matrix form as:

$$L=J\begin{Bmatrix} h \\ hu \\ hv \end{Bmatrix},\ M=J\begin{Bmatrix} hU \\ hUu+\xi_x\frac{gh^2}{2} \\ hUv+\xi_y\frac{gh^2}{2} \end{Bmatrix},$$

$$N=J\begin{Bmatrix} hV \\ hVu+\eta_x\frac{gh^2}{2} \\ hVv+\eta_y\frac{gh^2}{2} \end{Bmatrix},\ P=J\begin{Bmatrix} 0 \\ -gh\left(S_{ox}-S_{fx}\right) \\ -gh\left(S_{oy}-S_{fy}\right) \end{Bmatrix}$$

In the above relation J is determinant of the coordinate transformation jacobian, h is the depth of flow, u is the velocity component in x direction, v is velocity component in y direction, U is the velocity component in ξ direction, V is the velocity component in η direction, g is acceleration due to gravity, S_{ox} is bed slope in x direction, S_{oy} is bed slope in y direction, S_{fx} and S_{fy} are the friction slopes in x and y direction respectively and are calculated using Manning's equation. Following relations exist for the coordinate transformation matrices:

$$J=x_\xi y_\eta-y_\xi x_\eta,\ \xi_x=\frac{y_\eta}{J},\ \xi_y=-\frac{x_\eta}{J},\ \eta_x=-\frac{y_\xi}{J},\ \eta_y=\frac{x_\xi}{J} \tag{16.2}$$

The matrices x_ξ,x_η,y_ξ,y_η calculated using central finite difference approximations. The relation between velocities of computational and physical coordinates is given by the following equation.

$$\begin{Bmatrix} U \\ V \end{Bmatrix}=\begin{Bmatrix} \xi_x & \xi_y \\ \eta_x & \eta_y \end{Bmatrix}\begin{Bmatrix} u \\ v \end{Bmatrix} \tag{16.3}$$

Equation (16.1) presented above is solved using Beam and Warming implicit finite difference scheme [13] in the computational coordinate. Beam and Warming scheme is a second order accurate Alternate Direction Implicit (ADI) scheme. This scheme consists of two-step (double sweep) sequence, where each step involves the solution of a block tridiagonal matrix. For the efficient solution of the block tridiagonal system, Thomas algorithm [5] is used. The equations solved in ξ-direction and η-direction are as follows:

$$\left[I + \Delta t \frac{\theta}{1+\gamma} \frac{\partial}{\partial \xi} A^k\right] L^* = -\Delta t \frac{1}{1+\gamma} \left(\frac{\partial M}{\partial \xi} + \frac{\partial N}{\partial \eta} + P \right)^k + \frac{\gamma}{1+\gamma} \Delta_t L^k \tag{16.4}$$

$$\left[I + \Delta t \frac{\theta}{1+\gamma} \left(\frac{\partial}{\partial \eta} B + Q \right)^k \right] L^{**} = L^* \tag{16.5}$$

$$L^{k+1} = L^k + L^{**} \tag{16.6}$$

Where, I is identity matrix of size 3 by 3; Δt is the time step, θ and γ are parameters of Beam and Warming scheme, k is time step index, L^* is incremental value obtained for matrix L after one sweep, $\Delta_t L^k$ is the increment of matrix L from the previous time step, L^{**} is the final increment of the matrix L and A, B and Q are the jacobians of matrices M, N and P respectively. Following expressions are exists for the jacobians A, B and Q.

$$A = \begin{Bmatrix} 0 & \xi_x & \xi_y \\ gh\xi_x - Uu & U + u\xi_x & u\xi_y \\ gh\xi_y - Uv & v\xi_x & U + v\xi_y \end{Bmatrix}, \quad B = \begin{Bmatrix} 0 & \eta_x & \eta_y \\ gh\eta_x - Vu & V + u\eta_x & u\eta_y \\ gh\eta_y - Vv & v\eta_x & V + v\eta_y \end{Bmatrix},$$

$$Q = \begin{Bmatrix} 0 & 0 & 0 \\ -gS_{ox} - \frac{7}{3} g n^2 u^2 h^{-\frac{4}{3}} & 2gn^2 u h^{-\frac{4}{3}} & 0 \\ -gS_{oy} - \frac{7}{3} g n^2 v^2 h^{-\frac{4}{3}} & 0 & 2gn^2 v h^{-\frac{4}{3}} \end{Bmatrix}$$

In this work $\theta = 1$ and $\gamma = 0.5$ (Three point backward formulation) is considered. The primitive flow variables at each step are calculated as:

$$h_{i,j}^{k+1} = h_{i,j}^k + \overline{h}_{i,j}^{**}$$

$$u_{i,j}^{k+1} = \frac{uh_{i,j}^k + \overline{uh}_{i,j}^{**}}{h_{i,j}^{k+1}}$$

$$v_{i,j}^{k+1} = \frac{vh_{i,j}^k + \overline{vh}_{i,j}^{**}}{h_{i,j}^{k+1}} \tag{16.7}$$

Boundary Condition

For solution of the Eqs. (16.4), (16.5) and (16.6) presented above, boundary conditions are required at each computational boundary. For supercritical case, at upstream all the three variables are needed. No conditions are required at downstream boundary, as no characteristics enter into the domain from the downstream boundary. However for subcritical case, two boundary conditions are required at upstream and one is needed at downstream. When there is no need of boundary condition one some boundaries, then extrapolation technique is generally used to get the values of flow variables. Free slip boundary technique [26] is used for simulating the banks. Reflection procedure [13] is used for simulating the groyne. The sand bars are simulated using no slip boundary technique [29].

Courant-Friedrichs-Lewy Condition

Since the implicit scheme presented above is a time marching method, Δt must be specified tosatisfy the Courant-Friedrichs-Lewy (CFL) condition. Equation (16.8) gives the value of Δt.

$$\Delta t = CN * \min\left[\left|\frac{U}{\Delta\xi}\right| + \left|\frac{V}{\Delta\eta}\right| + \sqrt{gh}\left\{\left(\frac{\xi_x}{\Delta\xi} + \frac{\eta_x}{\Delta\eta}\right)^2 + \left(\frac{\xi_y}{\Delta\xi} + \frac{\eta_y}{\Delta\eta}\right)^2\right\}^{\frac{1}{2}}\right]^{-1} \tag{16.8}$$

Where, *CN* is the Courant number. Molls & Zhao [30] used Beam and Warming scheme and observed that the solution remains stable for a higher value of *CN* of 2.5 for their case. However for explicit scheme this value is limited to one only.

Artificial Viscosity

For the present numerical scheme, artificial viscosity is required to damp the high-frequency oscillations [13]. Following artificial viscosity (Eq. 16.9) is added to the right hand side of the Eq. (16.4).

$$AV = Ct \times \left[\left(L_{i,j+1} - 2L_{i,j} + L_{i,j-1}\right) + \left(L_{i+1,j} - 2L_{i,j} + L_{i-1,j}\right)\right] \tag{16.9}$$

Where, *Ct* is a dissipation coefficient and a small value of *Ct* which is necessary to damp the numerical oscillations should be used.

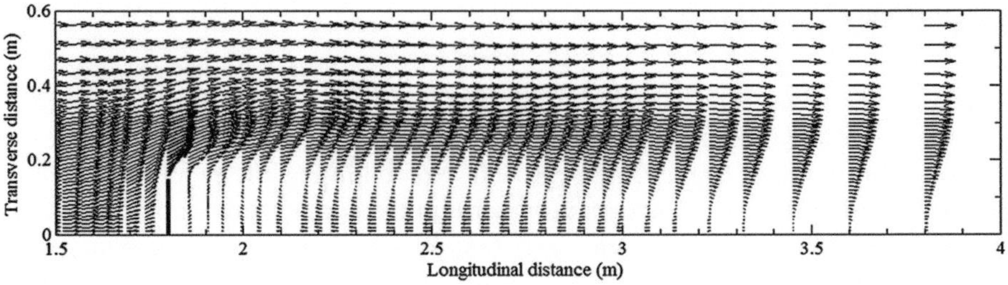

Fig. 16.1 Computed velocity vector by the present model for the Rajaratnam & Nwachukwu [35] experimental channel

Hydrodynamic Model Validation

For testing the present hydrodynamic model for flow simulation around groyne, the experimental data from Rajaratnam & Nwachukwu [35] are used. The experimental studies were done on a straight rectangular flume of 37 m long and 0.914 m wide. The groyne considered was an aluminum plate, of length 0.15 m. These experimental data were used by many researchers [21, 36] for validation of their numerical model. In all of the studies presented above, extent of the area that has been simulated mathematically was from 1.8 m upstream of the groyne to 3.6 m downstream of the groyne. In this present study also, same extent is considered. The study area was discretized into finite difference grid of size 97 by 45. Manning's n value is taken equal to 0.01, as the channel is said to be a smooth one. At the upstream boundary, u and v are given as 0.253 m/s and 0 m/s respectively. The value of h is extrapolated from interior domain. However at downstream boundary, the values of u and v are extrapolated from interior domain and h is given as 0.189 m. The courant number is set equal to 2.2 and Ct is considered as 0.01.

Figure 16.1 shows the velocity vector plot of the final solution. The recirculation zone developed at downstream of the groyne ends at 3.5 m. The value of the same in experimental case was 3.6 m. The resultant velocity (W) is non-dimensionalized with respect to free stream velocity of 0.253 m/s. Figure 16.2 shows the comparison of simulated and experimental velocity at different distances from the bank where groyne is attached. These results show very good agreement between the numerical and experimental data.

Optimization Model Formulation

The proposed linked simulation-optimization model minimizes the total construction cost of the groyne system for reducing the flow speed to a desired minimum value. Two different formulations of the optimization model are prepared as given below:

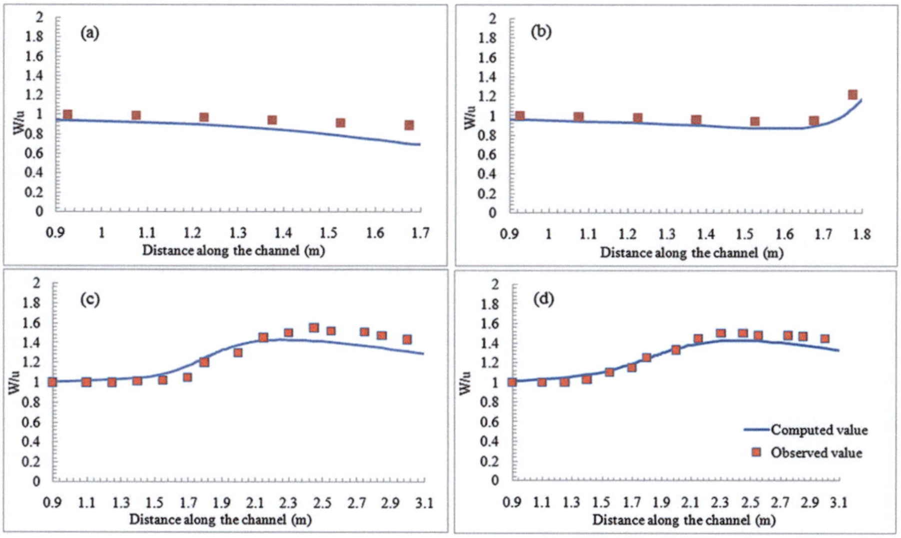

Fig. 16.2 Comparison of simulated and experimental speed for the Rajaratnam & Nwachukwu [35] experimental channel. (**a**) 0.15 m from the bank; (**b**) 0.225 m from the bank; (**c**) 0.4 m from the bank; (**d**) 0.6 m from the bank

Formulation I

The formulation I deals with determination of optimal number, position and length of the groynes on both the banks of a river to achieve a predefined target speed on a predefined area. The objective is to minimize the total construction cost of the project. The number of groynes having non-zero length is the required number of groynes. The optimization model can be formulated as given below.

$$\text{Minimize } C\,(Lrg, Llg) = \sum\nolimits_{Ng=1}^{N} (Lrg + Llg) \times C1 \tag{16.10}$$

Subject to,

$$UU \leq \Omega, Llg_{min} \leq Llg \leq Llg_{max}, Plg_{min} \leq Plg \leq Plg_{max},$$

$$Lrg_{min} \leq Lrg \leq Lrg_{max}, Prg_{min} \leq Prg \leq Prg_{max} \tag{16.11}$$

where C is total construction cost of groynes in i ₹, *Lrg* is length of groyne on the right bank in terms of grid number, *Llg* is length of groyne on the left bank in terms of grid number, *Prg* is position of groyne on the right bank in terms of grid number from upstream boundary, *Plg* is position of groyne on the left bank in terms of grid number from upstream boundary, N is the initial number of groynes considered, *Ng* is total number of groynes required, $C1$ is per meter construction cost of the groyne system and is taken equal to ₹100,000, UU is maximum speed on the specified areas in m/s and is obtained at every iteration of optimization model by simulating the hydrodynamic process, and Ω is target speed value in m/s, Llg_{min} to Llg_{max} is extreme limit of groyne length in terms of grid number on the left bank, Plg_{min} to Plg_{max} is extreme limit of groyne position on the left bank in terms of grid number from the upstream boundary within which the groynes are to be placed, Lrg_{min} to Lrg_{max} is extreme limit of groyne length in terms of grid number on the right bank, Prg_{min} to Prg_{max} is extreme limit of groyne position on the right bank in terms of grid number within which the groynes are to be placed.

Formulation II

In the second formulation, the positions of the groynes are fixed on both the banks. The present model determines the optimal length and number of the groynes for achieving a predefined target speed on a predefined area leading to minimum construction cost. The mathematical formulation is as follows.

$$\text{Minimize } C\,(Lrg, Llg) = \sum\nolimits_{Ng=1}^{N} (Lrg + Llg) \times C1 \tag{16.12}$$

Subject to,

$$UU \leq \Omega, Llg_{min} \leq Llg \leq Llg_{max}, Lrg_{min} \leq Lrg \leq Lrg_{max} \tag{16.13}$$

Solution of Linked Simulation-Optimization Model Using Genetic Algorithm

The different formulations of the proposed linked simulation-optimization model presented above are solved using genetic algorithm. As discussed earlier, the variables in this optimization problem are number, length and position of groynes. Since, values of all these parameters are integer; the variables are coded using binary strings with an accuracy of one. This leads to results of the variables as an integer. The string lengths used for coding the variables for all the formulations are presented in the subsequent sections. The selection operation is carried out using the tournament selection method with a tourney size of two. The crossover probability of 0.9, the mutation probability of 0.003 and the elitism fraction of 0.1 are employed in all the formulations.

Figure 16.3 shows the schematic representation of the proposed genetic algorithm based linked simulation-optimization model. The initial population is created randomly between specified lower and upper bounds, which are actually some

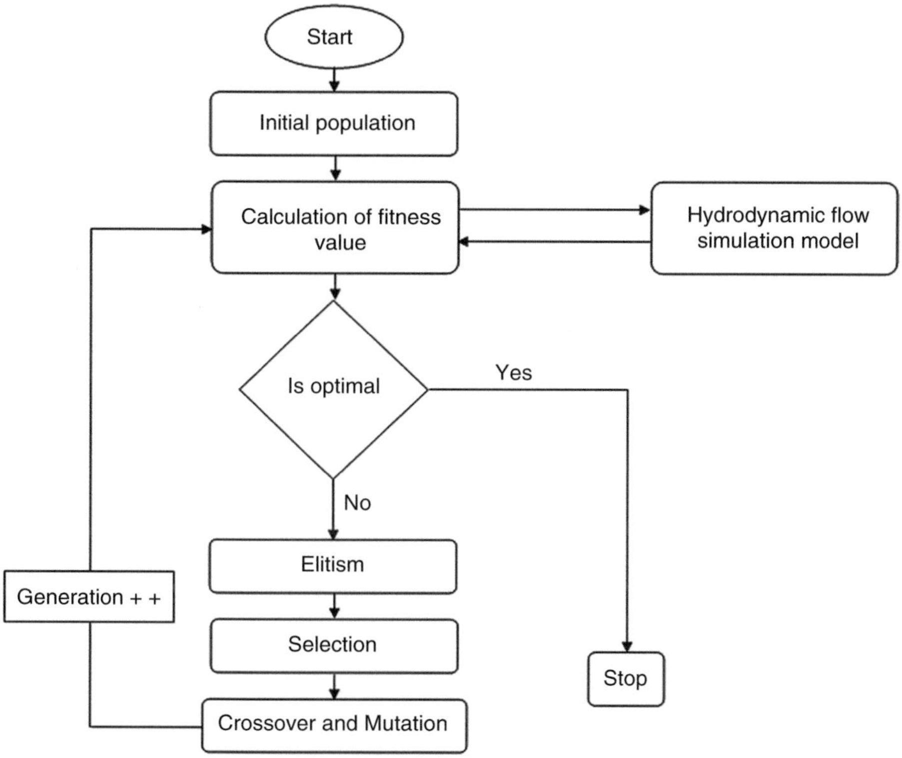

Fig. 16.3 Schematic representation of the proposed GA based linked simulation optimization model

strings (chromosome) representing the location and length of the groynes. For each chromosome, the model finds the fitness function value. The constraint function value is obtained by running the hydrodynamic model with the combination created by genetic algorithms. The population is then checked for termination criteria. If it is satisfied, the iteration terminates automatically. If not, the population again passes through the genetic operators, i.e. reproduction, crossover, mutation and elitism.

Application of the Proposed Methodology

The proposed model is tested for two different cases. Firstly, the model is applied for a hypothetical channel bend. For showing the field applicability of the proposed model, the model is also applied on a portion of the River Brahmaputra.

Case A: Hypothetical Channel Bend

Figure 16.4 shows the study area composed of two bends connected by a short tangent. The inner radiuses of the bends are 350 m each and the channel width is 240 m. The lengths of the approach channel, middle channel and the exit channel are 246 m, 123 m and 246 m respectively. The grid selected is 86 by 17. Figure 16.5 shows the finite difference grid along with proposed near bank low speed zones for this channel. The hydrodynamic model simulates a discharge of 1200 cumec. The values of *CN* and *Ct* are 1.2 and 0.24 respectively. Manning's roughness *n* is 0.031. Slope of the channel is 1:10000. Binary coded string of length 3 is used for coding the decision variable 'length of groyne' (*Llg* and *Lrg*). However, position of groyne (*Plg* and *Prg*) is coded using binary string of length 5.

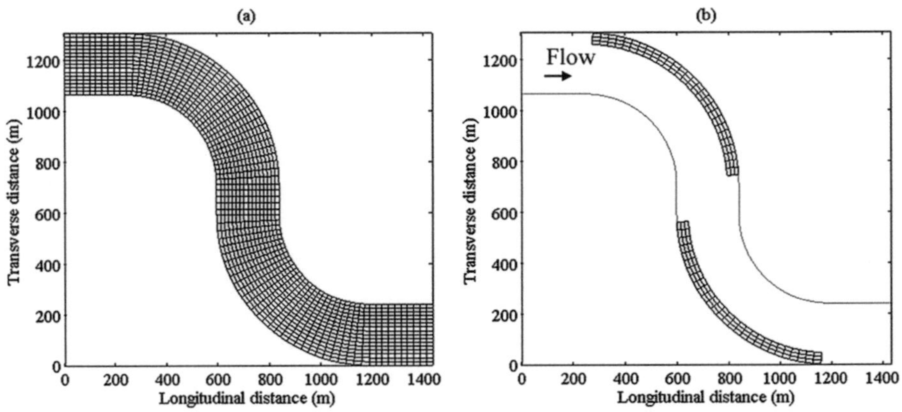

Fig. 16.4 Study area for the hypothetical channel bend. (**a**) Finite difference grid; (**b**) Area of speed reduction

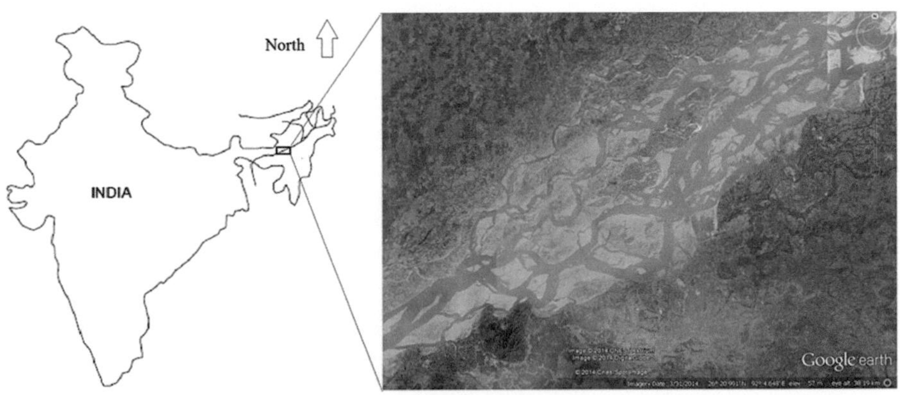

Fig. 16.5 Satellite image of the vulnerable reach of the River Brahmaputra

Case B: Application to River Brahmaputra

The River Brahmaputra and its tributaries constitute the major river system of northeastern part of India. Extensive bank erosion occurs particularly during flood in several reaches of River Brahmaputra. An area located on the south bank of River Brahmaputra in the Nagaon and Morigaon district of Assam in India is also facing the problem of extensive river bank erosion. Figure 16.5 shows the satellite image of the study area. To mitigate the erosion hazard of this area, the present linked simulation-optimization model was applied here to determine the optimal combination of groynes required to save this area. Figure 16.6 shows the generated grid of size 100 by 50 cells and the proposed low speed zone.

The hydrodynamic model simulates a high discharge of 40,000 cumec. The values of *CN*, *Ct*, and *n* are 1.2, 0.2 and 0.031 respectively. To have an idea of the present situation, the hydrodynamic model is run without considering any groyne. Figure 16.7 shows the velocity vector and speed contour map for the present condition. Velocity vectors (Fig. 16.7a) indicate that water is directed towards the south bank of the river in several locations. Speed contour (Fig. 16.7b) reveals high speed near the south bank indicating flow concentration at the south bank. These results lead to a conclusion that erosion is initiated on the southern bank primarily because of high flow speed near the bank.

As the erosion is near the south bank, the groynes are placed on the south bank only. Binary coding having string length 3 is used for coding the decision variable length of groyne (*Llg*). The decision variable, position of groyne (*Plg*) is coded using binary string of length 7.

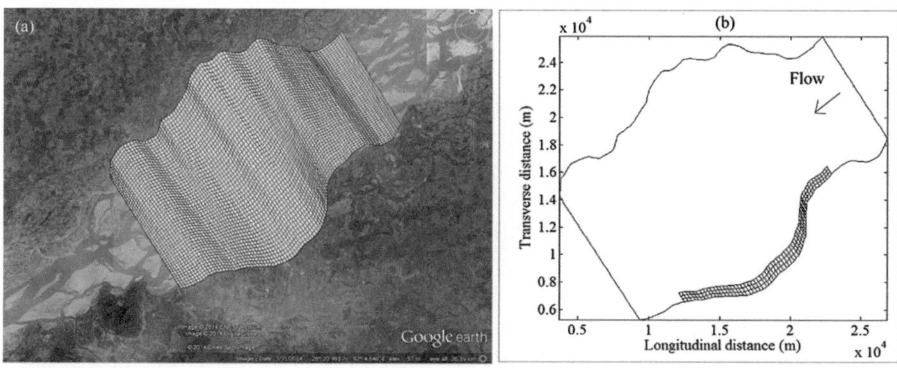

Fig. 16.6 Vulnerable reach of the River Brahmaputra. (**a**) Finite difference grid; (**b**) Area of speed reduction

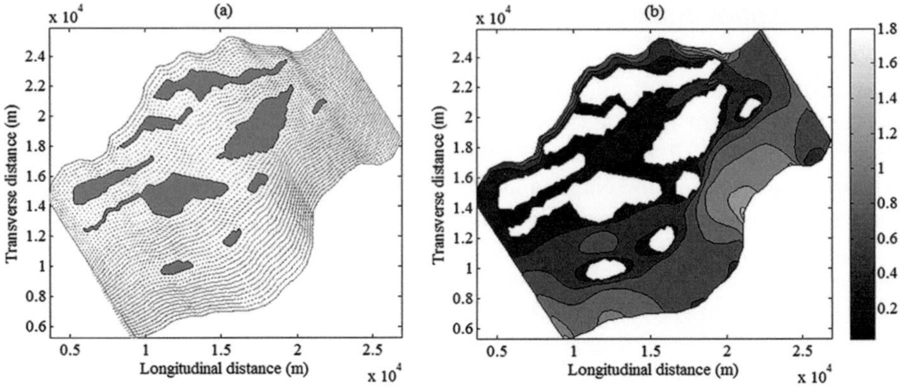

Fig. 16.7 Simulation of the present condition for the reach of River Brahmaputra. (**a**) Velocity vector; (**b**) Speed contour map

Results and Discussion

The proposed genetic algorithms based linked simulation-optimization model is run for both the channels with different formulations. The results obtained are presented in the subsequent sections.

Case A: Hypothetical Channel Bend

Formulation I minimizes the total construction cost of the groyne system for attaining a target speed value of 0.3 m/s on the predefined areas on both the banks. The value of Plg_{min} and Plg_{max} are grid numbers 11 and 41 respectively. Similarly, at the right bank values of Prg_{min} and Prg_{max} are 46 and 76 respectively. The values of Llg_{min} and Llg_{max} are 0 and 4 respectively. Similarly, value of Lrg_{min} and Lrg_{max} are also considered as 0 and 4 respectively. Number of possible groynes on each of the banks is initially assumed as 6 within the position limits. This leads to maximum of 12 possible groynes. The associated variables with all of these 12 groynes are the position of groyne and length of groyne. As such, the number of variables becomes 24. The population size is 240. The actual length of the groyne is calculated by multiplying the transverse grid spacing (15 m in this channel) with the integer form of groyne length obtained from the optimization model.

The present model results in 7 number ($Ng = 7$) of groynes for attaining the predefined target speed value. Out of these 7 groynes, 4 groynes having length of 45 m, 30 m, 15 m and 60 m are needed on the left bank at grid locations 13, 25, 28 and 41 respectively. The other 3 groynes of length 30 m, 15 m, and 15 m are to be placed on the right bank at locations 46, 60 and 75 respectively from the upstream boundary. Figure 16.8 shows the velocity vector plot and speed contour map with

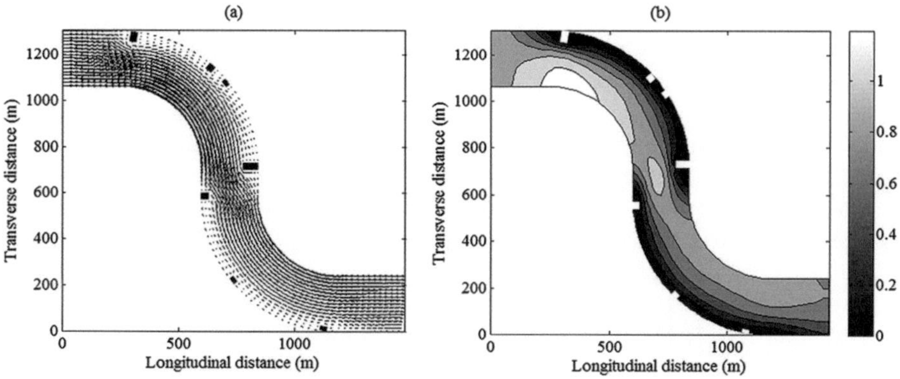

Fig. 16.8 Optimal combination of groynes for hypothetical channel bend by formulation I with target speed of 0.3 m/s. (**a**) Velocity vector; (**b**) Speed contour

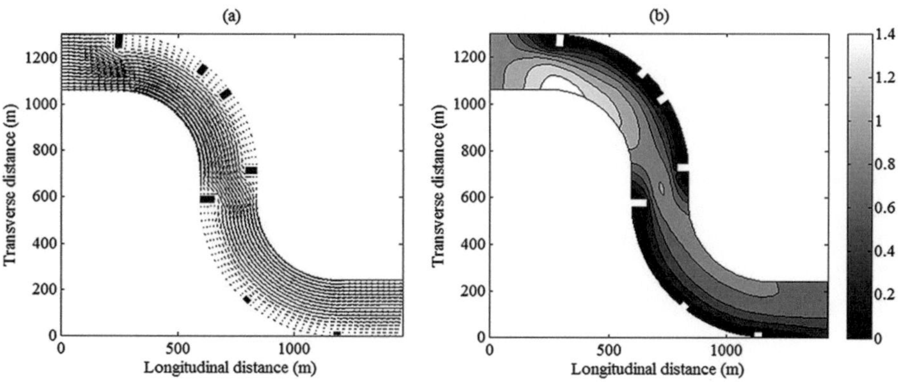

Fig. 16.9 Optimal combination of groynes for hypothetical channel bend by formulation I with target speed of 0.2 m/s. (**a**) Velocity vector; (**b**) Speed contour

the proposed combination of the groynes. Due to the presence of the groynes, the velocity vector shows a deflected pattern from the banks towards the central portion of the channel. In between the groynes, the lengths of the velocity vectors are very less representing very low speed on those areas. For this combination, the total construction cost is found as ₹2.10 $\times$ 10^7. Another experiment is conducted for achieving target speed value of 0.2 m/s on the predefined areas and the models results a need of 7 number of groynes ($Ng = 7$). Out of these 7 groynes, 4 groynes with length of 60 m, 45 m, 45 m and 45 m are needed on the left bank at the grid locations of 11, 24, 29 and 41 respectively. On the right bank, optimal lengths of the groynes are 60 m, 15 m and 15 m and the positions are 46, 62 and 76 respectively. Figure 16.9 shows the velocity vector and speed contour map. As a result of increase in length of the groynes, the total construction cost is also raised to ₹2.85 $\times$ 10^7.

In case of formulation II, the positions of the groynes are already fixed on both the banks. On the left bank, the uniformly distributed positions of groynes are grid number 11, 16, 21, 26, 31, 36 and 41. On the right bank, these positions are on grid number 46, 51, 56, 61, 66, 71 and 76. For this case, possible maximum number of groynes is 14, which is the summation of the number of groynes on both the banks. For each of the groyne, the associated variable is only length of groynes as positions of the groynes are already fixed. Thus, the number of variables is 14. The initial population is considered as 140 and the values of Llg_{min} and Llg_{max} are 0 and 4 respectively. Similarly the values of Lrg_{min} and Lrg_{max} are also 0 and 4 respectively.

The formulation II of the proposed model is run to find the required length of groynes on those specified 14 positions to achieve a target speed of 0.3 m/s on the predefined areas. The present model results in 7 numbers of groynes. Out of these, 4 groynes having lengths of 30 m, 15 m, 45 m and 60 m are needed on the left bank at grid numbers 11, 16, 26 and 41 respectively. Similarly, the optimal positions of the groynes on the right bank are 46, 61 and 76 with optimal lengths of 30 m, 15 m and 15 m respectively. The associated total construction cost is ₹2.10×10^7. Cost of construction for this case is similar to formulation I which shows that the problem has alternate optimal solutions. Figure 16.10 shows the velocity vector plot and speed contour map for the optimal combination of groynes. In order to reduce the target speed to 0.2 m/s also, 7 numbers of groynes is sufficient. However, position and length of the groynes are different than the previous case. Out of these 7 groynes, 4 groynes having length of 45 m each are needed on the left bank at positions of 11, 21, 31 and 41 respectively. The remaining 3 groynes are needed at the right bank at optimal positions of 46, 61 and 76 having length of 45 m, 15 m and 45 m respectively. The total construction cost for this groyne system is ₹2.85×10^7. In this case also, the construction cost is similar to formulation I. Figure 16.11 shows the velocity vector plot and speed contour map with this proposed combination of groynes. One important observation is that, length of the groynes required is more when the reduction of speed required is more. These results are intuitively as expected.

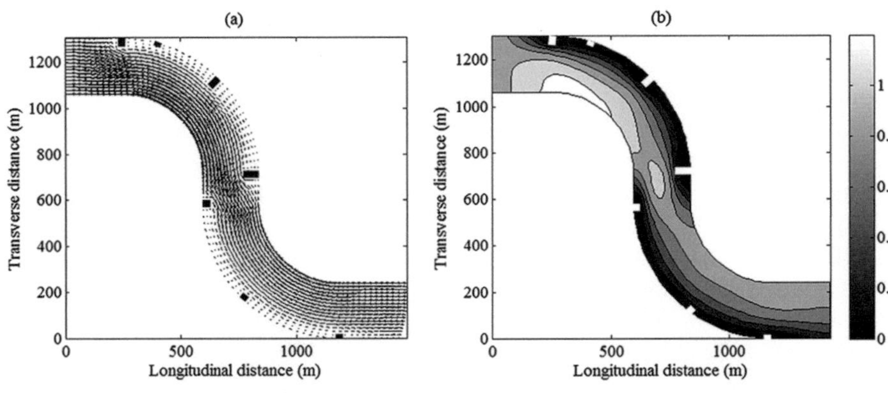

Fig. 16.10 Optimal combination of groynes for hypothetical channel bend by formulation II with target speed of 0.3 m/s. (**a**) Velocity vector; (**b**) Speed contour

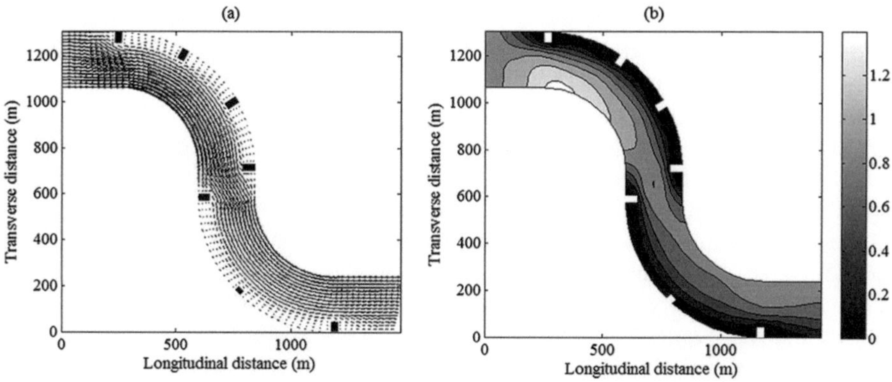

Fig. 16.11 Optimal combination of groynes for hypothetical channel bend by formulation II with target speed of 0.2 m/s. (**a**) Velocity vector; (**b**) Speed contour

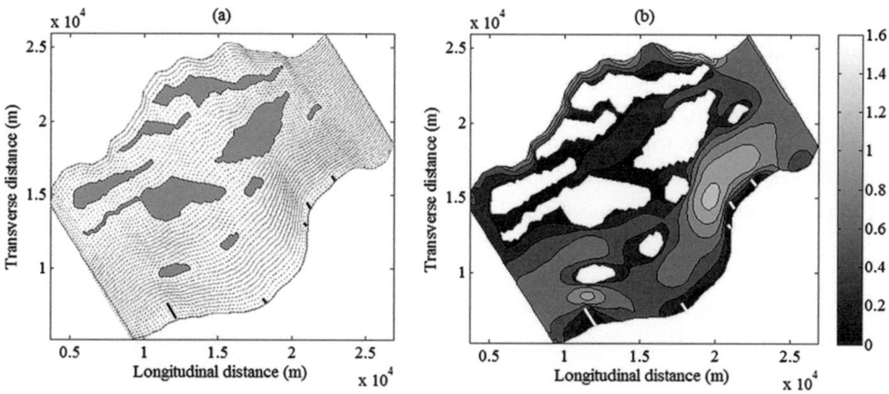

Fig. 16.12 Optimal combination of groynes for River Brahmaputra by formulation I with target speed of 0.6 m/s. (**a**) Velocity vector; (**b**) Speed contour

Case B: Application to River Brahmaputra

The main objective the model in this case is to find out the number of groynes which are to be placed on the reach so that the bank of the river can be protected. In the first case, flow speed limit is 0.6 m/sec. The possible maximum number of groynes is 7. Thus initial population is considered as 140. The values of Plg_{min} and Plg_{max} are grid numbers 18 and 90 respectively and the values of Llg_{min} and Llg_{max} are 0 and 5 respectively. The optimal solution shows that 5 numbers of groynes ($Ng = 5$) are necessary for attaining the predefined target speed value. The lengths of these groynes are 440 m, 500 m, 260 m, 320 m and 1250 m with corresponding positions on grid numbers 22, 32, 37, 59 and 87 respectively from the upstream boundary. For this combination, the total construction is ₹2.77×10^8. Figure 16.12 shows the velocity vector plot and speed contour map with the proposed combination of groyne for this case. The present model also results in 5 numbers of groynes, to achieve another target speed of 0.5 m/s on the predefined area. The lengths of these groynes are found as 1160 m, 1040 m, 320 m, 280 m and 1020 m with corresponding positions on grid numbers 26, 37, 61, 75 and 88 respectively from upstream boundary. Due to the increase in length of groynes, the total construction cost has increased to ₹3.82 × 108. Figure 16.13 shows the velocity vector plot and speed contour map with the proposed combination of groyne.

For formulation II, initially assumed groyne positions are grid numbers 8, 27, 36, 45, 54, 63, 72, 81 and 90 from upstream boundary. The initial population is 90. The values of Llg_{min} and Llg_{max} are 0 and 5 respectively. The proposed model is run to find the required lengths of groynes on those positions to achieve a target speed of 0.6 m/s. For this case, the present model results in 6 numbers of groynes. The optimal position of these 6 groynes are grid numbers 18, 27, 36, 72, 81 and 90, having optimal length of 200 m, 720 m, 800 m, 300 m, 500 m and 260 m respectively. The total construction cost is ₹2.78 × 108. Figure 16.14 shows the velocity vector plot and speed contour map for this combination. With a target speed value of 0.5 m/s on the predefined area, the present model results in 7 numbers of groynes. The optimal position of these groynes are 18, 27, 36, 54, 63, 81 and 90, having optimal lengths of 200 m, 960 m, 800 m, 340 m, 330 m, 1000 m and 260 m respectively. Due to the increase in the length of groynes, the total construction cost has also increased to ₹3.89 × 108. Figure 16.15 shows the velocity vector plot and speed contour map for this combination.

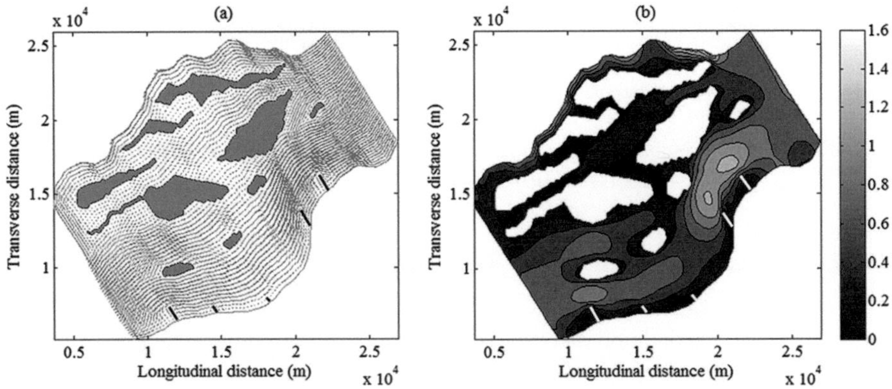

Fig. 16.13 Optimal combination of groynes for River Brahmaputra by formulation I with target speed of 0.5 m/s. (**a**) Velocity vector; (**b**) Speed contour

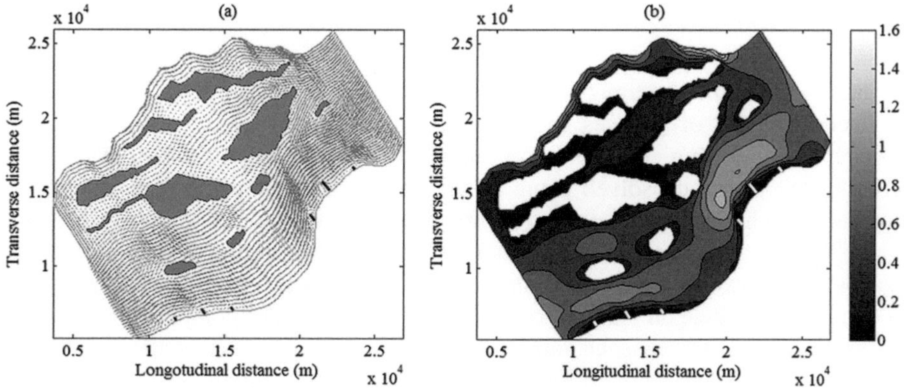

Fig. 16.14 Optimal combination of groynes for River Brahmaputra by formulation II with target speed of 0.6 m/s. (**a**) Velocity vector; (**b**) Speed contour

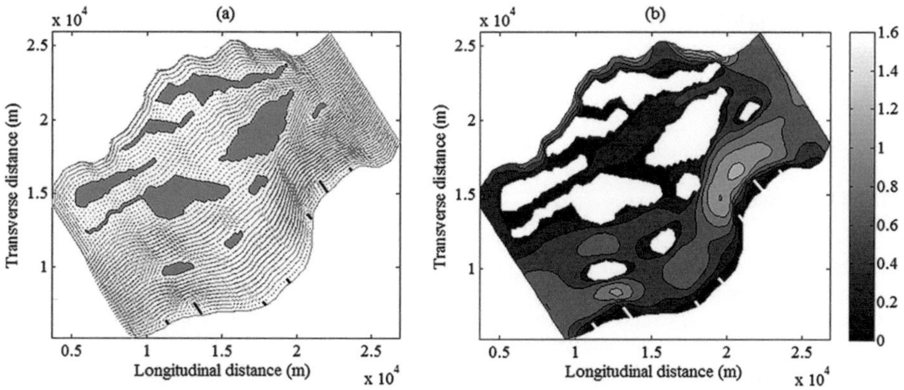

Fig. 16.15 Optimal combination of groynes for River Brahmaputra by formulation II with target speed of 0.5 m/s. (**a**) Velocity vector; (**b**) Speed contour

Computational Time Requirement

The example problems are run on a Windows 7 based Intel Core 2 Quad CPU with 2.66 GHZ Processor computer. The genetic algorithm based linked simulation-optimization model is solved using Matlab. For checking the computational efficiency of the present methodology, the computational time required for all the channels for different formulations are found out. The computational time required for formulation I in both the channels are very high, i.e. 180 h for the hypothetical channel bend and 200 h for the reach of River Brahmaputra. Computational time required by formulation II is lesser than formulation I (105 h and 128 h for the channel bend and River Brahmaputra respectively). This shows that the time required by the proposed model is basically dependent upon the size of the problem being considered.

Conclusions

A linked simulation-optimization methodology is developed for obtaining optimal combination of groynes for required training of a river. In the simulation model, the governing equations of unsteady free surface flow in a boundary fitted coordinate system are solved using Beam and Warming implicit finite difference method. The hydrodynamic model is linked with a binary coded genetic algorithm based optimization model. The limited evaluation made through the two example problems shows that the cost of the project is dependent to the amount of speed reduction to be done on that particular reach of the river. The comparison of results between different target speed values for a same formulation reveals that if the target speed value reduces, required length of the groyne also increases which eventually increases the construction cost of the groyne system. Though computational time required for such genetic algorithm based linked simulation-optimization model is high, this may not be considered as a limitation, as number of alternatives tested during the computational procedure is enormous. Test of such large alternatives would have otherwise taken much more time and effort as compared to the time required by the model.

References

1. Ahmed H, Tanaka N, Tamai N (2011) Flow modeling and analysis of compound channel in river network with complex floodplains and groynes. J Hydroinf 13(3):474–488
2. Ahmed JA, Sarma AK (2005) Genetic algorithm for optimal operating policy of a multipurpose reservoir. Water Resour Manag 19(2):145–161
3. Alauddin M, Tsujimoto T (2012) Optimum configuration of groynes for stabilization of alluvial rivers with fine sediments. Int J Sediment Res 27(2):158–167
4. Aljuboori M, Datta B (2018) Linked simulation-optimization model for optimum hydraulic design of water retaining structures constructed on permeable soils. Int J Geomate 14(44):39–46
5. Anderson DA, Tannehill JD, Pletcher RH (1984) Computational fluid mechanics and heat transfer. McGraw-Hill, New York
6. Ayvaz MT (2017) Optimal dewatering of an excavation site by using the linked simulation – optimization approaches. Water Sci Technol Water Supply. https://doi.org/10.2166/ws.2017.175
7. Beam RM, Warming RF (1976) An implicit finite difference algorithm for hyperbolic systems in conservation law form. J Comput Phys 22:87–110
8. Bellos CV, Soulis JV, Sakkas JG (1991) Computation of two-dimensional dam-break-induced flows. Adv Water Resour 14(1):31–41
9. Bellos V, Hrissanthou V (2011) Numerical simulation of dam break flood wave. Eur Water 33:45–53
10. Bhattacharjya RK, Datta B (2009) ANN-GA based multiple objective management of coastal aquifers. J Water Resour Plan Manag 135(4):314–322
11. Bhuiyan F, Hey RD, Wormleaton PR (2010) Bank-attached vanes for bank erosion control and restoration of river meanders. J Hydraul Eng 136(9):583–596
12. Celeste A, Suzuki K, Kadota A (2004) Genetic algorithms for real-time operation of multipurpose water resource systems. J Hydroinf 6(1):19–38

13. Chaudhry MH (2008) Open channel flow, 2nd edn. Prentice Hall, Englewood Cliffs
14. Damodaram C, Zechman EM (2013) Simulation-optimization approach to design low impact development for managing peak flow alterations in urbanizing watersheds. J Water Resour Plan Manag 139(3):290–298
15. Dehghani AA, Azamathulla HM, Najafi SAH, Ayyoubzadeh SA (2013) Local scouring around L-head groynes. J Hydrol 504:125–131
16. Fang T, Ball JE (2007) Evaluation of spatially variable control parameters in a complex catchment modeling system: a genetic algorithm application. J Hydroinf 9(3):163–173
17. Fazli M, Ghodsian M, Neyshabouri SAAS (2008) Scour and flow around a spur dike in a 90^0 bend. Int J Sediment Res 23(1):56–68
18. Fennema RJ, Chaudhry MH (1990) Explicit methods for 2D transient free-surface flows. J Hydraul Eng 116(8):1013–1034
19. Haleem DA, Kesserwani G, Caviedes-Voullieme D (2015) Haar wavelet-based adaptive finite volume shallow water solver. J Hydroinf 17(6):857–873
20. Jha A, Akiyama J, Ura M (1994) Modeling unsteady open-channel flows-modification to Beam and Warming scheme. J Hydraul Eng 120(4):461–476
21. Jia Y, Wang SSY (1999) Numerical model for channel flow and morphological change studies. J Hydraul Eng 125(9):924–933
22. Johns MB, Keedwell E, Savic D (2014) Adaptive locally constrained genetic algorithm for least-cost water distribution network design. J Hydroinf 16(2):288–301
23. Kalita HM, Sarma AK, Bhattacharjya RK (2014) Evaluation of optimal river training work using GA based linked simulation-optimization approach. Water Resour Manag 28(8):2077–2092
24. Kalita HM (2016) A new total variation diminishing predictor corrector approach for two-dimensional shallow water flow. Water Resour Manag 30(4):1481–1497
25. Kassem AA, Chaudhry MH (2005) Effect of bed armoring on bed topography of channel bends. J Hydraul Eng 131(12):1136–1140
26. Klonidis AJ, Soulis JV (2001) An implicit scheme for steady two-dimensional free-surface flow calculation. J Hydraul Res 39(3):1–10
27. Kuhnle RA, Alonso CV, Shields FD (1999) Geometry of scour holes associated with 90^0 spur dikes. J Hydraul Eng 125(9):972–978
28. Molls T, Chaudhry MH, Khan KW (1995) Numerical simulation of two dimensional flow near a spur dike. Adv Water Resour 18(4):221–236
29. Molls T, Chaudhry MH (1995) Depth-averaged open-channel flow model. J Hydraul Eng 121(6):453–465
30. Molls T, Zhao G (2000) Depth-averaged simulation of supercritical flow in channel with wavy sidewall. J Hydraul Eng 126(6):437–445
31. Mousavi SJ, Shourian M (2010) Capacity optimization of hydropower storage projects using particle swarm optimization algorithm. J Hydroinf 12(3):275–291
32. Ostfeld A, Salomons E, Lahav O (2011) Chemical water stability in optimal operation of water distribution systems with blended desalinated water. J Water Resour Plan Manag 137(6):531–541
33. Petaccia G, Natale L, Savi F, Velickovic M, Zech Y, Soares-Frazao S (2013) Flood wave propagation in steep mountain rivers. J Hydroinf 15(1):120–137
34. Rahimpour M, Tavakoli A (2011) Multi-grid Beam and Warming scheme for the simulation of unsteady flow in an open channel. Water SA 37(2):229–236
35. Rajaratnam N, Nwachukwu A (1983) Flow near groin like structure. J Hydraul Eng 109(3):463–480
36. Sarveram H, Shamsai A, Banihashemi MA (2012) Two-dimensional simulation of flow pattern around a groyne using semi-implicit semi Lagrangian method. Int J Phys Sci 7(20):2775–2783
37. Schwanenberg D, Harms M (2004) Discontinuous Galerkin finite-element method for trans-critical two-dimensional shallow water flows. J Hydraul Eng 130(5):412–421

38. Shourian M, Mousavi SJ, Menhaj M, Jabbari E (2008) Neural network-based simulation optimization model for optimal water allocation planning at basin scale. J Hydroinf 10(4):331–343
39. Simons F, Busse T, Hou J, Ozgen I, Hinkelmann R (2014) A model for overland flow and associated processes within the hydroinformatics modelling system. J Hydroinf 16(2):375–391
40. Singh RM, Datta B (2006) Identification of groundwater pollution sources using GA-based linked simulation optimization model. J Hydrol Eng 11(2):101–109
41. Vaghefi M, Ghodsian M, Neyshabouri SAAS (2012) Experimental study on scour around a T shaped spur dike in a channel bend. J Hydraul Eng 138(5):471–474
42. Wu W, Simpson AR, Maier HR, Marchi A (2012) Incorporation of variable-speed pumping in multi objective genetic algorithm optimization of the design of water transmission systems. J Water Resour Plan Manag 138(5):543–552
43. Zhao DH, Shen HW, Tabios GQ, Lai JS, Tan WY (1994) Finite volume two dimensional unsteady flow model for river basins. J Hydraul Eng 120(7):863–883

Chapter 17
A GA Based Iterative Model for Identification of Unknown Groundwater Pollution Sources Considering Noisy Data

Leichombam Sophia and Rajib Kumar Bhattacharjya

Abstract Genetic Algorithms have been applied in solving various complex engineering optimization problems. This chapter presented the application of Genetic Algorithms in identifying unknown groundwater pollution sources of an aquifer. The unknown groundwater pollution sources can be identified by using the inverse optimization model. The inverse optimization model minimizes the difference between the simulated and observed concentration at the observation locations for obtaining the unknown pollution sources. However, the model cannot be setup unless and until the number of pollutions sources are not known. As such, an iterative based methodology is used to obtain the exact number of pollution sources along with their source strength. Further, it is not always possible to accurately measure the concentration data in the field. As such an analysis has been carried out to evaluate the model performance when noisy data is used for the prediction of the sources. The performance of the model is evaluated using an illustrative study area.

Keywords Groundwater pollution source · Inverse optimization model · Linked simulation-optimization · Genetic algorithm

Introduction

Groundwater is considered to be an important resource of water as it supplies almost half of the drinking water to the world [29]. It can be also regarded as a long-term reservoir because the amount of freshwater stored in the earth by groundwater is around 30.1% whereas surface water stores only a meagre

L. Sophia (✉)
College of Food Technology, Central Agricultural University, Imphal, India

R. K. Bhattacharjya
Department of Civil Engineering, Indian Institute of Technology Guwahati, Guwahati, Assam, India
e-mail: rkbc@iitg.ac.in

F. Bennis, R. K. Bhattacharjya (eds.), *Nature-Inspired Methods for Metaheuristics Optimization*, Modeling and Optimization in Science and Technologies 16,
https://doi.org/10.1007/978-3-030-26458-1_17

0.3% in the forms like river, streams, lakes etc. However, the immense growth in the industrial and the agricultural fields has led to numerous adverse effects on the socio-economic condition of different communities of the world. Unwanted activities like the improper dumping of industrial waste, leachate from landfill, overuse of pesticides, septic tanks leakage etc. have led to an alarming rate of groundwater contamination. Above this, the contamination of the groundwater has also given birth to numerous waterborne diseases to the mankind. Therefore, the identification of the groundwater pollution sources becomes the initial steps for efficiently remediating the contaminated groundwater aquifer.

In most of the situation, groundwater pollution sources can be identified effectively using linked simulation-optimization approach. This technique is considered one of the best methods due to its ability to solve the inverse problem associated with the identification of pollution sources. The source identification approaches can be regarded as an inverse problem. The problem is considered an ill-posed problem because a unique solution can never be guaranteed. The response matrix approach adopted by Gorelick et al. [13] reflected the earlier technique of the simulation-optimization approach in identification of groundwater pollution sources. The application of this technique could be further seen when Datta et al. [10] developed an expert system methodology to identify groundwater pollution sources. The algorithm of the methodology is based on the optimal statistical pattern recognition technique. However, this linear response matrix approach has shown large computational error in the result when the source identification problem is solved for a heterogeneous aquifer. With the inability of the response matrix to solve highly non-linear problem, the embedded approach came into light which is capable of solving complex and non-linear problem. The applications of embedded approach were first presented by Aguado and Remson [1]. Later, Mahar and Datta [18–20] employed embedded technique for groundwater source identification and monitoring network design. They adopted this approach and embedded the physical processes of groundwater as constraints in the optimization model. Aral et al. [3] further used this technique and proposed a new combinatorial optimization approach using Genetic Algorithms which they presented as progressive genetic algorithm (PGA). Using this new approach, the release histories and the location of the contaminant sources were successfully identified. However, the embedded technique is limited to a small-scale problem. In order to solve the problem of pollution sources identification for a large-scale area, the linked simulation-optimization evolves to be a better option. In this technique, the groundwater simulation model is linked as an external module to the optimization algorithm. Various sophisticated aquifer simulation models like FEFLOW, MODFLOW, MT3DMS, SEAWAT, SUTRA etc. which have the capability to simulate the complex phenomenon of groundwater flow and transport processes can be linked to the optimization model for solving the groundwater source identification model or the groundwater management model. Datta et al. [11] solved the groundwater pollutant source identification problem by externally linking the simulation model SUTRA with a nonlinear optimization model. Datta et al. [12] further modified the problem by linking a classical nonlinear optimization model with the SUTRA. But the degree of complexity was much higher when applied to a large aquifer. As a result, many researchers have attempted different heuristic global search approach for solving the source identification problems.

With the advancement of simulation-optimization technique, different types of source identification model were developed based on various techniques like statistical pattern recognition [10], Tikhonov regularization [28], Genetic Algorithm [3, 5, 21, 25], Simulated Annealing [15, 16]. Jha and Datta [17] proposed Adaptive Simulated Annealing (ASA) to determine the source characteristics and compared with the results obtained by GA. They concluded that the proposed approach can converge to an optimal solution faster proving to be a computationally efficient technique but with a large number of simulation GA gives better results. Ayvaz [4] successfully identified all the characteristics of unknown pollution sources by integrating the user-friendly groundwater flow and transport models, MODFLOW and MT3DMS with the optimization model based on heuristic harmony search (HS) algorithm. Additionally, an iterative technique was also employed for identifying the exact number of pollution sources. An overview on the identification of groundwater pollution was also presented by Chadalavada et al. [8] and Amirabdollahian and Datta [2].

Another major concern in the study of groundwater source identification problem is the computational efficiency of the model. In order to overcome the computational burden, Bhattacharjya et al. [6] used the artificial neural network (ANN) as the surrogate model and replaced the density-dependent flow and transport processes in a coastal aquifer. Results indicated that the developed ANN model proved to be computationally efficient and require less CPU time when compared with the embedded technique. Furthermore Borah and Bhattacharjya [7] suggested that improvement in the computational time can be seen with hybrid optimization model. They initially solved the model by using surrogate ANN model and later using numerical simulation models available in Groundwater Modeling System (GMS). The optimization model was solved using GA. This technique drastically reduced the computational time and also could efficiently obtain the optimal solution. Focussing on the budgetary constraints, an effective monitoring network plays an important role in the identification of unknown groundwater pollution sources. The optimal placement of monitoring wells will monitor the contaminant concentration effectively and the accuracy of the source identification model can be further enhanced. Therefore, based on this notion many researchers have identified the unknown pollution sources effectively. Meyer and Brill [23] presented a method for locating the wells in the monitoring network using Monte Carlo technique. A facility well location model was also used to select a fixed number of well locations. Later, Dhiman and Datta [9] developed a model for designing a groundwater quality monitoring network by linking the groundwater pollution transport simulation model with chance-constrained optimization model. The model was solved using mixed integer programming. Prakash and Datta [24] used trained genetic programming models for developing a monitoring network design. The model identifies the pollution source fluxes using the candidate monitoring locations.

The review of the literature suggested that the inverse model can efficiently identify the unknown pollution sources. In this technique, the difference between the simulated and observed contaminant concentrations at the observation well locations is minimized by using an optimization model. The simulated concentrations can be obtained using groundwater simulation model whereas the observed concentrations are recorded in the field. Most of the earlier studies assumed that the information about the pollution source location is known to the problem. However, in real case scenario, the location of the pollution sources is not known. As such, an iterative search methodology has been used for identifying the location and flux of the pollution sources. In this approach, the number of pollution sources is subsequently increased based on the feedback of the function values until the optimal solutions are not achieved. Further, measurement error associated with the observed data has also play a significant role in determining the unknown pollution sources. As such, an analysis has also been carried out to evaluate the effect of different level of measurement error in the observed concentrations in identifying the unknown pollution sources. The groundwater flow and transport processes have been solved using MODFLOW-2000 and MT3DMS. The models are then externally linked with the GA based optimization model in MATLAB platform. The performance of the developed methodology is evaluated using an illustrative study area [20].

Methodology

As discuss above, for the real scenario, the number and the source characteristics are not known. For solving such problems, an iterative based process is used in this study. Figure 17.1 shows the schematic representation of the iterative based unknown source identification model. The search for the optimal locations and source fluxes is initiated considering two pollutant sources. The reason for initiating with two number of pollution sources is that it always requires a one dummy source to confirm the actual number of pollution sources. It is then continued with the successive number of pollutant source until all the sources are identified. The source locations, number and the magnitude of the source fluxes are treated as the unknown decision variables. The Genetic Algorithm is used for solving the optimization model. Thus the initial solutions are generated randomly. The initial solution is known as the population. The initial population comprises of the source locations and the fluxes. For each of the solution, the MT3DMS and the MODFLOW packages of the Groundwater Modeling System (GMS) is used for simulating observed concentrations. The observed concentration at observation location is used to get the objective function value or the fitness function value.

A large value of fitness function value (F, as given by Eq. 17.1) at the end of the optimization run means that the magnitude of the observed and simulated concentration does not match with each other. The recovered source location and the source flux are bound to be erroneous values as the concentration values at the respective observation wells are giving inaccurate results. Henceforth, it clearly shows the presence of more pollution sources in the aquifer. As such, another

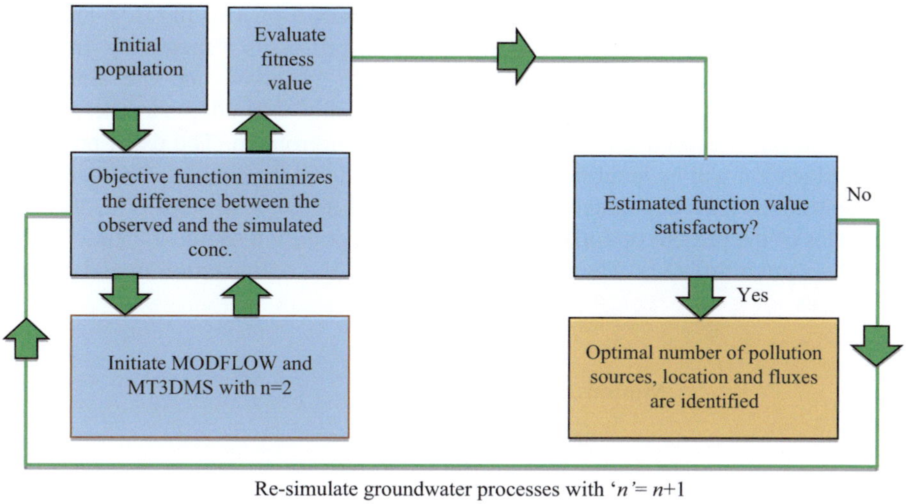

Fig. 17.1 Methodology of the iterative search model

pollution source is added to the existing one, i.e. n = n + 1. The model will again run and optimal fitness function will be calculated. If the fitness function value converges to a minimum value for n = 3, it corresponds to the best match between the observed and simulated contaminant concentration. Therefore, for further confirmation, another source is added (n = n + 1) to the prevailing one and now with n = n + 1, the present methodology is repeated. After repeating the whole process, if the fitness value converges to a minimum value with a dummy source (sources with negligible concentrations), this signifies that the actual number of pollution sources have been attained. The final confirmation of the exact number of pollution sources is based on the dummy source. It is because the retrieved source being a redundant source (dummy), does not have contaminant concentration. It suggests that there is no further pollution source in the aquifer. Hence, the dummy source can be discarded and the exact number of identified pollution source is given by n = n−1.

Source Identification Model

The groundwater pollution source identification problem can be solved using the linked simulation-optimization model by minimizing the difference between the simulated concentration and the observed concentration at the well locations. The observed concentration is the concentration measured at different observation locations in different time steps. The simulated concentration can be obtained by solving the groundwater flow and transport simulation model. In the present methodology, the modules MODFLOW and MT3DMS present in GMS are used in simulating the groundwater flow and transport processes respectively. In order to minimize the absolute difference between the observed and the simulated concentration in space and time, an optimization model is adopted. The genetic algorithm (GA) is used for solving the optimization problem. The exact pollution source location will be identified when the observed and the simulated contaminant concentrations perfectly match with each other at different time steps. The decision variables of the present optimization model are the pollution source location (X, Y) and the source flux (Sf). The objective function of the optimization model can be written as,

$$\text{Minimize } F_n = \sum_{i}^{M} \sum_{j}^{N} \left(C_{o,i}^{j} - C_{s,i}^{j} \right)^2 \tag{17.1}$$

Subject to

$$\mathrm{C} = \mathrm{f}\,(\mathrm{X}, \mathrm{Y}, \mathrm{Sf}) \tag{17.2}$$

$$\mathrm{Sf}_{min} \leq \mathrm{Sf} \leq \mathrm{Sf}_{max} \tag{17.3}$$

$$\mathrm{X}_{min} \leq \mathrm{X} \leq \mathrm{X}_{max} \tag{17.4}$$

$$\mathrm{Y}_{min} \leq \mathrm{Y} \leq \mathrm{Y}_{max} \tag{17.5}$$

Here, F_n is the objective function for the present optimization model with n number of pollution source; $C_{o,i}^{j}$ is the observed concentration at j^{th} time step for i^{th} well location; $C_{s,i}^{j}$ is the simulated concentration at j^{th} time step for i^{th} well location; M is the total number of observation wells and; N is the total number of time steps; C is the concentration vectors of the simulated concentration; Sf is the vector of the pollutant source fluxes such that $\mathrm{Sf} = [\mathrm{Sf}_1, \mathrm{Sf}_2, \mathrm{Sf}_{3,} \ldots, \mathrm{Sf}_\mathrm{n}]^\mathrm{T}$; X and Y are the source location vectors such that $\mathrm{X} = [\mathrm{x}_1, \mathrm{x}, \mathrm{x}_{3,} \ldots, \mathrm{x}_\mathrm{n}]^\mathrm{T}$ and $\mathrm{Y} = [\mathrm{y}_1, \mathrm{y}_2, \mathrm{y}_{3,} \ldots, \mathrm{y}_\mathrm{n}]^\mathrm{T}$; Sf_{min} and Sf_{max} are the lower and upper bounds of the source flux; X_{min}, X_{max}, Y_{min} and Y_{max} are the lower and upper bounds where the source location are expected. It can be noted that the bounds for source fluxes are taken as $\mathrm{Sf}_{min} = 0$ g/s and $\mathrm{Sf}_{max} =$ 100 g/s. It may be noted that in MT3DMS model the location of the pollutant sources should coincide with the centre of the discretized grid blocks [4]. Therefore, the lower and upper bounds for the source locations are provided on the basis of the maximum number of grids discretised for the study area in the GMS environment which will ultimately assist in reducing the search space.

Table 17.1 Genetic Algorithm parameters used in the present methodology

Parameter	Value	Function parameter	Adopted
Population size	200	Scaling function	Rank
Generations	1000	Selection function	Stochastic uniform
Crossover fraction	0.8	Mutation function	Constraint dependent
Elite count	0.5	Crossover function	Scattered

Optimization Algorithm

The Genetic Algorithms (GA) is employed to solve the simulation-optimization based inverse optimization model. The GA was first introduced by professor Holland [14] and is based on the theory of natural selection and genetics. IT is a heuristic search technique which mimics the natural selection processes of nature. Unlike the traditional search, GA opts for different path producing many directions for the optimal solution. For the source identification model, the algorithm starts by generating candidate solutions of source fluxes and the source locations. The set of candidate solutions is known as the population. The randomly generated population is then sent to the objective function routine for calculating the function value of each individual. If the calculated function value does not fulfil the termination criteria, the population will pass on to next successive generation following the three genetic operators i.e. selection, mutation and crossover for producing better offspring. Each of these successive steps will continue until the stopping criterion is reached. The genetic algorithm parameters used in the algorithm is shown in Table 17.1.

Simulation Model

The groundwater flow and transport processes have been simulated using MODFLOW and MT3DMS models. MODFLOW [22] is an executable program written in FORTRAN. It is based on finite difference method that can numerically solve a groundwater flow equation efficiently. MODFLOW consists of various packages which can be used in specifying various hydrogeological characteristics of an aquifer. The division of various packages enables the user to selects particular hydrologic conditions of the aquifer accordingly. After successfully simulating MODFLOW, different ASCII and binary files of the used packages gets saved in the disk. These files are then used by MT3DMS in simulating the groundwater transport equation. The groundwater flow equation used in MODFLOW can be written as,

$$\frac{\partial}{\partial x}\left(K_{xx}\frac{\partial h}{\partial x}\right)+\frac{\partial}{\partial y}\left(K_{yy}\frac{\partial h}{\partial y}\right)+\frac{\partial}{\partial z}\left(K_{zz}\frac{\partial h}{\partial z}\right)+W=S_s\frac{\partial h}{\partial t} \tag{17.6}$$

Where, K_{xx}, K_{yy} and K_{zz} are the hydraulic conductivity along the x, y and z directions (LT^{-1}); h is the hydraulic head (L); S_s is the specific storage coefficient; t is the time (T); W is the recharge flux per unit area (LT^{-1}).

MT3DMS [30] is a 3D modular transport model that can be used in the simulation of advection, dispersion and chemical reactions of dissolved constituents in the groundwater system. MT3DMS has a number of packages similar to MODFLOW that will deal the groundwater transport processes. MT3DMS works in conjunction with MODFLOW. MODFLOW compute the heads cell-by-cell during the flow simulation and are written in formatted files. Later MT3DMS read these files as the flow fields that are used in transport model simulation. Therefore, for performing MT3DMS simulation, one has to run the MODFLOW in order to obtain the flow field. The graphical interphase present in the GMS can be used for performing the MT3DMS transport simulation by undergoing pre-processing and post-processing steps. The inputs required for MT3DMS are generated in GMS, and then the files are saved in the disk. Later, the outputs from the MT3DMS are exported for post-processing in GMS. The transient groundwater transport equation can be written as

$$\frac{\partial(\theta C)}{\partial t} = \frac{\partial}{\partial x_i}\left(\theta D_{ij}\frac{\partial C}{\partial x_j}\right) - \frac{\partial(\theta v_i C)}{\partial x_j} + q_s C_s + \sum R_n \tag{17.7}$$

Where, C is the dissolved concentration in the groundwater (ML^{-3}); θ is the porosity of the subsurface medium; t is the time(T); x_i is the distance along the respective Cartesian co-ordinate axis (L); D_{ij} is the hydrodynamic dispersion coefficient tensor ($L^2\,T^{-1}$); v_i is the seepage or linear pore water velocity (LT^{-1}); q_s is the volumetric flow rate per unit volume of aquifer representing fluid sources (positive) and sinks (negative) (T^{-1}); C_s is the concentration of the source or sinks flux (ML^{-3}); $\sum R_n$ is the chemical reaction term ($ML^{-3}\,T^{-1}$).

Measurement Errors

As discussed earlier, the pollution source can be efficiently identified by minimizing the error function as stated in Eq. 17.1. In actual scenario, the observed contaminant concentrations data can be obtained from the field. However, for the hypothetical problem considered in this study, the numerical simulation model (MT3DMS) is used to generate the observed data. As some measurement errors always exist in the observed contaminant concentration during field measurement, a random error has been incorporated to the data. However, the extent of measurement error is not known to us. As such different level of error has been introduced in the observed concentrations to evaluate the effect of the measurement error. As reported, the effect of measurement error in the source identification problem was also performed by many researchers [5, 12, 20, 24, 27]. However, much emphasis was given on the impact it might have while determining the pollution source locations and magnitude of source fluxes. The effect of measurement error in the observed contaminant concentration can be shown by adding randomly generated error term in the simulated values of the observed concentrations. The following [20] the equation is used in the study.

$$PCo_i^j = Co_i^j \,(1 + err) \tag{17.8}$$

Where, PCo_i^j is the perturbed simulated concentration value; Co_i^j is the simulated concentration value; *err* is the error term to be introduce. The error term (*err*) can be further represented as

$$err = a \text{ x } \xi \tag{17.9}$$

Here, a signifies different level of error magnitude equal to ranging from 0.05 to 0.2 [26]. When $a< 0.1$, it signifies a low noise level, $0.1\leq a \leq 0.15$ denotes moderate level of noise and $a \geq 0.15$ signifies high level of noise [26]. ξ represent random fraction generated following normal distribution for mean 0 and standard deviation of 1.

If the *err* term is not introduced in the numerically simulated observed concentrations, then there is zero noise in the measured contaminant concentration. With no measurement error in the observed concentrations, the source identification model will give the best optimal solution. However, it does not support the realistic case as explained above. Therefore, introducing the error term will be regarded as one of the best approaches to check the effectiveness of the model. As such the perturbed concentrations will be used to identify the pollution source location and source fluxes using the linked simulation-optimization model. Thus the present approach will demonstrate the performance and robustness of the model in determining the characteristics of the pollution sources under the different level of measurement errors. If the location of the pollution sources and the magnitude of the pollution sources are efficiently identified under the given conditions, then the present source identification model will prove to be an effective one.

However, it may be noted that the present methodology also requires obtaining the exact number of pollution sources when it is assumed that very limited information is available in the affected aquifer. Therefore, determining the exact number of pollution sources also a primary objective of the study. As the number of pollution sources is completely unknown, an iterative approach has been adopted for determining the exact number of pollution sources [5]. As such in this study, a methodology has been presented for identification of the number and location of the pollution sources along with the magnitude of the source fluxes under the influence of different measurement error. Initially it is assumed that there is no measurement error in the observed contaminant concentrations, therefore a =0 for zero noise level. The same given steps are followed subsequently with different noise levels i.e. $a< 0.05$ for low noise level, $0.1\leq a \leq 0.15$ for moderate noise level and $a \geq 0.15$ for high noise level.

Performance Evaluation Criteria

The performance of the proposed simulation-optimization model is evaluated using different performance criteria. Analysis using the error free data and erroneous data can be performed using the normalized error (*NE*) [20] and relative error (*RE*) [7]. The normalized error (*NE*) can be represented as

$$NE = \frac{\sum_{p=1}^{ns}\sum_{r=1}^{nl}\left|Ef_{p,r} - Af_{p,r}\right|}{\sum_{p=1}^{ns}\sum_{r=1}^{nl}Af_{p,r}} \times 100 \tag{17.10}$$

The relative error (RE) criteria to evaluate each of the source flux can be computed as

$$RE = \frac{\left|Ef_{p,r} - Af_{p,r}\right|}{Af_{p,r}} \times 100 \tag{17.11}$$

Where $Ef_{p,\,r}$ and $Af_{p,\,r}$ are the estimated and the actual source fluxes respectively at r^{th} location and p^{th} stress period; *nl* is the total number of source location; *ns* is the total number of stress period.

Study Area

An illustrative study area with an area of 1.04 km^2 (Fig. 17.2) is adopted for evaluating the performance of the proposed model. The boundary conditions and the geometry of the study area have been considered as proposed by Mahar and Datta [20]. It is a homogeneous and isotropic aquifer with constant head boundary on the left and right-hand boundaries. The no-flow boundary condition has been considered on the upper and lower boundaries of the aquifer. The 2D study area has been discretized in 13 columns and 8 rows. The cell size is 100 m x 100 m. A total number of 12 observation wells are placed randomly around the suspected area of the aquifer. The hydrological parameters used in the study area are shown in Table 17.2. The observation wells are designated as W1, W2 . . . W12. The adopted study area can be regarded as a complex model due to the presence of a large number of pollutant source in the aquifer. A total number of four pollutant sources are present in the aquifer but the location and number of source pollutant are completely unknown to us and have to be identified using the present methodology. The groundwater flow and transport processes are simulated for 5 years (5 stress periods) at an interval of 3 months. It is assumed the pollution sources are active for 4-time steps i.e. the source releases pollutant for 1 year. The magnitudes of the

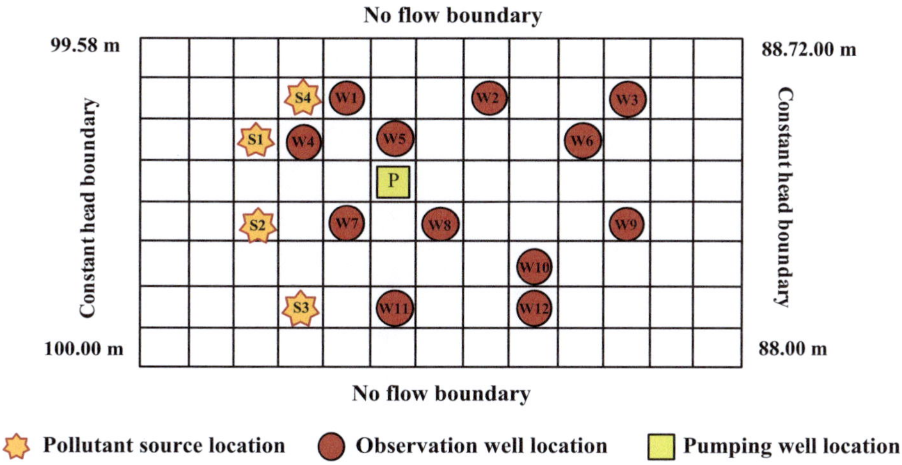

Fig. 17.2 Illustrative study area showing pollutant source locations, observation wells and pumping wells

Table 17.2 Hydrological parameters used in the study area

Parameters	Values
Hydraulic conductivity in x direction, K_{xx}(m/s)	0.0002
Hydraulic conductivity in y direction, K_{yy} (m/s)	0.0002
Porosity, ε	0.25
Thickness of the aquifer, b (m)	30.5
Longitudinal dispersivity, α_L (m)	40
Transverse dispersivity, α_T (m)	9.6
Time steps, Δt (months)	3

Table 17.3 Source fluxes for different time steps (g/s)

Sources	Time step 1	Time step 2	Time step 3	Time step 4
S1	30	58.5	0	35
S2	47	15	37	0
S3	41.26	0	14.40	16.88
S4	21.7	0	29.68	0

pollutant sources are shown in Table 17.3. There is a pumping well in the aquifer and the pumping rates for the twenty stress period are given in Table 17.4.

Results and Discussion

The present linked simulation-optimization model has searched for the optimal number, locations and source fluxes starting from one number of pollution sources. As such, the search has been initiated considering two pollution sources in the

Table 17.4 Pumping rates of the well at the pumping location of the aquifer (m^3/d)

Time step	Rate	Time step	Rate
1	273.02	11	163.29
2	163.29	12	327.45
3	327.45	13	273.02
4	163.29	14	163.29
5	273.02	15	381.02
6	327.45	16	217.72
7	163.29	17	273.02
8	273.02	18	163.29
9	381.02	19	327.45
10	217.72	20	217.72

Table 17.5 Estimated source flux, Final fitness (*F*) value and normalized error (*NE*) for different number of source considering no measurement error in the observed concentrations

			Estimated source flux (g/s)					
No. of sources	No. of decision variable	Estimated source location	Time step 1	Time step 2	Time step 3	Time step 4	Final function value *F*	*NE*
2	10	(3,3)	59.38	30.00	5.01	0.00	F2 = 5.38	101.21
		(4,4)	69.73	58.99	76.89	20.70		
3	15	(3,3)	43.89	72.49	8.99	20.22	F3 = 3.94	67.93
		(5,3)	70.54	28.01	10.45	0.07		
		(7,4)	68.96	47.95	16.03	8.05		
4	20	(3,3)	30.18	57.91	1.48	33.98	F4 = 0.072	0.054
		(5,3)	46.67	16.61	34.59	1.18		
		(7,4)	40.79	0.137	15.59	15.94		
		(2,4)	20.92	2.16	27.10	1.41		
5	25	(3,3)	30.58	56.81	3.026	33.36	F5 = 0.0201	1.32
		(5,3)	46.36	17.33	34.05	1.24		
		(7,4)	41.16	2.42	9.85	19.20		
		(2,4)	21.70	0.13	29.23	0.34		
		(1,10)	6.82	3.90	4.38	6.12		

aquifer. The number of pollution sources is then successively increased and objected function values have been noted down. Initially, it is assumed that there is no measurement error in the observed concentrations. As such the identification of the source location, the magnitude of the flux and the number of pollution sources are based on zero noise level. Table 17.5 shows the estimated source flux, Normalized Error (NE) and the final objective function values (F) for different number of pollution sources considering no measurement error in the observed concentration. In the first trial, it is assumed that only two pollutant sources are present in the affected aquifer. For the present scenario, the candidate solution for a two sources will have only 10 decision variables. Eight of these decision variables will represent the source fluxes while the remaining two will represent the source locations. The estimated source locations are (3, 3) and (4, 4), and the objective function value (F2) is 5.38 which is considered to be a large value. Moreover, the NE value is also found to be very large 101.21. This indicates the possibilities of more pollution sources. Henceforth, the simulation-optimization model is run further with three number of pollution sources.

The source locations predicted by the model for three number of pollution sources are (3, 3), (5, 3) and (7, 4). For this case, the F3 and NE values are reduced to 3.94 and 67.93 respectively. But both the values are considered to be high and are not enough to give an optimal solution. So, the iterative process is again repeated with four number of pollution sources and found the source locations as (3, 3), (5, 3), (7, 4) and (2, 4). The F4 value has been remarkably reduced to 0.072 and goes well with the decreased value of NE as 0.054. The evaluated values of the present scenario demonstrate an optimal solution but for further confirmation, an additional iteration is performed with five number of pollution sources. The final F5 value and NE are obtained as 0.0201 and 1.32 respectively. This shows that the actual number of pollution sources in the aquifer is four. Therefore, the results with four number of pollution sources are the optimal solution.

The trend at which the fitness values are obtained for different number of pollution sources can be seen in Fig. 17.3. It is also observed that the fitness function value slowly decreases with the increase in the number of pollution sources. Later, the fitness value converges to a minimum value when the number of pollution sources reaches four which is the actual number of pollution sources. Furthermore, when the fitness function is evaluated for five number of pollution sources. It is observed that the fitness function value for four and five number of pollution sources almost matches with each other with minimum values.

Even though the exact number of pollution sources is successfully identified, the concern lies in the proximity of the identified source fluxes. Figure 17.4 shows the comparison between the actual and the identified pollution sources. It clearly signifies the close resemblance of the source fluxes when no measurement error has

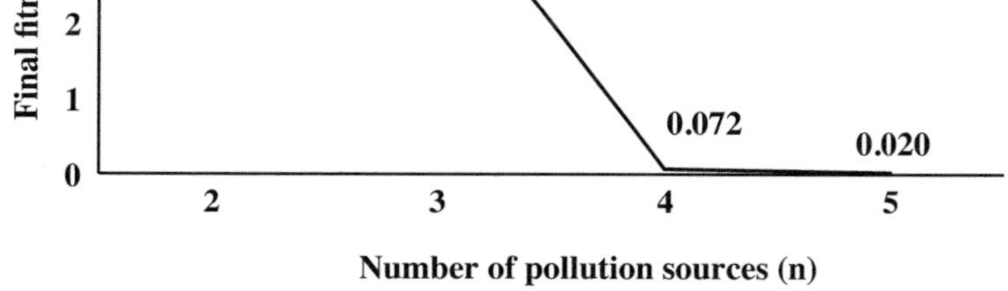

Fig. 17.3 Final fitness function for different number of pollution sources

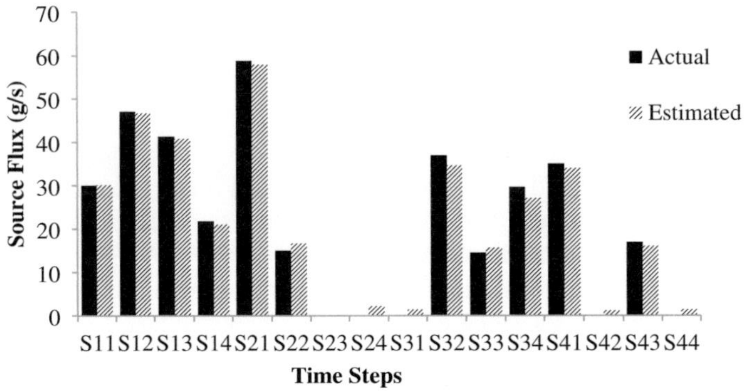

Fig. 17.4 Comparison between the actual and the estimated source fluxes

been considered. Presence of source fluxes could be seen on some of the inactive time steps (S24, S31, and S44) but as the range of the recovered source fluxes is found to be a minimal one and it can be neglected.

From the results obtained using the present methodology, it can be concluded that there is four number of pollution sources and the location and magnitude of source flux were efficiently identified by the proposed methodology. These optimal results were obtained considering that there is no measurement error in the observed concentration. Now to see the effect of measurement error, different noise levels ($a = 0.05$, $a = 1$, $a = 0.15$ and $a = 0.2$) are added to the observed concentration.

Table 17.6 shows the coordinates of the estimated source locations considering different noise levels. It is observed that when two pollution sources are considered, the simulation-optimization model could not predict the exact pollution source location with all the different noise levels except for source location (3, 3). For 3 numbers of pollution sources, the same location (3, 3) has been identified by all the different number of pollution sources. However, it may be noted that the other source locations estimated by the model for different level of noises are found to be very near to the actual ones. In case of 4 number of pollution sources, the model could identify the exact source i.e. (3,3), (5,3), (7,4) and (2,4) very precisely except for noise level a = 0.2. This shows that the different noise levels have less impact on identification of source locations. Similarly, for n = 5 also, the model could identify the actual source locations along with a dummy source.

Table 17.7 shows the estimated source fluxes for five number of pollution source under different noise levels. At zero noise level, the predicted source fluxes showed some similarity with the actual fluxes however some contradictory flux values are also observed. The actual magnitude of source flux for S1, S2 and S3 at second and third-time steps are 58.80 g/s, 37 g/s and 14.40 g/s but the model predicted as 56.81 g/s, 34.05 g/s and 9.29 g/s respectively. However, these differences do not

Table 17.6 Coordinates of the actual and estimated source location

Actual source location	No. of pollution sources	Estimated source location: With noise level $a = 0.05$	With noise level $a = 0.1$	With noise level $a = 0.15$	With noise level $a = 0.2$
(3,3)	2	(3,3)	(8,4)	(3,3)	(6,3)
		(4,4)	(3,3)	(5,5)	(3,3)
(5,3)	3	(3,3)	(3,3)	(1,4)	(6,3)
		(6,3)	(1,4)	(6,3)	(3,3)
		(4,4)	(6,3)	(3,3)	(1,3)
(7,4)	4	(3,3)	(3,3)	(3,3)	(3,3)
		(5,3)	(5,3)	(5,3)	(6,3)
		(7,4)	(7,4)	(7,4)	(7,4)
		(2,4)	(2,4)	(2,4)	(2,4)
(2,4)	5	3,3)	(3,3)	(3,3)	(3,3)
		(5,3)	(5,3)	(5,3)	(5,3)
		(7,4)	(7,4)	(7,4)	(7,4)
		(2,4)	(2,4)	(2,4)	(2,4)
		(1,11)	(1,10)	(1,11)	(1,11)

show any major impact on the other source fluxes of the respective locations. S5 being a dummy source, a negligible amount of source fluxes is predicted by the model. The reason for the presence of some source fluxes in dummy source is due to the effect of contaminant concentration from other pollution sources on it. With the rise in noise level, the magnitudes of source fluxes shows some variations. This indicates that the magnitude of the source fluxes are susceptible to erroneous results with the increase in noise levels.

Figure 17.5 presents the bar graph showing objective function values for the different number of pollution sources considering the different level of noises and its impact on the function value. The effect of the various noise levels (i.e. zero noise level, moderate noise level and high noise level) does not show much variation on the objective function value for each of the different number of pollution sources. It can be remarked from the bar graph that the final fitness function values are converging towards the best minimum value as it proceeds towards the actual number of pollution sources. The final fitness value with four and five number of pollutant sources almost matches with a minimum value of fitness value and thus confirming that no more pollution sources are available in the affected aquifer. This implies that with the presence of noise in observed data has a less significant impact in determining the optimal number of pollution sources. It clearly shows that the present model is capable of identifying the exact number of pollution sources even under different noise levels.

Table 17.7 Estimated source fluxes for $n = 5$ at different noise level

Time Steps	Source locations	Actual source fluxes (g/s)	Magnitude of estimated source fluxes (g/s)				
			At zero noise level	At noise level $a = 0.05$	At noise level $a = 0.1$	At noise level $a = 0.15$	At noise level $a = 0.2$
1	S1	30.00	30.58	30.75	28.41	27.79	29.82
	S2	47.00	46.36	45.61	46.83	45.35	43.19
	S3	41.26	41.16	40.62	43.32	39.65	37.59
	S4	21.70	21.70	24.62	24.33	24.78	27.29
	S5	x	3.82	1.17	3.20	7.83	7.42
2	S1	58.80	56.81	56.17	63.89	61.80	64.94
	S2	15.00	16.33	18.83	17.43	17.83	21.36
	S3	0.00	2.42	1.63	4.32	2.42	3.76
	S4	0.00	0.13	2.69	2.76	3.87	7.78
	S5	x	2.90	5.73	1.89	6.90	7.01
3	S1	0.00	2.03	4.18	6.19	3.02	5.82
	S2	37.00	34.05	32.92	29.92	29.34	23.05
	S3	14.40	9.29	13.63	6.73	9.85	14.56
	S4	29.68	29.23	29.16	25.91	19.76	20.90
	S5	x	4.38	1.39	3.60	5.37	10.01
4	S1	35.00	33.36	31.68	29.70	28.76	27.73
	S2	0.00	1.24	1.59	2.13	4.76	8.80
	S3	16.88	15.53	16.69	17.56	19.21	13.85
	S4	0.00	0.34	1.17	1.03	1.39	3.36
	S5	x	6.12	1.53	5.96	4.31	6.05

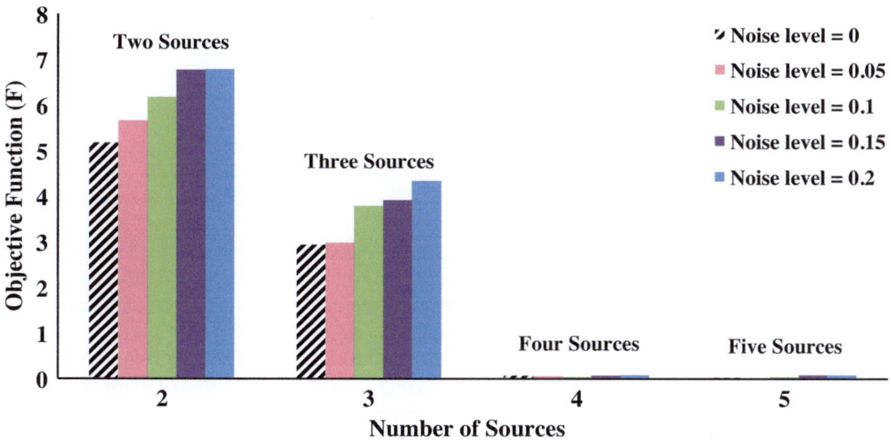

Fig. 17.5 Bar graph of different function values for different number of pollution sources considering different noise level

Conclusions

This chapter presented a GA based methodology for optimal identification of unknown pollution sources under different noise levels. It has been assumed that no information is available about the pollution sources. As such, an iterative based search technique is adapted for identifying the exact number of pollution sources and the source locations. The identification of the unknown groundwater pollution sources is carried out using the inverse optimization technique. For evaluating the methodology, an illustrative study area is taken. The application of the model to the study area shows that the GA based model is capable of identifying the exact location and source fluxes of the pollution sources when there is no noise in the observed data. Further, the model is also capable of finding the exact location of the pollution sources even when measurement errors are added to the observed source concentration. However, the accuracy in determining the source fluxes has been reduced when errors are added to the observed data.

References

1. Aguado E, Remson I (1974) Groundwater hydraulics in aquifer management. J Hydraul Div ASCE 100(1):103–118
2. Amirabdollahian M, Datta B (2013) Identification of contaminant source characteristics and monitoring network design in groundwater aquifers: an overview. J Environ Pro 4:26–41
3. Aral MM, Guan J, Maslia ML (2001) Identification of contaminant source location and release history in aquifers. J Hydrol Eng ASCE 6(3):225–234

4. Ayvaz MT (2010) A linked simulation-optimization model for solving the unknown groundwater pollution source identification problems. J Contam Hydrol 117(1–4):46–59
5. Ayvaz MT (2015) A new simulation-optimization approach for simultaneously identifying the spatial distribution and source fluxes of the areal groundwater pollution sources. In: 36th IAHR world congress. The Hague, pp 1–7
6. Bhattacharjya RK, Datta B, Satish MG (2005) Optimal management of coastal aquifer using linked simulation optimization approach. Water Resour Manag 19(3):295–320
7. Borah T, Bhattacharjya RK (2014) Solution of source identification problem by using GMS and MATLAB. J Hydrol Eng 19(3):297–304
8. Chadalavada S, Datta B, Naidu R (2011) Uncertainty based optimal monitoring network design for a chlorinated hydrocarbon contaminated site. Environ Monit Assess 173:929–940
9. Datta B, Dhiman SD (1996) Chance constrained optimal monitoring network design for pollutants in groundwater. J Water Resour Plann Manage 122(3):180–188
10. Datta B, Beegle JE, Kavvas ML, Orlob GT (1989) Development of an expert-system embedding Pat- tern-recognition techniques for pollution source identification. Technical Report, Department of Civil Engineering, California University, Davis
11. Datta B, Chakrabarty D, Dhar A (2009) Simultaneous identification of unknown groundwater pollution sources and estimation of aquifer parameters. J Hydrol 376(1):48–57
12. Datta B, Chakrabarty D, Dhar A (2011) Identification of unknown groundwater pollution sources using classical optimization with linked simulation. J Hydro Environ Res 5(1):25–36
13. Gorelick SM, Evans B, Remson I (1983) Identifying sources of groundwater pollution: an optimization approach. Water Resour Res 19(3):779–790
14. Holland JH (1975) Adaptation in natural and artificial systems. MIT Press, Cambridge, MA
15. Jha MK, Datta B (2011) Simulated Annealing based simulation-optimization approach for identification of unknown contaminant sources in groundwater aquifer. Desalin Water Treat 32(1–3):79–85
16. Jha M, Datta B (2013) Three-dimensional groundwater contamination source identification using adaptive simulated annealing. J Hydrol Eng 18:307–317. https://doi.org/10.1061/(ASCE)HE.1943-5584.0000624
17. Jha M, Datta B (2014) Linked simulation-optimization based dedicated monitoring network design for unknown pollutant source identification using dynamic time warping distance. Water Resour Manag 28(12):4162–4182
18. Mahar PS, Datta B (1997) Optimal monitoring network and groundwater pollution source identification. Water Resour Manag 123(4):199–207
19. Mahar PS, Datta B (2000) Identification of pollution sources in transient groundwater. Water Resour Manag 14(3):209–227
20. Mahar PS, Datta B (2001) Optimal identification of ground-water pollution sources and parameter estimation. J Water Resour Plan Manag 127(1):20–29
21. Mahinthakumar G, Sayeed M (2005) Hybrid genetic algorithm – local search methods for solving groundwater source identification inverse problems. J Water Resour Plan Manag 1(45):45–57. https://doi.org/10.1061/(ASCE)0733-9496(2005)131
22. McDonald MG, Harbaugh AW (1988) A modular three-dimensional finite difference groundwater flow model. USGS Report
23. Meyer PD, Bril ED (1988) A method for locating wells in a groundwater monitoring network under conditions of uncertainty. Water Resour Res 24(8):1277–1282
24. Prakash O, Datta B (2015) Optimal characterization of pollutant sources in contaminated aquifers by integrating sequential-monitoring-network design and source identification: methodology and an application in Australia. Hydro J 23(6):1089–1107
25. Singh R, Datta B (2006) Identification of groundwater pollution sources using GA-based linked simulation optimization model. J Hydrol Eng 11(2):101–109

26. Singh RM, Datta B (2007) Artificial neural network modeling for identification of unknown pollution sources in groundwater with partially missing concentration observation data. Water Resour Manage 21(3):557–572
27. Singh RM, Datta B, Jain A (2004) Identification of un- known groundwater pollution sources using artificial neural networks. Water Resour Plann Manag 130(6):506–514
28. Skaggs TH, Kabala ZJ (1994) Recovering the release history of a groundwater contaminant plume: method of quasi-reversibility. Water Resour Res 3(11):2669–2673
29. UN-WWAP (2009) United Nations world water assessment programme. The world water development report 3: water in a changing world. UNESCO, Paris
30. Zheng C, Wang PP (1999) MT3DMS: a modular three-dimensional multispecies transport model for simulation of advection, dispersion, and chemical reactions of contaminants in groundwater systems; documentation and user's guide. Alabama University

Chapter 18
Efficiency of Binary Coded Genetic Algorithm in Stability Analysis of an Earthen Slope

Rajib Kumar Bhattacharjya

Abstract The critical factor of safety and the corresponding slip circle of an earthen slope can be determined by using an optimization technique. This chapter evaluates the efficiency of genetic algorithms in locating critical slip circle of a homogeneous earthen slope. Genetic algorithm, a global search technique, is highly efficient in finding the global optimal solution of a problem, having highly irregular response surface. The evaluation of results shows that genetic algorithm is very robust in locating critical slip circle. On the other hand, the gradient-based classical optimization method is highly sensitive to the initial solution supplied to the problem. This implies that there are multiple local optimal solutions of the problem. As a result, the classical optimization techniques many times trap at local optimal solutions.

Keywords Slope stability · Method of slices · Optimization · Genetic algorithms

Introduction

The stability analysis of an earthen slope is one of the main exercises for the civil engineers working on various activities in hilly terrain. The factor of safety of the most critical slip circle is generally used to evaluate the stability of an earthen slope. The critical factor of safety and the corresponding slip circle can be obtained by using optimization techniques. The limiting equilibrium-based methods such as the method of slices [6], modified Bishop's method [2], Janbu's method [9], Sarma's method [12] etc. are generally used to calculate the factor of safety for an arbitrary slip circle. Generally, a series of trial slip circles are considered and the factor of

R. K. Bhattacharjya (✉)
Department of Civil Engineering, Indian Institute of Technology Guwahati, Guwahati, Assam, India
e-mail: rkbc@iitg.ernet.in

F. Bennis, R. K. Bhattacharjya (eds.), *Nature-Inspired Methods for Metaheuristics Optimization*, Modeling and Optimization in Science and Technologies 16,
https://doi.org/10.1007/978-3-030-26458-1_18

safety of each slip circle is calculated. The slip circle corresponding to the minimum factor of safety is then determined from the series of trial slip circles. The slip circle corresponding to the minimum factor of safety is known as the critical slip circle.

The classical optimization techniques were widely used to calculate the factor of safety of the critical slip circle. Some of the application of classical algorithms for locating the critical slip circle are found in Celestino and Duncan [4], Baker [1], Nguyen [11], Li and White [10], Chen and Shao [3], etc. The classical optimization methods generally use the gradient search techniques to find the optimal solution. The performance of the gradient-based classical optimization methods is not satisfactory when the response surface is highly irregular. In such a situation, it is very likely that the solutions obtained would be a local optimal solution. One possible remedy is to use multiple solution points as an initial solution to obtain the global optimal solution. Moreover, many times it may be difficult to calculate the gradient of the objective or the constraint functions for the real-world optimization problem.

There is another class of optimization algorithms which do not require the gradient information of the objective function or the constraint functions. However, these algorithms start from single initial solution. This class of algorithms is known as direct search method. Some of the applications of direct search technique are Yamagami and Ueta [13], Chen and Shao [3], Greco [7] etc. The direct search technique works by creating a set of search direction iteratively. This technique can be effectively used for solving nonlinear optimization problems. It is therefore considered as a robust tool for optimization. These algorithms are simple in concept and can easily be implemented for solving complex optimization problems. The main advantage of the algorithm is that it works without gradient information of the objective function and the constraint functions. The disadvantage of these algorithms is that it can yield only the local optimal solutions as it starts from a single initial solution and does not have any mechanism to overcome the local optimal solution. In case of non-convex optimization problems, the algorithm is also highly sensitive to the initial solution chosen to start the iteration. As an alternative to classical optimization methods and direct search algorithms, the global search methods such as Genetic Algorithm can be used to solve the non-convex, nonlinear optimization problems.

The Genetic Algorithms is a search technique based on the concept of natural selection inherent from natural genetics. This algorithm is relatively more efficient in obtaining global optimal solution even when the response surface is highly irregular. Unlike the classical algorithms, the search process is initiated from a set of solutions generated randomly using a distribution function. The Genetic Algorithm combines the law of 'survival of fittest' with the genetic operators abstracted from nature. One of the advantages of the Genetic Algorithm is that it does not require continuity or differentiability of either the objective function or the constraints. The main difference of genetic algorithm with classical methods is that the Genetic Algorithm works with the coding of the parameter set, and not the parameters themselves and it starts from a population of points rather than a single initial starting point [8].

The chapter presented here evaluates the performance the Genetic Algorithms in locating the critical slip circle of an earthen slope. The method of slices [6] is used to calculate the factor of safety for an arbitrary slip circle. An example problem taken from literature has been used to evaluate the relative efficiency of the Genetic Algorithms.

Optimization Model Formulation

The ordinary method of slices is one of the most frequently used iterative procedures to calculate the factor of safety of an earthen slope. Figure 18.1 shows a homogeneous earthen slope with a trial slip circle. Let BD is the trial slip circle with r, the radius, and $O(x, y)$ centre of the slip circle. $B(0, 0)$ is the toe of the slope. The factor of safety can be determined as

$$F = \frac{cr\delta + \tan\phi \sum (N_i - U_i)}{\sum T_i} \tag{18.1}$$

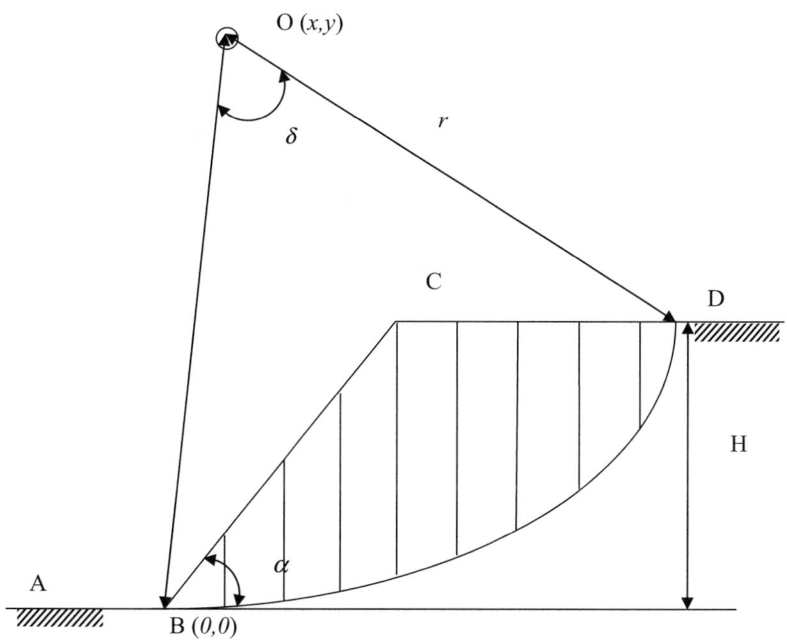

Fig. 18.1 Homogeneous earthen slope

Where, F is the factor of safety; c is the cohesion in kN/m^2; δ is the arc angle, c is cohesion, ϕ is the angle of internal friction in degree; i is the slice number; N_i is the normal component of weight; T_i is the tangential component of weight; U_i is the pore water pressure. For any arbitrary slip circle, the factor of safety can be determined using Eq. (18.1).

The parameters of the critical slip circle can be determined by minimizing factor of safety. The optimization model can be formulated as:

Minimize

$$F(x, y, r) = \frac{cr\delta + \tan\phi \sum (N_i - U_i)}{\sum T_i} \tag{18.2}$$

Subject to

$$x_L < x < x_U \tag{18.3}$$

$$y_L < y < y_U \tag{18.4}$$

$$r_L < r < r_U \tag{18.5}$$

Where, F is the factor of safety calculated using method of slices, x_L and x_U are the lower and upper limit on x; y_L and y_U are the lower and upper limit on y; and r_L and r_U are the lower and upper limit on r.

Genetic Algorithms

The Genetic Algorithms (GA) was introduced by Prof. John H. Holland, Professor of electrical engineering and computer science at the University of Michigan, Ann Arbor. Since its introduction, the algorithm has been applied in various fields of science and engineering for solving nonlinear non-convex optimization problems. It has now emerged as one of the most powerful and robust tools for function optimization.

GA is the search techniques that are motivated by the theory of natural genetics and natural selection. The basic techniques of GA are designed to simulate the mechanism of population genetics and natural rules of survival in pursuit of the ideas of adaptation. The GAs operators are borrowed from the natural genetics and applied artificially to search for the global optimal solution of an optimization problem. The genetic operators are reproduction, crossover, and mutation. The algorithms are computationally simple, but powerful in their search for improvement after each generation [8].

Working Principle of GA

The algorithm starts with the random generation of a set of initial solutions (chromosomes), called the population of possible solutions. Each solution of the population is known as a chromosome. These chromosomes are evaluated based on their objective function value. The population is then passed through the three basic genetic operators with an aim to produce better offspring in the next generation. These operators are reproduction, crossover, and mutation. Reproduction is a process in which individual strings are copied according to their fitness [8]. This operator eliminates the weaker chromosomes and makes multiple copies of the stronger chromosomes. Crossover is a process of exchange of gene between two chromosomes for producing their offsprings. In this operation, the algorithm picks up two chromosomes from the population to perform crossover at a randomly selected crossover site of the chromosome. A probability, known as 'crossover probability' is also used to control the crossover process. The aim of this operator is to search for better solution near the current solution. Therefore, this operator performs the local search. Mutation is the occasional introduction of a new feature into the population pool to maintain diversity in the population. The main purpose of this operator is to come out of the location optimal solution and try to explore the other areas of the search space. Random bit-by-bit mutation is used in this study to generate new chromosomes. These three genetic operators are designed in such a way that it would produce better solution after each generation and the search process would converge towards the global optimal solution of the function. Sometimes, crossover and mutation may produce inferior solutions. But these solutions will not be able to propagate further and the selection process will eliminate these solutions. Moreover, to increase the efficiency of the genetic algorithms, it is also necessary to preserve some better solutions for the next generation. This process is known as elitism. Generally, some percentage (say 5%) of the population is preserved for the next generation without applying crossover and mutation operator. As such these solutions will propagate to the next generation without going through the crossover and mutation operators. Elitism is important since it allows the solutions to get better over time. These procedures would continue till the better individual is better enough to suit the objective of the problem. Figure 18.2 shows the working principle of GAs in the form of a schematic diagram.

Representation of a Solution String

The Genetic Algorithms represent the solution in binary code. Each solution string contains the binary representation of all the variables. If there are two variables in the solution string with x bits to code each variable, then the overall string length will be $2x$. Figure 18.3 shows a binary solution string with two variables. For this string, each variable is coded by 6 bits. Therefore, the total string length of the string

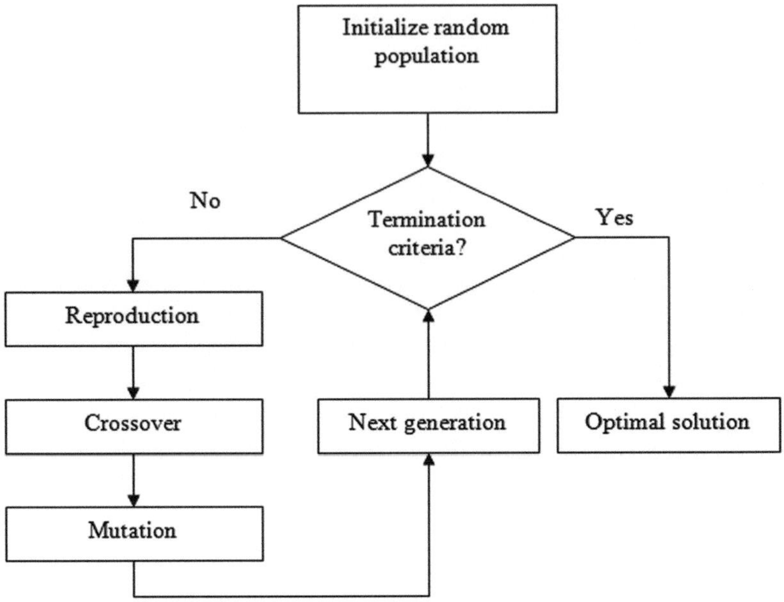

Fig. 18.2 Simple GA

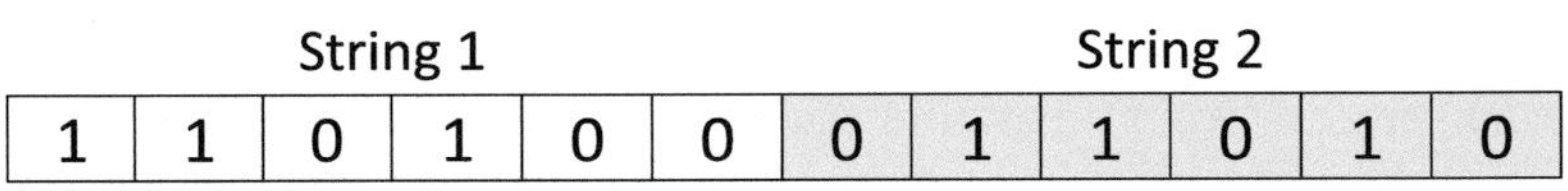

Fig. 18.3 Binary coded string

is 12 in this case. The solution string is also called a chromosome and each bit of the chromosome is called a gene. The mapping between real variable and the binary variable can be done by the following equation [5].

$$x_i = x_i^{\min} + \frac{x_i^{\max} - x_i^{\min}}{2^{l_i} - 1} DV\left(s_i\right) \tag{18.6}$$

Here, $x_i^{\min}$ and $x_i^{\max}$ are the lower bound and the upper bound of the variable x_i; l_i is the string length for the variable x_i; $DV(s_i)$ is the decoded value of the string s_i. Any arbitrary precision can be achieved by using Eq. (18.6).

Fitness of a Solution String

The search technique of GA is based on the principle of natural selection and natural genetics. GA does not use any gradient or auxiliary problem information to guide the search for evolving at an optimal solution. The search technique of GAs is guided by the fitness of a solution string. The fitness of a string is an assigned value which is a function of the objective function value. The fitness function determines the relative quality of the solution strings. The solution string having better fitness would receive more emphasis and would have a better probability of survival for the next generation.

Reproduction Operator

The main objective of reproduction operator is to identify better solutions and also to make multiple copies of better individuals. Inferior solutions are also eliminated by this operator. The operator is designed in such a way that the total population is remaining same after eliminating the inferior solutions, and after making multiple copies of the better solutions. Some of the popular methods for reproduction are proportionate selection, tournament selection, rank selection, etc. Proportionate selection selects the better individual in proportion to their fitness function. The solution having a higher fitness value will have multiple copies in the next generation according to their proportion. The solution with a lower fitness value will be eliminated. In binary tournament selection, tournaments are played between two or more solutions, selected randomly from the population pool and the winner of the tournament is selected for the next generation. For tournament size of two, each solution of the population will play the tournament for two times. Therefore, the best solution in the population pool will win the tournament both the times and will always have two copies in the next generation. The solution with the worst fitness value will lose the tournament both the times and will not have a copy in the next generation. The other solutions will have one or two copies in the next generation. Higher tournament size can also be taken. However, in that case, the population will be saturated with the better individuals only. This is not a healthy situation and exploration may end prematurely. The algorithm for the rank selection operator is similar to that of roulette wheel selection. However, the selection probability of a solution is proportional to relative fitness of the solution rather than the absolute fitness.

Crossover Operator

The crossover operator is used to create new solutions from the existing solutions available in the mating pool after applying selection operator. This operator exchanges the gene information between the solutions in the mating pool. There are several crossover operators proposed by various researchers. The most popular crossover selects any two solution strings randomly from the mating pool and a portion of the first string is exchanged with the corresponding portion of the other string. The crossover site, i.e. the exchange point is selected randomly. A probability of crossover (P_c) is also used to give freedom to an individual solution string to determine whether the particular solution would go for crossover or not. A biased coin with a probability (P_c) towards head is tossed. If it is head, the individual solution would go for crossover, and if it is tail the individual solution would go to the next generation without crossover. Therefore, (P_c) percent of the total population will participate in crossover operation and ($1 - P_c$) percent of the total population will move to the next generation without going through crossover operator. The crossover operator produces two children for the next generation. The children may be better than their parents. However, the crossover operator may also produce inferior solutions. In this case, the inferior solutions will be eliminated by the selection operator in the next iteration. Figure 18.4 shows a single point crossover operation. In the figure, (P_1) and (P_2) are the parent solutions. Similarly, (C_1) and (C_2) are the child solutions created after the crossover operators. Figure 18.5 shows two points crossover operation.

Mutation Operator

Mutation is the occasional introduction of new features into a solution string of the population. This operator is used to maintain diversity in the population. Though crossover has the main responsibility to search for the optimal solution, mutation is also used for this purpose. Mutation operator changes a '1' to '0' or otherwise, with a mutation probability of (P_m). The mutation probability is generally kept low for steady convergence. A high value of mutation probability would search here and there like a random search technique. Figure 18.6 shows the mutation operation where fourth bit has been changed from '1' to '0'.

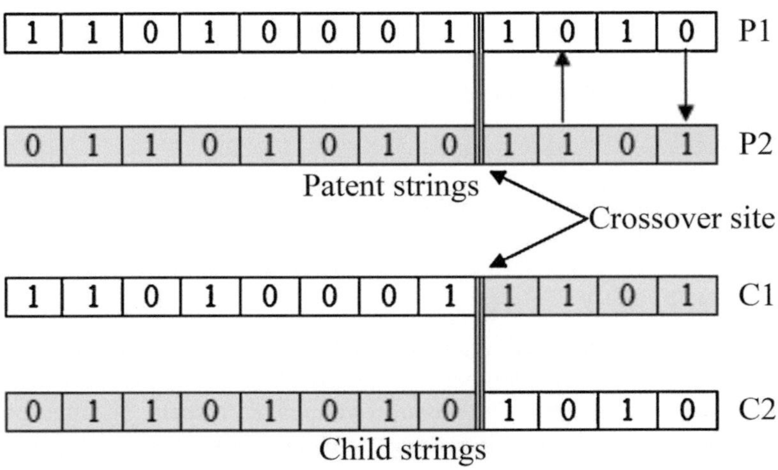

Fig. 18.4 Single-point crossover operation

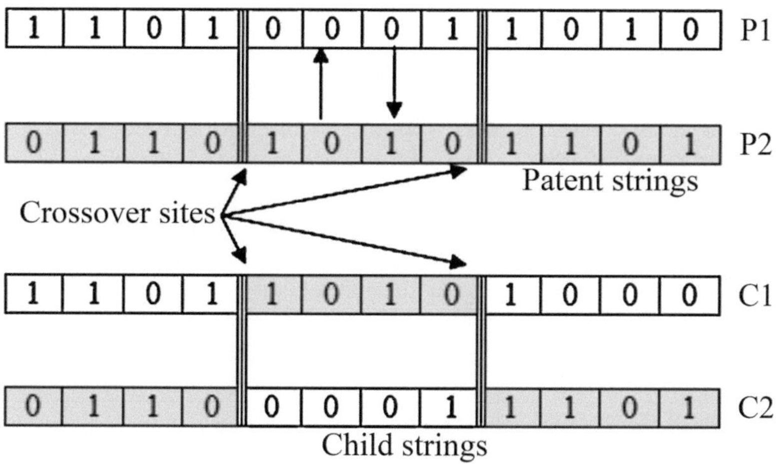

Fig. 18.5 Two-point crossover operation

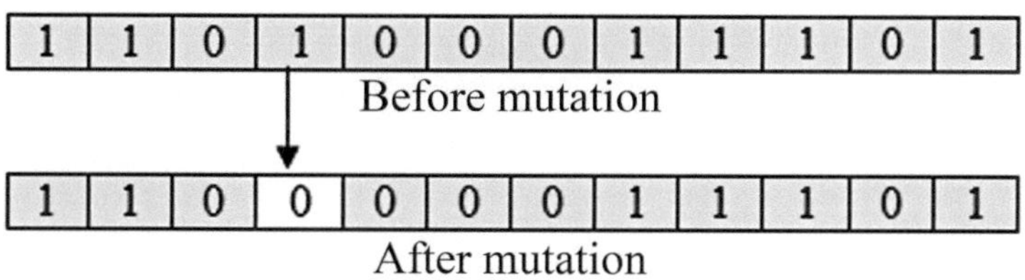

Fig. 18.6 Mutation operation

Elitism

The genetic operators, i.e. crossover and mutation are designed in such a way that it would produce better solution after each generation. However, sometime these operators may produce inferior solutions. But these solutions would die out when passing through the selection operator next time. Moreover, to increase the efficiency of the GA, it is necessary to preserve some better solutions for the next generation without applying crossover and mutation operator. This process is known as elitism. In this process, the better 5–10% of the total population is preserved for the next generation without applying crossover and mutation operator. Elitism is important since it allows the solutions to get better over time. If the elitism percentage is high, GA would lose its diversity to search for better solutions. The population will be saturated with the better individuals and it will converge at the local optimal solution.

Results and Discussion

A graphical interface based model in MATLAB (Fig. 18.7) is developed to calculate the factor of safety of an arbitrary slip circle. The developed MATLAB model calculates the factor of safety along with the parameters of the critical slip circle for a given earthen slope. The input parameters to the model are slope angle (α), the height of the slope (H), cohesion of the soil (c), the angle of shearing resistance (ϕ), unit weight of the soil (γ), and the pore pressure ratio (r_u). The developed MATLAB model is then linked with the genetic algorithm code to find critical slip circle of the slope along with the corresponding value of the factor of safety. Therefore, the output from the model is the critical factor of safety (F), the coordinate of the critical slip circle (x, y) and the radius (r) of the critical slip circle. The Genetic Algorithms parameters used in this study are listed in Table 18.1.

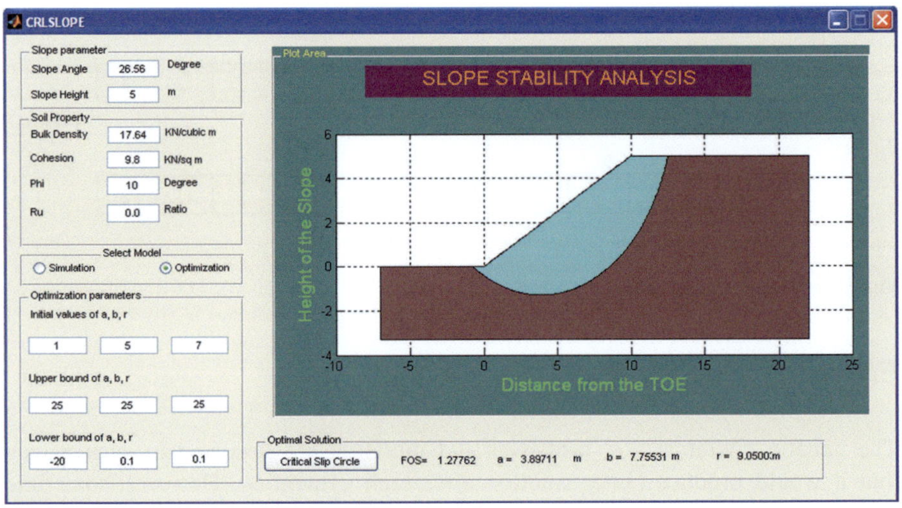

Fig. 18.7 GUI based model to calculate factor of safety

Table 18.1 Genetic algorithm parameters

Genetic parameters	Value
Population size (p)	100
Crossover probability (p_c)	0.85
Mutation probability (p_m)	0.001
Generation (g)	500
Elitism size (E)	5% of p

Table 18.2 Range F obtained using different optimization techniques

Method	Range of factor of safety
Yamagami and Ueta [13]	
BFGS	1.338
DFP	1.338
Powel	1.338
Simplex	1.339–1.348
Greco [7]	
Pattern search	1.327–1.33
Monte Carlo	1.327–1.333
The present study	
Genetic algorithm	1.276
Gradient search	1.335

Example Problem

A 5 m high slope with an inclination of 26.56° be comprised of soil whose cohesion $c = 9.8$ kPa, the angle of internal friction $\phi = 10°$, unit weight of soil $\gamma = 17.64$ kN/m^3. The range of factor of the safety calculated by Yamagami and Ueta [13], Greco [7], and the present study are shown in Table 18.2. The factor of safety obtained using genetic algorithm is 1.276. The factor of safety is also obtained by using the gradient search method and the value obtained is 1.335. This value is similar to the factor of safety calculated Yamagami and Ueta [13] and [7]. The factor of safety calculated using genetic algorithm is more critical than the factor of safety calculated using other methods. It has been observed that classical optimization techniques are highly sensitive to the initial solution, which implies that there may have several local optimal solutions of the problem. Therefore, the genetic algorithms produce better solution that the classical optimization algorithms. The evaluation of the results shows that the genetic algorithm is very robust in locating critical slip circle and the corresponding factor of safety of an earthen slope.

Conclusion

The study presents the efficiency of genetic algorithm in locating critical slip circle of an earthen slope. The method of slices is used to calculate the factor of safety of an arbitrary slip circle. The evaluation of results shows that the result derived using genetic algorithm is consistence and better than the other search algorithms. The gradient based search methods are highly sensitive to initial solution supplied and many times they trapped in local optimal solutions. This implies that the optimization problem is nonlinear and there may have several local optimal solutions.

References

1. Baker R (1980) Determination of the critical slip surface in slope stability computations. Int J Numer Anal Methods Geomech 4:333–359
2. Bishop AW (1955) The use of slip circle in the stability analysis of slopes. Geotechnique London 5:7–17
3. Chen Z-Y, Shao C-M (1988) Evolution of minimum factor of safety in slope stability analysis. Canadian Geotech J Ottawa 25:735–748
4. Celestino TB, Duncan JM (1981) Simplified search for noncircular slip surface. In: Proceedings of the 10th international conference on SMFE, pp 391–394
5. Deb K (1999) An introduction to genetic algorithms. Sadhana 24(4):293–315
6. Fellenius W (1936) Calculation of the stability of earth dams. Trans, of 2nd congress on Large Dams, vol 4, pp 445–459
7. Greco VR (1996) Efficient Monte Carlo technique for locating critical slip surface. J Geotech Eng ASCE 122(7):517–525
8. Goldberg DE (1989) Genetic algorithms in search, optimization, and in machine learning. Addison Wiley Longman, Inc, Boston
9. Janbu N (1973) Slope stability computations. In: Hirschfield E, Poulos S (eds) Embankment dam engineering, Casagrande memorial volume. Wiley, New York, pp 47–86
10. Li KS, White W (1987) Rapid evaluation of the critical slip surface in slope stability problems. Int J Numer Anal Methods Geomech 11:449–473
11. Nguyen VU (1985) Determination of critical slope failure surface. J Geotech Eng ASCE 111(2):238–250
12. Sarma SK (1979) Stability analysis of embankments and slopes. J Geotech Eng ASCE 105(12):1511–1524
13. Yamagami T, Ueta Y (1988) Search for noncircular slip surface by Morgenstern-price method. In: Proceedings of the 6th international conference numerical methods in geomechanics, pp 1219–12223

Chapter 19
Corridor Allocation as a Constrained Optimization Problem Using a Permutation-Based Multi-objective Genetic Algorithm

Zahnupriya Kalita and Dilip Datta

Abstract The corridor allocation problem (CAP) seeks the optimum arrangement of given facilities along two sides of a central corridor. The CAP is so far handled as an unconstrained optimization problem without imposing any restriction on the placement of the facilities. In practice, however, some facilities may need to satisfy certain constraints on their placement. Accordingly, a constrained bi-objective CAP (cbCAP) model is proposed here, where some facilities are restricted to fixed, same and/or opposite rows. Realizing the difficulties to any algorithm for handling such a combinatorial problem, a cbCAP specific permutation-based genetic algorithm (cbCAP-pGA) with specialized operators is also proposed for solving the cbCAP model by generating only feasible solutions. In the numerical experimentation, the cbCAP-pGA is found capable in searching promising solutions even for a set of large-size benchmark instances.

Keywords Combinatorial optimization · Corridor allocation problem · Genetic algorithm

Introduction

The well-known corridor allocation problem (CAP) seeks the effective placement of given facilities along the two sides of a straight central corridor, so as to optimize some objective functions. The CAP studied by Amaral [2], Ghosh and Kothari [8] and Ahonen et al. [1] is a variant of the double row facility layout problem (DRFLP), where the placement of the facilities along the two rows can be started from any level and some physical gap between two adjacent facilities of a row is also allowed

Z. Kalita · D. Datta (✉)
Department of Mechanical Engineering, Tezpur University, Tezpur, Assam, India
e-mail: zk@tezu.ernet.in; ddatta@tezu.ernet.in

F. Bennis, R. K. Bhattacharjya (eds.), *Nature-Inspired Methods for Metaheuristics Optimization*, Modeling and Optimization in Science and Technologies 16,
https://doi.org/10.1007/978-3-030-26458-1_19

without any restriction on the length of the corridor [4]. Further, the DRFLP is simplified to a linear problem by neglecting the width of the corridor. However, Kalita and Datta [9] claimed that the DRFLP and CAP may find applications mainly in planning machines in workshops or factories, but they are not applicable to many problems, such as the placement of office rooms in an administrative building or shops in a supermarket, where the placement of the facilities is to be started from a common level along the corridor of non-zero width. Further, any physical gap between two adjacent facilities of a row is not allowed and equal lengths of the two rows are preferred, even if such restrictions increase the material handling cost among the facilities. Accordingly, Kalita and Datta [9] proposed the bi-objective corridor allocation problem (bCAP) for minimizing simultaneously both the overall material handling cost among the facilities and the required length of the corridor. It is to be mentioned that Zuo et al. [14] and Wang et al. [13] also used non-zero corridor width in their DRFLP formulations. Zuo et al. [14] minimized the material flow cost and the layout area, while Wang et al. [13] minimized the material flow and re-arrangement costs.

In all the above cases, optimal layouts were searched considering arbitrary placement of the facilities along the two sides of a corridor. However, it is often found that some restrictions are preferred on the placement of the facilities, particularly in the layout design of service industries. Such restrictions were imposed by Motaghi et al. [12] in a hospital layout, where the rooms of patients were placed on one side and the technical rooms on the other side of the corridor. Lin et al. [11] also stated that the sterile storage room and operating rooms in operation theaters should always be placed on the same side of a corridor. According to Ebster [7], in order to promote sales rather than minimizing material handling costs, the checkouts and registers in supermarkets or shopping malls should be located on one side of the entrance (corridor), while the other side should be used for the exposure of products.

Motivated by above, a constrained bi-objective CAP (cbCAP) model is proposed here for minimize simultaneously the material handling cost among the facilities and the length of the corridor by imposing constraints on the placement of some facilities on the two rows. In this model, some facilities are restricted to a fixed row, some facilities in the same row, while some others in opposite rows. A permutation based genetic algorithm (pGA) was proposed by Datta et al. [5] for solving SRFLP, which was modified by Kalita and Datta [9] and Kalita et al. [10] for handling the bCAP. But the pGA alone is not adequate for solving the cbCAP effectively due to the presence of the constraints. Hence, for solving the cbCAP by generating feasible solutions only, the pGA is further modified here as the cbCAP-pGA with problem-specific operators. In the numerical experimentation, the potentiality of the cbCAP-pGA is demonstrated by solving a set of large-size instances in the range of [60, 80].

The article is divided into different Sections. The proposed cbCAP model is described in Sect. 19, followed by the cbCAP-pGA in Sect. 19. Presenting the computational experiments and discussion in Sect. 19, the article is finally concluded in Sect. 19.

The Proposed cbCAP Model

The cbCAP model requires an optimal arrangement of given facilities in two parallel rows along a central corridor by imposing some constraints on the placement of the facilities, so as to minimize the material handling cost among them and the length of the corridor. As in the bCAP model, the placement of the facilities in the cbCAP is also to be started from the same level on both the rows and no physical gap is allowed between two adjacent facilities placed in the same row. Additionally, the following constraints are imposed to the cbCAP:

1. Fixed row constraint: Some facilities are to be placed in the specified fixed row.
2. Same row constraint: Some pairs of facilities are to be placed in the same row.
3. Opposite row constraint: Some pairs of facilities are to be placed in opposite rows.

Accordingly, the cbCAP model can be stated as to minimize simultaneously the material handling cost among n number of facilities and the length of the corridor by arranging some facilities in the first row and remaining facilities in the second row with t_1 number of facilities in fixed rows, t_2 pairs of facilities in the same row and t_3 pairs of facilities in opposite row. The problem related parameters and variables are defined as follows (constraint related parameters are denoted in uppercase and variables are superscribed with π):

Indices

i, j : Indices of facilities
π : Index of a permutation of given facilities

Fixed parameters

l_i : Length of the ith facility ($i = 1, 2, \ldots, n$)
c_{ij} : Flow cost between the ith and jth facilities ($i, j = 1, 2, \ldots, n;\ i \neq j$)
w : Width of the corridor
$\prod_n$: Set of all permutations of the given n facilities
X_i : Index of placing the ith facility in fixed row $X_i \in$ {Y (yes), N (no)}
F_i : Required fixed row of the ith facility ($F_i \in \{1, 2\}$)
S_{ij} : Index of placing the ith facility in the same row with the jth facility $S_{ij} \in$ {Y (yes), N (no)}
Z_{ij} : Index of placing the ith facility in opposite row with the jth facility $Z_{ij} \in$ {Y (yes), N (no)}

Variables

n_p^π : Number of facilities placed in the pth row ($p \in \{1, 2\};\ n_1^\pi + n_2^\pi = n,\ \forall \pi \in \prod_n$)
B_i^π : Index of the row of the given ith facility
r_{pq}^π : Index of the qth facility of the pth row

$x^{\pi}_{r_{pq}}$: Centroidal distance of facility r^{π}_{pq} along the corridor

Objective functions

f_1^{π} : Overall material handling cost among all the given n facilities

f_2^{π} : Difference between the required lengths of the two rows

In terms of the above notations, the objective functions of the cbCAP can be formulated mathematically as in Eqs. (19.1)–(19.5) [10], while the imposed constraints as in Eqs. (19.6)–(19.8).

$$\text{Minimize} \quad f_1^{\pi} \equiv \sum_{i=1}^{n_1^{\pi}-1} \sum_{j=i+1}^{n_1^{\pi}} c_{r^{\pi}_{1,i} r^{\pi}_{1,j}} \left| x^{\pi}_{r_{1,i}} - x^{\pi}_{r_{1,j}} \right| + \sum_{k=1}^{n_2^{\pi}-1} \sum_{m=k+1}^{n_2^{\pi}} c_{r^{\pi}_{2,k} r^{\pi}_{2,m}} \left| x^{\pi}_{r_{2,k}} - x^{\pi}_{r_{2,m}} \right| + \sum_{i=1}^{n_1^{\pi}} \sum_{k=1}^{n_2^{\pi}} c_{r^{\pi}_{1,i} r^{\pi}_{2,k}} \left\{ \left(x^{\pi}_{r_{1,i}} - x^{\pi}_{r_{2,k}} \right)^2 + w^2 \right\}^{\frac{1}{2}} \tag{19.1}$$

$$\text{Minimize} \quad f_2^{\pi} \equiv \frac{L^{\pi} - L^{\text{th}}}{L^{\text{th}}} \times 100\% \tag{19.2}$$

$$\text{where,} \quad L^{\pi} = \max \left\{ \left(x^{\pi}_{r_{1,n_1}} + \frac{1}{2} l_{r^{\pi}_{1,n_1}} \right), \left(x^{\pi}_{r_{2,n_2}} + \frac{1}{2} l_{r^{\pi}_{2,n_2}} \right) \right\} \tag{19.3}$$

$$L^{\text{th}} = \frac{1}{2} \sum_{i=1}^{n} l_i \tag{19.4}$$

$$x^{\pi}_{r_{p,q}} = \frac{1}{2} l_{r^{\pi}_{p,1}} + \frac{1}{2} \sum_{t=2}^{q} \left(l_{r^{\pi}_{p,t-1}} + l_{r^{\pi}_{p,t}} \right) \tag{19.5}$$

$$\text{Subject to,} \quad B_i^{\pi} = F_i \; ; \quad \text{if} \quad X_i = \text{Y} \; ; \quad i = 1, 2, \cdots, n \tag{19.6}$$

$$B_i^{\pi} = B_j^{\pi} \; ; \quad \text{if} \quad S_{ij} = \text{Y} \; ; \quad i, j = 1, 2, \cdots, n \; ; \; i \neq j \tag{19.7}$$

$$B_i^{\pi} \neq B_j^{\pi} \; ; \quad \text{if} \quad Z_{ij} = \text{Y} \; ; \quad i, j = 1, 2, \cdots, n \; ; \; i \neq j \tag{19.8}$$

The first and second terms of f_1^{π} in Eq. (19.1) are the material flow costs among the facilities placed respectively in the first row and second rows, while the third term is the material flow cost among the facilities placed in opposite rows. The centroidal distance of a facility along the length of the corridor from the starting level of the placement of facilities, used in Eq. (19.1), is expressed by Eq. (19.5). The distance between two facilities is simply the difference between their such centroidal distances if they are placed in the same row (as used in the first two terms of f_1^{π}), while the same is computed using the Pythagoras' rule if the facilities are placed in opposite rows (as used in the third term of f_1^{π}). The distance between the jth

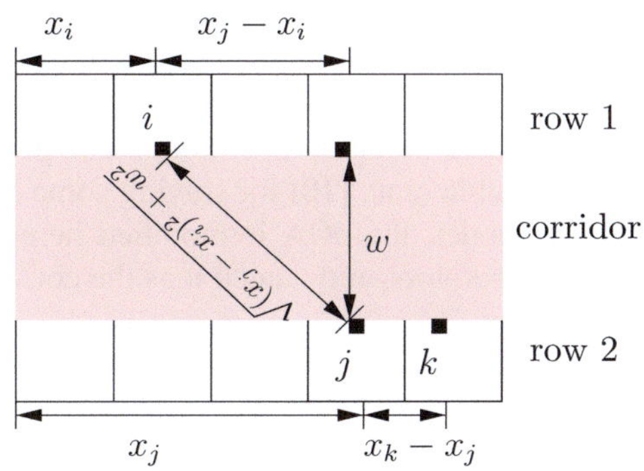

Fig. 19.1 Computation of the distances between pairs of facilities placed in the same row and in opposite rows

and kth facilities placed in the same row, and that between the ith and jth facilities placed in opposite rows, are illustrated in Fig. 19.1, where w is the width of the corridor.

In order to minimize the length of the corridor, f_2^{π} in Eq. (19.2) is defined as the percentage deviation of the longest row expressed by Eq. (19.3) from the theoretical minimum length of the corridor expressed by Eq. (19.4) as one half of the total length of all the n facilities placed in the two rows.

The constraint in Eq. (19.6) ensures that facilities are placed in their fixed rows, if required, where $X_i = \mathrm{Y}$ indicates that the given ith facility will occupy a fixed row and F_i is that fixed row. Similarly, the constraint in Eq. (19.7) ensures that both the ith and jth facilities are placed in the same row if $S_{ij} = \mathrm{Y}$, while they are not allowed in the same row if $Z_{ij} = \mathrm{Y}$ as shown by the constraint in Eq. (19.8).

The Proposed Genetic Algorithm for the cbCAP Model

Genetic Algorithm (GA) is a stochastic optimization technique developed based on the concept of biological evolution. The basic element of a GA is an individual that represents a solution of a problem. A set of individuals, known as a GA population, is evolved gradually towards the optima of a problem by repeated applications of mainly three operators, namely the selection, crossover and mutation operators. A selection operator identified some good individuals from the current GA population, a crossover operator generates new individuals by exploiting those identified by the selection operator, and a mutation operator is employed for exploring the neighborhood of the new individuals generated by the crossover operator. The process of evolution is continued until some termination criteria are met, such as the desired optimum is obtained or the predefined number of generations (i.e., iterations) are completed.

A permutation based genetic algorithm (pGA), suitable for solving single-row facility layout problems (SRFLPs), was introduced by Datta et al. [5], where an individual of the pGA represents a permutation of given facilities (the pGA is a modification of the nondominated sorting genetic algorithm II NSGA-II, a widely applied multi-objective GA proposed by Deb et al. [6], with permutation based crossover and mutation operators). The pGA was customized by Kalita and Datta [9] and Kalita et al. [10] for solving some bCAP models. For solving the proposed cbCAP model, the pGA is modified here further by introducing some problem-specific operators, and named it as the cbCAP-pGA.

Individual Representation and Initialization

In the pGA introduced by Datta et al. [5], an individual can be any permutation of given facilities as a feasible solution of the SRFLP or bCAP. However, any arbitrary permutation of the facilities may not be a feasible solution of the proposed cbCAP model, but it has to satisfy the constraints imposed to the cbCAP. Hence, in order to generate a feasible solution for the cbCAP model, an individual initialization technique is introduced in the proposed cbCAP-pGA, where the constrained facilities are arranged first and then the remaining facilities are arranged arbitrarily in vacant positions. Still all the constraints may not get satisfied until the permutation of the facilities represented by a cbCAP-pGA individual is split into two rows, which is explained in Sects. 19 and 19.

Splitting an Individual into Two Rows

An individual is first initialized with an arbitrary permutation of the facilities by the random initialization technique (refer Datta et al. [5] for detail). In the next step, the individual is split into two parts in order to form a cbCAP solution with the facilities of the two parts in two rows of the solution. The individual is split in a way to obtain a solution with the minimum possible length of the corridor, i.e., the individual is to be split after the n_1th element by minimizing the deviation of the first row from the theoretical minimum corridor length L^{th} given by Eq. (19.4). Accordingly, n_1 can be determined through Eq. (19.9).

$$n_1 = \begin{cases} s\,; & \text{if} \;\; L_1 = L^{\text{th}} \\ s\,; & \text{if} \;\; L^{\text{th}} - L_1 \leqslant L_2 - L^{\text{th}}\,;\; L_1 < L^{\text{th}}\,;\; L_2 > L^{\text{th}} \\ s+1\,; & \text{if} \;\; L_2 - L^{\text{th}} < L^{\text{th}} - L_1\,;\; L_1 < L^{\text{th}}\,;\; L_2 > L^{\text{th}} \end{cases} \tag{19.9}$$

$$\text{where,} \quad L_1 = \sum_{i=1}^{s} l_i\,; \qquad L_2 = L_1 + l_{s+1}$$

```
subroutine splitting_element(n,l,P,n1)
  L^th := 0
  for (i = 1 to n) do   L^th := L^th + l_i   end for
  L1 := 0
  for (i = 1 to n) do
  |  if (L1 + l_i ⩽ L^th) do
  |  |  L1 := L1 + l_i
  |  |  s := i
  |  end if
  end for
  L2 := L1 + l_{s+1}
  if (L1 = L^th) do   n1 := s   end if
  if (L1 < L^th and L2 > L^th and L^th − L1 ⩽ L2 − L^th) do   n1 := s   end if
  if (L1 < L^th and L2 > L^th and L2 − L^th < L^th − L1) do   n1 := s+1   end if
  return n1
end subroutine
```

Fig. 19.2 Pseudo-code implementing Eq. (19.9) for finding the splitting element in a cbCAP-pGA individual for forming a cbCAP solution

The pseudo-code, implementing Eq. (19.9) for finding the splitting element in a cbCAP-pGA individual in the process of forming a cbCAP solution, is presented in Fig. 19.2 as subroutine `splitting_element`(n, l, P, n_1), where n is the number of given facilities, l is a vector containing lengths of the facilities, P is the cbCAP-pGA individual, and n_1 is the splitting element to be returned by the subroutine.

Forming cbCAP Individual

A cbCAP solution is formed by satisfying the constraints imposed on the placement of the facilities. The constraints are satisfied based on the decreasing order of their complexities. The fixed row constraint is satisfied first, followed by the same row constraint and then the opposite row constraint. Hence, a constraint satisfying procedure for the cbCAP model is developed here. For formulating the procedure, the following additional parameters and notations are used:

P : Array of the facilities representing a feasible cbCAP solution
P' : Temporary array of the facilities before satisfying the constraints
I : Array of the positions of the facilities in P
n' : Number of available positions in array P
e : Array of the vacant positions of P

In the above notations, P is a feasible cbCAP solution, i.e., an array where the facilities are stored after satisfying the constraints imposed on their placement. P' is a temporary array of the facilities placed serially, i.e., as $1, 2, \cdots, n$. The constrained facilities are first picked up from P' one by one and placed in proper positions in P. Each time a facility is shifted, P' is updated by eliminating

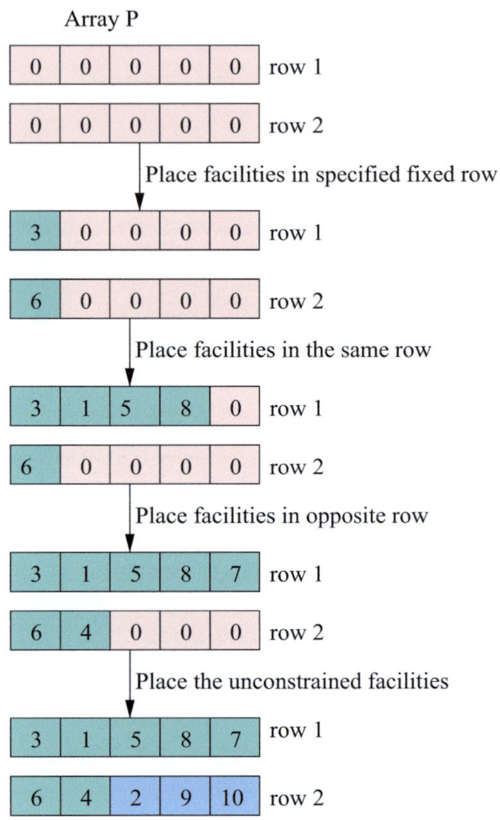

Fig. 19.3 Illustration of the procedure for forming a feasible cbCAP solution by satisfying the imposed constraints step by step

that facility from it. Once the placement of the constrained facilities is over, the unconstrained facilities are shifted from P' to arbitrary vacant positions of P. For illustrating the procedure, consider a small example of 10 facilities, where the third and sixth facilities are to be placed respectively in the first and second rows. The pair of the first and third facilities as well as the pair of the fifth and eighth are to be placed in the same row, while the pair of the third and fourth facilities as well as the pair of the sixth and seventh facilities are to placed in opposite rows. The formation of P by satisfying the constraints step by step is demonstrated in Fig. 19.3, where the constrained facilities are shown in cyan color and the unconstrained ones in light blue color.

Formulation and satisfaction procedures of the three types of constraints imposed to the cbCAP model are presented below:

1. *Fixed row constraint*: If the ith facility requires a fixed row in P, it is placed in the vacant position b of that row through Eq. (19.10), where b is obtained randomly using Eq. (19.11) and $I_i = 0$ means that the ith facility is not yet placed in P.

$$\left.\begin{array}{l} P_b = i \\ I_i = b \\ P'_j = P'_{j+1} \end{array}\right\} ; \quad \text{if} \quad \begin{array}{l} X_i = \text{Y} \; ; \; I_i = 0 \\ i = 1, 2, \cdots, n \\ j = i, i+1, \cdots, n-1 \end{array} \tag{19.10}$$

$$\text{where,} \quad b = e_t \; ; \quad t \in \{1, n'\} \; ; \; e_t \begin{cases} \leqslant n_1 \; ; & \text{if} \quad F_i = 1 \\ > n_1 \; ; & \text{if} \quad F_i = 2 \end{cases} \tag{19.11}$$

The self-explanatory pseudo-code, implementing Eqs. (19.10) and (19.11) for placing facilities in the specified rows in P is shown in Fig. 19.4 as subroutine `fix_row`(n, P, P', n_1, n_2),where subroutine `random_positions`$(1, n_1, n', e, P)$ find the vacant positions in P and subroutine `random_integer`$(1, n')$ returns a random integer in the range of $[1, n']$.

2. *Same and opposite row constraints*: Each of the same/opposite row constraints involves a pair of facilities. Hence, two cases may arise while satisfying such constraints, either one facility of a pair is already placed in P or both the facilities are yet to be placed.

   (a) If one facility of a pair of the same/opposite row facilities is already placed in P, the other one can be placed through Eq. (19.12) in the vacant position b of the appropriate row, where b is obtained randomly using Eq. (19.13).

$$\left.\begin{array}{l} P_b = i \\ I_i = b \\ P'_j = P'_{j+1} \end{array}\right\} ; \quad \begin{array}{l} \text{if} \;\; \left(S_{iy} = \text{Y or } Z_{iy} = \text{Y}\right) \; ; \; I_y \neq 0 \; ; \; I_i = 0 \\ j = i, i+1, \cdots, n-1 \\ i = 1, 2, \cdots, n \end{array} \tag{19.12}$$

```
subroutine fix_row(n,P,P',n1,n2)
  for (i = 1 to n) do                                         // Loop of the facilities
    if (X_i = Y and I_i = 0) do                               // Facility i requires a fixed row
      if (F_i = 1 ) do call random_positions(1,n1,n',e,P)     // Vacant positions in 1st row of P
      else call random_positions(n1 + 1,n2,n',e,P)            // Vacant positions in 2nd row of P
      end if
      if (n' ≠ 0) do t := call random_integer(1,n') end if    // A random integer in (1,n')
      b := e_t
      P_b := i                                                // Place i in P
      I_i := b                                                // Position of i in P
      for (j = i+1 to n − 1) do P'_j := P'_{j+1} end for      // Update P'
    end if
  end for
  return P,P'
end subroutine
```

Fig. 19.4 Pseudo-code implementing Eqs. (19.10) and (19.11) for placing facilities in fixed rows in cbCAP solution where, $b = e_t \; ; \; t \in \{1, n'\}$

$$e_t \in \begin{cases} \{1, n_1\} \; ; & \text{if } S_{iy} = \mathrm{Y} \; ; \; I_y \leqslant n_1 \\ \{n_1+1, n_2\} \; ; & \text{if } S_{iy} = \mathrm{Y} \; ; \; n_1 < I_y \leqslant n_2 \\ \{n_1+1, n_2\} \; ; & \text{if } Z_{iy} = \mathrm{Y} \; ; \; I_y \leqslant n_1 \\ \{1, n_1\} \; ; & \text{if } Z_{iy} = \mathrm{Y} \; ; \; n_1 < I_y \leqslant n_2 \end{cases} \tag{19.13}$$

In Eq. (19.13), $S_{iy} = \mathrm{Y}$ means that the ith and yth facilities are to be placed in the same row, while $Z_{iy} = \mathrm{Y}$ means that they are to be placed in opposite rows. Further, $I_y \leqslant n_1$ means that the yth facility is placed in the first row, while $n_1 < I_y \leqslant n_2$ if it is placed in the second row.

The self-explanatory pseudo-code, implementing Eqs. (19.12)–(19.13) for placing one facility of a pair in the same/opposite row with the other facility, which is already placed in P is shown in Fig. 19.5 in the subroutine `same_opp_row_cons_fixed`(n, P, P', n_1, n_2). Here, subroutines `random_positions`$(1, n_1, n', e, P)$ and `random_positions`$(n_1 + 1, n_2, n', e, P)$ find the vacant positions in the first row and second row of P, respectively. The subroutine `random_integer`$(1, n')$ returns a random integer in the range of $[1, n']$.

(b) If both the same/opposite row facilities of a pair are yet to be placed in P, they can be placed through Eq. (19.14) in the vacant positions a and b of the appropriate row(s), where a and b are obtained randomly using Eq. (19.15) or (19.16).

```
subroutine same_opp_row_cons_fixed(n,P,P',n1,n2)
  for (i = 1 to n) do                                    // Loop of the facilities
    if (Iy ≠ 0 and Ii = 0) do                            // y is in P not i
      if (Siy = Y ) do                                   // i required in same row with y
        if (Iy ≤ n1 ) do call random_positions(1,n1,n',e,P)     // Vacant positions in 1st row of P
        else call random_positions(n1 + 1,n2,n',e,P)            // Vacant positions in 2nd row of P
      end if
      if (Ziy = Y ) do                                   // i required in opposite row with y
        if (Iy ≤ n1 ) do call random_positions(n1 + 1,n2,n',e,P) // Vacant positions in 2nd row of P
        else call random_positions(1,n1,n',e,P)                 // Vacant positions in 1st row of P
      end if
      if (n' ≠ 0) do t := call random_integer(1,n') end if      // A random integer in (1,n')
      b := e_t
      P_b := i                                           // Place i in P
      I_i := b                                           // Position of i in P
      for (j = i+1 to n−1) do P'_j := P'_{j+1} end for   // Update P'
    end if
  end for
  return P,P'
end subroutine
```

Fig. 19.5 Pseudo-code implementing Eqs. (19.12) and (19.13) for placing one facility of each pair in the same/opposite rows with the other facility already placed in the cbCAP individual

$$\left.\begin{aligned} P_a &= i \\ P_b &= y \\ I_i &= a \\ I_y &= b \\ P'_j &= P'_{j+1} \\ P'_k &= P'_{k+1} \end{aligned}\right\} ; \quad \begin{aligned} &\text{if } \left(S_{iy} = \text{Y or } Z_{iy} = \text{Y}\right) \; ; \; I_y = 0 \; ; \; I_i = 0 \\ &j = i, i+1, \cdots, n-1 \\ &k = y, y+1, \cdots, n-1 \\ &i = 1, 2, \cdots, n \end{aligned} \tag{19.14}$$

$$\text{where,} \quad a = e_u \; ; \; b = e_t \; ; \; u, t \in \{1, n'\}$$

$$e_u, e_t \in \{1, n_1\} \text{ or } \{n_1 + 1, n_2\} \; ; \quad \text{if } S_{iy} = \text{Y} \tag{19.15}$$

$$e_u \in \begin{cases} \{1, n_1\} \; ; & \text{if} \quad Z_{iy} = \text{Y} \; ; \; e_t \in \{n_1 + 1, n_2\} \\ \{n_1 + 1, n_2\} \; ; & \text{if} \quad Z_{iy} = \text{Y} \; ; \; e_t \in \{1, n_1\} \end{cases} \tag{19.16}$$

In Eq. (19.15), $S_{iy} = \text{Y}$ means that the ith and yth facilities are to be placed in the same row, while $Z_{iy} = \text{Y}$ in Eq. (19.16) means that they are to be placed in opposite rows.

The self-explanatory pseudo-code, implementing Eqs. (19.14)–(19.16) for placing both facilities of a pair in the same/opposite rows, such that none of them are already in P is shown in Fig. 19.6 in the subroutine `same_opp_row_cons_free` (n, P, P', n_1, n_2).

Here, subroutines `random_positions` $(1, n_1, n', e, P)$ and `random_positions` $(n_1 + 1, n_2, n', e, P)$ find the vacant positions in the first row and second row of P, respectively. The subroutine `random_integer` $(1, n')$ returns a random integer in the range of $[1, n']$.

Selection Operation

Once the population of the cbCAP-pGA is initialized as explained in Sect. 19, its each individual is evaluated first through Eqs. (19.1)–(19.5) and then a mating pool is formed by applying the crowded tournament selection operator [6] to the population. The operator picks up two random individuals at a time from the mating pool and a copy of the best individual, based on the non-dominated ranks and crowding distances of the individuals, is stored in the mating pool. This is the same selection operator used in the original NSGA-II [6].

```
subroutine same_opp_row_cons_free(n,P,P',n1,n2)
  for (i = 1 to n) do                                          // Loop of the facilities
    if (I_y = 0 and I_i = 0) do                                // i and y not in P
      if (S_iy = Y ) do                                        // i required in same row with y
        call random_positions(1,n1,n',e,P)                     // Vacant positions in 1st row of P
        if (n' ≠ 0) do u := call random_integer(1,n') end if   // A random integer in (1,n')
        call random_positions(1,n1,n',e,P)                     // Vacant positions in 1st row of P
        if (n' ≠ 0) do t := call random_integer(1,n') end if   // A random integer in (1,n')
        if (u = 0 or t = 0) do                                 // No vacant positions in 1st row
          call random_positions(n1+1,n2,n',e,P)                // Vacant positions in 2nd row of P
          if (n' ≠ 0) do u := call random_integer(1,n') end if // A random integer in (1,n')
          call random_positions(n1+1,n2,n',e,P)                // Vacant positions in 2nd row of P
          if (n' ≠ 0) do t := call random_integer(1,n') end if // A random integer in (1,n')
        end if
      end if
      if (Z_iy = Y ) do                                        // i required in opposite row with y
        call random_positions(1,n1,n',e,P)                     // Vacant positions in 1st row of P
        if (n' ≠ 0) do u := call random_integer(1,n') end if   // A random integer in (1,n')
        call random_positions(n1+1,n2,n',e,P)                  // Vacant positions in 2nd row of P
        if (n' ≠ 0) do t := call random_integer(1,n') end if   // A random integer in (1,n')
        if (u = 0 or t = 0) do                                 // No vacant positions
          call random_positions(n1+1,n2,n',e,P)                // Vacant positions in 2nd row of P
          if (n' ≠ 0) do u := call random_integer(1,n') end if // A random integer in (1,n')
          call random_positions(1,n1,n',e,P)                   // Vacant positions in 1st row of P
          if (n' ≠ 0) do t := call random_integer(1,n') end if // A random integer in (1,n')
        end if
      end if
      a := e_u
      b := e_t
      P_a := i                                                 // Place i in P
      P_b := y                                                 // Place y in P
      I_i := a                                                 // Position of i in P
      I_y := b                                                 // Position of y in P
      for (j = i+1 to n-1) do P'_j := P'_{j+1} end for
      for (k = y+1 to n-1) do P'_k := P'_{k+1} end for         // Update P'
    end if
  end for
  return P,P'
end subroutine
```

Fig. 19.6 Pseudo-code implementing Eqs. (19.14)–(19.16) for placing both facilities of a pair in the same/opposite rows in the cbCAP solution

Crossover Operation

In the cbCAP-pGA, a cbCAP specific crossover operator is proposed, which generates only feasible cbCAP solutions. Taking two random parent individuals at a time from the mating pool formed by the selection operator, the crossover operator generates a child (new) individual by importing facilities, as many as possible, from one of the parent individuals. Importing means that a facility is placed in the same position of the child individual as it was occupying in the parent individual where from it is imported. The operator works first on importing facilities in the order of facilities requiring fixed rows, pairs of facilities requiring the same/opposite rows, and unconstrained facilities. The facilities, that could not be imported from any of the parent individuals due to the pre-occupancy of their specific positions in the child individual by other facilities, are first stored in reserve during the process of importing and finally they are placed randomly in the vacant positions of the child individual following the same procedure used in Sect. 19 for initializing an individual. For explaining the procedure of importing facilities from the parent individuals, the following additional notations are defined:

C : To be generated child individual
$I^{(c)}$: Array of the positions of the facilities in C
$I^{(1)}$: Array of the positions of the facilities in first parent individual
$I^{(2)}$: Array of the positions of the facilities in second parent individual

In terms of the above notations and those defined earlier, the importing of a facility requiring a fixed row in the child individual is expressed by Eq. (19.17), where the ith facility is imported to the vacant bth position of the child individual if it were occupying the bth position in one of the parent individuals.

$$\left.\begin{array}{ll} \left.\begin{array}{r} C_b = i \\ I_i^{(c)} = b \end{array}\right\}; & \text{if} \quad X_i = \mathrm{Y}\ ;\ I_i^{(c)} = 0 \\ & \quad\;\; i = 1, 2, \cdots, n \\ \text{where,} \quad b = I_i^{(1)} \text{ or } b = I_i^{(2)} & \end{array}\right\} \tag{19.17}$$

The self-explanatory pseudo-code, implementing Eq. (19.17) for importing the fixed-row facilities to the child individual C from any one parent individual is shown in Fig. 19.7 in the subroutine `crossover_fix_row`$(n, I^{(c)}, I^{(1)}, I^{(2)}, C)$. Here, the parent individual is chosen based on a random number, represented by r, ranging from [0, 1]. If $r < 0.5$, then the first parent is chosen, otherwise the second parent is chosen.

After importing fixed-row facilities as many as possible, pairs of facilities requiring the same/opposite rows are imported through Eq. (19.18), where such a

```
subroutine crossover_fix_row(n,I^(c),I^(1),I^(2),C)
  for (i = 1 to n) do                          // Loop of the facilities
    if (X_i = Y and I_i^(c) = 0) do            // i in fixed row and not in C
      r := rreal(0,1)                          // A random number in (0,1)
      if (r < 0.5) do b := I_i^(1)             // Select parent based on the value of r
      else b := I_i^(2)
      end if
      C_b := i                                 // Place i in C
      I_i^(c) := b                             // Position of i in C
    end if
  end for
  return C
end subroutine
```

Fig. 19.7 Pseudo-code showing the procedure used for placing facilities in fixed rows in the child individual by the crossover operator of the cbCAP-pGA pair of the ith and yth facilities are imported respectively to the ath and bth positions in the same/opposite rows of the child individual if they were occupying the same positions in one of the parent individuals.

$$\left.\begin{array}{l}\left.\begin{array}{r} C_a = i \\ C_b = y \\ I_i^{(c)} = a \\ I_y^{(c)} = b \end{array}\right\} ; \quad \begin{array}{l} \text{if } \left(S_{iy} = \text{Y or } Z_{iy} = \text{Y}\right) \; ; \; I_y^{(c)} = 0 \; ; \; I_i^{(c)} = 0 \\ i = 1, 2, \cdots, n \end{array} \\ \text{where, } \left(a = I_i^{(1)} \text{ and } b = I_y^{(1)}\right) \text{ or } \left(a = I_i^{(2)} \text{ and } b = I_y^{(2)}\right) \end{array}\right\} \tag{19.18}$$

The self-explanatory pseudo-code, implementing Eq. (19.17) for placing facilities in same/opposite rows in the child individual C form any one parent individual is shown in Fig. 19.8 in the subroutine `crossover_same_opp_row` $(n, I^{(c)}, I^{(1)}, I^{(2)}, C)$. The parent individual is selected randomly, with the similar procedure used for importing the fixed-row facilities.

Then, unconstrained facilities are imported using Eq. (19.19), where the ith facility is imported to the vacant bth position of the child individual, if it were occupying the bth position in one of the parent individuals and not paired with any facility (shown as the yth and zth facilities) for placing in the same/opposite rows.

$$\left.\begin{array}{l}\left.\begin{array}{r} C_b = i \\ I_i^{(c)} = b \end{array}\right\} ; \quad \begin{array}{l} \text{if } \; X_i \neq \text{Y} \; ; \; S_{iy} \neq \text{Y} \; ; \; Z_{iz} \neq \text{Y} \; ; \; I_i^{(c)} = 0 \\ y, z \in \{1, 2, \cdots, n\} \; ; \; y \neq i \; ; \; z \neq i \\ i = 1, 2, \cdots, n \end{array} \\ \text{where, } \; b = I_i^{(1)} \text{ or } b = I_i^{(2)} \end{array}\right\} \tag{19.19}$$

```
subroutine crossover_same_opp_row(n,I(c),I(1),I(2),C)
  for (i = 1 to n) do                                   // Loop of the facilities
    if (S_iy = Y or Z_iy = Y and I_i(c) = 0 and I_y(c) = 0) do   // i in same/opp. row with y not in C
      r := rreal(0,1)                                   // A random number in (0,1)
      if (r < 0.5) do                                   // Select parent based on the value of r
        a := I_i(1)
        b := I_y(1)
      else
        a := I_i(2)
        b := I_y(2)
      end if
      C_a := i                                          // Place i in C
      I_i(c) := a                                       // Position of i in C
      C_b := y                                          // Place y in C
      I_y(c) := b                                       // Position of y in C
    end if
  end for
  return C
end subroutine
```

Fig. 19.8 Pseudo-code showing the procedure used for placing facilities in same/opposite rows in the child individual by the crossover operator of the cbCAP-pGA

```
subroutine crossover_uncons(n, I^(c), I^(1), I^(2), C)
  for (i = 1 to n) do                                  // Loop of the facilities
    if (X_i ≠ Y or S_iy ≠ Y or Z_iz ≠ Y and I_i^(c) = 0) do   // i and y are unconstrained and not in C
      r := rreal(0,1)                                  // A random number in (0, 1)
      if (r < 0.5) do b := I_i^(1)                     // Select parent based on the value of r
      else b := I_i^(2)
      end if
      C_b := i                                         // Place i in C
      I_i^(c) := b                                     // Position of i in C
    end if
  end for
  return C
end subroutine
```

Fig. 19.9 Pseudo-code showing the procedure used for placing unconstrained facilities in the child individual by the crossover operator of the cbCAP-pGA

The self-explanatory pseudo-code, implementing Eq. (19.19) for importing each unconstrained facility to the child individual C from any one parent individual chosen randomly, as mentioned above, is shown in Fig. 19.9 in the subroutine crossover_uncons(n, $I^{(c)}$, $I^{(1)}$, $I^{(2)}$, C).

Finally, as stated earlier, the facilities, that could not be placed in the child individual through any of Eqs. (19.17)–(19.19), are placed randomly in the vacant positions of the child individual following the same procedure used in Sect. 19 for initializing an individual.

Mutation Operation

A mutation operator explores the neighborhood of a child individual generated by a crossover operator. The cbCAP specific mutation operator incorporated in the cbCAP-pGA chooses two random elements at a time from a child individual and their values are interchanged if no constraint imposed to the cbCAP model is violated from such swapping.

Elite Preserving Mechanism

Like those in any other variants of GA, the cbCAP specific crossover and mutation operators, incorporated in the proposed cbCAP-pGA, also do not guarantee the generation of children individuals better than their parent individuals, even the former could be worse than the latter pushing the search opposite to the optima of a problem. Hence, the elite preserving mechanism, proposed in the NSGA-II [6], is applied in the cbCAP-pGA also for carrying over the elite individuals of one generation to the next generation. The mechanism first combines the current population and the newly generated children population, and then the population for the next generation is formed with the best 50% of the combined population based on the nondominated ranks and crowding distances of the individuals (refer Deb et al. [6] for detail).

Computational Experiment and Discussion

The proposed cbCAP-pGA is coded in C programming language and executed in Fedora 15 Linux environment. It is tested for 20 large-size instances in the range of [60, 80], which were introduced by Anjos et al. [3] for testing SRFLP models.

The cbCAP involves the satisfaction of the constraints imposed on the placement of some facilities. For easy recognition, such constraints are imposed on the same set of facilities in each of the studied 20 instances. The fixed-row constraint is imposed on facilities 13 and 40, requiring the placement of facility 13 in the first row and facility 40 in the second row. The pair of facilities 15 and 24, as well as that of facilities 40 and 42, are constrained to be placed in the same rows. On the other hand, the constraint of opposite rows is imposed to the pair of facilities 13 and 18, as well as that of facilities 27 and 39.

It is a known fact that the performance of stochastic algorithms including the proposed cbCAP-pGA may be influenced by their algorithmic parameter settings. However, without going through any procedure for fine tuning such parameters, the cbCAP-pGA related parameters for each of the studied 20 instances are set simply based on some trial runs as follows: a population of size 60 is evolved over 10,000 generations with crossover probability of 75% and mutation probability of 5%. Further, 30 independent runs for each instance are performed with different sets of initial solutions and the obtained best results are reported here.

Since the proposed cbCAP is a bi-objective optimization problem, its result is not a single solution but a set of trade-off solutions known as the Pareto front. In a Pareto front of cbCAP, its solutions are conflicting with each other in terms of the objective functions f_1 (material handling cost) and f_2 (deviation of the obtained corridor length from its theoretical minimum length) as expressed by Eqs. (19.1) and (19.2), respectively. As an illustration, the best Pareto front obtained for instance Anjos60_1 [3] is shown in Fig. 19.10, where the extreme solutions of the Pareto front are best either in the material handling cost (f_1^{best}) or deviation in corridor length (f_2^{best}).

As shown in Fig. 19.10, the objective values of the extreme solutions of the Pareto fronts obtained for the studied 20 instances (i.e., the solutions containing f_1^{best} and f_2^{best}), are shown in Table 19.1.

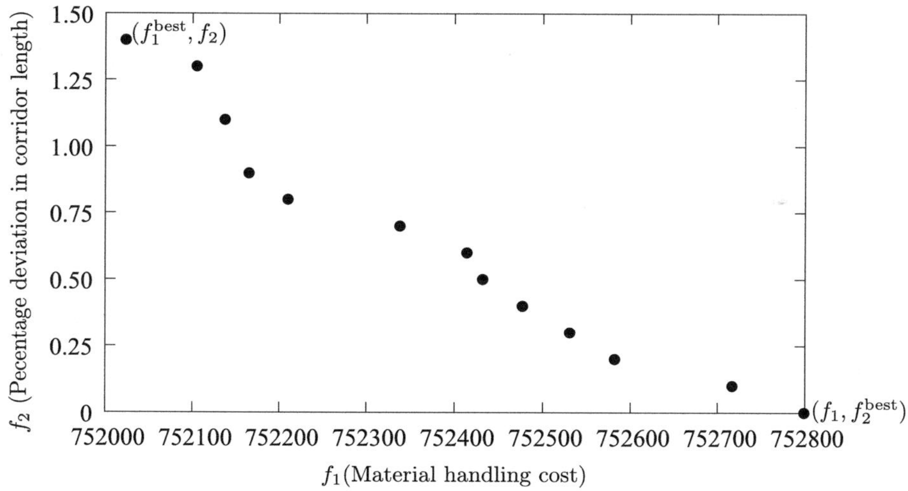

Fig. 19.10 Pareto front for instance Anjos60_1 [3]

Since no work could be found in the specialized literature studying any CAP model with constraints, the acceptability of the results obtained in the present work could not be verified. Hence, in order to have some coarse estimate, the cbCAP results for the studied instances presented in Table 19.1 are compared with the bCAP results reported by Kalita et al. [10] for those instances. Such comparison is also presented in Table 19.1 marking the best values in boldface fonts. It seen that the values of f_1^{best} for the instances under the cbCAP model are higher than those under the bCAP model, which is obvious as some facilities under the cbCAP model missed the best possible positions as they are constrained to be placed in some other positions. However, it is observed that their relative deviations (presented in percentage in Table 19.1) are marginal in all the studied 20 instances. On the other hand, an interesting observation is that the corridor lengths (presented in Table 19.1 in the forms of f_2 and f_2^{best}) are reduced in the case of some instances under the cbCAP model. Hence, it may be concluded that the proposed cbCAP model can be applied to practical applications even if the constraints on the facilities increase the material flow cost by some amount.

The arrangements of the facilities in the two rows obtained by the proposed cbCAP model for the two scenarios (i.e., f_1^{best} and f_2^{best}) are shown in Tables 19.2 and 19.3, respectively. For easy identification, the constrained facilities are shown in boldface fonts.

Table 19.1 Objective values of the extreme solutions of the Pareto fronts obtained for the studied 20 instances and their comparison with those of bCAP model reported by Kalita et al. [10] (the best values are shown in bold fonts)

			Solutions with f_1^{best}					Solutions with f_2^{best}				
			cbCAP model		bCAP model		Increase in f_1^{best} (in %)	cbCAP model		bCAP model		Increase in f_2^{best} (in %)
SN	Instance ID	Size	f_1^{best}	f_2	f_1^{best}	f_2		f_1	f_2^{best}	f_1	f_2^{best}	
1	Anjos60_1	60	752023.0	1.45	**739516.0**	**0.10**	1.69	752799.0	0.00	**739582.0**	0.00	1.79
2	Anjos60_2	60	428807.0	**1.76**	**421074.0**	1.87	1.84	429127.0	0.05	**421419.0**	0.05	1.83
3	Anjos60_3	60	330315.5	**1.55**	**324931.5**	1.88	1.66	330488.5	0.05	**325222.5**	0.05	1.62
4	Anjos60_4	60	202678.0	0.18	**199434.0**	**0.09**	1.63	202766.0	0.00	**199441.0**	0.00	1.67
5	Anjos60_5	60	164394.0	2.98	**159652.0**	**1.39**	2.97	164615.0	0.06	**159870.0**	0.07	2.97
6	Anjos70_1	70	773603.0	**1.16**	**765123.0**	1.17	1.11	773651.0	0.03	**765711.0**	0.04	1.04
7	Anjos70_2	70	729819.0	1.49	**721156.0**	**0.97**	1.20	729888.0	0.00	**721251.0**	0.00	1.20
8	Anjos70_3	70	771258.5	0.97	**760263.5**	**0.60**	1.45	771849.5	0.00	**760469.5**	0.00	1.50
9	Anjos70_4	70	493014.0	0.32	**484746.0**	**0.26**	1.71	493090.0	0.03	**484848.0**	0.03	1.70
10	Anjos70_5	70	2146937.5	1.64	**2109741.5**	**0.78**	1.76	2148070.5	0.00	**2109845.5**	0.00	1.81
11	Anjos75_1	75	1214707.5	0.86	**1197910.5**	**0.13**	1.40	1214820.5	0.00	**1198083.5**	0.00	1.40
12	Anjos75_2	75	2178628.0	1.53	**2162182.0**	**0.53**	0.76	2179350.0	0.03	**2162581.0**	0.03	0.78
13	Anjos75_3	75	637176.0	1.13	**625118.0**	**0.99**	1.93	637425.0	0.00	**625317.0**	0.00	1.94
14	Anjos75_4	75	1989668.5	0.85	**1972587.5**	**0.41**	0.87	1990027.5	0.03	**1972897.5**	0.03	0.87
15	Anjos75_5	75	915945.0	2.64	**896016.0**	**0.45**	2.22	918182.0	0.00	**896127.0**	0.00	2.46
16	Anjos80_1	80	1050883.5	0.12	**1035134.5**	0.12	1.52	1050999.5	0.00	**1035150.5**	0.00	1.53
17	Anjos80_2	80	977673.0	0.28	**961338.0**	**0.09**	1.70	977673.0	0.28	**961420.0**	**0.03**	1.69
18	Anjos80_3	80	1651956.0	1.29	**1626934.0**	**1.23**	1.54	1652273.0	0.00	**1627508.0**	0.00	1.52
19	Anjos80_4	80	1888011.0	**0.15**	**1874656.0**	0.57	0.71	1888127.0	0.03	**1875113.0**	0.03	0.69
20	Anjos80_5	80	807499.0	0.73	**795089.0**	**0.03**	1.56	807633.0	0.03	**795089.0**	0.03	1.58

Table 19.2 Arrangement of the facilities obtained by the cbCAP model in the solutions presented in Table 19.1 against f_1^{best} of the studied 20 instances (constrained facilities are presented in boldface fonts)

SN	Instance ID	Size	Arrangements of the facilities in the solutions with f_1^{best}
1	Anjos60_1	60	**27** 11 **24** **15** 28 21 25 37 6 46 9 36 35 32 8 26 2 14 **13** 49 23 57 31 45 53 59 19 60 20 48 34 56 **18** 55 58 3 **40** 51 54 7 12 44 **42** 16 33 4 17 22 41 47 29 38 5 43 50 10 52 30 **39** 1
2	Anjos60_2	60	20 22 **15** 4 23 3 7 38 58 60 16 **27** 51 59 36 46 17 1 **24** 25 37 28 5 21 34 14 12 9 30 **13** 10 53 11 57 **42** 19 2 48 49 47 6 43 55 **18** 26 45 54 **39** 52 **40** 56 33 32 31 8 50 29 41 35 44
3	Anjos60_3	60	4 **24** 59 20 44 1 **27** 37 52 34 51 12 **15** 56 35 46 43 26 17 25 58 9 55 **13** 57 54 22 36 16 19 5 6 53 49 47 21 **40** 32 31 10 **39** 23 **18** 33 30 50 8 60 45 28 7 2 14 41 11 48 29 38 3 **42**
4	Anjos60_4	60	20 12 **13** 10 7 **24** 22 11 36 6 17 46 23 5 53 3 50 14 35 16 54 **27** 26 15 32 30 56 52 38 4 41 25 31 19 57 55 **39** 44 **40** 33 21 37 59 45 **18** 2 58 8 60 29 51 **42** 47 48 9 49 1 28 34 43
5	Anjos60_5	60	33 53 49 52 11 41 19 1 43 29 23 51 **24** 57 45 **27** 25 5 **15** 60 6 28 14 30 35 46 **13** 54 34 16 12 **39** 44 37 59 38 55 26 58 22 32 31 56 **42** **40** 21 50 8 48 10 36 **18** 20 17 9 2 4 47 3 7
6	Anjos70_1	70	50 35 17 48 25 45 34 66 14 59 49 58 **13** 21 36 23 12 **27** **24** 52 67 44 61 64 68 51 31 63 32 62 2 **15** 47 **39** 37 57 29 43 46 4 3 38 20 7 6 54 11 60 55 5 30 19 **42** **18** 33 26 10 70 56 41 16 1 69 9 8 28 65 22 **40** 53
7	Anjos70_2	70	50 25 47 21 36 1 57 19 58 9 45 60 14 68 44 43 **24** **15** 54 22 **27** 16 34 55 12 17 69 70 48 59 64 52 28 **13** 53 10 26 7 3 67 23 29 30 62 11 63 **18** 6 **40** 49 8 4 **42** 2 20 46 5 37 32 31 35 38 65 33 66 51 41 61 **39** 56
8	Anjos70_3	70	**15** 44 52 17 64 33 **27** 12 **13** 55 62 60 65 36 5 28 25 53 14 70 37 68 4 11 6 41 23 32 21 **24** 51 30 16 31 57 34 69 46 38 59 48 45 35 3 22 **39** **18** 66 49 **40** 56 10 50 **42** 2 19 7 61 26 9 47 67 1 20 58 29 43 54 63 8
9	Anjos70_4	70	7 **15** 47 9 54 22 61 29 **24** 12 63 26 60 58 53 2 49 50 23 **27** **13** 44 69 57 70 17 67 62 43 3 10 31 4 8 35 66 16 51 30 52 6 64 **42** 11 14 33 **18** 68 46 5 34 **40** 36 37 1 28 56 59 19 21 55 20 48 32 45 65 **39** 38 25 41
10	Anjos70_5	70	28 38 68 23 31 51 58 22 6 10 16 63 **27** 8 61 14 **24** 35 12 11 29 34 3 **13** 26 59 45 **15** 57 52 33 64 44 67 19 32 1 62 **40** 2 37 50 54 65 49 41 43 21 56 20 **42** 70 5 36 30 4 7 53 **39** 60 46 66 55 25 48 47 **18** 69 9 17

(continued)

Table 19.2 (continued)

SN	Instance ID	Size	Arrangements of the facilities in the solutions with f_1^{best}
11	Anjos75_1	75	47 48 **13** 51 21 17 35 75 62 **15** **24** 43 68 66 34 12 10 19 20 4 55 44 7 65 61 **27** 33 26 52 28 73 54 69 60 38 14 16 72 5 56 8 **39** 11 6 22 25 71 53 37 70 **40** **42** 74 31 1 3 32 2 41 23 30 49 50 45 **18** 59 36 57 63 9 29 67 46 64 58
12	Anjos75_2	75	49 64 32 72 28 44 14 50 33 1 58 5 52 53 71 67 12 8 62 **27** 22 **15** 68 55 **13** 17 9 19 10 69 **24** 45 30 70 2 37 43 25 51 61 47 21 6 63 60 7 57 **42** 59 4 20 75 65 3 34 36 11 23 16 38 46 74 48 31 73 41 29 54 **18** 35 **40** 66 **39** 56 26
13	Anjos75_3	75	47 54 69 55 8 72 26 **27** 38 2 68 44 62 59 32 6 **13** 58 75 9 **24** 45 21 73 65 11 30 71 61 37 60 41 43 63 14 33 10 **15** 50 1 17 25 **42** 29 70 4 **39** 66 **18** 20 48 3 34 49 5 67 28 35 **40** 7 52 57 36 16 31 12 56 74 64 23 46 51 53 19 22
14	Anjos75_4	75	60 **15** 9 14 7 75 10 37 30 57 22 29 56 46 55 64 16 28 **13** 73 32 1 2 **24** 35 63 66 59 54 71 4 11 61 25 6 68 **27** 17 36 5 49 **42** 50 62 8 70 47 20 41 **40** 33 12 3 65 52 44 53 **18** 34 67 45 19 48 58 26 23 43 **39** 31 74 51 38 69 72 21
15	Anjos75_5	75	20 37 9 63 2 43 44 53 16 22 69 56 45 34 57 54 61 25 17 3 55 29 12 5 49 **15** 23 73 32 47 1 35 64 52 **13** **24** 33 **27** 26 19 6 58 50 60 14 74 21 75 4 **18** 59 67 48 11 28 30 36 **42** 70 8 10 46 65 7 31 66 38 41 71 68 **39** 62 72 **40** 51
16	Anjos80_1	80	78 25 80 67 **27** 65 29 46 43 19 52 14 75 51 45 34 47 54 41 32 58 9 31 36 **15** 76 74 56 17 37 72 **13** 69 **24** 59 3 64 1 71 60 11 55 70 2 44 8 **40** 21 16 63 53 61 12 68 49 **39** 62 66 48 20 10 50 23 6 22 33 7 **42** 73 26 5 30 35 77 38 28 4 **18** 79 57
17	Anjos80_2	80	22 10 58 46 56 **27** 12 47 **15** 29 **24** 1 7 **13** 28 35 20 36 34 8 65 62 75 59 68 76 70 79 37 31 64 80 69 49 48 45 17 11 57 55 5 **18** 14 32 74 54 41 16 33 77 72 23 25 67 53 3 6 4 44 38 9 **40** 2 50 **39** **42** 26 21 19 30 73 71 52 78 60 51 63 43 61 66
18	Anjos80_3	80	74 16 33 49 46 47 43 56 **24** 67 28 6 66 38 9 75 2 11 73 **15** 59 **27** 21 80 44 78 51 77 **13** 76 22 35 48 70 26 34 36 54 63 37 58 68 61 62 64 45 5 4 65 **42** 79 12 31 60 55 71 1 19 7 41 23 14 17 3 **39** 8 52 50 10 **18** 25 30 72 32 **40** 69 29 20 53 57
19	Anjos80_4	80	69 50 75 38 54 30 28 **24** 80 43 74 76 70 6 67 2 16 20 72 31 33 48 52 **15** 79 77 51 5 **27** 11 **13** 23 25 34 46 36 57 22 47 60 64 71 35 78 63 32 66 12 **42** 14 29 62 17 4 3 7 **18** 19 26 1 53 37 8 **39** 49 65 9 59 45 68 73 55 10 44 21 58 61 **40** 56 41
20	Anjos80_5	80	2 47 52 61 10 67 **24** 22 **15** 1 9 76 33 6 54 30 34 75 71 66 32 80 43 62 63 50 5 28 14 64 29 72 **27** **13** 60 58 48 25 7 78 38 73 21 68 37 74 70 31 19 36 44 **39** 53 35 55 59 **40** 65 3 4 77 20 **18** 79 45 23 46 57 11 56 26 16 8 51 49 12 69 **42** 41 17

Table 19.3 Arrangement of the facilities obtained by the cbCAP model in the solutions presented in Table 19.1 against f_2^{best} of the studied 20 instances (constrained facilities are presented in boldface fonts)

SN	Instance ID	Size	Arrangements of the facilities in the solutions with f_2^{best}
1	Anjos60_1	60	**27** 11 **24** **15** 28 21 25 37 6 46 12 36 35 33 8 26 2 14 **13** 49 23 57 31 43 10 59 19 60 20 48 34 56 **18** 55 58 3 **40** 51 54 7 9 44 **42** 16 32 4 17 22 41 29 47 38 5 45 50 53 52 30 **39** 1
2	Anjos60_2	60	20 57 **15** 4 23 3 7 38 58 60 16 **27** 51 55 36 46 17 1 **24** 25 37 28 5 21 34 14 12 9 41 **13** 10 53 11 22 **42** 19 2 48 49 47 6 43 59 **18** 26 45 54 **39** 52 **40** 56 33 32 31 8 50 29 30 35 44
3	Anjos60_3	60	4 **24** 59 20 44 1 **27** 37 52 34 51 12 **15** 56 35 46 43 26 17 60 58 9 55 **13** 57 41 11 36 16 19 5 6 53 49 47 21 **40** 32 31 10 **39** 23 **18** 33 30 50 8 25 45 28 7 2 14 54 22 48 29 38 3 **42**
4	Anjos60_4	60	31 12 **13** 10 7 **24** 22 36 11 6 17 46 23 5 53 3 50 14 35 16 54 **27** 26 **15** 32 30 56 48 28 1 41 25 20 19 57 55 **39** 44 **40** 33 21 37 59 45 **18** 2 58 8 60 29 51 **42** 47 52 9 49 38 4 34 43
5	Anjos60_5	60	33 53 49 52 38 41 19 1 43 29 23 51 **24** 57 45 **27** 25 5 **15** 60 10 28 14 30 35 46 **13** 54 34 16 7 **39** 44 37 59 11 55 26 58 22 32 31 56 **42** **40** 21 50 8 48 6 36 **18** 20 17 9 2 4 47 3 12
6	Anjos70_1	70	50 35 17 48 25 45 34 66 14 59 49 58 **13** 21 36 23 12 **27** **24** 52 67 44 61 64 68 51 31 63 32 28 2 **15** 53 **39** 37 57 29 43 46 4 3 38 20 7 6 54 11 60 5 55 30 19 **42** **18** 33 26 10 70 56 41 16 1 69 9 8 62 65 22 **40** 47
7	Anjos70_2	70	50 25 47 21 36 1 57 19 58 9 45 60 14 6 44 43 **24** **15** 54 22 **27** 16 34 55 12 17 69 66 48 59 64 56 28 **13** 53 10 26 7 3 67 23 29 30 62 11 63 **18** 68 **40** 49 8 4 **42** 2 20 46 5 37 32 31 35 38 65 33 70 51 41 61 **39** 52
8	Anjos70_3	70	**15** 44 52 17 64 33 **27** 12 **13** 55 62 60 65 5 28 49 25 53 14 68 70 7 4 11 6 41 23 32 21 **24** 51 30 16 31 57 34 69 46 38 59 48 45 35 3 22 **39** **18** 36 66 **40** 10 50 56 **42** 2 37 19 61 26 9 47 67 1 20 58 43 29 54 63 8
9	Anjos70_4	70	7 **15** 47 9 54 22 61 29 **24** 12 63 26 60 58 53 2 49 50 23 **27** **13** 44 69 57 70 21 67 62 43 3 10 31 4 8 35 41 16 30 51 52 6 64 **42** 11 14 33 **18** 68 46 5 34 **40** 36 37 1 28 56 59 19 17 55 20 48 32 45 65 **39** 38 25 66
10	Anjos70_5	70	28 38 68 23 31 51 58 22 6 10 16 63 **27** 8 61 70 **24** 35 12 11 29 34 3 **13** 26 59 45 **15** 57 52 33 64 17 67 19 32 1 62 **40** 2 37 50 54 65 49 41 43 21 56 20 **42** 14 5 30 36 4 7 53 **39** 60 46 66 55 25 48 47 **18** 69 9 44

(continued)

Table 19.3 (continued)

SN	Instance ID	Size	Arrangements of the facilities in the solutions with f_2^{best}
11	Anjos75_1	75	47 48 **13** 51 21 17 35 75 62 **15 24** 43 68 66 34 12 10 19 20 4 55 44 7 65 61 **27** 33 26 52 28 73 54 29 60 38 14 16 72 5 56 8 **39** 11 6 22 25 71 53 37 70 **40 42** 74 31 1 3 32 2 41 23 30 49 50 45 **18** 59 36 57 63 9 67 69 46 64 58
12	Anjos75_2	75	49 64 32 72 28 44 63 50 33 1 58 5 52 53 71 67 12 8 62 **27** 22 **15** 68 55 **13** 17 9 19 10 69 **24** 45 66 70 2 37 26 25 51 61 47 21 6 14 60 7 57 **42** 59 4 20 75 65 3 34 36 11 23 16 38 46 74 48 31 73 41 29 54 **18** 35 **40** 30 **39** 56 43
13	Anjos75_3	75	47 54 69 55 8 72 26 **27** 38 2 68 44 62 59 32 6 **13** 58 9 75 **24** 28 21 52 65 16 71 30 61 37 60 41 43 63 14 19 10 **15** 50 1 17 25 **42** 29 70 4 **39** 66 **18** 20 48 3 34 49 5 67 45 35 **40** 7 73 57 36 11 31 12 56 74 64 23 46 51 53 33 22
14	Anjos75_4	75	36 **15** 9 14 7 75 10 37 30 57 22 29 56 46 55 3 16 28 **13** 73 32 1 2 **24** 35 63 26 59 54 71 4 11 61 25 6 68 **27** 21 60 5 49 **42** 50 62 8 70 47 20 41 **40** 33 12 64 65 52 44 53 **18** 34 67 45 19 48 58 66 23 43 **39** 31 74 51 38 69 72 17
15	Anjos75_5	75	20 37 9 63 2 43 44 53 16 22 69 56 45 34 57 54 61 11 17 28 30 36 12 5 49 **15** 23 73 32 47 1 35 64 52 **13 24** 33 **27** 26 19 6 58 50 60 14 74 21 75 4 **18** 59 67 48 25 3 55 29 **42** 70 8 10 46 65 7 31 38 66 41 71 68 **39** 62 72 **40** 51
16	Anjos80_1	80	78 25 80 67 **27** 65 21 29 43 19 52 14 75 51 45 47 34 54 41 32 58 9 31 36 **15** 6 74 56 17 37 72 **13** 69 **24** 59 3 64 1 71 60 11 55 70 2 44 8 **40** 16 46 63 53 61 12 68 49 **39** 62 66 48 20 50 10 23 76 22 33 7 **42** 73 26 5 30 35 77 38 28 4 **18** 79 57
17	Anjos80_2	80	22 10 58 46 56 **27** 12 47 **15** 29 **24** 1 7 **13** 28 35 20 36 34 8 65 62 75 59 68 76 70 79 37 31 64 80 69 49 48 45 17 11 57 55 5 **18** 14 32 74 54 41 16 33 77 72 23 25 67 53 3 6 4 44 38 9 **40** 2 50 **39 42** 26 21 19 30 73 71 52 78 60 51 63 43 61 66
18	Anjos80_3	80	37 26 69 64 72 30 76 67 77 3 8 78 44 11 **27** 59 19 **15** 1 55 52 38 31 79 13 65 6 **24** 5 45 43 34 49 68 33 16 54 57 53 36 62 32 29 48 22 25 **42** 56 10 50 51 17 80 21 14 23 7 41 73 75 **39** 2 71 66 **18** 60 9 12 28 4 35 46 47 70 **40** 61 20 58 63 74
19	Anjos80_4	80	69 60 64 21 78 54 30 28 66 80 43 76 29 9 **27** 17 49 2 3 7 20 52 33 26 53 13 51 **15** 70 11 68 74 23 63 36 **24** 71 22 50 41 47 75 35 38 46 **42** 34 32 12 14 59 62 6 4 67 79 77 16 31 72 48 19 37 **18** 1 8 5 65 45 25 **39** 73 55 44 10 58 57 61 56 **40**
20	Anjos80_5	80	2 47 52 61 10 67 **24** 22 **15** 1 9 76 33 6 54 30 34 75 71 66 32 80 43 62 63 57 5 28 14 64 51 72 **27 13** 60 58 48 25 7 78 38 73 21 68 37 74 70 31 19 36 44 39 53 35 55 59 **40** 65 3 4 77 20 **18** 79 45 23 46 50 11 56 26 16 8 29 49 12 69 **42** 41 17

Conclusion

Owing practical applications, the bi-objective corridor allocation problem (bCAP), which requires to arrange given facilities along two sides of a central corridor, is presented here as a constrained bCAP (cbCAP) by restricting some facilities to be placed in fixed, same, and/or opposite rows. Satisfying the imposed constraints, the facilities are to be arranged starting in both the rows from a common level along the length of the corridor and without allowing any physical gap between two adjacent facilities of a row, so as to minimize simultaneously the material handling cost among the facilities and the required length of the corridor. Since it would be difficult for any general-purpose optimizer to handle such a constrained permutation problem, a cbCAP specific permutation based genetic algorithm (cbCAP-pGA) is also proposed for handling the problem. The specialty of the cbCAP-pGA is that it generates only feasible solutions for the cbCAP model through its problem-specific initialization technique and crossover and mutation operators.

Applying to a set of benchmark instances in the numerical experimentation, it is observed that the cbCAP-pGA could search promising solutions with marginally increased material handling costs in comparison to those of the bCAP model. It is also observed that the cbCAP-pGA could even improve the required corridor lengths for studied some instances. The proposed cbCAP model and the cbCAP-pGA procedure may be applied in future to real-life problems in order to investigate their practical applicability.

References

1. Ahonen H, de Alvarenga AG, Amaral ARS (2014) Simulated annealing and tabu search approaches for the corridor allocation problem. Eur J Oper Res 232:221–233
2. Amaral ARS (2012) The corridor allocation problem. Comput Oper Res 39(12):3325–3330
3. Anjos MF, Kennings A, Vannelli A (2005) A semidefinite optimization approach for the single-row layout problem with unequal dimensions. Discrete Optim 2:113–122
4. Chung J, Tanchoco JMA (2010) The double row layout problem. Int J Prod Res 48(3):709–727
5. Datta D, Amaral ARS, Figueira JR (2011) Single row facility layout problem using a permutation-based genetic algorithm. Eur J Oper Res 213(2):388–394
6. Deb K, Agarwal S, Pratap A, Meyarivan T (2002) A fast and elitist multi-objective genetic algorithm: NSGA-II. IEEE Trans Evol Comput 6(2):182–197
7. Ebster C (2011) Store design and visual merchandising: creating store space that encourages buying. Business Expert Press, New York
8. Ghosh D, Kothari R (2012) Population heuristics for the corridor allocation problem. Technical report W.P. No. 2012-09-02, Indian Institute of Management, Ahmedabad, India
9. Kalita Z, Datta D (2014) Solving the bi-objective corridor allocation problem using a permutation-based genetic algorithm. Comput Oper Res 52:123–134
10. Kalita Z, Datta D, Palubeckis G (2019) Bi-objective corridor allocation problem using a permutation-based genetic algorithm hybridized with a local search technique. Soft Comput 23(3):961–986
11. Lin QL, Liu HC, Wang DJ, Liu L (2015) Integrating systematic layout planning with fuzzy constraint theory to design and optimize the facility layout for operating theatre in hospitals. J Intell Manuf 26(1):87–95

12. Motaghi M, Hamzenejad A, Riahi L, Kashani MS (2011) Optimization of hospital layout through the application of heuristic technique (diamond algorithm) in Shafa Hospital. Int J Manag Bus Res 1(3):133–138
13. Wang S, Zuo X, Liu X, Zhao X, Li J (2015) Solving dynamic double row layout problem via combining simulated annealing and mathematical programming. Appl Soft Comput 37:303–310
14. Zuo X, Murray CC, Smith AE (2014) Solving an extended double row layout problem using multiobjective tabu search and linear programming. IEEE Trans Autom Sci Eng 11(4):1122–1132

Chapter 20
The Constrained Single-Row Facility Layout Problem with Repairing Mechanisms

Zahnupriya Kalita and Dilip Datta

Abstract The single-row facility layout problem (SRFLP), which deals with the placement of some facilities along a row by minimizing the overall material flow cost among them, is usually studied as an unconstrained problem allowing the placement of the facilities arbitrarily without any restriction. But a practical SRFLP instance may require to respect certain constraints imposed on the arrangement of its facilities. Such an SRFLP model, which can be termed as a *constrained* SRFLP (cSRFLP), is studied here by requiring to place some facilities in fixed locations, and/or in predefined orders with/without allowing the arrangement of any other facility in between two ordered facilities. The handling of such a complex problem generally requires a specialized algorithm incorporating some problem-specific information for intelligent search. But the development of an algorithm needs expertise, from which practitioners often suffer. Hence, it is shown here how the cSRFLP can be tackled using a general-purpose algorithm with some repairing mechanisms outside the algorithm for forcibly satisfying the constraints of the problem. Employing a permutation-based genetic algorithm for this purpose, the potentiality of the proposed procedure is demonstrated by applying it to a set of cSRFLP instances of different sizes.

Keywords Combinatorial optimization · Facility layout problem · Constraint · Repairing mechanism · Genetic algorithm

Introduction

The single-row facility layout problem (SRFLP) seeks the placement of some facilities along a row on one side of a corridor, so as to minimize the overall material flow cost among the facilities. It has applications in different domains, such as

Z. Kalita · D. Datta (✉)
Department of Mechanical Engineering, Tezpur University, Tezpur, Assam, India
e-mail: zk@tezu.ernet.in; ddatta@tezu.ernet.in; datta_dilip@rediffmail.com

F. Bennis, R. K. Bhattacharjya (eds.), *Nature-Inspired Methods for Metaheuristics Optimization*, Modeling and Optimization in Science and Technologies 16,
https://doi.org/10.1007/978-3-030-26458-1_20

installation of machines in flexible manufacturing system [12], or arrangement of stores in supermarkets, departments in office buildings, and laboratories in hospitals, among others [24, 29].

The SRFLP is a combinatorial optimization problem, which has been investigated by both exact and heuristic procedures. Studied exact methods include branch-and-bound [29], dynamic programming [24], semidefinite programming [3], and mixed-integer linear programming [1, 13, 20]. Various heuristic and metaheuristic procedures, investigated to obtain approximate solutions for large-size instances in a lesser computational time, include simulated annealing [11], constructive greedy heuristic [17], metaheuristics [6], ant colony algorithm [30], scatter search [18], particle swarm optimization [28], tabu search [27], and genetic algorithm [5], among others.

The SRFLP has been studied so far as an unconstrained problem, where a facility is allowed to be placed in any location in a permutation of the facilities. However, many real-life problems might require to satisfy certain positioning and ordering constraints on the placement of some facilities, such as placing a facility in a fixed location, two facilities together, a facility before another, and so on. In the hospital layout problem investigated by Padgaonkar [23], the departments involving more inter-floor movement were arranged closer to elevators and those involving more inter-departmental movement were arranged on the same/next floor. Considering the possibility of more sale if a customer passes through more items, Ozgormus [22] proposed in the design of a grocery store layout to disperse basic commodities at the end of the store. In regard of ordering, Monma [21] found that duplicate billing in subsequent months can be prevented by preferring the job of recording the payments of customers to precede the job of preparing the bills. In manufacturing systems, Gapp et al. [9] studied precedence constraints requiring a given operation to precede another operation (e.g., drilling a hole in a work-piece before tapping it) and the contiguity constraints requiring a given set of operations to be performed together irrespective of their intra-ordering (e.g., performing all the milling operations together). It is also found in sequence flexibility based manufacturing systems that the ordering of operations can be altered by performing some operations in sequence and others arbitrarily [4, 19, 26]. Such restrictions in sequencing operations are common in many cases as a manufacturing system is neither very rigid like the well-known job-shop problem which requires strict ordering of the operations nor very flexible like the open-shop problem where the operations can be performed arbitrarily [14, 25].

In view of above, a constrained SRFLP model, naming as the *constrained single-row facility layout problem* (cSRFLP), is studied in the present work considering positioning and ordering constraints on the arrangement of some given facilities. Since it would be computationally very expensive for a general-purpose search technique to fulfill such permutation-based constraints on its own, the development of a specialized algorithm by incorporating certain problem-specific components becomes essential for intelligent search. However, practitioners and academia working on application areas often suffer from expertise required for the development of specialized algorithms, and hence they prefer some alternative techniques to get their works done through some existing general-purpose algorithms or commercial software packages. Such a procedure is presented here by developing some repairing mechanisms, which can be applied along with a general-purpose algorithm for forcibly satisfying the constraints of the cSRFLP outside the algorithm. Each time the algorithm sends a solution to an outside subroutine for evaluation, the solution is first checked and steered to the feasible region by forcibly satisfying the violated constraints, if any, and then the repaired solution along with its objective value is returned to the algorithm. Taking the permutation-based genetic algorithm (pGA), employed by Datta et al. [5] for the unconstrained SRFLP, as the general-purpose algorithm, the potentiality of the proposed procedure is demonstrated in the numerical experimentation by solving of a set of cSRFLP instances of sizes in the range of [64,100].

Note that prior to formulating the general SRFLP model by Datta et al. [5] as an unconstrained permutation-based optimization problem, it was studied with different constraints on the placement of given facilities in a row of minimum possible length by avoiding the overlapping of the facilities. The positioning and ordering constraints imposed in the present cSRFLP model should not to be confused with those length based and overlapping constraints imposed in the exact formulation of the general SRFLP model.

The remaining of the present article is organized as follows: presenting the cSRFLP model in Sect. 20, the proposed repairing mechanisms are explained in Sect. 20. The pGA developed for the unconstrained SRFLP model is presented briefly in Sect. 20, and its performance to some cSRFLP instances in Sect. 20. Finally, the article is summarized in Sect. 20.

The cSRFLP Formulation

As stated in Sect. 20, the cSRFLP model is studied in the present work incorporating the following two types of constraints:

1. *Positioning constraints*: Some given facilities are to be placed in predefined fixed positions.
2. *Ordering constraints*: Some given pairs of facilities are to be placed in predefined orders, with/without allowing to place other facilities in between such an ordered pair.

The studied cSRFLP is explained here through a small instance of 10 facilities. Say, the eighth and seventh positions are reserved for placing facilities 3 and 4, respectively. Also, facility 2 is required to be placed on the right side of facility 4, facility 6 is prior to facility 5 and facility 5 is prior to facility 9. Further, facility 10 is required to be placed prior to facility 6 and facility 8 prior to facility 7, but without allowing the placement of any other facility in between the two facilities of a pair, i.e., each pair is required to come together. In terms of the two types of constraints mentioned above, the same can be stated as facilities 3 and 4 are

Fig. 20.1 An illustrative solution of the cSRFLP model showing the constrained facilities in cyan color and others in gray color

required to be placed in predefined fixed positions, facilities 10 and 6 as well as facilities 8 and 7 are required to be ordered in pairs without allowing the placement of other facilities in between a pair, while facilities 6 and 5 as well as facilities 5 and 9 are required to be ordered in pairs allowing the placement of other facilities in between a pair. A possible solution of the instance is shown in Fig. 20.1 by marking the constrained facilities with cyan color and the unconstrained one with gray color. Note that the total material flow cost (i.e., the objective value) in a constrained solution (i.e., a cSRLFP solution) in general would be higher than that in an unconstrained solution (i.e., a SRFLP solution).

According to above, the cSRFLP can be defined as a problem requiring the effective placement of n number of facilities along a row with some facilities in predefined fixed positions and some pairs of facilities in predefined orders with/without allowing the placement of other facilities in between such an ordered pair. Hence, the following notations can be defined for formulating the problem (parameters related constraints are defined by uppercase alphabets):

Indices

i, j, k : Indices to represent facilities and positions
π : Index of a permutation of given facilities

Parameters

n : Number of given facilities to be arranged along a row
l_i : Length of facility i in the direction of the row
c_{ij} : Material flow cost between facilities i and j $(i \neq j)$
T_i : Indicator of the predefined fixed position for facility i $(T_i \in \{\text{Y (yes), N (no)}\})$
P_i : predefined fixed position of facility i (positioning constraint)
O_i : Indicator for ordering facility i on the left side another facility ($O_i \in \{\text{Y (yes), N (no)}\}$ ordering constraint)
R_i : Facility to be placed on the right side of facility i (ordering constraint)
A_{ij} : Indicator to allow the placement of other facilities in between the ordered facilities i and j $(A_{ij} \in \{\text{Y (yes), N (no)}\})$

Variables

r_i^π : Facility placed at position i of permutation π
$x_{r_i^\pi}$: Distance to the centre point of facility r_i^π from the left end of the row
f^π : Overall material flow cost among all the facilities of π

For demonstration, the constraints posed in the instance of Fig. 20.1 are shown in Table 20.1 in terms of the notations T_i–A_{ij} defined above. The value 'Y' to T_i indicates that facility i is required to be placed at a predefined fixed position,

Table 20.1 Representation of the constraints of the cSRFLP instance shown in Fig. 20.1 in a matrix form

Facility (i)	1	2	3	4	5	6	7	8	9	10
Whether to be arranged in predefined position (T_i)	N	N	Y	Y	N	N	N	N	N	N
Predefined position (P_i)	–	–	8	7	–	–	–	–	–	–
Whether to be ordered on the left side (O_i)	N	N	N	Y	Y	Y	N	Y	N	Y
Ordered facility on the right side ($R_i = j$)	–	–	–	2	9	5	–	7	–	6
Other facilities allowed within the pair (A_{ij})	–	–	–	Y	Y	Y	–	N	–	N

which is marked by P_i. On the other hand, values to R_i and A_{ij} are required against facility i only if it is paired with another facility for placing on its right side. Such pairing of facility i is denoted by $O_i = Y$ (otherwise $O_i = N$ if the facility does not have such pairing).

In terms of the notations defined above, the cSRFLP model in a general form for permutation π can be expressed mathematically by Eqs. (20.1)–(20.4).

$$\text{Minimize} \qquad f^{\pi} \equiv \sum_{i=1}^{n-1} \sum_{j=i+1}^{n} c_{r_i^{\pi}, r_j^{\pi}} \left| x_{r_j^{\pi}} - x_{r_i^{\pi}} \right| \tag{20.1}$$

$$\text{Subject to} \qquad P_{r_i^{\pi}} = k \; ; \qquad \text{if} \quad T_{r_i^{\pi}} = \text{Y} \; ; \; k \in \{1, 2, \ldots, n\} \; ; \; i = 1, 2, \ldots, n \tag{20.2}$$

$$j \begin{cases} = i+1 \; ; & \text{if} \quad O_{r_i^{\pi}} = \text{Y} \; ; \; R_{r_i^{\pi}} = r_j^{\pi} \; ; \; A_{r_i^{\pi}, r_j^{\pi}} = \text{N} \\ \geqslant i+1 \; ; & \text{if} \quad O_{r_i^{\pi}} = \text{Y} \; ; \; R_{r_i^{\pi}} = r_j^{\pi} \; ; \; A_{r_i^{\pi}, r_j^{\pi}} = \text{Y} \\ & i, j = 1, 2, \ldots, n \; ; \; i \neq j \end{cases} \tag{20.3}$$

$$\text{where,} \qquad x_{r_i^{\pi}} = \frac{l_{r_1^{\pi}}}{2} + \frac{1}{2} \sum_{j=2}^{i} \left(l_{r_{j-1}^{\pi}} + l_{r_j^{\pi}} \right) \tag{20.4}$$

For permutation π, f^{π} expressed by Eq. (20.1) is the objective function. In Eq. (20.2), $P_{r_i^{\pi}}$ bears the predefined fixed position of r_i^{π} if it is subject to positioning constraint. Similarly, j given by Eq. (20.3) denotes ordering constraints with $R_{r_i^{\pi}} = r_j^{\pi}$ indicating that facilities r_i^{π} and r_j^{π} are to be ordered by placing r_j^{π} right to r_i^{π}. Further, '$A_{r_i^{\pi}, r_j^{\pi}} = \text{N}$' in Eq. (20.3) says that facilities r_i^{π} and r_j^{π} are required to be ordered without allowing other facilities in between them, while '$A_{r_i^{\pi}, r_j^{\pi}} = \text{Y}$' relaxes that restriction. On the other hand, $x_{r_i^{\pi}}$ in Eq. (20.4) indicates the centroidal distance of facility r_i^{π} from the left end of the row of placement of the facilities of permutation π.

The Repairing Mechanisms

Because of the increasing complexity with the rigidity of the constraints, they should be repaired based upon their decreasing complexities. Accordingly, the positioning constraints will be repaired prior to the ordering constraints. Further, three cases are possible with an ordering constraint of a pair of two facilities, which in decreasing order of their complexities are as follows: one facility is in a fixed position, no other facility is allowed to be placed in between the facilities, and other facilities are allowed to be placed in between the facilities. Hence, the imposed positioning and ordering constraints will be repaired in a sequence of a total of four cases (the positioning constraints and the three cases of the ordering constraints).

In order to formulate the repairing mechanisms for the above four cases, the following notations are also defined in addition to those defined in Sect. 20:

u_i : Facility at position i before repairing
v_i : Facility at position i after repairing
p_i^{r} : Position of facility i before repairing
p_i^{v} : Position of facility i after repairing
z : Right side facility of an ordered pair ($z = R_i$ if $O_i = \mathrm{Y}$)
a : Position of the left side facility of an ordered pair after repairing
b : Position of the right side facility of an ordered pair after repairing
n' : Number of available positions for placing a particular facility
e_q : qth available position for placing a particular facility

In the above notations, v is an array in which the repaired facilities will be stored, while u is another array that will start as a copy of the current permutation r and will delete a facility from it once it (the facility) is placed in v after repairing. Both u and v will act globally until all the constraints are repaired. Before starting the repairing of any constraint, u will be initialized with r and v with zero (i.e., $u_i = r_i$ and $v_i = 0\ ;\ i = 1, 2, \ldots, n$). On the other hand, after placing the facilities in v by repairing each type of constraints, r and u will be updated accordingly before going to repair another type of constraints (r will be updated by copying the repaired facilities from v and filling the vacant positions by the remaining facilities taking in order from u). It is to be mentioned that u is mandatory only if any facility appears in two or more ordering constraints, otherwise (if no facility appears in more than one ordering constraint) a facility can be deleted directly from r once it is repaired and placed in v.

While repairing ordered facilities, the array p^{r} will keep record of the positions of the facilities in r before repairing, and p^{v} those of the facilities placed in v after repairing. Both p^{r} and p^{v} will act locally during the repairing of each type of ordering constraints. Before starting the repairing of any type of ordering constraints, p^{r} will be initialized with the positions of the facilities in r (i.e., $p_{r_i}^{\mathrm{r}} = i\ ;$ $i = 1, 2, \ldots, n$), while p^{v} with the positions of those facilities already placed in v and with zero for the remaining facilities (i.e., $p_{v_i}^{\mathrm{v}} = i$ if $v_i \neq 0$ and $p_{v_i}^{\mathrm{v}} = 0$ if $v_i = 0\ ;\ i = 1, 2, \ldots, n$).

The array e will be used to report the vacant positions for placing a facility in v and n' is the number of such vacant positions. On the other hand, only for handling convenient, a and b as well as z will be used, respectively, as shorter notations of the positions of the two facilities of an ordered pair and the right side facility of that ordered pair.

Positioning Constraints

As the very first step of repairing the constraints, irrespective of any other requirement, Eq. (20.5) is applied for placing the facilities of r in their specified fixed positions in v, if any, and subsequently for removing them from u.

$$\left.\begin{aligned} v_k &= r_i \\ u_i &= 0 \end{aligned}\right\} ; \quad \text{if} \quad T_{r_i} = \mathrm{Y} \; ; \; P_{r_i} = k \; ; \; i = 1, 2, \ldots, n \tag{20.5}$$

Ordering Constraints With a Facility of a Pair in a Fixed Position

For an ordered pair with one facility already placed in a specified fixed position in v using Eq. (20.5), the other facility is to be placed in another suitable position in v, if not yet placed.

(i) If the left facility r_i is already placed in its specified fixed position P_{r_i} in v, Eqs. (20.6a) and (20.6b) can be used to place the right facility z in another suitable position of v and subsequently to update p^{v} and u.

$$\left.\begin{aligned} v_b &= z \\ p^{\mathrm{v}}_z &= b \\ u_{p^{\mathrm{r}}_z} &= 0 \end{aligned}\right\} ; \quad \text{if} \quad O_{r_i} = \mathrm{Y} \; ; \; T_{r_i} = \mathrm{Y} \; ; \; T_z = \mathrm{N} \; ; \; p^{\mathrm{v}}_z = 0 \tag{20.6a}$$

where, $a = P_{r_i}$

$$b = \begin{cases} a + 1 \; ; & \text{if} \quad A_{r_i,z} = \mathrm{N} \; ; \; v_{a+1} = 0 \text{ or } z \\ p^{\mathrm{r}}_z \; ; & \text{if} \quad A_{r_i,z} = \mathrm{Y} \; ; \; v_{p^{\mathrm{r}}_z} = 0 \; ; \; p^{\mathrm{r}}_z > a \\ e_t \; ; & \text{if} \quad A_{r_i,z} = \mathrm{Y} \; ; \; b = 0 \; ; \; e_t > a \; ; \; t \in \{1, n'\} \end{cases} \tag{20.6b}$$

$i = 1, 2, \ldots, n$.

For placing z in v, the first condition in Eq. (20.6b) for b checks the availability of the adjacent position on the right side of the fixed position a ($= P_{r_i}$) in v (i.e., whether the position $a + 1$ is vacant or already occupied by z only) if r_i and z are to be placed without allowing other facilities in between them. On the other hand, when r_i and z are allowed to be placed with other facilities in between them, the second condition in Eq. (20.6b) checks if the position p_z^{r} of z in r is on the right side of a and its corresponding position in v is vacant (i.e, $v_{p_z^{\mathrm{r}}} = 0$), otherwise the third condition finds a random vacant position in v on the right side of a ($b = 0$ in the third condition means that a suitable value for b could not be obtained by satisfying the previous condition). Then, Eq. (20.6a) places z in position b of v if it is not yet placed in v (i.e., $p_z^{\mathrm{v}} = 0$).

(ii) If the right facility z is already placed in its specified fixed position P_z in v, Eqs. (20.6c) and (20.6d) can be used to place the left facility r_i in another suitable position of v and subsequently to update p^{v} and u.

$$\left.\begin{aligned} v_a &= r_i \\ p_{r_i}^{\mathrm{v}} &= a \\ u_i &= 0 \end{aligned}\right\} ; \quad \text{if} \quad O_{r_i} = \mathrm{Y} \; ; \; T_{r_i} = \mathrm{N} \; ; \; T_z = \mathrm{Y} \; ; \; p_{r_i}^{\mathrm{v}} = 0 \tag{20.6c}$$

where, $b = P_z$

$$a = \begin{cases} b-1 \; ; & \text{if} \;\; A_{r_i,z} = \mathrm{N} \; ; \; v_{b-1} = 0 \text{ or } r_i \\ i \; ; & \text{if} \;\; A_{r_i,z} = \mathrm{Y} \; ; \; v_i = 0 \; ; \; i < b \\ e_t \; ; & \text{if} \;\; A_{r_i,z} = \mathrm{Y} \; ; \; a = 0 \; ; \; e_t < b \; ; \; t \in \{1, n'\} \end{cases} \tag{20.6d}$$

$$i = 1, 2, \ldots, n \; .$$

For placing r_i in v, the first condition in Eq. (20.6d) for a checks the availability of the adjacent position on the left side of the fixed position b ($= P_z$) in v (i.e., whether the position $b - 1$ is vacant or already occupied by r_i only) if r_i and z are to be placed without allowing other facilities in between them. On the other hand, when r_i and z are allowed to be placed with other facilities in between them, the second condition in Eq. (20.6d) checks if the position i of r_i in r is on the left side of b and its corresponding position in v is vacant (i.e, $v_i = 0$), otherwise the third condition finds a random vacant position in v on the left side of b ($a = 0$ in the third condition means that a suitable value for a could not be obtained by satisfying the previous condition). Then, Eq. (20.6c) places r_i in position a of v if it is not yet placed in v (i.e., $p_{r_i}^{\mathrm{v}} = 0$).

Ordering Constraints with a Pair of Facilities in Two Adjacent Positions

For an ordered pair of facilities r_i and z, none of which requires a specified fixed position and no other facility is allowed in between them, Eq. (20.7) can be used to place them in two adjacent positions of v and subsequently to update p^{v} and u.

$$\left.\begin{aligned} v_a &= r_i \\ p^{\mathrm{v}}_{r_i} &= a \\ u_i &= 0 \end{aligned}\right\} ; \quad \text{if} \quad O_{r_i} = \mathrm{Y}\,;\ A_{r_i,z} = \mathrm{N}\,;\ T_{r_i} = \mathrm{N}\,;\ T_z = \mathrm{N}\,;\ p^{\mathrm{v}}_{r_i} = 0 \qquad (20.7\mathrm{a})$$

$$\left.\begin{aligned} v_b &= z \\ p^{\mathrm{v}}_{z} &= b \\ u_{p^{\mathrm{r}}_z} &= 0 \end{aligned}\right\} ; \quad \text{if} \quad O_{r_i} = \mathrm{Y}\,;\ A_{r_i,z} = \mathrm{N}\,;\ T_{r_i} = \mathrm{N}\,;\ T_z = \mathrm{N}\,;\ p^{\mathrm{v}}_{z} = 0 \qquad (20.7\mathrm{b})$$

$$\text{where,}\ a = \begin{cases} i\,; & \text{if} \quad v_i = 0 \text{ or } r_i\,;\ v_{i+1} = 0 \text{ or } z \\ i-1\,; & \text{if} \quad a = 0\,;\ v_{i-1} = 0 \text{ or } r_i\,;\ v_i = 0 \text{ or } z \\ p^{\mathrm{r}}_z\,; & \text{if} \quad a = 0\,;\ v_{p^{\mathrm{r}}_z} = 0 \text{ or } r_i\,;\ v_{p^{\mathrm{r}}_z+1} = 0 \text{ or } z \\ p^{\mathrm{r}}_z - 1\,; & \text{if} \quad a = 0\,;\ v_{p^{\mathrm{r}}_z-1} = 0 \text{ or } r_i\,;\ v_{p^{\mathrm{r}}_z} = 0 \text{ or } z \\ e_t\,; & \text{if} \quad a = 0\,;\ v_{e_t+1} = 0\,;\ t \in \{1, n'\} \end{cases} \qquad (20.7\mathrm{c})$$
$$b = a + 1$$
$$i = 1, 2, \ldots, n\ .$$

In this case, two adjacent vacant positions a and b in v, in which r_i and z can be placed, are first obtained by satisfying any of the five conditions for a given by Eq. (20.7c). The first condition checks if positions i and $i+1$ of v are vacant or already filled up respectively with r_i and z, while the second condition checks positions $i-1$ and i of v for the same. Similarly, the third and fourth conditions check positions in v for the same, corresponding to the pairs of positions (P^{r}_z, P^{r}_z+1) and (P^{r}_z-1, P^{r}_z), respectively. If two suitable adjacent positions in v, corresponding to the present position of r_i or z in r, are not found, then the fifth condition is applied, which checks for two suitable random adjacent positions in v. Note that $a = 0$ in each condition, starting from the second condition, means that its previous condition was not satisfied.

Once two suitable adjacent positions a and b are found in v by satisfying one of the five conditions given by Eq. (20.7c), facility r_i is placed in a using Eq. (20.7a) if it is not yet placed in v (i.e., $p^{\mathrm{v}}_{r_i} = 0$) and z is placed in b using Eq. (20.7b) if it is not yet placed in v (i.e., $p^{\mathrm{v}}_z = 0$). Finally, the positions and facilities of p^{v} and u, respectively, are updated as per the requirement.

Ordering Constraints Allowing Other Facilities in Between a Pair of Facilities

For an ordered pair of facilities r_i and z, none of which requires a specified fixed position and other facilities are also permitted in between them, Eq. (20.8) can be used to place them in two adjacent positions of v and subsequently to update p^{v} and u.

$$\left.\begin{aligned} v_a &= r_i \\ p^{\mathrm{v}}_{r_i} &= a \\ u_i &= 0 \end{aligned}\right\} ; \quad \text{if} \quad O_{r_i} = \mathrm{Y}\, ;\ A_{r_i,z} = \mathrm{Y}\, ;\ T_{r_i} = \mathrm{N}\, ;\ T_z = \mathrm{N}\, ;\ p^{\mathrm{v}}_{r_i} = 0 \tag{20.8a}$$

$$\left.\begin{aligned} v_b &= z \\ p^{\mathrm{v}}_{z} &= b \\ u_{p^{\mathrm{r}}_z} &= 0 \end{aligned}\right\} ; \quad \text{if} \quad O_{r_i} = \mathrm{Y}\, ;\ A_{r_i,z} = \mathrm{Y}\, ;\ T_{r_i} = \mathrm{N}\, ;\ T_z = \mathrm{N}\, ;\ p^{\mathrm{v}}_{z} = 0 \tag{20.8b}$$

$$\text{where, } (a,b) = \begin{cases} (i, p^{\mathrm{r}}_z)\, ; & \text{if} \quad v_i = 0 \text{ or } r_i\, ;\ v_{p^{\mathrm{r}}_z} = 0 \text{ or } z\, ;\ i < p^{\mathrm{r}}_z \\ (p^{\mathrm{r}}_z, i)\, ; & \text{if} \quad a = b = 0\, ;\ v_i = 0 \text{ or } z\, ;\ v_{p^{\mathrm{r}}_z} = 0 \text{ or } r_i\, ;\ i > p^{\mathrm{r}}_z \\ (i, e_{t_2})\, ; & \text{if} \quad a = b = 0\, ;\ v_i = 0 \text{ or } r_i\, ;\ p^{\mathrm{v}}_z = 0\, ;\ p^{\mathrm{r}}_z < i\, ;\ e_{t_2} > i \\ (e_{t_1}, p^{\mathrm{r}}_z)\, ; & \text{if} \quad a = b = 0\, ;\ p^{\mathrm{v}}_{r_i} = 0\, ;\ i > p^{\mathrm{r}}_z\, ;\ v_{p^{\mathrm{r}}_z} = 0 \text{ or } z\, ;\ e_{t_1} < p^{\mathrm{r}}_z \\ (e_{t_1}, e_{t_2})\, ; & \text{if} \quad a = b = 0\, ;\ p^{\mathrm{v}}_{r_i} = 0\, ;\ p^{\mathrm{v}}_z = 0\, ;\ e_{t_1} < e_{t_2}\, ;\ t_1, t_2 \in \{1, n'\} \end{cases}$$
$$i = 1, 2, \ldots, n\ . \tag{20.3}$$

This constraint is similar with the previous one of Sect. 20, with the only difference that r_i and z can be placed here in any two suitable positions in v allowing the placement of other facilities in between them. In this case, two suitable vacant positions a and b ($a < b$) in v, in which r_i and z can be placed, are first obtained by satisfying any of the five conditions for a and b given by Eq. (20.3). The first condition checks if positions in v corresponding to positions i and p^{r}_z of r are vacant or already filled up respectively with r_i and z when $i < p^{\mathrm{r}}_z$, while the second condition checks the same in reverse order when $i > p^{\mathrm{r}}_z$. In the third condition, the position in v corresponding to position i of r is either vacant or filled up with r_i, but z is not yet placed in v and its position in r is on the left side of r_i (i.e., $p^{\mathrm{r}}_z < i$), and hence, a suitable random position e_{t_2} ($> i$) is obtained in v as b for placing z. The forth condition is similar to the third condition. In this case, the position in v corresponding to position p^{r}_z of r is either vacant or filled up with z, but r_i is not yet placed in v and its position in r is on the right side of z (i.e., $i > p^{\mathrm{r}}_z$), and hence, a suitable random position e_{t_1} ($< p^{\mathrm{r}}_z$) is obtained in v as a for placing r_i. On the other hand, the fifth condition says that none of r_i and z is yet placed in v and the positions in v corresponding to their positions in r are also not vacant, and hence, two suitable random positions e_{t_1} and e_{t_2} ($e_{t_1} < e_{t_2}$) are obtained in v as a and b for placing r_i and z, respectively. Note that $a = b = 0$ in each condition, starting from the second condition, means that its previous condition was not satisfied.

Once two suitable positions a and b ($a < b$) are found in v by satisfying one of the five conditions given by Eq. (20.3), facility r_i is placed in a using Eq. (20.8a) if it is not yet placed in v (i.e., $p^{\mathrm{v}}_{r_i} = 0$) and z is placed in b using Eq. (20.8b) if it is not yet placed in v (i.e., $p^{\mathrm{v}}_z = 0$). Finally, the positions and facilities of p^{v} and u, respectively, are updated as per the requirement.

Illustration of the Repairing Mechanisms

The instance considered in Sect. 20 involves all the four types of constraints introduced in Sects. 20, 20, 20, and 20, detail of which is given in Table 20.1. Starting with a permutation arranging the 10 facilities serially, the step-wise repairing of various constraints using Eqs. (20.5)–(20.8), leading to the final solution of Fig. 20.1, is shown in Fig. 20.2. Initially all the facilities are shown in gray color. Thereafter, once a facility is repaired and placed in a new position, its color is changed to cyan. At the very first step, facilities 3 and 4 are placed at their fixed positions of 8 and 7, respectively (positioning constraints explained in Sect. 20), and their original positions are filled up by shifting the un-repaired facilities 5–8 towards left. In the next step, facilities 4 and 2 are ordered (ordering constraint explained in Sect. 20), in which facility 2 is placed at position 9 and its original position is filled up by shifting facilities 5–9 towards left. The third step repairs the ordering constraints addressed in Sect. 20, in which facility 10 is first placed on the immediate left side of facility 6 by shifting facility 5, 7, 8 and 9 towards right (to positions 4–6 and 10, respectively), and then facilities 7 and 8 are interchanged so as to order them together. Although the final step is responsible for repairing the ordering constraints of Sect. 20, practically nothing is required to do as both the pairs (facilities 6 and 5, and facilities 5 and 9) are already ordered.

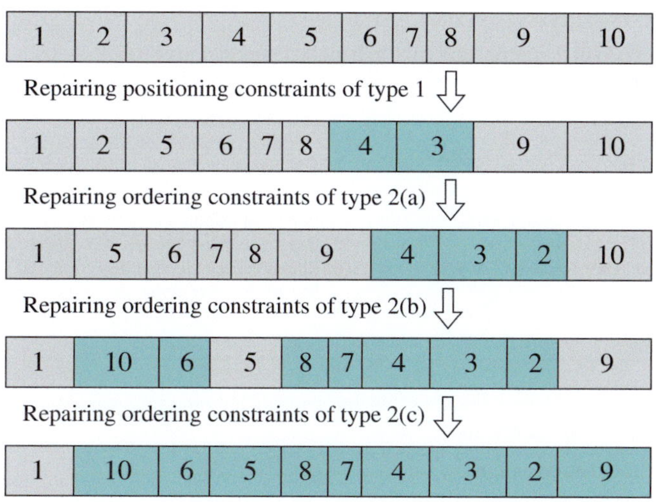

Fig. 20.2 Illustration of the step-wise repairing of various constraints imposed to the proposed cSRFLP model (repaired facilities are shown in cyan color and others in gray color)

Pseudo-Codes of the Repairing Mechanisms

It can be observed in Eqs. (20.5)–(20.8), as well as in the illustration in Fig. 20.2, that the repairing of the considered positioning and ordering constraints could be complicated to understand and implement. Therefore, the pseudo-codes for the four types of repairing mechanisms are also provided through Tables 20.2, 20.3, 20.4, 20.5, 20.6, and 20.7 by implementing Eqs. (20.5)–(20.8).

The subroutine repair_cSRFLP(), shown in Table 20.2, first initializes the global vectors u and v, then calls other subroutines for repairing different types of constraints, and finally returns the completely repaired permutation r.

The positioning constraints are repaired by implementing Eq. (20.5) in the subroutine cSRFLP_position() as shown in Table 20.3. It first places all the fixed-position facilities in the specified positions in v, then updates r and u by calling the subroutine cSRFLP_update(), and finally returns u, v and r.

The subroutine cSRFLP_order_fix() in Table 20.4 implements Eq. (20.6) for repairing the first type of ordering constraints, in which one facility of a pair is already placed in a fixed position by the subroutine cSRFLP_position() shown in Table 20.3. It first initializes p^{r} and p^{v} locally by calling the subroutine cSRFLP_repair_initialize(). Next, a suitable position b in v is obtained for z using Eq. (20.6b) when r_i is in a specified fix position, or a is obtained for r_i using Eq. (20.6d) when z is in a specified fix position. Then, r_i and z are arranged in v

Table 20.2 Pseudo-code for repairing the constraints imposed to cSRFLP

```
subroutine repair_cSRFLP(n, r)
  for (i = 1 to n) do                      // Initialize global u and v
  |  v_i := 0
  |  u_i := r_i
  end for
  call cSRFLP_position(n, u, v, r)         // Repair positioning constraints expressed by Eq. (5)
  call cSRFLP_order_fix(n, u, v, r)        // Repair ordering constraint expressed by Eq. (6)
  call cSRFLP_order_together(n, u, v, r)   // Repair ordering constraint expressed by Eq. (7)
  call cSRFLP_order_free(n, u, v, r)       // Repair ordering constraint expressed by Eq. (8)
  return r
end subroutine
```

Table 20.3 Pseudo-code implementing the mechanism for repairing positioning constraints as expressed by Eq. (20.5) in Sect. 20

```
subroutine cSRFLP_position(n, u, v, r)
  for (i = 1 to n) do
  |  if (T_{r_i} = Y) do
  |  |  v_k := r_i                         // r_i is placed at position k (= P_{r_i}) of v
  |  |  u_i := 0                           // r_i is removed from u
  |  end if
  end for
  call cSRFLP_update(n, v, r, u)           // Update the status of r and u
  return u, v, r
end subroutine
```

Table 20.4 Pseudo-code implementing the mechanism for repairing ordering constraints with a facility of a pair in a fixed position as expressed by Eq. (20.6) in Sect. 20

```
subroutine cSRFLP_order_fix(n,u,v,r)
  call cSRFLP_repair_initialize(n,r,v,p^r,p^v)                    // Initialize p^r and p^v
  for (i = 1 to n) do
  | if (O_{r_i} = N) continue   end if                           // r_i is not ordered on left side
  | z := R_{r_i}                                                  // Right hand side facility
  | if (T_{r_i} = N, T_z = N) continue   end if                  // No facility in fixed position
  | a := b := 0
  | if (T_{r_i} = Y) do                                           // r_i in a fixed position
  | | a := P_{r_i}                                                // Fixed position of r_i in v
  | | if (A_{r_i,z} = N, a+1 ⩽ n, v_{a+1} = 0 or z)  do  b := a+1  end if   // 1st condition of Eq. (6b)
  | | if (A_{r_i,z} = Y, v_{p^r_z} = 0, p^r_z > a) do  b := p^r_z  end if  // 2nd condition of Eq. (6b)
  | | | if (A_{r_i,z} = Y, b = 0) do                              // 3rd condition of Eq. (6b)
  | | | | call random_positions(v,a+1,n,n',e)                     // Vacancy in (a+1,n) in v
  | | | | | if (n' ≠ 0) do                                        // Vacant position(s) available
  | | | | | | call t := random_integer(1,n')                      // Random integer in (1,n')
  | | | | | | b := e_t                                            // Random position for z in v
  | if (T_z = Y) do                                               // z in a fixed position (not r_i)
  | | b := P_z                                                    // Fixed position of z in v
  | | if (A_{r_i,z} = N, b > 1, v_{b-1} = 0 or r_i) do  a := b-1  end if   // 1st condition of Eq. (6d)
  | | if (A_{r_i,z} = Y, v_i = 0, i < b) do  a := i  end if      // 2nd conditions of Eq. (6d)
  | | if (A_{r_i,z} = Y, a = 0) do                                // 3rd conditions of Eq. (6d)
  | | | call random_positions(v,1,b-1,n',e)                       // Vacancy in (1,b-1) in v
  | | | if (n' ≠ 0) do                                            // Vacant position(s) available
  | | | | t := call random_integer(1,n')                          // Random integer in (1,n')
  | | | | a := e_t                                                // Random position for r_i in v
  | if (a ≠ 0, b ≠ 0, (v_a = 0 or r_i), (v_b = 0 or z)) do
  | | call cSRFLP_arrange(a,b,i,p^r_z,r_i,z,u,v,p^v)              // Arrange r_i and z in v
  | else  call cSRFLP_constraint_error(n,v,r,u,r_i,z)            // Arrangement not possible
  | end else if
  end for
  call cSRFLP_update(n,v,r,u)                                     // Update r and u
  return u,v,r
end subroutine
```

by calling the subroutine cSRFLP_arrange(). Finally, u, v and r are returned after updating r and u through the subroutine cSRFLP_update().

Equation (20.7) is implemented in the subroutine cSRFLP_order_together(), as shown in Table 20.5, for repairing the second type of ordering constraints, in which the two facilities of an ordered pair are to be placed in two adjacent positions, i.e., without allowing any other facility in between them. The subroutine first initializes p^{r} and p^{v} locally by calling the subroutine cSRFLP_repair_initialize(). Next, two adjacent vacant positions a and b are obtained for r_i and z using Eq. (20.7c). Then, r_i and z are arranged in v by calling the subroutine cSRFLP_arrange(). Finally, u, v and r are returned after updating r and u through the subroutine cSRFLP_update().

Next, the pseudo-code implementing Eq. (20.8) is presented through the subroutine cSRFLP_order_free() in Table 20.6 for repairing the third type of ordering constraints, which allow the placement of other facilities in between the two facilities of an ordered pair. The subroutine first initializes p^{r} and p^{v} locally by calling the subroutine cSRFLP_repair_initialize(). Then, r_i and z are arranged in v by calling the subroutine cSRFLP_arrange() after obtaining there two suitable

Table 20.5 Pseudo-code implementing the mechanism for repairing ordering constraints with the two facilities of an ordered pair in two adjacent positions as expressed by Eq. (20.7) in Sect. 20

```
subroutine cSRFLP_order_together(n, u, v, r)
  call cSRFLP_repair_initialize(n, r, v, p^r, p^v)                       // Initialize p^r and p^v
  for (i = 1 to n) do
    if (O_{r_i} = N) continue  end if                                    // r_i is not ordered on left side
    z := R_{r_i}                                                         // Right hand side facility
    if (T_{r_i} = Y or T_z = Y) continue  end if                         // One facility in fixed position
    if (A_{r_i,z} = Y) continue  end if                                  // Adjacent placement not required
    a := b := 0
    if (v_i = 0 or r_i, i+1 <= n, v_{i+1} = 0 or z) do  a := i  end if   // 1st condition of Eq. (7c)
    if (a = 0, i-1 >= 1, v_{i-1} = 0 or r_i, v_i = 0 or z) do  a := i-1  end if   // 2nd condition of Eq. (7c)
    if (a = 0, v_{p^r_z} = 0 or r_i, p^r_z+1 <= n, v_{p^r_z+1} = 0 or z) do  a := p^r_z  end if   // 3rd condition of Eq. (7c)
    if (a = 0, p^r_z-1 >= 1, v_{p^r_z-1} = 0 or r_i, v_{p^r_z} = 0 or z) do  a := p^r_z-1  end if   // 4th condition of Eq. (7c)
    if (a = 0) do                                                        // 5th condition of Eq. (7c)
      call random_positions(v, 1, n, n', e)                              // Vacancy in (1, n) in v
      if (n' >= 2) do                                                    // Vacant positions available
        repeat
          t_1 := call random_integer(1, n')                              // A random integer in (1, n')
          t_2 := call random_integer(1, n')                              // Another random integer in (1, n')
        while |t_1 - t_2| != 1                                           // Not adjacent positions
        if (t_1 < t_2) do  a := e_{t_1}  end if
        if (t_1 > t_2) do  a := e_{t_2}  end if
      if (a != 0) do  b := a+1  end if                                   // a and b are required positions
    if (a != 0, b != 0, (v_a = 0 or r_i), (v_b = 0 or z)) do
      call cSRFLP_arrange(a, b, i, p^r_z, r_i, z, u, v, p^v)            // Arrange r_i and z
    else  call cSRFLP_constraint_error(n, v, r, u, r_i, z)               // Arrangement not possible
    end else if
  end for
  call cSRFLP_update(n, v, r, u)                                         // Update r and u
  return u, v, r
end subroutine
```

vacant positions a and b using Eq. (20.3). Finally, r and u are updated through the subroutine cSRFLP_update(), and then u, v and r are returned.

On the other hand, the five auxiliary subroutines called in the above mentioned four repairing subroutines are presented in Table 20.7, which are cSRFLP_repair_initialize(), random_positions(), cSRFLP_arrange(), cSRFLP_update() and cSRFLP_constraint_error(). The positioning subroutine cSRFLP_position() in Table 20.3 calls cSRFLP_update() only, while all the five auxiliary subroutines are called in the three ordering subroutines presented in Tables 20.4, 20.5, and 20.6. At the very beginning of each of the three ordering subroutines, the auxiliary subroutine cSRFLP_repair_initialize() is called for initializing the vectors p^{r} and p^{v} locally with the positions of the facilities arranged in r and v, respectively. The auxiliary subroutine random_positions() is called for obtaining the vacant positions in v in a given range, in one of which a particular facility is to be placed. The auxiliary subroutine cSRFLP_arrange() is engaged for placing the facilities of a pair if suitable positions for them can be obtained in v, otherwise the auxiliary subroutine cSRFLP_constraint_error() reports the inability to repair the ordering pair. Upon successful repairing of each type of constraints, the auxiliary subroutine cSRFLP_update() is called for updating the global vectors r and u.

Table 20.6 Pseudo-code implementing the mechanism for repairing ordering constraints with the permission for placing other facilities in between the two facilities of an ordered pair as expressed by Eq. (20.8) in Sect. 20

```
subroutine cSRFLP_order_free(n, u, v, r)
  call cSRFLP_repair_initialize(n, r, v, p^r, p^v)                 // Initialize p^r and p^v
  for (i = 1 to n) do
    if (O_{r_i} = N) continue  end if                               // r_i is not ordered on left side
    z := R_{r_i}                                                    // Right hand side facility
    if (T_{r_i} = Y or T_z = Y) continue  end if                    // One facility in fixed position
    if (A_{r_i,z} = N) continue  end if                             // Facilities to be placed together
    a := b := 0
    if (v_i = 0 or r_i, v_{p^r_z} = 0 or z, i < p^r_z) do  a := i, b := p^r_z  end if   // 1st condition of Eq. (8c)
    if (a = b = 0, v_i = 0 or z, v_{p^r_z} = 0 or r_i, i > p^r_z) do  a := p^r_z, b := i  end if   // 2nd condition of Eq. (8c)
    if (a = b = 0, v_i = 0 or r_i, p^v_z = 0, p^r_z < i) do          // 3rd condition of Eq. (8c)
      call random_positions(v, i+1, n, n', e)                       // Vacancy in (i+1, n) in v
      if (n' > 0) do                                                // Vacant position(s) available
        t_2 := call random_integer(1, n')                           // A random integer in (1, n')
        a := i, b := e_{t_2}                                        // i and e_{t_2} are positions in v
    if (a = b = 0, p^v_{r_i} = 0, i > p^r_z, v_{p^r_z} = 0 or z) do  // 4th condition of Eq. (8c)
      call random_positions(v, 1, p^r_z − 1, n', e)                 // Vacancy in (1, p^r_z) in v
      if (n' > 0) do                                                // Vacant position(s) available
        t_1 := call random_integer(1, n')                           // A random integer in (1, n')
        a := e_{t_1}, b := p^r_z                                    // e_{t_1} and p^r_z are positions in v
    if (a = b = 0, p^v_{r_i} = 0, p^v_z = 0) do                     // 5th condition of Eq. (8c)
      call random_positions(v, 1, n, n', e)                         // Vacancy in (1, n) in v
        if (n' ⩾ 2) do                                              // Vacant positions available
          repeat
            t_1 := call random_integer(1, n')                       // A random integer in (1, n')
            t_2 := call random_integer(1, n')                       // Another random integer in (1, n')
          while t_1 = t_2
          if (t_1 < t_2) do  a := e_{t_1}, b := e_{t_2}  end if
          if (t_1 > t_2) do  a := e_{t_2}, b := e_{t_1}  end if
    if (a ≠ 0, b ≠ 0, (v_a = 0 or r_i), (v_b = 0 or z)) do
      call cSRFLP_arrange(a, b, i, p^r_z, r_i, z, u, v, p^v)        // Arrange r_i and z
    else  call cSRFLP_constraint_error(n, v, r, u, r_i, z)          // Arrangement not possible
    end else if
  end for
  call cSRFLP_update(n, v, r, u)                                    // Update r and u
  return u, v, r
end subroutine
```

Implementation of the Repairing Mechanisms

The repairing mechanisms, expressed through Eqs. (20.5)–(20.8) and pseudo-codes in Tables 20.3, 20.4, 20.5, and 20.6, can be implemented outside of an optimizer (a general algorithm or a commercial software package). However, the optimizer should have the provision to support an external subroutine evaluating a solution (a permutation in the present study) and also to accept the solution modified in the subroutine. Such a procedure is shown in Fig. 20.3. In this case, the optimizer starts by initializing a solution, which is sent to the external subroutine for evaluation. The external subroutine first repairs the solution if required, and then evaluates (i.e., computes the objective value) and returns the repaired solution to the optimizer. The optimizer improves the repaired solution towards the optimum and then sends it to the external subroutine for evaluation. In this way, the process of improving and evaluating the solution is continued until some given termination criteria are fulfilled.

Table 20.7 Five auxiliary subroutines called in other subroutines presented in Tables 20.3, 20.4, 20.5, and 20.6

```
subroutine cSRFLP_repair_initialize(n, r, v, p^r, p^v)
  for (i = 1 to n) do   p^v_i := 0 end for
  for (i = 1 to n) do
  |  p^r_{r_i} := i                                         // Position of r_i
  |  if (v_i ≠ 0) do   p^v_{v_i} := i end if                // Position of v_i
  return p^r, p^v
end subroutine

subroutine random_positions(v, a, b, n', e)
  n' := 0
  for (i = a to b) do
  |  if (v_i = 0) do   n' = n' + 1, e_{n'} := i   end if    // Vacant positions in v found
  return n', e
end subroutine

subroutine cSRFLP_arrange(a, b, l^p, r^p, l^f, r^f, u, v, p^v)
  if (p^v_{l^f} = 0) do                                     // Facility l^f to be arranged in v
  |  p^v_{l^f} := a                                         // Facility l^f placed at a of v
  |  v_a := l^f
  |  u_{l^p} := 0                                           // Facility l^f removed from l^p of u
  if (p^v_{r^f} = 0) do                                     // Facility r^f to be arranged in v
  |  p^v_{r^f} := b                                         // Facility r^f placed at b of v
  |  v_b := r^f
  |  u_{r^p} := 0                                           // Facility r^f removed from r^p of u
  return u, v, p^v
end subroutine

subroutine cSRFLP_update(n, v, r, u)
  s := 1
  for (i = 1 to n) do                                       // Updating r
  |  if (v_i ≠ 0) do   r_i := v_i
  |  else do
  |  |  for (j = s to n) do
  |  |  |  if (u_j ≠ 0) do
  |  |  |  |  r_i := u_j
  |  |  |  |  s := j + 1
  |  |  |  |  break
  for (i = 1 to n) do                                       // Updating u
  |  if (v_i ≠ 0) do   u_i := 0
  |  else do           u_i := r_i
  return r, u
end subroutine

subroutine cSRFLP_constraint_error(n, v, r, u, l^f, r^f)
  write "Facilities l^f and r^f could not be ordered" quit  // Error message
end subroutine
```

Genetic Algorithm for Optimizing the cSRFLP Model

Genetic algorithm (GA) is a stochastic technique that mimics the Darwin's evolution mechanisms based on the "survival of the fittest" concept. The primary component of a GA is an individual representing a complete solution of a problem, where the individual is usually codified as an array of the decision variables of the problem. A set of individuals is taken as a GA population, which is gradually evolved towards the optimum of the problem through repeated application of three operators similar to those from natural evolution, namely selection, crossover, and

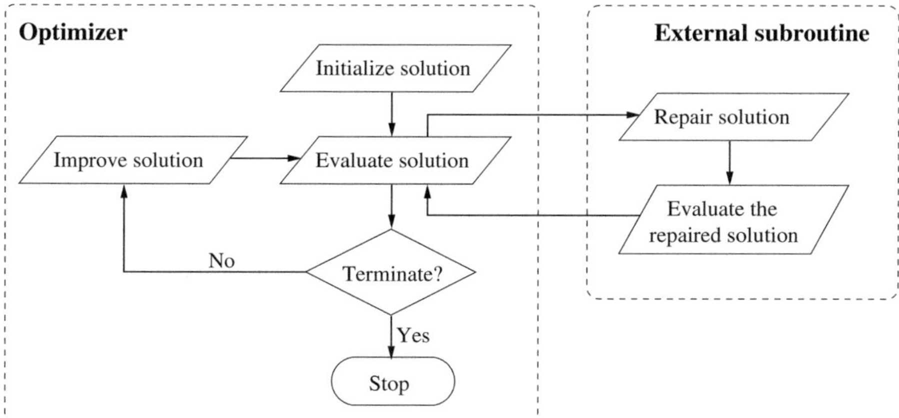

Fig. 20.3 Implementation of an external subroutine in an optimizer for repairing and evaluating a solution

mutation operators. The selection operator identifies the better individuals from the current GA population, the crossover operator generates new individuals by exploiting those better individuals identified by the selection operator, and finally the mutation operator is engaged to explore the neighborhood of the new individuals generated by the crossover operator. The process of evolution of the GA population is continued in this way until some termination criteria are fulfilled, usually until a given maximum number of generations are performed (see Deb [7], Goldberg [10] for further detail).

The permutation-based genetic algorithm (pGA), investigated by Datta et al. [5] for solving the unconstrained SRFLP model, is adopted here as the general-purpose algorithm for handling the cSRFLP model, which will call an external subroutine for repairing and evaluating its individuals. The main working procedure of the pGA, in the context of the proposed cSRFLP, is presented here in brief (refer Datta et al. [5], Kalita and Datta [15] for detail of the pGA).

Individual Initialization

Each individual of the pGA is considered to be an array to represent a permutation of a given number (in the present work, the number of facilities to be arranged). Unlike in real and integer-valued problems, a random initialization process is likely to generate an invalid individual for a permutation-based problem by repeating some values while omitting others. Therefore, a greedy technique is applied in the pGA for initializing an individual with a valid permutation. In this process, a temporary array is taken first to represent a permutation of n in ascending order starting from one. Then, the elements of an individual are assigned one by one with the values of random positions of the temporary array. After assigning a value to an element of the individual, the temporary array is updated by deleting that value from it. The process is continued until the temporary array is exhausted, i.e., the pGA individual is assigned with a valid permutation.

Individual Evaluation

An individual of the pGA, as initialized in Sect. 20 or generated by the crossover and mutation operators in Sects. 20 and 20 respectively, represents a valid permutation of a given number. However, the cSRFLP seeks the placement of its facilities subject to the positioning and ordering constraints imposed to some facilities as given by Eqs. (20.2) and (20.3), respectively, which means that an arbitrary permutation of the given facilities may not be a feasible solution of the cSRFLP model. Therefore, as shown in Fig. 20.3, each pGA individual is sent to an external subroutine, where it is first repaired by applying the repairing mechanisms addressed in Sect. 20 and then evaluated for the objective function expressed by Eq. (20.1).

Selection Operator

For forming a mating pool, the binary tournament selection operator [7] is used to select better individuals from the current pGA population. The operator picks up at a time two random individuals from the pGA population and updates the mating pool with a copy of the best one based on their objective values. In order to reduce the algorithmic complexity of a GA, the size of the mating pool is usually made equal to that of the GA population by repeating the process of selecting and copying of better individual.

Crossover Operator

A problem-based crossover operator is used in the pGA for generating a new feasible individual (i.e., a valid permutation of the given facilities) by exploiting two random parent individuals picked up from the mating pool generated by the selection operator. The crossover operator generates a new individual with a prefixed crossover probability as follows: each element of the new individual is assigned the value of the corresponding element of one of the selected two parent individuals with equal probability. If an element of the new individual is not vacant, the value selected against it is kept temporarily in reserve. Finally, all the values kept in reserve are assigned randomly to the vacant elements of the new individual.

Mutation Operator

The pGA employs a very simple mutation operator, which chooses two elements randomly from an individual, generated in the crossover operation. Then the values of the elements are swapped with a low mutation probability. Since none of the values is altered, but those are swapped only, the mutation operator forms a new individual by exploring successfully the neighborhood of an existing individual.

Elite Preserving Mechanism

Although GA is expected to converge to the optimum because of the exploitation of the previous individuals by the crossover operator and exploration of the search space by the mutation operator, the stochastic nature may cause it to fail in generating better individuals at some generations, or even to push the search opposite to the optimum of upon generation of new individuals worse than those of the present population. For preventing such situations, it is generally suggested to preserve some elite individuals of the population over generations. In such an attempt, the elite preserving mechanism proposed by Deb et al. [8] is employed in the pGA. This mechanism first combines the current and new populations of a generation, and then the best 50% individuals of the combined population is taken to form the population for the next generation 20.

Computational Experiment

The repairing mechanisms proposed in Sect. 20 and the pGA presented in Sect. 20 are coded in C programming language and executed in a Linux environment.

The set of 20 instances in the range of [60,80], introduced by Anjos et al. [2] for evaluating their SRFLP formulation, are taken in the present study for evaluating the cSRFLP model by imposing positioning and ordering constraints to some of the facilities. For easy recognition, the two types of constraints are imposed to the same facilities of all the instances. Facilities 1 and 60 are made to subject to the positioning constraints requiring to place them, respectively, at the first and sixtieth positions of a permutation. On the other hand, for demonstrating the ordering constraints, the pair of facilities 20 and 30 are made to be ordered without allowing other facilities in between them, while the pair of facilities 25 and 10 are made to be ordered allowing to place other facilities in between them.

In the numerical experiment, the number of facilities requiring predefined positions, as well as the indices and required predefined positions for those facilities are taken as the input parameters for the positioning constraints. For the ordering constraints, the input parameters include the number of facility pairs required to be ordered, the indices of the left side and right side facilities of each ordering pair. Further input parameter for the ordering constraints is the option for allowing or not allowing the placement of other facilities in between the facilities of a given pair. For applying the repairing mechanisms expressed by Eqs. (20.5)–(20.8), whose pseudocodes are given in Tables 20.2, 20.3, 20.4, 20.5, 20.6, and 20.7, all the facilities can first be arranged in a matrix form as shown in Table 20.1.

It is a known fact that the performance of a stochastic optimizer, like a GA, may greatly be influenced by the chosen algorithmic parameter values, particularly in the case of combinatorial optimization problems including the facility layout problem. Hence, researchers often conduct some transition rule based experiments for obtaining a suitable set of algorithmic parameter values for effectively tackling a problem at hand. However, since the present work is not aimed at developing an algorithm but to demonstrate a procedure for handling the cSRFLP problem, based on some trial runs a pGA population of size 60 is considered to be evolved for a maximum of 500 generations with fixed crossover probability of 75% and mutation probability of 1%. With those parameter values, the pGA is executed for 30 times for each problem instance with different sets of initial solutions.

The obtained best results in terms of the objective values (i.e., the overall material flow cost among the given facilities) for the studied set of problem instances are shown in Table 20.8, where the constrained facilities are shown in boldface fonts for identifying easily. Since it is an introductory work on the cSRFLP, the results obtained here could not be compared with any existing work. Therefore, the best known results, obtained by Kothari and Ghosh [16] studying the instances under their unconstraint SRFLP model, are also shown in Table 20.8 for comparing with the same obtained for the cSRFLP model. Due to the obvious reasons upon imposing the positioning and ordering constraints, the results of the cSRFLP instances are found inferior to those of the SRFLP instances by some amount.

Conclusion

So far, the single-row facility layout problem (SRFLP) is studied as an unconstrained optimization problem for minimizing the overall material flow cost among a given number of facilities, i.e., the facilities can be placed freely at any position, thus making any permutation of the facilities as a feasible solution of the problem. However, a practical problem might face some restrictions on the arrangement of some facilities in random positions, instead requiring a facility to be placed at a predefined fixed position or a particular pair of facilities to be arranged in a given order with or without allowing the placement of other facilities in between them. Accordingly, a constrained SRFLP model (cSRFLP in short) is introduced here by imposing positioning and ordering constraints to some given facilities.

Since a general-purpose optimizer may face a huge computational burden on satisfying such permutation-based constraints, an intelligent search will require the development of a specialized algorithm by incorporating certain problem-

Table 20.8 The best cSRFLP solutions obtained for the 20 instances of Anjos et al. [2] and their objective values (i.e., the total material flow costs) along with the best known corresponding SRFLP values reported by Kothari and Ghosh [16]

		Material flow cost			
ID	Instance	SRFLP	cSRFLP	% increase	Permutation of the facilities (cSRFLP model)
1	60-01	1477834.0	1499318.0	1.45	**1** 27 11 48 56 28 24 15 21 3 40 **25** 6 51 54 46 7 57 12 47 36 44 16 13 35 4 32 33 8 26 2 17 14 42 22 49 41 9 5 38 29 23 45 31 43 37 50 53 **10** 52 59 19 58 55 **20 30** 39 18 34 **60**
2	60-02	841776.0	858864.0	2.03	**1** 4 5 19 6 59 24 53 49 47 27 21 32 52 **20 30** 37 2 57 7 51 **25 10** 12 23 18 56 40 35 46 39 17 50 43 8 33 26 9 45 28 58 15 55 34 31 14 44 41 11 54 13 22 48 36 38 29 16 42 3 **60**
3	60-03	648337.5	696937.5	7.50	**1** 4 5 19 6 59 24 53 49 47 27 21 32 52 **20 30** 37 2 57 7 51 **25 10** 12 23 18 56 40 35 46 39 17 50 43 8 33 26 9 45 28 58 15 55 34 31 14 44 41 11 54 13 22 48 36 38 29 16 42 3 **60**
4	60-04	398406.0	457737.0	14.89	**1** 13 12 43 **25** 55 34 4 41 28 38 **20 30** 49 9 52 48 51 32 47 54 37 42 26 29 16 8 14 27 18 58 15 50 35 2 45 5 3 53 59 40 17 46 6 23 21 11 56 44 33 22 24 36 7 39 **10** 57 19 31 **60**
5	60-05	318805.0	345453.0	8.36	**1** 33 12 49 44 52 34 38 3 11 26 43 29 32 14 57 23 22 18 5 51 15 58 31 45 24 28 56 42 21 50 **25 10** 8 48 6 27 40 **20 30** 36 35 17 46 2 55 13 9 54 4 41 19 47 59 16 7 37 53 39 **60**
6	70-01	1528537.0	1596183.0	4.43	**1** 53 47 40 65 2 39 28 62 8 22 15 32 9 63 51 69 68 **25** 16 64 61 41 38 56 **10** 67 44 70 26 14 19 33 42 **20 30** 59 55 11 23 21 24 36 18 27 13 54 5 12 58 6 49 52 7 66 3 34 4 31 **60** 45 46 43 48 17 29 57 35 37 50
7	70-02	1441028.0	1456777.0	1.09	**1** 28 39 53 52 **25** 36 21 **10** 26 7 3 19 13 69 58 9 67 33 23 45 29 62 18 63 14 11 6 44 43 24 42 49 20 30 8 40 4 2 46 54 5 16 22 27 15 37 34 68 31 55 32 35 12 65 38 17 70 66 **60** 51 48 47 41 59 61 57 64 50 56
8	70-03	1518993.5	1561147.5	2.78	**1** 8 63 51 **20 30** 27 54 29 21 43 58 32 23 24 67 41 47 6 9 26 4 22 65 61 7 70 19 2 42 37 68 53 14 **25 10** 50 49 28 18 56 36 66 39 5 3 40 11 35 62 13 55 45 48 12 59 33 46 38 **60** 15 64 69 34 17 31 57 52 44 16

(continued)

Table 20.8 (continued)

ID	Instance	Material flow cost			Permutation of the facilities (cSRFLP model)
		SRFLP	cSRFLP	% increase	
9	70-04	968796.0	1033612.0	6.69	**1** 41 16 25 35 8 38 31 45 47 43 52 63 48 **20 30** 3 6 40 67 17 55 23 21 19 70 69 57 56 27 34 50 37 28 59 44 18 5 49 2 13 46 36 14 26 24 12 53 68 62 58 33 42 61 11 29 22 64 9 **60** 54 **10** 65 32 39 51 4 15 7 66
10	70-05	4218002.5	4275606.5	1.37	**1** 28 19 38 32 68 23 62 31 51 2 40 58 37 50 22 39 6 54 16 65 63 27 49 8 26 41 43 61 21 56 14 35 24 70 **20 30** 5 12 36 4 53 11 7 29 34 46 3 42 59 66 15 **25 10** 55 45 48 57 47 **60** 52 13 18 33 69 9 64 44 17 67
11	75-01	2393456.5	2489354.5	4.01	**1** 38 46 16 29 9 63 72 54 28 51 5 13 33 21 8 45 17 22 **25** 75 11 39 62 **20 30** 43 68 70 3 4 66 **10** 32 31 12 34 74 19 2 42 40 24 37 53 55 71 41 44 6 35 7 23 50 49 65 15 56 18 **60** 61 26 59 27 52 36 57 73 69 67 48 14 47 64 58
12	75-02	4321190.0	4415170.0	2.17	**1 25** 49 26 56 37 61 47 72 28 66 39 18 35 7 24 33 54 **10** 4 19 53 5 75 46 65 71 12 36 67 22 8 62 23 34 27 15 3 11 38 16 74 68 13 17 55 48 9 31 52 73 58 41 69 29 59 **20 30** 57 **60** 50 42 45 40 63 14 70 6 44 2 21 32 64 43 51
13	75-03	1248423.0	1251980.0	0.28	**1** 22 50 54 14 63 53 55 **25** 8 72 18 70 66 38 2 37 44 **20 30** 31 62 65 3 59 34 52 49 6 32 7 40 58 9 75 24 21 28 35 45 67 5 73 57 11 16 48 71 13 27 36 12 56 61 64 74 39 4 26 60 68 23 29 46 41 51 43 17 15 33 19 **10** 42 69 47
14	75-04	3941816.5	4205911.5	6.70	**1** 21 17 9 72 38 69 68 27 **25** 6 51 61 74 11 31 4 71 43 54 59 23 26 58 66 63 56 48 19 2 35 67 32 13 45 24 18 34 73 53 44 28 52 16 65 3 64 12 55 46 39 33 40 29 41 **20 30** 57 22 **60** 47 70 37 8 62 **10** 42 75 7 50 15 14 5 49 36

15	75-05	1791408.0	1852782.0	3.43	**1** 24 37 72 52 13 35 50 43 58 **20** **30** 68 71 44 47 32 22 73 66 65 7 31 23 46 42 49 8 38 3 17 28 12 70 5 55 **25** 29 **10** 59 45 34 36 11 57 48 54 15 67 4 56 21 75 69 39 74 16 61 41 **60** 18 53 14 40 64 62 6 2 51 19 63 9 26 33 27
16	80-01	2069097.5	2108989.5	1.93	**1** 78 **25** 11 79 57 71 4 28 64 8 77 38 3 35 59 5 26 73 72 69 24 37 33 13 7 17 74 6 18 23 22 76 36 **20** **30** 31 9 15 58 42 32 50 **10** 41 48 54 66 56 62 34 51 49 75 45 47 68 52 53 **60** 12 61 39 43 14 63 19 29 46 16 21 40 65 27 44 67 80 2 70 55
17	80-02	1921136.0	1990790.0	3.63	**1** 57 24 55 58 46 14 49 15 74 47 56 32 77 12 54 41 7 16 28 72 27 **25** 23 76 67 21 **20** **30** 19 13 6 4 38 36 34 39 8 65 42 2 44 50 9 40 62 75 53 68 3 35 26 59 33 29 73 37 71 79 60 52 70 31 78 64 80 51 69 63 43 45 5 48 61 66 17 **10** 11 18 22
18	80-03	3251368.0	3487874.0	7.24	**1** 57 54 37 53 36 68 34 69 29 **20** **30** 32 13 48 72 22 **25** 56 76 **10** 77 31 50 52 51 8 78 71 17 27 80 44 21 18 11 14 23 41 7 73 59 2 19 15 75 55 38 39 3 9 66 12 79 6 28 65 42 24 **60** 67 4 45 5 35 43 64 47 46 70 26 62 49 61 33 16 58 40 63 74
19	80-04	3746515.0	4009518.0	7.02	**1** 69 64 75 71 35 38 78 54 46 28 32 42 12 80 74 43 14 **20** **30** 76 29 9 62 27 6 17 67 4 18 3 2 16 7 72 31 19 15 26 52 48 33 79 37 53 77 13 49 8 51 65 70 5 11 39 45 68 23 **25** **60** 59 66 73 63 34 24 55 **10** 44 36 21 58 57 22 56 61 40 50 41 47
20	80-05	1588885.0	1636438.0	2.99	**1** **25** 2 48 7 47 52 38 61 73 10 68 21 37 72 49 74 70 11 8 22 19 76 31 36 79 62 9 33 75 34 35 54 6 44 71 32 53 **20** **30** 55 59 65 3 39 66 77 18 43 80 40 4 45 23 63 46 50 57 5 **60** 26 56 24 28 15 14 64 67 16 51 29 12 13 27 42 78 69 58 41 17

specific components in it. However, many practitioners and academia working on application areas suffer from expertise required for the development of such a specialized algorithm. Hence, they prefer some alternative techniques to get their works done through some existing general-purpose algorithms or commercial software packages. Such a procedure is presented here by developing some repairing mechanisms, which can be applied along with a general-purpose algorithm for forcibly satisfying the constraints of the cSRFLP outside the algorithm.

Taking a permutation-based genetic algorithm (name the pGA) as the general-purpose algorithm, the potentiality of the proposed procedure is demonstrated by applying it to a set of instances of sizes in the range of [60,80], which could be taken as benchmark instances in future studies on constrained SRFLP models.

References

1. Amaral ARS (2006) On the exact solution of a facility layout problem. Eur J Oper Res 173:508–518
2. Anjos MF, Kennings A, Vannelli A (2005) A semidefinite optimization approach for the single-row layout problem with unequal dimensions. Discrete Optim 2:113–122
3. Anjos MF, Vannelli A (2008) Computing globally optimal solutions for single-row layout problems using semidefinite programming and cutting planes. INFORMS J Comput 20:611–617
4. Benjaafar S, Ramakrishnan R (1996) Modelling, measurement and evaluation of sequencing flexibility in manufacturing systems. Int J Prod Res 34(5):1195–1220
5. Datta D, Amaral ARS, Figueira JR (2011) Single row facility layout problem using a permutation-based genetic algorithm. Eur J Oper Res 213(2):388–394
6. de Alvarenga AG, Negreiros-Gomes FJ, Mestria M (2000) Metaheuristic methods for a class of the facility layout problem. J Intell Manuf 11:421–430
7. Deb, K. (2001). Multi-objective optimization using evolutionary algorithms. Wiley, Chichester
8. Deb K, Agarwal S, Pratap A, Meyarivan T (2002) A fast and elitist multi-objective genetic algorithm: NSGA-II. IEEE Trans Evol Comput 6(2):182–197
9. Gapp W, Mankekar PS, Mitten LG (1965) Sequencing operations to minimize in-process inventory costs. Manage Sci 11(3):476–484
10. Goldberg DE (1989) Genetic algorithms in search, optimization, and machine learning. Addison-Wesley, New Jersey
11. Heragu SS, Alfa AS (1992) Experiment Analysis of simulated annealing based algorithms for the layout problem. Eur J Oper Res 57(2):190–202
12. Heragu SS, Kusiak A (1988) Machine layout problem in flexible manufacturing systems. Oper Res 36:258–268
13. Heragu SS, Kusiak A (1991) Efficient models for the facility layout problem. Eur J Oper Res 53:1–13
14. Huang J, Lu X, Zhang G, Qu J (2014) Study on the rheological, thermal and mechanical properties of thermoplastic polyurethane/poly (butylene terephthalate) blends. Polym Test 36:69–74
15. Kalita Z, Datta D (2014) Solving the bi-objective corridor allocation problem using a permutation-based genetic algorithm. Comput Oper Res 52:123–134
16. Kothari R, Ghosh D (2014) A scatter search algorithm for the single row facility layout problem. J Heuristics 20(2):125–142
17. Kumar KR, Hadjinicola GC, Lin TL (1995) A heuristic procedure for the single-row facility layout problem. Eur J Oper Res 87:65–73

18. Kumar S, Asokan P, Kumanan S, Varma B (2008) Scatter search algorithm for single row layout problem in FMS. Adv Prod Eng Manag 3:193–204
19. Lafou M, Mathieu L, Pois S, Alochet M (2016) Manufacturing system flexibility: sequence flexibility assessment. In: 49th CIRP Conference on Manufacturing System (CIRP CMS2016), Stuttgart
20. Love RF, Wong JY (1976) On solving a single row space allocation problem with integer programming. INFOR 14:139–143
21. Monma CL (1981) Sequencing with general precedence constraints. Discrete Appl Math 3(2):137–150
22. Ozgormus E (2015) Optimization of block layout for grocery stores. PhD thesis, Auburn University, USA
23. Padgaonkar AS (2004) Modelling and analysis of the hospital facility layout problem. Master's thesis, Department of Industrial and Manufacturing Engineering, New Jersey Institute of Technology, New Jersey
24. Picard J, Queyranne M (1981) On the one dimensional space allocation problem. Oper Res 29(2):371–391
25. Rachamadugu R, Nandkeolyar U, Schriber T (1993) Scheduling with sequencing flexibility. Decis Sci 24(2):315–342
26. Rachamadugu R, Schriber TJ (1990) Performance of dispatching rules under perfect sequencing flexibility. In: 22nd conference on winter simulation. IEEE Press, Piscataway, pp 653–658
27. Samarghandi H, Eshghi K (2010) An efficient tabu algorithm for the single row facility layout problem. Eur J Oper Res 205:98–105
28. Samarghandi H, Taabayan P, Jahantigh FF (2010) A particle swarm optimization for the single row facility layout problem. Comput Ind Eng 58:529–534
29. Simmons DM (1969) Single row space allocation: an ordering algorithm. Oper Res 17(5):812–826
30. Solimanpur M, Vrat P, Shankar R (2005) An ant algorithm for the single row layout problem in flexible manufacturing systems. Comput Oper Res 32:583–598

Chapter 21
Geometric Size Optimization of Annular Step Fin Array for Heat Transfer by Natural Convection

Abhijit Deka and Dilip Datta

Abstract Although a substantial amount of research is available on annular step fin, only a few works are dedicated to the study of annular step fin array. In the present work, an annular fin array consisting of two-stepped rectangular cross-sectional identical fins is modeled as a multi-objective optimization problem for simultaneously maximizing the total heat transfer rate from the fin array and minimizing the total volume of the fins. Maximization of the surface efficiency and augmentation factor of the fin array are also studied as two additional objective functions for further assessment of the fin array. Considering a constant base temperature and one-dimensional heat flow along the radial direction of the fins, the cross-sectional half-thicknesses and outer radii of the two steps of a fin as well as the total number of fins in the fin array are taken as five design variables. The hybrid spline difference method is used to solve the one-dimensional heat transfer equations and then NSGA-II is applied for approximating the Pareto-optimal fronts for different cases. In order to investigate the influence of various design variables on the objective functions, a Pareto-optimal sensitivity analysis is also carried out. The proposed procedure should aid designers in adopting suitable fin array configurations as per their requirements and practicability.

Keywords Annular fin array · Step fin profile · Convection heat transfer · Multi-objective optimization · Genetic algorithm

---

A. Deka · D. Datta (✉)
Department of Mechanical Engineering, School of Engineering, Tezpur University, Tezpur, Assam, India
e-mail: adeka13@tezu.ernet.in; ddatta@tezu.ernet.in

F. Bennis, R. K. Bhattacharjya (eds.), *Nature-Inspired Methods for Metaheuristics Optimization*, Modeling and Optimization in Science and Technologies 16,
https://doi.org/10.1007/978-3-030-26458-1_21

Nomenclature

A_b	Area of the base of an individual fin of the fin array, m^2
A_s	Heat transfer surface area of an individual fin of the fin array, m^2
A_{sp}	Fin inter-spacing base area, m^2
Bi	Biot number; $Bi = hr_b/k_a$
f_1	Heat transfer rate through the fin array, W
f_2	Fin volume of the array, m^3
f_3	Surface efficiency
f_4	Augmentation factor
g	Acceleration due to gravity, m/s^2
h	Convective heat transfer coefficient, W/m^2K
k	Thermal conductivity of the fin material, W/mK
k_a	Thermal conductivity of the fin material at ambient temperature, W/mK
n_{fin}	Number of fins
R	Non-dimensional radial coordinate; $R = \frac{r}{r_2}$
R_1	Non-dimensional inner radius of the fin; $R_1 = \frac{r_b}{r_2}$
R_2	Non-dimensional radius of the fin at the point of step change in thickness; $R_2 = \frac{r_1}{r_2}$
r	Radial coordinate for the entire fin, m
r_b	Inner radius of the fin, m
r_1	Radius of the fin at the point of step change in thickness ($r_1 > r_b$), m
r_2	Outer radius of the fin ($r_2 > r_1$), m
Ra	Rayleigh number
s_b	Fin inter-spacing at the base, m
s_m	Mean fin inter-spacing, m
T_b	Base temperature of the fin, K
T_1	Local temperature in the radial direction within the first step of the fin, K
T_2	Local temperature in the radial direction within the second step of the fin, K
T_∞	Ambient temperature, K
t	Thickness ratio; $t = \frac{t_2}{t_1}$
t_1	Half-thickness of the first step of the fin, m
t_2	Half-thickness of the second step of the fin ($t_2 < t_1$), m
Z_0, Z_1	Non-dimensional fin parameters defined in Eq. (21.6)
W	Length of the primary cylinder, m

Greek Symbols

α	Non-dimensional variable thermal conductivity parameter; $\alpha = (T_b - T_\infty)\,\beta$
α'	Thermal diffusivity, m/s^2
β	Parameter describing the variation of the thermal conductivity, K^{-1}

β' Thermal expansion coefficient, K^{-1}
δ Non-dimensional temperature factor; $\delta = \frac{T_\infty}{T_b - T_\infty}$
θ Non-dimensional temperature in the first step of the fin; $\theta = \frac{T_1 - T_\infty}{T_b - T_\infty}$
ν Kinematic viscosity, m/s^2
ξ Fin aspect ratio; $\xi = \frac{t_1}{r_b}$
ϕ Non-dimensional temperature in the second step of the fin; $\phi = \frac{T_2 - T_\infty}{T_b - T_\infty}$

Introduction

An annular fin array is used for augmenting heat transfer from a cylindrical surface to the surroundings. This is done primarily by increasing the heat transfer surface area. However, the addition of fins is associated with additional design material and fabrication costs, thus raising the total production cost. Hence, efforts are being made on size and shape optimization as well as the fin inter-spacing optimization of an annular fin array for effective utilization of the fin material. Designers need to identify a fin array configuration that would dissipate maximum heat through minimum fin volume. Although an annular fin array with fins of rectangular cross-section is the simplest one from the manufacturing point of view, its material at the tip of the fins is not utilized efficiently. Because of that, annular fins of different profiles, such as triangular, trapezoidal, parabolic and hyperbolic, are found in literature [7]. In order to achieve the maximum heat transfer rate with the lowest production cost, designers strive to optimize the fin profile as well as the fin inter-spacing of a fin array, so as to dissipate either the maximum amount of heat from a fixed fin volume or a specified amount of heat from the minimum fin volume. However, because of the complexity involved with the manufacturing processes, the practical applicability of complex fin profiles (shape optimization) is still limited. As an alternate approach, researchers also attempt to optimize the dimensions (size optimization) of the fin profiles of predefined shapes. Therefore, there lies immense scope in modifying the geometry of a rectangular fin profile, such that the fin material is used effectively in dissipating heat as well as reducing the complexity of the fabrication processes [11]. Since the shape of the profile of a fin with step change in thickness is similar to that of a constant thickness fin, its fabrication is simple. Also, the heat dissipation rate per unit volume of a stepped fin is always higher than that of a constant thickness fin. Even if the design criteria of fin arrays differ from application to application, the primary concerns are the maximization of the rate of heat transfer from an fin array and minimization of the total volume of the fins. Thus, the optimization of a fin array based on these two requirements is highly desirable.

An excellent comprehensive review of the existing literature on extended surface heat transfer was published by Kraus et al. [7]. Analyzing the performance of a stepped annular fin under dehumidifying surface conditions, Kundu [8] observed that the stepped annular fin profiles are better than the concentric annular disc fin profiles from the point of view of the heat dissipation rate under the identical surface conditions. Some more studies are found, where similar observations as in Kundu [8] were made after carrying out thermal analysis and optimization of annular rectangular stepped fins [11, 14]. With the application of Bessel functions, Brown [2] optimized the dimensions of a constant thickness radial fin. Extending the work of Brown [2] for determining the efficiency and optimum dimensions of annular fins of various linear and nonlinear profiles (rectangular, triangular, parabolic and hyperbolic) under constant heat transfer coefficient, Ullmann and Kalman [25] observed that the fin of parabolic profile outperformed the fins of all other considered profiles. Arslanturk [1] proposed an analytical approach for maximizing the heat transfer rate from an annular fin of constant volume rectangular profile, where heat transfer only by convection was considered under a constant thermal conductivity of the fin material. Based on differential transformation, Yu and Chen [31] presented an analytical method for thermal analysis and constant volume optimization of a rectangular profile annular fin with variable convective heat transfer coefficient and variable thermal conductivity. Sharqawy and Zubair [24] analyzed the performance and optimization issues of annular fin related to simultaneous heat and mass transfers under various dehumidifying conditions. Based on a variable separation method, Kang [6] presented an optimization strategy for a rectangular annular fin of fixed height. Optimizing eccentric and elliptical annular disc fins with restriction of space in one or both sides of the fluid carrying tube, Kundu and Das [10, 12] observed that for a given volume, eccentric and elliptical annular disc fins can dissipate more heat in comparison to concentric annular disc fins. Other works optimizing annular fins can be found in [4, 17, 20–22].

All of the above works were concerned with the optimization of a single annular fin isolated from the surrounding. However, application of a single fin for heat dissipation is rarely found in practice. Instead, a number of fins in a row, known as a fin array, is generally used for augmenting heat transfer. In that case, optimization of a fin profile in combination with fin inter-spacing becomes essential for efficiently dissipating heat from a fin array. Still, only a limited number of works in that direction have been reported in literature. Kundu and Das [13] developed an analytical model for performance analysis and design optimization of fins attached to flat and curved primary surfaces. In another study, Kundu et al. [15] carried out thermal analysis of a step annular disc fin array considering heat dissipation from the fin surface by convection only. Other works dealing with the design of annular fin arrays were reported by Kundu and Barman [9], Lai et al. [16]. Many such optimization processes are carried out considering fins of constant thickness or constant inter-spacing [13, 15, 16]. However, an appreciable saving in the fin material can be achieved by modifying the fin profile and fin inter-spacing of a fin array.

Though the maximization of heat dissipation from fins is a primary requirement, there exist some other objectives to be satisfied, thus leading the fin design to be a multi-objective optimization problem for optimizing all such conflicting objective functions simultaneously. As for example, the total heat dissipation rate from a fin array conflicts with both total volume and surface efficiency of the fin array, which may result in an optimized solution with respect to one of those objectives (single-objective optimization) unacceptable in terms of the other objectives. The multi-objective optimization process gives a set of trade-off solutions, where an optimized solution satisfies all the objectives with a certain level of acceptance without being influenced by any other solution. However, scrutinizing the present literature, it is observed that no work designing an annular step fin array as a multi-objective optimization problem has been reported so far. The only work in that direction, found in specialized literature, studied a single annular step fin as a multi-objective optimization problem [5].

Motivated by the research gaps as stated above, the present study proposes a complete methodology for the multi-objective optimization of an annular fin array consisting of two-stepped rectangular cross-sectional identical fins. Taking the cross-sectional half thicknesses and outer radii of the two steps of a fin as well as the number of fins in the fin array as five design variables, they are tuned through an optimization tool for optimizing some fin performance related objective functions under various thermal conditions. The chosen main objective functions are the maximization of the total heat transfer rate and minimization of the fin volume of the array. For further assessment of the performance of the fin array, the maximization of the surface efficiency and augmentation factor of the fin array are taken as two more objective functions. Finally, in order to analyze the level of influence of the design variables on the objective functions, a Pareto-optimal sensitivity analysis is also carried out. In the solution procedure, the temperature field is first evaluated by solving the one-dimensional differential heat transfer equations using the hybrid spline difference method (HSDM), a highly accurate numerical method proposed by Wang et al. [29]. Then, the heat transfer rate and subsequently the surface efficiency and augmentation factor are evaluated using the known temperature field. On the other hand, the objective functions are optimized using the non-dominated sorting genetic algorithm II (NSGA-II), a well-known and widely used multi-objective genetic algorithm proposed by Deb et al. [3].

The main objective of the present study is to put forward the trade-off optimal scenarios based on various conflicting objective functions along with the level of influence of different design variables on the objective functions, so that designers can enjoy the freedom to choose favourable solutions as per their requirements depending upon the availability and practicability of information and resources.

Thermal Modeling of Annular Stepped Fin

The present study analyzes an annular fin array consisting of identical fins with two-stepped rectangular cross-sectional area, which is attached to a heat exchanger of cylindrical primary surface with uniform fin inter-spacing. The thickness of a fin is considered to be very small, so that the temperature difference in its lateral direction would become negligible and the flow of heat through the fin can be treated as one-dimensional. Since the heat transfer by radiation can be neglected for a low temperature difference [18], the heat loss from the fin surface as well as from the fin inter-spacing is considered to be taken place by natural convection only. Steady state heat transfer without internal heat generation is made another assumption.

The schematic diagram of such an annular fin array is shown Fig. 21.1a. The fin array is taken in the horizontal orientation so as to make the gravity forces parallel to the fins. In Fig. 21.1a, r_b and r_2 are respectively the inner and outer radii of a fin with r_1 as the radius at its point of step change in thickness, t_1 and t_2 are respectively the cross-sectional half-thickness at the base and tip of the fin, s_b is the fin inter-spacing at the base, W is the total length of the primary surface, and n_{fin} is the number of fins in the fin array.

Note in Fig. 21.1a that the attachment of the fin array on the primary surface is actually a repetitive surface containing one fin and some spacing (fin inter-spacing). Hence, the thermal analysis of the whole fin assembly can be done by analyzing a repetitive symmetric sector (heat transfer module) only. A schematic diagram of such a module is shown in Fig. 21.1b.

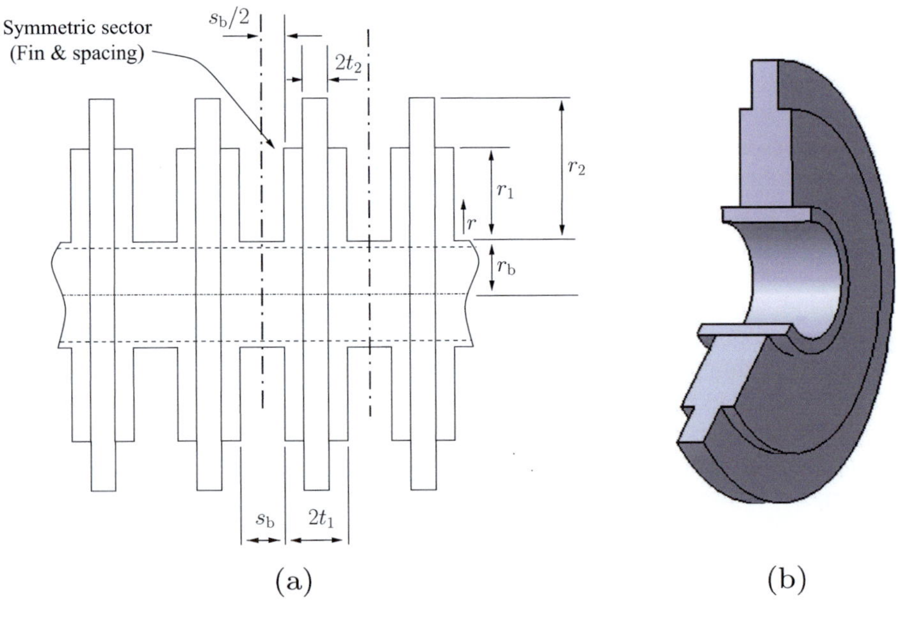

Fig. 21.1 Array of identical annular fins with profile of step change in thickness and equal fin inter-spacing. (**a**) Schematic diagram of the fin array. (**b**) Transverse section of a fin (heat transfer module)

Formulation of the Thermal Model

Since the thermal conductivity (k) of most of the engineering materials varies linearly with temperature (T), the simple relationship expressed by Eq. (21.1) is adopted here, where T_∞ is the ambient temperature and k_a is the thermal conductivity of the fin material at T_∞, β is a predefined parameter for governing the variation in k with T_i (the steps of the fin are denoted by i with $i = 1$ for the first step and $i = 2$ for the second step).

$$k = k_a \{1 + \beta (T_i - T_\infty)\} \; ; \quad i \in \{1, 2\} \tag{21.1}$$

The average natural convective heat transfer coefficient (h) for air flowing through two adjacent fins can be estimated by the correlation expressed by Eq. (21.2) [23], where s_m is the mean fin inter-spacing and Ra is the Rayleigh number in the laminar range of $(5, 10^8)$ with properties evaluated at the film temperature of $\frac{T_b + T_\infty}{2}$.

$$\left.\begin{aligned} h &= \frac{k}{s_m}\left\{c_0 + c_1 Ra^{a_0}\left(\frac{r_2}{r_b}\right)^{a_1} + c_2 Ra^{a_2} + c_3\left(\frac{r_2}{r_b}\right)^{a_3}\right\} \\ \text{where,} \quad Ra &= \frac{g\beta'(T_b - T_\infty)s_m^4}{2\nu\alpha' r_2} \\ c_0 &= -3.827\;; \quad c_1 = 0.047\;; \quad c_2 = 1.039\;; \quad c_3 = 2.548 \\ a_0 &= \;\;0.348\;; \quad a_1 = 0.173\;; \quad a_2 = 0.175\;; \quad a_3 = 0.009 \end{aligned}\right\} \tag{21.2}$$

It is to be mentioned that the correlation given by Eq. (21.2) was proposed by Senapati et al. [23] for uniform thickness annular fin array. Due to the non-availability of a similar correlation for annular stepped fin array, it is adopted here considering s_m as the mean fin inter-spacing.

Having Eqs. (21.1) and (21.2), the steady state energy balance governing equation for an individual fin of the fin array can be expressed by Eq. (21.3).

$$\left.\begin{aligned} &\frac{d}{dr}\left[r\{1 + \beta(T_i - T_\infty)\}\frac{dT_i}{dr}\right] - \frac{hr}{k_a t_i}(T_i - T_\infty) = 0 \\ \text{where,} \quad & r_b \le r \le r_1\;; \quad \text{for} \quad i = 1 \\ & r_1 \le r \le r_2\;; \quad \text{for} \quad i = 2 \end{aligned}\right\} \tag{21.3}$$

In this study, it is considered that the temperature at the base of a fin (i.e., T_b) is constant and heat is transferred from the tip of the fin to the surrounding by natural convection only. Also, there must be a continuity of temperature as well as energy balance at the interface of the two steps of a fin. Hence, Eq. (21.3) will be subjected to the boundary conditions given by Eq. (21.4).

$$T_1 = T_b\;; \qquad \text{at} \quad r = r_b \tag{21.4a}$$

$$-k_a \{1 + \beta (T_2 - T_\infty)\} \frac{dT_2}{dr} = h (T_2 - T_\infty) \; ; \qquad \text{at} \quad r = r_2 \tag{21.4b}$$

$$T_1 = T_2 \; ; \qquad \text{at} \quad r = r_1 \tag{21.4c}$$

$$-t_1 k_a \{1 + \beta (T_1 - T_\infty)\} \frac{dT_1}{dr} = -t_2 k_a \{1 + \beta (T_2 - T_\infty)\} \frac{dT_2}{dr} + h (t_1 - t_2)(T_1 - T_\infty) \; ; \quad \text{at} \quad r = r_1 \tag{21.4d}$$

Non-dimensional Formulation of the Thermal Model

In order to normalize the temperature distribution and fin dimensions shown in Fig. 21.1a, some non-dimensional parameters are defined as given by Eq. (21.5).

$$\left.\begin{array}{llll} R_1 = \frac{r_b}{r_2} & R_2 = \frac{r_1}{r_2} & R = \frac{r}{r_2} & \\ t = \frac{t_2}{t_1} & \xi = \frac{t_1}{r_b} & \mathrm{Bi} = \frac{h r_b}{k_a} & \\ \theta = \frac{T_1 - T_\infty}{T_b - T_\infty} & \phi = \frac{T_2 - T_\infty}{T_b - T_\infty} & \delta = \frac{T_\infty}{T_b - T_\infty} & \alpha = (T_b - T_\infty)\,\beta \end{array}\right\} \tag{21.5}$$

Finally, in terms of the non-dimensional parameters of Eq. (21.5), the thermal model governing Eqs. (21.3) and (21.4) are normalized in dimensionless forms as expressed by Eqs. (21.6) and (21.7).

$$\left.\begin{array}{ll} & (1 + \alpha\kappa)\frac{d^2\kappa}{dR^2} + \left\{\alpha \frac{d\kappa}{dR} + \frac{1}{R}(1 + \alpha\kappa)\right\} \frac{d\kappa}{dR} - Z^2\kappa = 0 \\ \text{where,} & \kappa = \theta \,,\; Z = \frac{Z_0}{R_1} \; ; \quad \text{if} \quad R_1 \le R \le R_2 \\ & \kappa = \phi \,,\; Z = \frac{Z_1}{\sqrt{t}} \; ; \quad \text{if} \quad R_2 \le R \le 1 \\ & Z_0 = \sqrt{\frac{\mathrm{Bi}}{\xi}} \\ & Z_1 = \frac{Z_0}{R_1} \end{array}\right\} \tag{21.6}$$

$$\theta = 1 \; ; \qquad \text{at} \quad R = R_1 \tag{21.7a}$$

$$-(1 + \alpha\phi)\frac{d\phi}{dR} = \frac{\mathrm{Bi}}{R_1}\phi \; ; \qquad \text{at} \quad R = 1 \tag{21.7b}$$

$$\theta = \phi \; ; \qquad \text{at} \quad R = R_2 \tag{21.7c}$$

$$R_1 (1 + \alpha\theta) \frac{d\theta}{dR} = t R_1 (1 + \alpha\phi) \frac{d\phi}{dR} - \mathrm{B}i\,(1 - t)\,\theta \; ; \quad \text{at} \quad R = R_2 \tag{21.7d}$$

Optimization Modeling

As stated in Sect. 21, the performance of the annular fin array having fins of step profile is evaluated in different combination of the four parameters, which are the total heat transfer rate from the fin array ($\dot{Q}$), total fin volume (V), surface efficiency (η_s) and augmentation factor (ϵ_s) of the fin array. The configuration of the fin array as shown in Fig. 21.1a can be defined in terms of five independent parameters, which are the radius of an individual fin at the point of step change in thickness (r_1), outer radius of the fin (r_2), cross-sectional half thickness of the thick (first) step of the fin (t_1), cross-sectional half thickness of the thin (second) step of the fin (t_2) and the total number of fins in the fin array (n_{fin}). Any change in the values of any of these five independent parameters will give rise to a new fin array configuration with new values of the four functions considered for measuring the performance of the array. Hence, the present problem at hand can be formulated as a multi-objective optimization problem as expressed by Eq. (21.8) by treating the five independent parameters as the design variables and the four performance parameters as the objective functions.

$$\left.\begin{array}{rl}
\text{Determine} & \boldsymbol{x} = (r_1, r_2, t_1, t_2, n_{\text{fin}})^{\mathrm{T}} \\
\text{to maximize} & f_1(\boldsymbol{x}) = \dot{Q} \\
\text{minimize} & f_2(\boldsymbol{x}) = V \\
\text{maximize} & f_3(\boldsymbol{x}) = \eta_s \\
\text{maximize} & f_4(\boldsymbol{x}) = \epsilon_s \\
\text{subject to} & g_1(\boldsymbol{x}) \equiv r_1 > r_{\text{b}} \\
 & g_2(\boldsymbol{x}) \equiv r_2 > r_1 \\
 & g_3(\boldsymbol{x}) \equiv t_1 \leqslant \frac{W}{4} \\
 & g_4(\boldsymbol{x}) \equiv t_1 > t_2 \\
 & g_5(\boldsymbol{x}) \equiv \left\lceil \frac{W}{2t_1+s_{\max}} \right\rceil \leqslant n_{\text{fin}} \leqslant \left\lfloor \frac{W}{2t_1+s_{\min}} \right\rfloor \\
 & r_1, r_2, t_1, t_2 \geq 0 \ .
\end{array}\right\} \quad (21.8)$$

In Eq. (21.8), constraints $g_1(\boldsymbol{x})$, $g_2(\boldsymbol{x})$ and $g_4(\boldsymbol{x})$ are related to the geometry of the individual fins, while constraints $g_3(\boldsymbol{x})$ and $g_5(\boldsymbol{x})$ are related to the configuration of the fin array. Constraint $g_1(\boldsymbol{x})$ ensures the existence of the fins by making the radius at step change in thickness (r_1) greater than the predefined radius at the base (r_{b}), and constraint $g_2(\boldsymbol{x})$ ensures the existence of two steps in a fin by making the outer radius (r_2) greater than radius at step change in thickness (r_1) while constraint $g_4(\boldsymbol{x})$ ensures that the inner step of the fin is thicker than its outer step. On the other hand, constraint $g_3(\boldsymbol{x})$ restricts the fin half-thickness at the base (t_1) to such a value that an array of fins can be formed by accommodating at least two fins within the limited predefined length (W) of the primary surface, and constraint $g_5(\boldsymbol{x})$ forms the fin array with the lower limit for two fins and the upper limit avoiding the excess number of fins over the length (W) of the primary surface (($s_{\min}$, $s_{\max}$) is the allowable range of fin inter-spacing at base). The last line in Eq. (21.8) makes the design variable non-negative.

The maximization of the heat transfer rate ($\dot{Q}$), surface efficiency (η_s) and augmentation factor (ϵ_s) would enhance the overall thermal performance of the fin array, while the minimization of the total fin volume (V) of the array will reduce the fin material cost. These objective functions, in terms of the notations and formulations of the thermal model of the fin array presented in Sect. 21, can be expressed by Eq. (21.9).

$$\dot{Q} = n_{\text{fin}} \times \left\{ -kA_{\text{b}} \left. \frac{dT_1}{dr} \right|_{r=r_{\text{b}}} + hA_{\text{sp}} \left(T_{\text{b}} - T_{\infty}\right) \right\} \tag{21.9a}$$

$$V = n_{\text{fin}} \times 2\pi \left\{ t_1 \left(r_1{}^2 - r_{\text{b}}{}^2 \right) + t_2 \left(r_2{}^2 - r_1{}^2 \right) \right\} \tag{21.9b}$$

$$\eta_s = \frac{1}{n_{\text{fin}}} \times \frac{f_1(\boldsymbol{x})}{hA_{\text{s}} \left(T_{\text{b}} - T_{\infty}\right) + hA_{\text{sp}} \left(T_{\text{b}} - T_{\infty}\right)} \tag{21.9c}$$

$$\epsilon_s = \frac{1}{n_{\text{fin}}} \times \frac{f_1(\boldsymbol{x})}{hA_{\text{b}} \left(T_{\text{b}} - T_{\infty}\right) + hA_{\text{sp}} \left(T_{\text{b}} - T_{\infty}\right)} \tag{21.9d}$$

where,

$$\begin{aligned} A_{\text{b}} &= 4\pi r_{\text{b}} t_1 \\ A_s &= 2\pi \left(r_2{}^2 - r_{\text{b}}{}^2 \right) + 4\pi \left\{ r_1 \left(t_1 - t_2 \right) + r_2 t_2 \right\} \\ A_{sp} &= 2\pi r_{\text{b}} s_{\text{b}} \end{aligned} \tag{21.9e}$$

Solution Procedure

The multi-objective optimization model of the fin array design problem, formulated in Eq. (21.8), is solved by using a very popular and widely applied multi-objective genetic algorithm, namely the non dominated sorting genetic algorithm II (NSGA-II) proposed by Deb et al. [3].

Constraints Handling Through Variable Bounds

Though the design of the studied fin array is formulated in Eq. (21.8) as a constrained optimization problem, it can easily be handled as an unconstrained optimization problem.

Since r_{b} (radius of a fin at its base) is a predefined fixed parameter, constraints $g_1(\boldsymbol{x})$ and $g_2(\boldsymbol{x})$ can be made satisfied by generating two values in the range of $(r_{\text{b}}, r_{\text{max}}]$ and then sorting them in ascending order as the values of r_1 (radius at step change in thickness) and r_2 (outer radius of the fin), respectively ($r_{\max}$ is the allowable upper limit of r_1 and r_2). Similarly, constraints $g_3(\boldsymbol{x})$ and $g_4(\boldsymbol{x})$ can be made satisfied by generating two values in the range of $\left[t_{\min}, \frac{W}{4}\right]$ and then sorting them in descending order as the values of t_1 (fin half-thickness at the base) and t_2 (fin half-thickness at the tip), respectively ($t_{\min}$ is the allowable lower limit of t_1 and t_2). Note that the maximum half-thickness of a fin could be $\frac{W}{4}$, but it may be taken to be $t_{\max} \ll \frac{W}{4}$ if the heat flow through the fin is to be treated as one-dimensional.

The process for satisfying constraint $g_5(\boldsymbol{x})$ is slightly different. In this case, the range $[s_{\min}, s_{\max}]$ for the fin inter-spacing at the base is to be so chosen that the air flowing through two consecutive fins would maintain the Rayleigh number in the laminar range of $\left[5, 10^8\right]$. Accordingly, the range $[s_{\min}, s_{\max}]$ may be obtained through a reserve calculation. Since the Rayleigh number (Ra) as expressed in Eq. (21.2) is a function only of the fin mean inter-spacing (s_m) and the fin outer radius (r_2) keeping all other terms constant for a given scenario, s_m can be computed as $s_{\min}$ by replacing r_2 by $r_{\max}$ and Ra by its lower limit of 5. Similarly, s_m can be computed as $s'_{\max}$ (maximum fin inter-spacing at the tip) by replacing r_2 by $r_{\min}$ ($r_{\min}$ is the allowable minimum value of r_2, i.e., $r_{\min} > r_b$) and Ra by its upper limit of 10^8, and then to estimate $s_{\max}$ by deducting $2(t_{\max} - t_{\min})$ from $s'_{\max}$.

Once all the five constraints are made satisfied automatically as above, the equal fin inter-spacing (s_b) at base for all the adjacent pairs of fins can be obtained using Eq. (21.10).

$$s_b = \frac{W}{n_{\text{fin}}} - 2t_1 \tag{21.10}$$

Evaluation of Objective Functions

In every iteration of the employed optimizer, the values of the four objective functions given by Eqs. (21.9a)–(21.9d) will be required, which are to be evaluated numerically. The critical one is the heat transfer rate $\dot{Q}$ given by Eq. (21.9a), which is to be evaluated by solving the problem governing Eq. (21.6) along with its initial and boundary conditions given by Eq. (21.7). For this, the hybrid spline difference method (HSDM) is used, which is a highly accurate numerical method proposed by Wang et al. [26, 27, 28, 29]. This method is based on a discretization scheme as given in Eq. (21.11), where Δr, N, n and p represent the grid size, number of grids, grid index of space, and spline parameter, respectively.

$$\kappa_n = \frac{p_{n-1}^{(i)} + 10p_n^{(i)} + p_{n+1}^{(i)}}{12} \tag{21.11a}$$

$$\kappa'_n = \frac{p_{n+1}^{(i)} - p_{n-1}^{(i)}}{2\Delta r} - \Delta\kappa'_n \tag{21.11b}$$

$$\kappa_n'' = \frac{p_{n-1}^{(i)} - 2p_n^{(i)} + p_{n+1}^{(i)}}{\Delta r^2} \tag{21.11c}$$

$$\text{where,} \quad \Delta\kappa_n' = \begin{cases} \frac{\left(-3\kappa_n''+4\kappa_{n+1}''-\kappa_{n+2}''\right)\Delta r}{24}; & n = 0 \\ \frac{\left(\kappa_{n+1}''-\kappa_{n-1}''\right)\Delta r}{24}; & n = 1, 2, \cdots, N^{(i)} - 1 \\ \frac{\left(3\kappa_n''-4\kappa_{n-1}''+\kappa_{n-2}''\right)\Delta r}{24}; & n = N^{(i)} \end{cases} \tag{21.11d}$$

$$i \in \{1, 2\}$$

The following is the detail procedure for evaluating $\dot{Q}$ through the HSDM scheme expressed by Eq. (21.11):

(a) Discretize Eqs. (21.6) and (21.7) using Eq. (21.11).
(b) Evaluate p_n ($n = 0, 1, \ldots, N$) by solving the discretized forms of Eqs. (21.6) and (21.7), which can be done through the variant of the Thomas algorithm proposed by Martin and Boyd [19].
(c) At all the grid points, evaluate the dimensionless temperature distribution and their derivatives up to the second order (i.e., θ_n, θ_n' and θ_n'') using the values of p_n in Eq. (21.11).
(d) Evaluate the temperature gradient at the base of the fin, i.e., $\left.\frac{dT_1}{dr}\right|_{r=r_b}$, using the dimensionless temperature gradient θ_o' at the base of the fin.
(e) Finally, evaluate the heat transfer rate from the fin array using the value of $\left.\frac{dT}{dr}\right|_{r=r_b}$ in Eq. (21.9a).

Numerical Experimentation and Discussion

An annular fin array with identical fins of rectangular cross-section having a step change in thickness is taken up in the present study. It is assumed that the temperature at the base of the fins is constant, thermal conductivity of the fin material varies linearly with temperature, and heat is dissipated from the fin array by natural convection only.

The operating condition, the thermal properties of the fin material, and the fin array configuration with reference to Fig. 21.1a, considered for numerical experimentation, are listed in Table 21.1, while the opted user-defined algorithmic parameter settings for NSGA-II (the applied optimizer) are given in Table 21.2. With those input parameters, NSGA-II is applied to Eq. (21.8) for studying the problem at hand under two scenarios. In the first scenario, the objective functions f_1–f_4 given by Eq. (21.9) are optimized in different pairs, while all the four objective functions are optimized simultaneously in the second scenario.

Table 21.1 Operating conditions, fin material properties, and fin array geometry

Parameter	Value/range
Ambient temperature, T_∞	300 K
Fin temperature at the base, T_b	373 K
Thermal conductivity of the fin material at T_∞, k_a	186 W/mK
Parameter for variable thermal conductivity, β	$-0.00018\ K^{-1}$
Fin base radius, r_b	2.0 cm
Outer radii of two steps of a fin, $[r_{min}, r_{max}]$ for r_1 and r_2	[2.5–6.0] cm
Half-thickness of two steps of a fin, $[t_{min}, t_{max}]$ for t_1 and t_2	[0.01–0.2] cm
Length of the primary cylinder, W	40.0 cm
Fin inter-spacing at base, $[s_{min}, s_{max}]$	[0.36, 18.0] cm

Table 21.2 User-defined parameter setting in the context of NSGA-II

Parameter	Value/range of value
Population size	100
Number of generations (iterations)	400
Crossover probability	90%
Mutation probability	(0–1)%

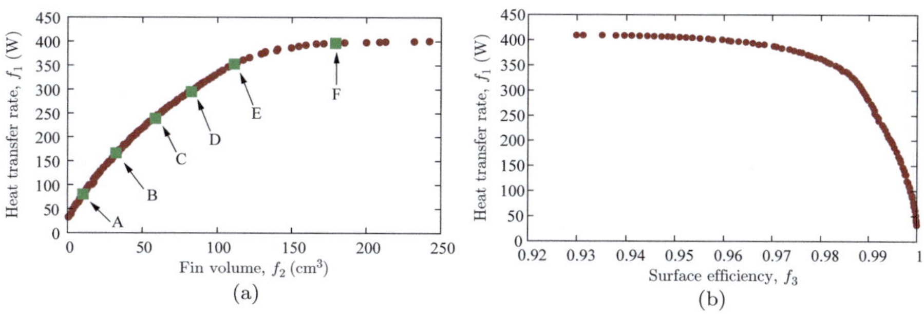

Fig. 21.2 Pareto fronts of f_1 separately paired with f_2 and f_3. (**a**) Pareto front in terms of f_1 and f_2. (**b**) Pareto front in terms of f_1 and f_3

Scenario I

At the first instance, the fin array design problem is studied for maximizing the total heat dissipation rate ($f_1 \equiv \dot{Q}$) from the fin array and simultaneously minimizing the total fin volume ($f_2 \equiv V$) of the array. Figure 21.2a shows the obtained Pareto optimal front containing a set of trade-off solutions in terms of f_1 and f_2, which clearly depicts the conflicting nature between the two objective functions, i.e., one cannot be improved without degrading the other at least by some amount. The efficient fin geometries corresponding to six selective trade-off solutions marked by A–F in Fig. 21.2a are shown in Fig. 21.3, where the variations in the pattern of the individual fins and the total number of fins in the fin array are most noticeable.

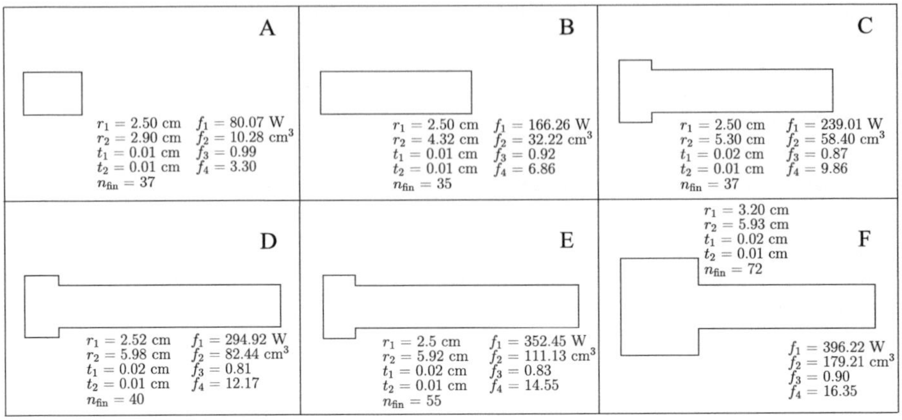

Fig. 21.3 Six selective efficient fin geometries corresponding to trade-off solutions (**A**)–(**F**) of Fig. 21.2a

Apart from the optimum values of the five design variables (r_1, r_2, t_1, t_2 and n_{fin}) and optimized two objective functions (f_1 and f_2), the corresponding values of the surface efficiency ($f_3 \equiv \eta_s$) and augmentation factor ($f_4 \equiv \epsilon_s$) are also computed and shown in Fig. 21.3.

Notice in Fig. 21.3 that the pattern of variation of f_3 with respect to those of f_1 and f_2 is not very clear. Hence, the fin design problem is studied in the second step for maximizing the heat dissipation rate f_1 from the fin array and simultaneously maximizing the surface efficiency f_3 of the fin array. The obtained Pareto front is shown in Fig. 21.2b, where it is seen that f_1 conflicts with f_3 also, i.e., an improvement in f_1 degrades f_3 by some amount and *vice-versa*.

Computing the augmentation factor values of the trade-off solutions of the Pareto front of Fig. 21.2b, they are plotted against the surface efficiency values. The plot shown separately in Fig. 21.4, where the surface efficiency and augmentation factor is found conflicting with each other, i.e., an increase in the augmentation factor decreases the efficiency of utilization of the fin material. Note that the surface efficiency of the fin array is the ratio of the actual heat transfer rate from the fin array to the heat transfer rate when all the surfaces of the fin array is at the base temperature, while the augmentation factor of the fin array is the ratio of the actual heat transfer rate from the fin array to the heat transfer rate from the base surface of the fin array when there are no fins.

Scenario II

In Sect. 21, studying the considered four objectives functions of the problem at hand in pairs, it is observed that their variations are arbitrary leading to no common pattern of the heat transfer rate. Hence, in order to arrive at a general conclusion,

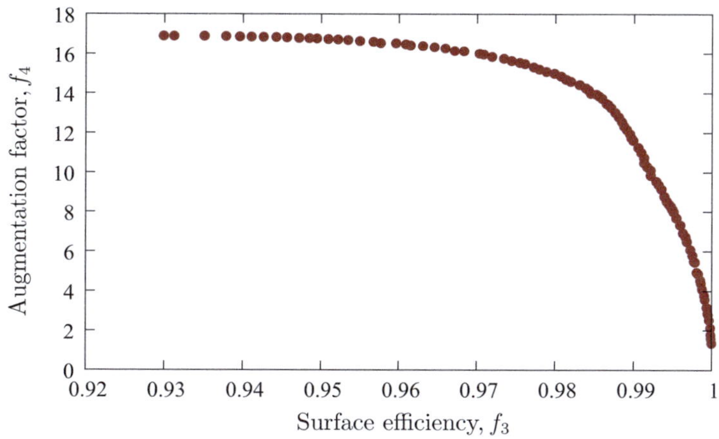

Fig. 21.4 Plot of the surface efficiency (f_3) and augmentation factor (f_4) of the trade-off solutions of Fig. 21.2b

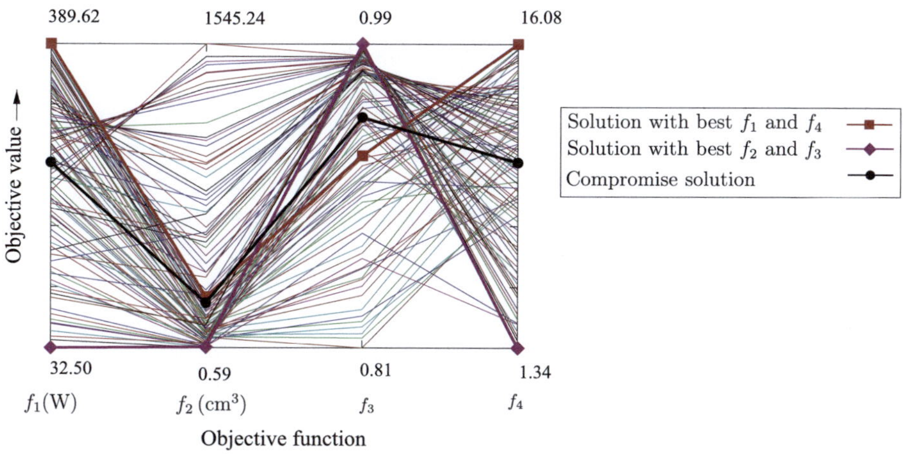

Fig. 21.5 Visualization of the four-dimensional Pareto in the parallel coordinate system

all the four objective functions, i.e., f_1–f_4 given by Eq. (21.9), are optimized here simultaneously. Since it is difficult to analyse a plot of more than two dimensions on a two dimensional-page, the obtained four-dimensional Pareto front is visualized in a parallel coordinate system (refer [30]) as shown in Fig. 21.5, where the objective functions are marked on the horizontal axis and their values corresponding to the trade-off solutions of the Pareto front are plotted on a series of parallel vertical coordinate axes of equal length. Connecting by a crossing line the objective values of each solution of the Pareto front plotted on different vertical axes in Fig. 21.5, the conflicting nature among all the optimized objective functions could be observed clearly. As an example, the solution corresponding to the highest heat transfer rate from the fin array have a moderately low fin volume and a moderately high surface efficiency, while the highest augmentation factor. With such information, it is now up to a designer to adopt suitable solution(s) based upon the availability and accessibility of resources at hand. One such compromise solution is shown in Fig. 21.5 by a thick crossing line.

Pareto Optimal Sensitivity Analysis

In order to study the influence of the design variables on the heat transfer rate, a Pareto optimal sensitivity analysis is performed. For this, an intermediate solution is chosen, whose various values are as follows: $r_1^\dagger = 3.35\,\text{cm}$, $r_2^\dagger = 5.11\,\text{cm}$, $t_1^\dagger = 0.05\,\text{cm}$, $t_2^\dagger = 0.01\,\text{cm}$, $n_{\text{fin}}^\dagger = 20$, $f_1^\dagger = 178.46\,\text{W}$ and $f_2^\dagger = 3.36\,\text{cm}^3$. The problem is solved here five times, each time allowing a design variable to vary $\pm 15\%$ from the chosen value while keeping the other four design variable fixed. Figure 21.6 shows the plots of the deviations of the heat transfer rate from the fin array against the corresponding variations in the design variables, where it is observed that the outer radius (r_2) of the fins has more influence on the heat transfer rate, followed by the number of fins (n_{fin}) in the fin array. On the other hand, the influences of the cross-sectional half-thickness (t_1 and t_2) and the radius of the step change in thickness (r_1) of the fins are comparatively very less. With this information at hand, a designer can adjust the design variables of the fin array in order to achieve the desired heat transfer effect based upon the availability and accessibility of information and resources.

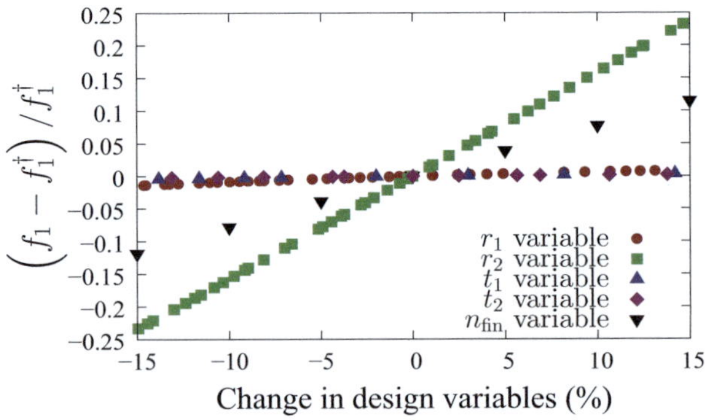

Fig. 21.6 Pareto optimal sensitivity analysis of the heat transfer rate in terms of the design variables

Conclusion

A complete procedure for multi-objective optimization of the design of fin arrays is proposed in this study. Considering the fin array to be consisting of identical fins with rectangular profile and having step change in thickness, steady state one-dimensional heat transfer is considered with heat dissipation from the fin array surface to the surrounding by natural convection only. Taking various geometric parameters of the fin array configuration as the design variables, Pareto fronts (sets of trade-off optimal solutions) are approximated by simultaneously optimizing the heat dissipation rate, fin volume, fin array surface efficiency and fin array augmentation factor. In the computational process, the hybrid spline difference method (HSDM) is used for evaluating heat transfer rate numerically and the nondominated sorting genetic algorithm-II (NSGA-II) is applied for optimizing objective functions. Finally, a comparative analysis on the degree of influences of the design variables on the heat transfer rate is also studied. It is observed from the numerical experimentation that the heat transfer rate from the fin array conflicts with fin volume and surface efficiency of the array, i.e., the heat transfer rate cannot be improved without degrading at least one of the other objectives, i.e., fin volume and surface efficiency by some amount and *vice-versa*. Further, it is also observed that different design variables influence the heat transfer rate differently. Such information can be exploited in designing economic annular stepped fin array with industrial viability.

References

1. Arslanturk C (2004) Performance analysis and optimization of a thermally non-symmetric annular fin. Int Commun Heat Mass Trans 31(8):1143–1153
2. Brown A (1965) Optimum dimensions of uniform annular fins. Int J Heat Mass Trans 8(4):655–662
3. Deb K, Pratap A, Agarwal S, Meyarivan T (2002) A fast and elitist multi-objective genetic algorithm: NSGA-II. IEEE Trans Evol Comput 6(2):182–197
4. Deka A, Datta D (2017) B-spline curve based optimum profile of annular fins using multiobjective genetic algorithm. J Therm Stresses 40:733–746
5. Deka A, Datta D (2017) Geometric size optimization of annular step fin using multi-objective genetic algorithm. J Therm Sci Eng Appl 9:0210131–0210139
6. Kang H-S (2009) Optimization of a rectangular profile annular fin based on fixed fin height. J Mech Sci Technol 23(11):3124–3131
7. Kraus AD, Aziz A, Welty J (2001) Extended surface heat transfer. John Wiley & Sons, New York
8. Kundu B (2009) Analysis of thermal performance and optimization of concentric circular fins under dehumidifying conditions. Int J Heat Mass Trans 52(11–12):2646–2659
9. Kundu B, Barman D (2011) An analytical prediction for performance and optimization of an annular fin assembly of trapezoidal profile under dehumidifying conditions. Energy 36(5):2572–2588
10. Kundu B, Das P (1999) Performance analysis and optimization of eccentric annular disk fins. ASME J Heat Trans 121(1):128–135

11. Kundu B, Das P (2001) Performance analysis and optimization of annular fin with a step change in thickness. ASME J Heat Trans 123(3):601–604
12. Kundu B, Das P (2007) Performance analysis and optimization of elliptic fins circumscribing a circular tube. Int J Heat Mass Trans 50(1–2):173–180
13. Kundu B, Das P (2009) Performance and optimum design analysis of convective fin arrays attached to flat and curved primary surfaces. Int J Refrig 32:430–443
14. Kundu B, Lee K-S (2014) Analytical tools for calculating the maximum heat transfer of annular stepped fins with internal heat generation and radiation effects. Energy 76:733–748
15. Kundu B, Lee K-S, Campo A (2012) Exact and approximate analytic methods to calculate maximum heat flow in annular fin arrays with a rectangular step profile. Int J Thermophys 33:1314–1333
16. Lai C-Y, Kou H-S, Lee J-J (2006) Optimum thermal analysis of annular fin heat sink by adjusting outer radius and fin number. Appl Therm Eng 26:927–936
17. Laor K, Kalman H (1996) Performance and optimum dimensions of different cooling fins with a temperature-dependent heat transfer coefficient. Int J Heat Mass Trans 39(9):1993–2003
18. Leung C, Probert S (1989) Heat exchanger performance: effect of orientation. Appl Energy 33:235–252
19. Martin A, Boyd ID (2010) Variant of the Thomas algorithm for opposite-bordered tridiagonal systems of equations. Int J Numer Method Biomed Eng 26(6):752–759
20. Nagarani N, Mayilsamy K, Murugesan A (2012) Fin effectiveness optimization of elliptical annular fins by genetic algorithm. Procedia Eng 38:2939–2948
21. Nemati H, Samivand S (2015) Performance optimization of annular elliptical fin based on thermo-geometric parameters. Alex Eng J 54(4):1037–1042
22. Pashah S, Moinuddin A, Zubair SM (2016) Thermal performance and optimization of hyperbolic annular fins under dehumidifying operating conditions-analytical and numerical solutions. Int J Refrig 65:42–54
23. Senapati JR, Dash SK, Roy S (2016) Numerical investigation of natural convection heat transfer over annular finned horizontal cylinder. Int J Heat Mass Trans 96:330–345
24. Sharqawy MH, Zubair SM (2007) Efficiency and optimization of an annular fin with combined heat and mass transfer–an analytical solution. Int J Refrig 30(5):751–757
25. Ullmann A, Kalman H (1989) Efficiency and optimized dimensions of annular fins of different cross-section shapes. Int J Heat Mass Trans 32(6):1105–1110
26. Wang C-C, Chen-Hung H, Yang D-J (2011) Hybrid spline difference method for steady-state heat conduction. Numer Heat Trans B 60(6):472–485
27. Wang C-C, Chao L-P, Liao W-J (2012) Hybrid spline difference method (HSDM) for transient heat conduction. Numer Heat Trans B 61(2):129–146
28. Wang C-C, Liao W-J, Hsu Y-S (2012) Hybrid spline difference method for the burgers' equation. Appl Math Comput 219(3):1031–1039
29. Wang C-C, Liao W-J, Yang C-Y (2013) Hybrid spline difference method for heat transfer and thermal stresses in annular fins. Numer Heat Trans B 64(1):71–88
30. Wegman EJ (1990) Hyperdimensional data analysis using parallel coordinates. J Am Stat Assoc 85:664–675
31. Yu L-T, Chen C-K (1999) Optimization of circular fins with variable thermal parameters. J Franklin Inst 336(1):77–95

Chapter 22
Optimal Control of Saltwater Intrusion in Coastal Aquifers Using Analytical Approximation Based on Density Dependent Flow Correction

Selva B. Munusamy and Anirban Dhar

Abstract Pumping well management in coastal aquifers required to account for the saltwater intrusion problem. The prevention saltwater contamination of pumping wells should be considered along with the objective of maximum groundwater withdrawal. Saltwater intrusion constraint can be based on (1) sharp interface model (2) density-dependent transport model. Sharp interface models are preferable in the case of limited computation cost available and density-dependent transport models are preferable for accuracy. The correction factor introduced to account for the density-dependent dispersion by Pool and Carrera (Water Resour Res 47(5):W05506, 2011) vastly improves the sharp interface solution. In this present study, the application of the modified sharp interface solution based on the density-dependent correction factor for the pumping optimization is demonstrated for a regional scale aquifer in Nellore, Andhra Pradesh, India. The proposed optimization model sought to maximize the total pumping and minimize the landward toe intrusion from the sea.

Keywords Coastal aquifer · Analytical solution · Pumping optimization · Density-dependent flow · Multi-objective optimization

Introduction

Coastal aquifers are important for the water demands of humans since they are heavily populated. The higher population density in the coastal region leads to over-exploitation of groundwater for the needs of domestic, agricultural and industrial use. The seawater in the coast is denser compared to freshwater in the coastal

S. B. Munusamy · A. Dhar (✉)
Department of Civil Engineering, Indian Institute of Technology Kharagpur, Kharagpur, West Bengal, India
e-mail: anirban@civil.iitkgp.ac.in

F. Bennis, R. K. Bhattacharjya (eds.), *Nature-Inspired Methods for Metaheuristics Optimization*, Modeling and Optimization in Science and Technologies 16,
https://doi.org/10.1007/978-3-030-26458-1_22

aquifer. The density difference results in the intrusion of denser saltwater in the coastal aquifer. There is a freshwater flux flows towards coast due to the head gradient between inland and coastal sea level. The saltwater intruded due to density gradient is in balance with the freshwater flux. However, the pumping of groundwater makes this balance vulnerable and excessive pumping leads to the further intrusion of saltwater due to negative gradient. Once the coastal aquifer is contaminated with saltwater, the available remedial measures based on engineering solutions are costly most of the time [25]. Thus the planning of pumping well system locations, pumping schedule and pumping rates should consider the possibility of saltwater contamination in the coastal aquifers [5]. Apart from drinking water purpose, coastal aquifer acts as a conduit for the contaminants to reach the sea, which is a habitat for large biodiversity. More details on the saltwater intrusion can be found in Cooper et al. [3], Bear et al. [1], Diersch and Kolditz [11], Simmons [21], and Werner et al. [25]. Optimization techniques are widely used for the pumping well planning and management [2, 4, 9, 16, 18, 20, 22].

The saltwater intrusion problems are solved either by considering sharp interface (Fig. 22.1) between freshwater and saltwater or diffused mixing region with the concentration of the mixing zone varies from freshwater to saltwater. However, sharp interface assumption overestimates the toe location of the saltwater interface [6]. Sharp interface approach is more preferable in computational costs point of view, whereas diffused interface method is better for the accurate mapping of the field aquifer. However, sharp interface approach can be applied for the coastal aquifers with narrow mixing zones [12]. Pool and Carrera [19] demonstrated that the sharp interface approximation with a non-dimensional correction factor $[1 - (\alpha_T/b)^{1/6}]$ with transverse dispersivity α_T and aquifer thickness b can fairly

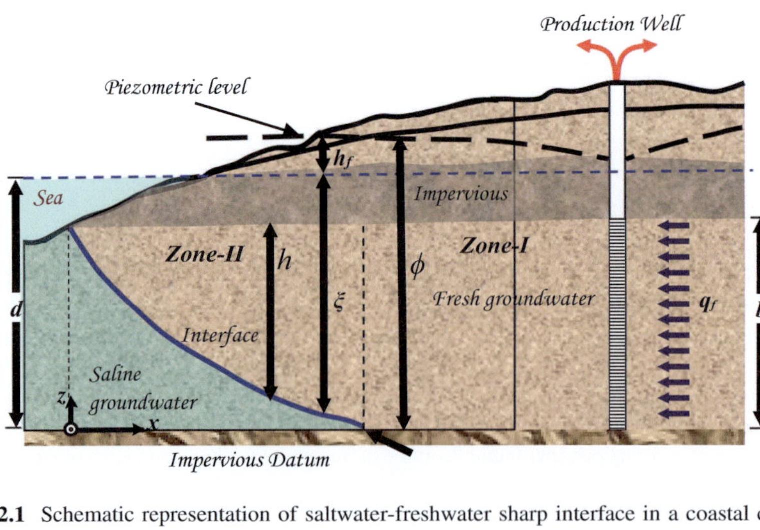

Fig. 22.1 Schematic representation of saltwater-freshwater sharp interface in a coastal confined aquifer [10]

estimate the density dependent flow equivalent interface location for a confined aquifer under pumping from a fully penetrated pumping well. The non-dimensional correction factor is obtained by minimizing the difference between Strack [23] analytical solution and density dependent numerical solution. The sharp interface is assumed to be equivalent to the 0.1% isochlor of saltwater contours corresponding to the dispersive mixing zone numerical solution. Pool and Carrera [19] study was based on constant recharge inland boundary condition, which is applicable for the infinitely large model domain [13]. The correction factor can be applied to modify the classical Ghyben-Herzberg sharp interface relationship based on the buoyancy factor. The Pool and Carrera correction factor is applicable for steady state, flux-controlled system. Lu et al. [13] analyzed the applicability of the correction factor given by Pool and Carrera [19] for the constant head inland boundary condition and concluded that the correction factor is applicable. The constant head boundary condition assumes sufficiently large model domain instead of an infinitely large model domain. Lu et al. [13] found that the difference between the maximum pumping rate corresponding to constant flux inland boundary and the constant head is less than 2.5%, if the well is located at a distance less than one-fifth of the domain length from the coastal boundary. Lu and Werner [14] demonstrated that for a head controlled system without pumping wells, $[1 - (\alpha_T/b)^{1/4}]$ can be used as a correction factor corresponding to isochlor of 0.1 saltwater concentration contour. Koussis et al. [15] proposed a composite correction by combining correction for freshwater outflow gap and correction for transverse dispersivity effects [19]. Outflow gap correction is derived to consider the freshwater outflow region at the coastal boundary (outcrop). Werner [26] used $(\alpha_T/b)^{1/4}$ to derive correction for the freshwater head of a river in a freshwater lens set-up in a saline aquifer. The correction factor is used to improve the freshwater-saltwater interface in freshwater lens. Freshwater lenses exist in saline aquifers near freshwater river, where the saline water is flowing towards the river.

The objective of the present work is to find out optimal pumping well management solution for multiple well system in a coastal aquifer. The multi-objective optimization approach uses the sharp interface solution modified to account for density-dependent correction [19] to prevent saltwater intrusion. The proposed sharp interface based optimization model reduces computational time compared to density-dependent numerical model. The multi-objective optimization is demonstrated for field conditions of coastal aquifer in Nellore, Andhra Pradesh, India.

Strack's Analytical Solution for Saltwater Intrusion

Strack [23] derived analytical solution for salt water intrusion in coastal aquifers under pumping well conditions. The derivation is based on Ghyben-Herzberg solution by using Strack's single potential. A semi-infinite confined aquifer with homogeneous and isotropic hydraulic conductivity is considered. Initially, the derivation of the Strack's single potential is explained. The vertical component of the velocity of flow is considered to be negligible as per Dupuit-Forchheimer assumption. The thickness of the aquifer is considered as linear function of the head. The average velocity vector components for the above assumptions can be obtained by Darcy law as following,

$$\begin{aligned} q_x &= -K\frac{\partial \phi}{\partial x} \\ q_y &= -K\frac{\partial \phi}{\partial y} \\ q_z &= 0 \end{aligned} \tag{22.1}$$

where K is hydraulic conductivity of the aquifer, x and y are the longitudinal and lateral directions in the horizontal plan, z is the vertical direction, q_x, q_y, and q_z are Darcy velocity (specific discharge) components in x, y, and z directions respectively. The components of discharge per unit distance can be written by multiplying the Darcy velocity with cross-sectional area. Discharge components can be given as,

$$\begin{aligned} Q_x &= -Kh\frac{\partial \phi}{\partial x} \\ Q_y &= -Kh\frac{\partial \phi}{\partial y} \\ Q_z &= 0 \end{aligned} \tag{22.2}$$

The aquifer thickness can be given as following by considering it as a linear function of head,

$$h = c_1\,\phi + c_2 \tag{22.3}$$

where c_1 and c_2 are constants. For unconfined aquifer if the potential is measured by considering aquifer impervious bottom as a datum, the thickness of the flow region is same as the potential head. Thus, $c_1 = 1$ and $c_1 = 0$ for flow in unconfined aquifer. For confined aquifer, the thickness of the flow region is equal to the distance between confining layers, which is a constant. Thus, $c_1 = 0$ and $c_2 = b$ for flow in unconfined aquifer, where b is the thickness between confining layers. By substituting the relationship for thickness of the aquifer Eq. (22.3), in the expression for discharge Eq. (22.2) and simplifying as following,

For case $c_1 \neq 0$,

$$\begin{aligned} Q_x &= -\frac{\partial}{\partial x}\left[\frac{1}{2}K\,c_1\left(\phi + \frac{c_2}{c_1}\right)^2\right] \\ Q_y &= -\frac{\partial}{\partial y}\left[\frac{1}{2}K\,c_1\left(\phi + \frac{c_2}{c_1}\right)^2\right] \end{aligned} \tag{22.4}$$

For case $c_1 = 0$,

$$Q_x = -\frac{\partial}{\partial x}(K\ c_2\ \phi)$$
$$Q_y = -\frac{\partial}{\partial y}(K\ c_2\ \phi) \tag{22.5}$$

The vertical components and unconfined aquifer cases are omitted since the problem considered Dupuit-Forchheimer assumptions and confined aquifer. The terms inside the differentials can be considered as potentials. In same aquifer there can be different zones of flow occurs, for example, the flow in saltwater zone, freshwater zone in saltwater intrusion problem. Strack introduced a constant to make the potential to be single valued function for all zones of flow in an aquifer. Thus, the Strack's single-valued potential for different cases can be written as,

For case $c_1 \neq 0$,

$$\Phi = \frac{1}{2}K\ c_1\left(\phi + \frac{c_2}{c_1}\right)^2 + C \tag{22.6}$$

For case $c_1 = 0$,

$$\Phi = K\ c_2\ \phi + C \tag{22.5}$$

By substituting Eq. (22.5) and Eq. (22.6) in Eq. (22.4), the components of discharge vector transform to following expressions,

$$Q_x = -\frac{\partial \Phi}{\partial x}$$
$$Q_y = -\frac{\partial \Phi}{\partial y} \tag{22.7}$$

The continuity equation which represents the conservation of mass equation for the groundwater equation can be written as,

$$\frac{\partial Q_x}{\partial x} + \frac{\partial Q_y}{\partial y} = N \tag{22.8}$$

where, N is the influx to the confined aquifer which can be recharge due to rainfall or any other source/sink term. By substituting Q_x and Q_y in terms of Strack's potential, continuity equation becomes,

$$\frac{\partial^2 \Phi}{\partial x^2} + \frac{\partial^2 \Phi}{\partial y^2} = -N \tag{22.9}$$

In coastal aquifers saltwater intrusion phenomenon happens due to the density difference between the denser saltwater from the seaside and freshwater in the aquifer. The saltwater and freshwater are in delicate balance, which is affected by the excessive pumping for water usage. In field due to the mixing between saltwater and freshwater a zone of diffusion or mixing zone forms. The salt concentration in the zone of diffusion changes from salt water concentration (ρ_s) which is maximum, to freshwater concentration (ρ_f) which is almost zero. To derive the analytical solution, the problem is simplified to certain assumptions as following: (1) The width of mixing zone or zone of diffusion is negligible. Thus, it is considered that the freshwater and saltwater separated by a sharp interface. (2) The saltwater flow inside saltwater zone is negligible when compared with the freshwater flow. Thus, saltwater is static. (3) Dupuit-Forchheimer assumption is valid. That is the vertical gradients of potential are negligible. (4) The coastal confined aquifer confining layers (both top and bottom) are horizontal. Thus the thickness of the aquifer (b) is uniform for full extent. (5) The extent of the confined aquifer is semi-infinite (6) The mean sea level above aquifer bottom is d (7) The pumping well is penetrated to the full depth of the aquifer (8) Ghyben-Herzberg theory is applicable. Ghyben-Herzberg theory is used to get the relation between the head above the mean sea level (h_f) to the depth of freshwater-saltwater interface from the mean sea level (ξ). The relationship can be obtained by equating the expressions for pressure at either side of the interface as following,

$$\begin{aligned} \rho_f g h_f + \rho_f g \xi &= \rho_s g \xi \\ \xi = \left(\frac{\rho_f}{\rho_s - \rho_f}\right) h_f &= \frac{h_f}{\varepsilon} \end{aligned} \tag{22.10}$$

where ε is buoyancy factor defined as $(\rho_s - \rho_f)/\rho_f$. The flow in the confined aquifer can be divided into the two zones: (1) Zone 1 in which only freshwater exists. The thickness of flow region is equal to the thickness of the confined aquifer (b), (2) Zone 2 in which the freshwater-saltwater interface exists. The thickness of the flow region ($\xi + h_f$) is above the saltwater interface since the flow in saltwater region is neglected. The piezometric head in zone 2 if aquifer bottom impermeable layer is considered as datum can be expressed as,

$$\phi = d + h_f \tag{22.11}$$

Thickness of the flow region can be expressed as,

$$h = \xi - (d - b) \tag{22.12}$$

By substituting Eq. (22.10) and Eq. (22.11) in Eq. (22.12), the thickness of the flow regime can be written as,

$$\begin{aligned} h &= \left(\frac{\rho_f}{\rho_s-\rho_f}\right)\phi - \left(\frac{\rho_s}{\rho_s-\rho_f}\right)d + b \\ &= \frac{\phi}{\varepsilon} - \left(\frac{1+\varepsilon}{\varepsilon}\right)d + b \end{aligned} \tag{22.13}$$

By substituting the expression for thickness of flow region as a linear function of piezometric head potential Eq. (22.2) into the Strack potential for confined aquifer Eq. (22.4), the equation for Strack potential in zone-2 can be obtained as,

$$\begin{aligned} \Phi &= \frac{1}{2}K\left(\frac{\rho_f}{\rho_s-\rho_f}\right)\left[\phi - d\frac{\rho_s}{\rho_f} + b\frac{\rho_s-\rho_f}{\rho_f}\right]^2 + C_{zone-2} \\ &= \frac{1}{2}K\varepsilon[\xi - (d-b)]^2 + C_{zone-2} \end{aligned} \tag{22.14}$$

In the confined aquifer flow part zone-1, the thickness of the flow region is constant,

$$h = b \tag{22.15}$$

By substituting Eq. (22.15) in the Strack's potential Eq. (22.5) for $c_1 = 0$,

$$\Phi = Kb\phi + C_{zone-1} \tag{22.16}$$

The potential of zone-1 and zone-2 should be equal at the point of tip of the saltwater interface to make the Strack potential be continuous for the full extent of aquifer. By enforcing this condition by equating Eq. (22.14) and Eq. (22.15),

$$C_{zone-1} - C_{zone-2} = \frac{1}{2}K\varepsilon[\xi - (d-b)]^2 - Kb\phi \tag{22.17}$$

At the tip of the interface, the thickness of the flow region in zone-2 h is equal to b. By using this, Eq. (22.13) reduces to,

$$\phi_{toe} = \frac{\rho_s}{\rho_f}d = (1+\varepsilon)\,d \tag{22.18}$$

By assuming $C_{zone-2} = 0$ and substituting Eq. (22.11) and Eq. (22.18) in Eq. (22.17),

$$C_{zone-1} = \frac{1}{2}K\varepsilon[\xi - (d-b)]^2 - Kb\,(1+\varepsilon)\,d \tag{22.19}$$

The Strack potential for both confined flow (zone-1) and confined-interface flow (zone-2) can be written as,

$$\Phi = \begin{cases} K\, b\, \varepsilon\, \left[\xi - \left(d - \frac{b}{2}\right)\right] \text{ for zone 1} \\ \frac{1}{2} K\, \varepsilon [\xi - (d - b)]^2 \text{ for zone 2} \end{cases} \tag{22.20}$$

To get the Strack potential at the toe of the location (Φ_{toe}), substituting $\xi = d$ in any one of the expressions in Eq. (22.20),

$$\Phi_{toe} = \frac{1}{2} K \varepsilon\, b^2 \tag{22.21}$$

The problem of saltwater intrusion in coastal aquifer in terms of Strack potential in Eq. (22.9) should be solved using coastal head boundary condition. Strack potential along the coastal boundary can be obtained by using $h = 0$ in Eq. (22.13) and Eq. (22.14),

$$\Phi = 0, x = 0, -\infty < y < \infty \tag{22.22}$$

The partial differential Eq. (22.9) is with harmonic routes. The solution can be following if the freshwater towards sea is q_f and the well with discharge rate Q is located at (x_w, y_w),

$$\Phi = q_f\ x + \frac{Q}{4\pi} ln \left[\frac{(x - x_w)^2 + (y - y_w)^2}{(x + x_w)^2 + (y - y_w)^2} \right] \tag{22.23}$$

In the above equation, the value of Φ_{toe} can be used in place of Φ to find out the toe location (x_{toe}, y_{toe}),

$$\Phi_{toe} = q_f\ x_{toe} + \frac{Q}{4\pi} ln \left[\frac{(x_{toe} - x_w)^2 + (y_{toe} - y_w)^2}{(x_{toe} + x_w)^2 + (y_{toe} - y_w)^2} \right] \tag{22.24}$$

For multiple wells, the toe location can be derived using superposition of the previous equation [2],

$$\Phi_{toe} = q_f\ x_{toe} + \sum_{i=1}^{n_w} \frac{Q_i}{4\pi} ln \left[\frac{(x_{toe} - x_i)^2 + (y_{toe} - y_i)^2}{(x_{toe} + x_i)^2 + (y_{toe} - y_i)^2} \right] \tag{22.25}$$

where n_w is the total number of wells, Q_i is the pumping rate of i-th well, (x_i, y_i) is the coordinate of i-th well.

Modified Ghyben-Herzberg Theory Based Analytical Solution of Saltwater Intrusion

Strack [23] solution is based on the sharp interface assumption. Strack solution is relatively accurate when the width of the mixing zone is narrow. Strack sharp interface solution does not consider the flow in the saltwater wedge and mixing between freshwater and saltwater. The sharp interface approach overestimates the toe location of the interface compared to dispersive interface approach [6, 8, 24]. Pool and Carrera [19] analyzed the error introduced in the saltwater intrusion under pumping well conditions by sharp interface assumption in confined aquifers by comparing the Strack's sharp interface solution with numerical solution of coupled flow and transport equation. Pool and Carrera [19] defined the critical pumping rate as maximum pumping rate for which the saltwater mixing with freshwater is less than 0.1% of maximum salt concentration (seawater concentration) at the pumping well. By analyzing the non-dimensional parameters, Pool and Carrera [19] found that the ratio of transverse dispersivity (α_T) to the thickness of the confined aquifer dictates the diffused interface flow mechanism. Based on regression analysis, Pool and Carrera [19] introduced a correction factor for the sharp interface analytical solution. Pool and Carrera correction for the Ghyben-Herzberg sharp interface relation for the depth of the interface from the mean sea level is given as $\xi = h_f/(\varepsilon\,\chi_{PC})$, where χ_{PC} is the Pool-Carrera correction factor. Pool-Carrera correction factor (χ_{PC}) is defined as,

$$\chi_{PC} = \left[1 - \left(\frac{\alpha_T}{b}\right)^{1/6}\right] \tag{22.26}$$

The toe location corresponding to Pool-Carrera correction model can be considered as tip location (x_{tip}, y_{tip}). The equation for tip location based on Strack's solution can be written as,

$$\frac{1}{2}K\,\varepsilon\,b^2\,\chi_{PC} = q_f\,x_{tip} + \sum_{i=1}^{n_w}\frac{Q_i}{4\pi}ln\left[\frac{\left(x_{tip}-x_i\right)^2 + \left(y_{tip}-y_i\right)^2}{\left(x_{tip}+x_i\right)^2 + \left(y_{tip}-y_i\right)^2}\right] \tag{22.27}$$

Optimization Formulation and Application

The objective of the optimization is to get the maximum total discharge from the number of wells in the field without contaminating any well with saltwater. So, one objective function is the maximization of the sum of pumping rates of all wells,

$$f_1 \equiv Maximization \sum_{i=1}^{n_w} Q_i \tag{22.28}$$

The second objective is that the saltwater front should not reach any of the pumping well. This can be represented as minimization of the maximam absolute difference of the x coordinate between the tip location of the interface for each well and seaside boundary.

$$f_1 \equiv Minimization \underset{\forall i}{\text{Max}} \left| x_i^t - x_{ref} \right| \tag{22.29}$$

where x_i^t is the location of the toe corresponding the i-th well and x_{ref} represents the reference location which is the shore location. The Strack's analytical solution with Pool and Carrera [19] correction factor for the tip location will act as a binding constraint,

$$\begin{aligned} &\tfrac{1}{2} K \varepsilon b^2 \chi_{PC} \\ &= q_f x_j^t + \sum_{i=1}^{n_w} \tfrac{Q_i}{4\pi} ln \left[\frac{\left(x_j^t - x_i\right)^2 + \left(y_j^t - y_i\right)^2}{\left(x_j^t + x_i\right)^2 + \left(y_j^t - y_i\right)^2} \right], \forall j \in \{1, 2, \ldots, n_w\} \end{aligned} \tag{22.30}$$

Moreover, upper and lower bounds of pumping values are specified to as constraints to make certain that the pumping values remain in technically feasible range.

$$Q_i^L \leq Q_i \leq Q_i^U, \forall j \in \{1, 2, \ldots, n_w\} \tag{22.31}$$

where Q_i^U and Q_i^L are the upper and lower limits of discharge rates from the wells.

The location of tip due to the effects of pumping wells can be obtained by solving the Eq. (22.30). The optimization problem can be solved by using nondominated sorting genetic Algorithm-II (NSGA-II) and evolutionary multi-objective optimization algorithm (EMO). NSGA-II algorithm for multi-objective optimization was developed by Deb [7]. NSGA-II can even handle discontinuous Pareto front for highly nonlinear problems. NSGA-II uses random population generation strategy to generate the initial population [9]. The generated spatial pumping value sets are utilized in the analytical solution to obtain the corresponding tip location. Then, the tip locations obtained from the analytical solutions are used by the NSGA-II algorithm to calculate the second objective function for the populated pumping values. The procedure is repeated till the iterations reach the specified termination criterion. The stepwise description of the algorithm is presented below:

Step 1: Generate initial population $(t = 0)P_t$ of size N using Latin Hypercube Sampling Strategy [17].
Step 2: Evaluate all objective functions and constraints.
Step 3: Classify initial population P_t into F fronts (1 is the best level) and calculate crowding distance.
Step 4: Perform tournament selection using population of size N.
Randomly pick a pair of population member.
Select the member which is having better rank, if the members are having equal rank then select the member having higher value of crowding distance.
Copy this member to O_t.
Repeat till O_t has N members.
Step 5: Perform crossover on O_t.
Step 6: Perform mutation on O_t.
Step 7: Evaluate all objective functions and constraints.
Step 8: Combine parent and offspring populations $(R_t = P_t \cup O_t)$
Step 9: Classify combined population P_t into F fronts (1 is the best level) and calculate crowding distance.
Step 10: Set $P_{t+1} = \Psi$, null set and $i = 1$.
Step 11: Perform $P_{t+1} = P_{t+1} \cup F_i$, till $|P_{t+1}| + |F_i| < N, i = i + 1$
Step 12: Choose $N - |P_{t+1}|$ widely spread solutions $F_i^{'} \subset F_i$,
$P_{t+1} = P_{t+1} \cup F_i^{'}$ based on crowding distance.
Step 13: Stop if termination criteria are satisfied, otherwise $t = t + 1$, go to Step 4.

The proposed methodology for saltwater intrusion problem is applied to the coastal aquifer in Nellore district (Fig. 22.2) of Andhra Pradesh, India. The costal aquifer is in the Penna river delta which is along the Bay of Bengal. The location of the aquifer extent is: Latitude: 14°35′ 24″ – 14°48′ 36″, Longitude: 79°57′ 00″-80°10′ 12″. Two following Mandals (administrative units) of the Nellore district considered for the model: Allur and Vidavalur, which have 15 villages (197 square km) and 10 villages (158 square km) in their divisions respectively. The population of Allur is 52,990 and the same for Vidavalur is 46,793. The pumping wells (55 pumping wells) are used extensively for the needs of agriculture, aquaculture, and domestic water utilization. Due to the excessive pumping, the groundwater levels are already below the mean sea level. The reliance on aquaculture for the revenue led to excessive water use, but still, the proliferation of the aquaculture has not stopped. The cropping pattern which is heavily dependent on water intensive paddy as a major crop is also puts additional pressure to water requirements. The deficient amount of rainfall (Allur: 113.3 cm/year and Vidavalur: 114.1 cm/year) is not sufficient for the requirement of aquifer recharge. The optimization algorithm NSGA-II is run for 500 generations, population size 24, lower limit of discharge $Q_i^L = 0\ \text{m}^3/\text{d}$, and upper limit of discharge $Q_i^U = 300\ \text{m}^3/\text{d}$. The parameter values

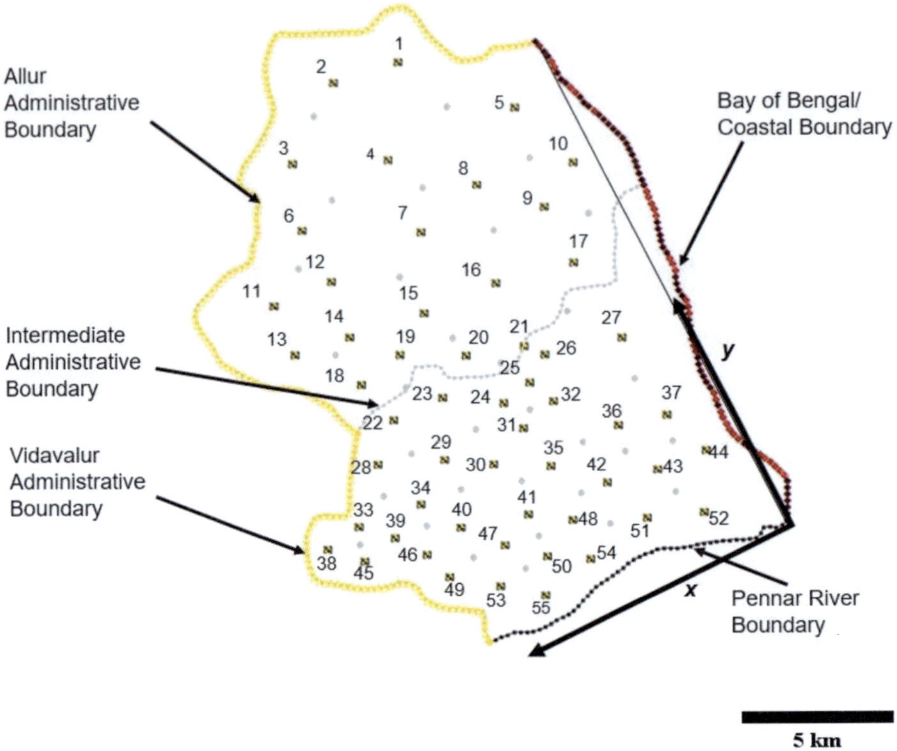

Fig. 22.2 Map representing the well locations in the study area: Allur and Vidavalur Mandals, Nellore District, Andhra Pradesh, India [10]

Table 22.1 Parameter values used for pumping optimization of Nellore aquifer

Parameter	Value
Freshwater flux from inland, q_f	$1.0\,m^2/d$, $0.5\,m^2/d$
Hydraulic conductivity, K	10 m/d, 20 m/d, 30 m/d, 100 m/d
Transverse dispersivity, α_t	1 m
Aquifer thickness, b	30 m
Freshwater density, ρ_f	$1000\,kg/m^3$
Saltwater density, ρ_s	$1025\,kg/m^3$

used for the optimization are given in Table 22.1. The Figs. 22.3, 22.4, 22.5, 22.6, 22.7, 22.8, 22.9 and 22.10 shows the comparison of optimization plots between the objective functions f_1 and f_2 for sharp interface without density correction ($\chi_{PC} = 1.00$) and with density correction ($\chi_{PC} = 0.43$). The optimization plots compared for four different hydraulic conductivity K values of 10 m/d, 20 m/d, 30 m/d, and 100 m/d, and two different freshwater flux q_f values of $1.00\,m^2/d$, and $0.50\,m^2/d$. The comparison plot for $K = 10\,m/d$, $q_f = 1.00\,m^2/d$ (Fig. 22.3) shows

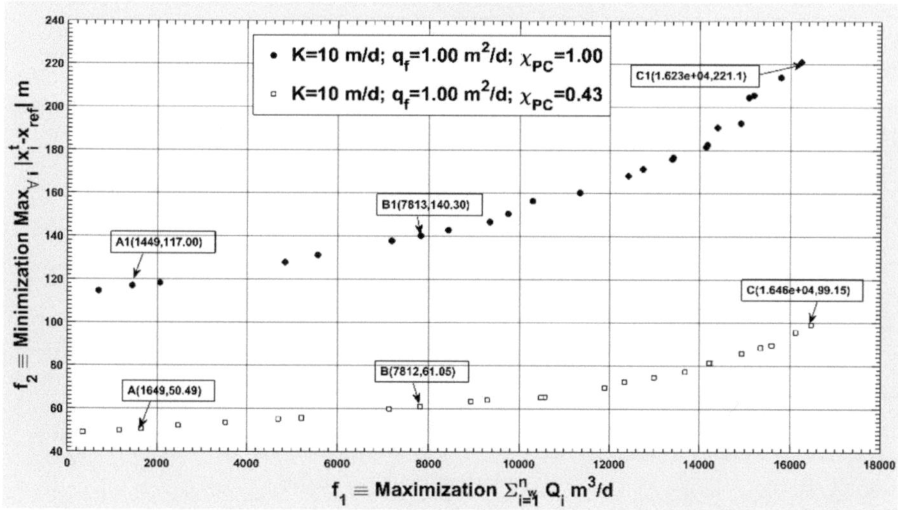

Fig. 22.3 Plot between the between the two objective functions for different iterations of NSGA-II algorithm $q_f = 1\,\text{m}^2/\text{d}$, $K = 10\,\text{m/d}$, $\alpha_t = 1\,\text{m}$, $b = 30\,\text{m}$, $\rho_f = 1000\,\text{kg/m}^3$, $\rho_s = 1025\,\text{kg/m}^3$

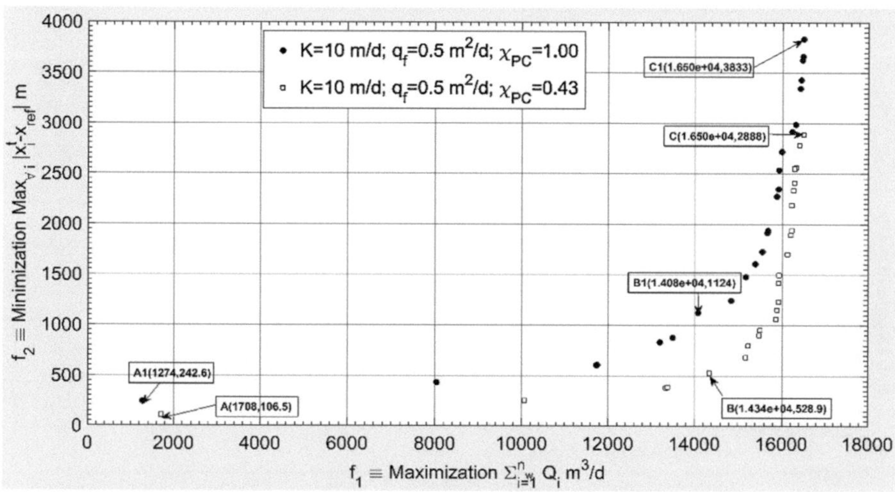

Fig. 22.4 Plot between the between the two objective functions for different iterations of NSGA-II algorithm $q_f = 0.5\,\text{m}^2/\text{d}$, $K = 10\,\text{m/d}$, $\alpha_t = 1\,\text{m}$, $b = 30\,\text{m}$, $\rho_f = 1000\,\text{kg/m}^3$, $\rho_s = 1025\,\text{kg/m}^3$

that the saltwater intrusion is more inland for the case without correction when compared to the density correction based on transverse dispersivity. Three points on the plot from lower pumping rate, medium pumping rate and higher pumping rates are highlighted in Fig. 22.3 (for all other comparison plots also). The lower pumping region of the optimization plot with correction $A(1649\,\text{m}^3/\text{d}, 50.49\,m)$, for

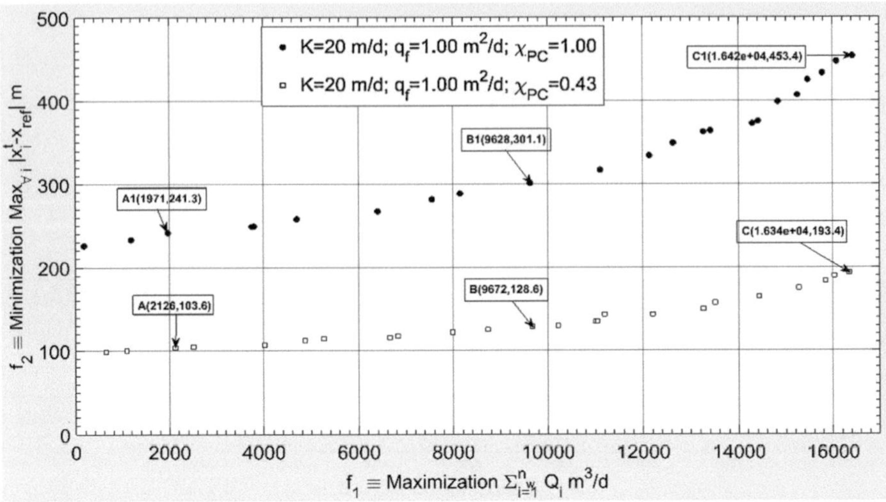

Fig. 22.5 Plot between the between the two objective functions for different iterations of NSGA-II algorithm $q_f = 1.0\,\text{m}^2/\text{d}$, $K = 20\,\text{m/d}$, $\alpha_t = 1\,\text{m}$, $b = 30\,\text{m}$, $\rho_f = 1000\,\text{kg/m}^3$, $\rho_s = 1025\,\text{kg/m}^3$

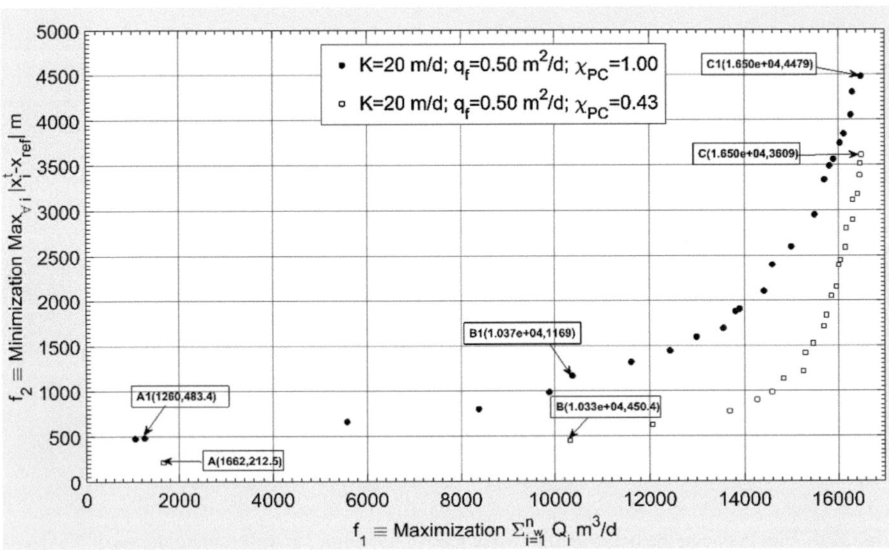

Fig. 22.6 Plot between the between the two objective functions for different iterations of NSGA-II algorithm $q_f = 0.5\,\text{m}^2/\text{d}$, $K = 20\,\text{m/d}$, $\alpha_t = 1\,\text{m}$, $b = 30\,\text{m}$, $\rho_f = 1000\,\text{kg/m}^3$, $\rho_s = 1025\,\text{kg/m}^3$

the total pumping rate of $1649\,\text{m}^3/\text{d}$ the saltwater tip intrudes to 50.49 m. However, for the saltwater interface without correction the intrusion is higher ($117.0\,m$) for the lower pumping rate of $1449\,\text{m}^3/\text{d}$ [point $A1(1449\,\text{m}^3/\text{d}, 117.0\,m)$ of Fig. 22.3]. The trend is similar for comparison of points B-B1and C-C1. The saltwater tip intrusion

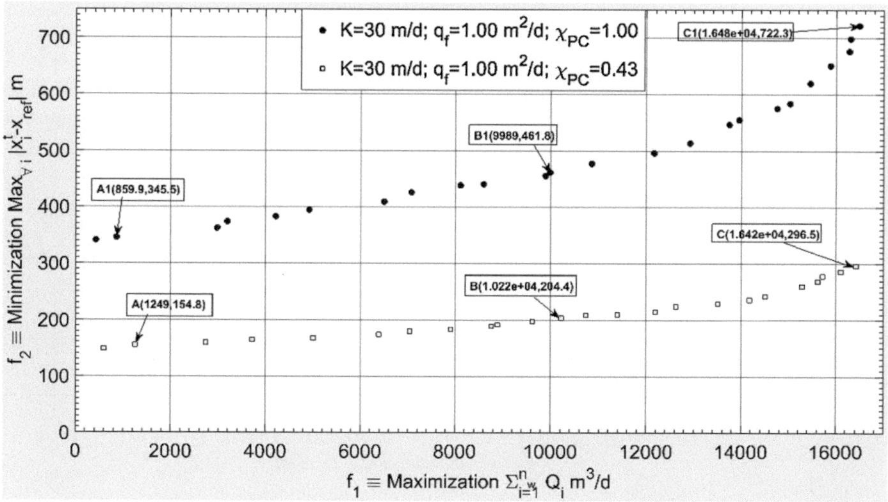

Fig. 22.7 Plot between the between the two objective functions for different iterations of NSGA-II algorithm $q_f = 1.0\,\mathrm{m^2/d}$, $K = 30\,\mathrm{m/d}$, $\alpha_t = 1\,\mathrm{m}$, $b = 30\,\mathrm{m}$, $\rho_f = 1000\,\mathrm{kg/m^3}$, $\rho_s = 1025\,\mathrm{kg/m^3}$

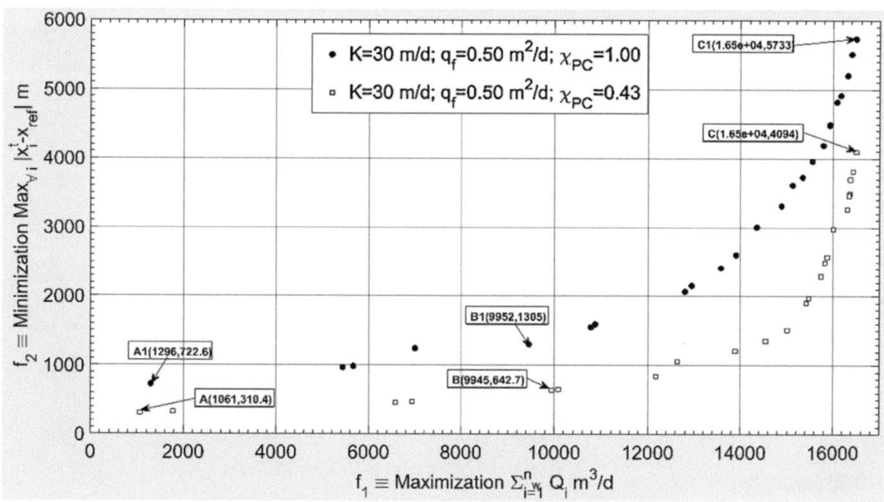

Fig. 22.8 Plot between the between the two objective functions for different iterations of NSGA-II algorithm $q_f = 0.5\,\mathrm{m^2/d}$,$K = 30\,\mathrm{m/d}$, $\alpha_t = 1\,\mathrm{m}$, $b = 30\,\mathrm{m}$, $\rho_f = 1000\,\mathrm{kg/m^3}$, $\rho_s = 1025\,\mathrm{kg/m^3}$

is doubled for the halving of the freshwater flux to $q_f = 0.50\,m^2/d$ in the lower pumping region [Fig. 22.4, $A(1708\,\mathrm{m^3/d}, 106.5\,m)$ and $A1(1274\,\mathrm{m^3/d}, 242.6\,m)$]. However, with the increasing total pumping rate, the intrusion corresponding to $q_f = 0.50\,m^2/d$ is rapid compared to $q_f = 1.00\,m^2/d$. The intrusion for highest total

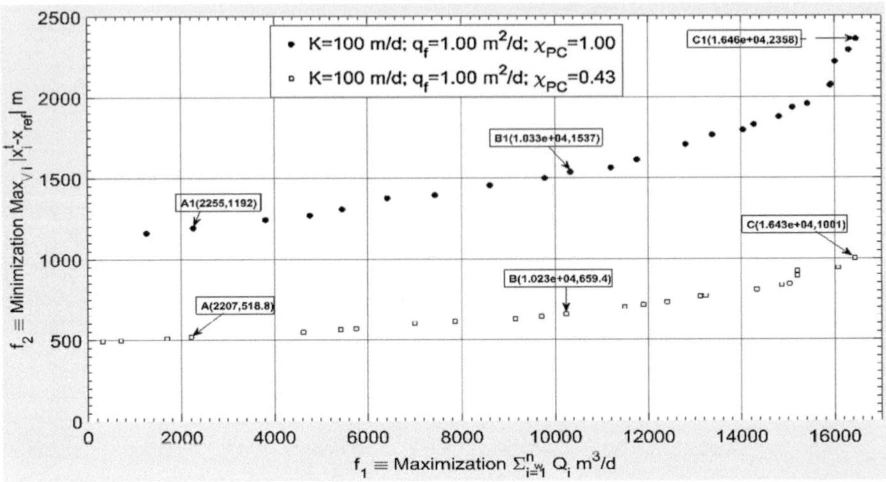

Fig. 22.9 Plot between the between the two objective functions for different iterations of NSGA-II algorithm $q_f = 1.0\,\mathrm{m^2/d}$, $K = 100\,\mathrm{m/d}$, $\alpha_t = 1\,\mathrm{m}$, $b = 30\,\mathrm{m}$, $\rho_f = 1000\,\mathrm{kg/m^3}$, $\rho_s = 1025\,\mathrm{kg/m^3}$

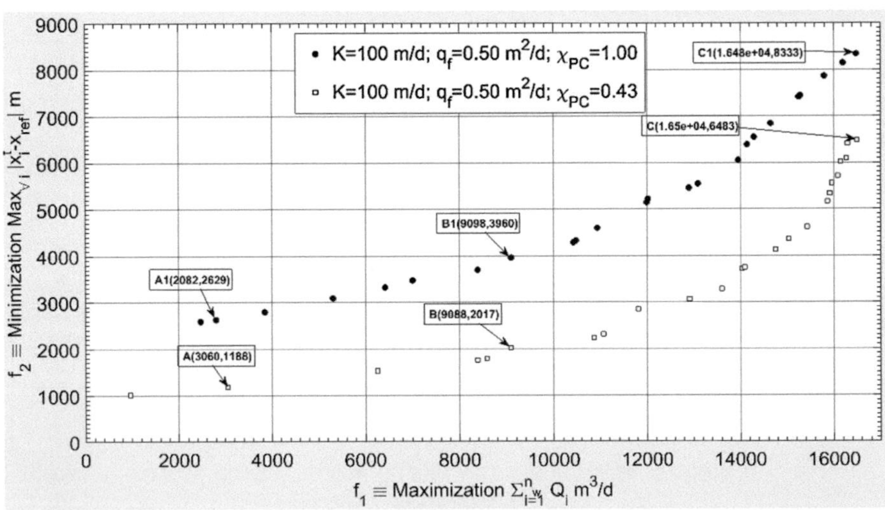

Fig. 22.10 Plot between the between the two objective functions for different iterations of NSGA-II algorithm $q_f = 0.5\,\mathrm{m^2/d}$, $K = 100\,\mathrm{m/d}$, $\alpha_t = 1\,\mathrm{m}$, $b = 30\,\mathrm{m}$, $\rho_f = 1000\,\mathrm{kg/m^3}$, $\rho_s = 1025\,\mathrm{kg/m^3}$

pumping rate region for $q_f = 1.00\,m^2/d$ and $q_f = 0.50\,m^2/d$ are 221.1 m and 3833 m respectively for the case with density correction.

The optimization plots for higher hydraulic conductivity values, i.e.,$K = 20\,m/d$, $q_f = 1.00\,m^2/d$ and $K = 20\,m/d$, $q_f = 0.50\,m^2/d$ are shown in Figs. 22.5 and 22.6 respectively. The hydraulic conductivity is double the value of previous plots. The saltwater intrusion also approximately doubled. This is consistent to the proportional relationship between hydraulic conductivity and coordinate x_j^t from Eq. (22.30). The decrease in freshwater flux between Figs. 22.5 and 22.6 shows similar trend as hydraulic conductivity $K = 10\,m/d$ cases [Figs. 22.3 and 22.4]. The Figs. 22.7 and 22.8 are corresponding to hydraulic conductivity $K = 30\,m/d$, and Figs. 22.9 and 22.10 are corresponding to hydraulic conductivity $K = 100\,m/d$. From the optimization plots it can be observed that the saltwater intrusion is very sensitive to regional freshwater flux. From the above points taken from the optimization front, it is evident that the two objective functions are conflicting objectives: with the increasing total pumping rate the minimization objective of saltwater intrusion is increasing.

Conclusions

A coastal aquifer pumping well management system framework based on multi-objective optimization is developed. The proposed framework links the saltwater intrusion in coastal aquifer with the total pumping rate constraint. The saltwater intrusion analytical solution is based on modified sharp interface solution considering density-dependent flow correction. Two conflicting objectives of maximizing the total pumping production from the number of pumping wells and minimizing the maximum toe intrusion from seashore are considered as objective functions. The modified sharp interface solution is used as a binding constraint. The developed framework is applied to Nellore coastal aquifer to test the performance. The regional freshwater flux plays important role in determining the maximum pumping rate. These nondominated fronts can provide initial management plan for solving the full-scale density-dependent flow based framework. The framework can be easily modified for the application of unconfined aquifer pumping optimization based on Koussis et al. [15] solution.

References

1. Bear J, Cheng AHD, Sorek S, Ouazar D, Herrera I (1999) Seawater intrusion in coastal aquifers: concepts, methods and practices, vol 14. Springer Science & Business Media, Dordrecht
2. Cheng AH-D, Halhal D, Naji A, Ouazar D (2000) Pumping optimization in saltwater-intruded coastal aquifers. Water Resour Res 36(8):2155–2165
3. Cooper HH Jr, Kohout FA, Henry HR, Glover RE (1964) Sea water in coastal aquifers: US Geological Survey Water-Supply Paper, 1613-C, C28
4. Datta B, Vennalakanti H, Dhar A (2009) Modeling and control of saltwater intrusion in a coastal aquifer of Andhra Pradesh, India. J Hydro Environ Res 3(3):148–159
5. Datta B, Dhar A (2011) Density dependent flows in saltwater intrusion and management, In: Aral MA, Taylor SW (eds) Groundwater quantity and quality management. Groundwater Management Technical Committee of the Groundwater Council of EWRI Environmental and Water Resources Institute (EWRI) of the American Society of Civil Engineers, pp 394–429

6. Dausman AM, Langevin C, Bakker M, Schaars F (2010) A comparison between SWI and SEAWAT – the importance of dispersion, inversion and vertical anisotropy. In: 21st Salt water intrusion meeting, pp 271–274
7. Deb K (2001) Multi-objective optimization using evolutionary algorithms. Wiley, Chichester
8. Dentz M, Tartakovsky DM, Abarca E, Guadagnini A, Sanchez-Vila X, Carrera J (2006) Variable-density flow in porous media. J Fluid Mech 561:209–235
9. Dhar A, Datta B (2009) Saltwater intrusion management of coastal aquifers. I: linked simulation-optimization. J Hydrol Eng 14(12):1263–1272
10. Dhar A, Munusamy SB (2014) Modified Ghyben-Herzberg theory based modelling and control of saltwater intrusion in coastal aquifers. In: 7th international symposium on environmental hydraulics ISEH 2014, Singapore, pp 355–358
11. Diersch H-JG, Kolditz O (2002) Variable-density flow and transport in porous media: approaches and challenges. Adv Water Resour 25(8–12):899–944
12. Llopis-Albert C, Pulido-Velazquez D (2014) Discussion about the validity of sharp-interface models to deal with seawater intrusion in coastal aquifers. Hydrol Process 28:3642–3654
13. Lu C, Chen Y, Luo J (2012) Boundary condition effects on maximum groundwater withdrawal in coastal aquifers. Ground Water 50(3):386–393
14. Lu C, Werner AD (2013) Timescales of seawater intrusion and retreat. Adv Water Resour 59:39–51
15. Koussis AD, Mazi K, Riou F, Destouni G (2015) A correction for Dupuit-Forchheimer interface flow models of seawater intrusion in unconfined coastal aquifers. J Hydrol 525:277–285
16. Mantoglou A (2003) Pumping management of coastal aquifers using analytical models of saltwater intrusion. Water Resour Res 39(12):1335
17. McKay MD, Conover WJ, Beckman RJ (1979) A comparison of three methods for selecting values of input variables in the analysis of output from a computer code. Technometrics 21:239–245
18. Park C-H, Aral M (2004) Multi-objective optimization of pumping rates and well placement in coastal aquifers. J Hydrol 290(1–2):80–99
19. Pool M, Carrera J (2011) A correction factor to account for mixing in Ghyben-Herzberg and critical pumping rate approximations of seawater intrusion in coastal aquifers. Water Resour Res 47(5):W05506. https://doi.org/10.1029/2010WR010256
20. Shamir U, Bear J, Gamliel A (1984) Optimal annual operation of a coastal aquifer. Water Resour Res 20(4):435–444
21. Simmons CT (2005) Variable density groundwater flow: from current challenges to future possibilities. Hydrogeol J 13(1):116–119
22. Sreekanth J, Datta B (2010) Multi-objective management of saltwater intrusion in coastal aquifers using genetic programming and modular neural network based surrogate models. J Hydrol 393(3–4):245–256
23. Strack ODL (1976) A single-potential solution for regional interface problems in coastal aquifers. Water Resour Res 12(6):1165–1174
24. Volker RE, Rushton KR (1982) An assessment of the importance of some parameters for seawater intrusion in aquifers and a comparison of dispersive and sharp-interface modelling approaches. J Hydrol 56(3–4):239–250
25. Werner AD, Bakker M, Post VEA, Vandenbohede A, Lu C, Ataie-Ashtiani B, Barry DA (2013) Seawater intrusion processes, investigation and management: recent advances and future challenges. Adv Water Resour 51:3–26
26. Werner AD (2017) Correction factor to account for dispersion in sharp-interface models of terrestrial freshwater lenses and active seawater intrusion. Adv Water Resour 102:45–52

Chapter 23
Dynamic Nonlinear Active Noise Control: A Multi-objective Evolutionary Computing Approach

Apoorv P. Patwardhan, Rohan Patidar, and Nithin V. George

Abstract Evolutionary-computing-algorithm-based nonlinear active noise control (ANC) removes the requirement of secondary path modeling, which is essential for proper functioning of a conventional gradient-descent-approach based ANC system. However, the noise mitigation capability of such algorithms is largely dependent on the proper selection of the agent count as well as on the number of sound samples processed by an agent in a given iteration. In order to alleviate this dependency, we propose a dynamic nonlinear ANC (DNANC) system, which adapts its parameters in accordance with the acoustic scenario under consideration. The nonlinear ANC (NANC) problem has been formulated as a multi-objective optimization problem in this chapter. We have used the non-domination sorting genetic algorithm II (NSGA-II) for solving the optimization task. The conflicting objectives employed in this chapter are the ensemble mean-square error and the computation time. The proposed DNANC system has been shown to adapt itself to several ANC scenarios in a dynamic manner, wherein, the controller structure has been optimized for the situation considered.

Keywords Nonlinear active noise control · Functional-link artificial neural network · Non-domination sorting genetic algorithm II · Particle swarm optimization · Differential evolution · Cuckoo search algorithm

Introduction

The field of noise cancellation can be broadly divided into two categories: passive and active. Active noise control (ANC), which is based on the principle of destructive superposition of sound waves, has seen enhanced interest in the recent

A. P. Patwardhan · R. Patidar · N. V. George (✉)
Department of Electrical Engineering, Indian Institute of Technology Gandhinagar, Gandhinagar, Gujarat, India
e-mail: apoorv@btech2011.iitgn.ac.in; rohanpatidar@btech2011.iitgn.ac.in; nithin@iitgn.ac.in

F. Bennis, R. K. Bhattacharjya (eds.), *Nature-Inspired Methods for Metaheuristics Optimization*, Modeling and Optimization in Science and Technologies 16,
https://doi.org/10.1007/978-3-030-26458-1_23

past owing to advances in control theory as well as the availability of improved processing capability. In its basic form, a feed-forward ANC system consists of a reference microphone which senses the noise, an active loudspeaker which produces the necessary anti-noise, and an error microphone to measure the level of noise mitigation attained [14]. The active loudspeaker is driven by an adaptive controller, which has a transfer function $W(z)$.

Figure 23.1 shows the block diagram of a filtered-x least-mean-square (FxLMS) based ANC system. In the figure, $P(z)$ is the transfer function of the primary path (the acoustic path from the reference microphone to the error microphone), $S(z)$ is the transfer function of the secondary path (the electroacoustic path from the output of the adaptive controller to the output of the error microphone), $\hat{S}(z)$ is the transfer function of a model of the secondary path, $W(z)$ is the transfer function of the adaptive controller, $x(n)$ is the reference signal, $d(n)$ is the primary path output, $y(n)$ is the output of the adaptive controller, $\hat{d}(n)$ is the output of the secondary path, and $e(n) = d(n) - \hat{d}(n)$ is the error signal. The weight vector $\mathbf{w}(n)$ of the adaptive controller is updated using the FxLMS algorithm as

$$\boldsymbol{w}(n+1) = \boldsymbol{w}(n) + \mu e(n)\boldsymbol{x}'(n) \tag{23.1}$$

where μ is the learning rate and $\boldsymbol{x}'(n)$ is the reference signal vector filtered through $\hat{S}(z)$.

The basic FxLMS algorithm, which uses a finite-impulse-response (FIR) filter as the adaptive controller, is designed on the assumption that the primary and secondary paths are linear. In the actual implementation of an ANC system, the primary and secondary paths may offer nonlinearities, and the FxLMS-based ANC system can fail to effectively cancel the primary noise. Several nonlinear ANC (NANC) systems have been proposed in the recent past to achieve noise mitigation in a nonlinear environment. A Volterra-filter-based ANC system has been proposed

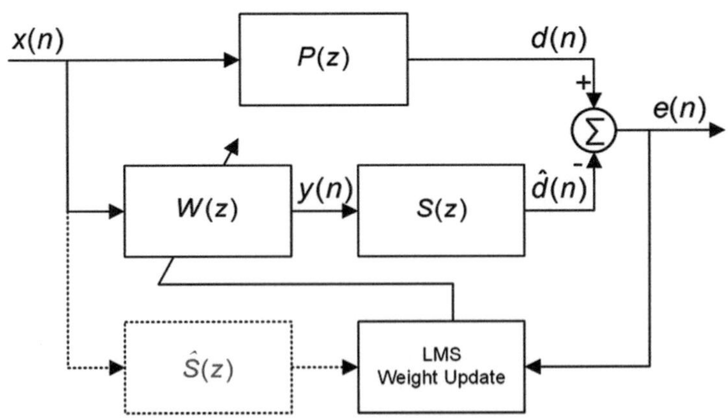

Fig. 23.1 Block diagram of an FxLMS algorithm based ANC system

in [21]. Das and Panda [7] have recently developed a nonlinear ANC system based on functional-link artificial neural networks (FLANNs). The same authors have also proposed a filtered-s least-mean-square (FsLMS) algorithm [7] to update the weights of the FLANN. Several newer versions of the FLANN-based ANC system have been reported in the literature that aims to further enhance noise cancellation capability [12, 15, 26].

All the ANC systems discussed so far require an accurate modeling of the secondary path and its model is kept fixed during the operation of the system. Imperfect modeling of the secondary path can cause the ANC system to diverge [3]. In a real-life scenario, the secondary path may vary with time and can affect the performance of the ANC system. Several online secondary-path modeling schemes have been proposed to overcome this limitation [1, 4, 22]. But, most of these schemes require the injection of an additional white noise in order to achieve the modeling [25]. This additional noise adds to the final residual noise and also increases the computational complexity. Some of the other online secondary path modeling methods require additional adaptive filters to achieve the modeling task, and in some cases, the order of the secondary path needs to be fixed *apriori* [2, 24]. Fixing the order *apriori* may lead to under or over estimation of the secondary path.

It has been lately reported that the use of meta-heuristic algorithms in ANC systems can remove the requirement of modeling secondary paths. A genetic algorithm (GA) based ANC system has been proposed in [6]. Rout et al. [16, 17] have recently proposed an ANC scheme based on particle-swarm optimization (PSO), wherein the authors design a de-multiplexer/multiplexer arrangement for real-time implementation. A nonlinear ANC system based on a bacterial-foraging algorithm has been reported in [13]. George and Panda [11] extended the approach to nonlinear multichannel ANC systems and introduced a decentralized version that offers flexibility in scaling up a multichannel ANC system. The presence of non-linear secondary paths as well as the use of infinite impulse response filters as the controller may result in a non-quadratic error cost function. The traditional gradient descent algorithms may fall into local minima in such cases and may lead to non-optimal noise cancellation. In addition to avoiding the requirement of secondary path modeling, meta-heuristic-based ANC systems can also avoid this local minima problem. Another gradient-free approach towards ANC has been presented in [19].

The success of the meta-heuristic-based ANC system is largely dependent on the proper selection of the switching quotient Q (the count of sound samples after which switching occurs in a de-multiplexer/multiplexer arrangement) and the population count L. In most of the available literature, these values have been computed experimentally for each of the experiments considered. In a practical implementation of an ANC system, the primary and secondary paths may vary, and fixing Q and L may lead to improper noise mitigation. In order to overcome this limitation of meta-heuristic-based ANC systems, this chapter proposes a dynamic nonlinear ANC (DNANC) system where switching in the de-multiplexer/multiplexer arrangement, as well as the population count of the meta-heuristic algorithm, are adaptively selected based on the ANC scenario. The selection of Q as well as L will be formulated as a multi-objective optimization problem, with the computation time and the ensemble mean-square error (EMSE) as two conflicting objective functions. The non-domination sorting genetic algorithm II (NSGA-II) [9] has been used as the multi-objective optimization algorithm in this work. In addition, the chapter also develops two novel meta-heuristic-based ANC systems, one based on a cuckoo search algorithm (CSA) [23] and the other based on differential evolution (DE) [20], which offer improved noise control over other meta-heuristic-based ANC systems available in literature. All the proposed schemes also avoid the local minima problem existing in ANC systems trained using traditional gradient descent algorithms.

The rest of the chapter is organized as follows. The meta-heuristic-based ANC scenario is discussed in Sect. 1.2. The dynamic nonlinear ANC system proposed in this chapter is designed in Sect. 1.3, which also discusses the various single- and multi-objective meta-heuristic algorithms used in the chapter. An extensive simulation study has been carried out in Sect. 1.4 and concluding remarks are drawn in Sect. 1.5.

Meta-Heuristic-Based NANC System

As discussed in the previous section, the FxLMS-based linear ANC system may fail to effectively mitigate noise in the presence of non-linearities. FLANN-based nonlinear ANC systems have gained significant attention in the recent past owing to the low computational complexity offered in comparison with other nonlinear-filter-based ANC schemes. In a FLANN-based nonlinear controller, the reference signal $x(n)$ is functionally expanded before it is input to an adaptive weight network. The functional expansion can be trigonometric, Legendre, or Chebyshev. Trigonometric FLANNs are among the most popular type.

In an FsLMS-based NANC system, the tap-delayed input signal vector $\boldsymbol{x}(n) = [x(n), x(n-1), \ldots, x(n-M+1)]^{\mathrm{T}}$ of length M is trigonometrically expanded into N terms as

$$\begin{aligned} \boldsymbol{X}(n) = \{&x(n), \sin\left[\pi \boldsymbol{x}(n)\right], \cos\left[\pi \boldsymbol{x}(n)\right], \sin\left[2\pi \boldsymbol{x}(n)\right], \\ &\cos\left[2\pi \boldsymbol{x}(n)\right], \ldots, \sin\left[\alpha\pi \boldsymbol{x}(n)\right], \cos\left[\alpha\pi \boldsymbol{x}(n)\right]\}^{T} \end{aligned} \tag{23.2}$$

with α as the order of the FLANN filter and $N = M\,(2\alpha + 1)$. The adaptive weights $\boldsymbol{W}(n)$ are updated as

$$\boldsymbol{W}\,(n+1) = \boldsymbol{W}\,(n) + \mu e(n)\boldsymbol{X}'(n) \tag{23.3}$$

where μ is the learning rate, $e(n)$ is the error signal and $\boldsymbol{X}'(n)$ is $\boldsymbol{X}(n)$ filtered through a model of the secondary path. The requirement of secondary-path modeling in an FsLMS-based ANC system can be avoided by using a meta-heuristic algorithm for weight update, which minimizes the objective function $E(e^2)$, where $E(\cdot)$ is the expectation operator. The block diagram of the meta-heuristic-based ANC scheme will be similar to that in Fig. 23.1, except that the dotted block will not be part of the final structure.

The controller structure in a meta-heuristic-based ANC system is similar to the primary meta-heuristic algorithm block in Fig. 23.2. The adaptive de-multiplexer/multiplexer arrangement has to be replaced by a regular de-multiplexer/multiplexer. A meta-heuristic algorithm utilizes a number of agents having different attributes. In the context of an NANC system, the attribute of any particular agent relates to the weight vector, which multiplies the expanded FLANN vector $\boldsymbol{X}(n)$. These attributes help in distinguishing one agent's performance from that of the other. In conventional meta-heuristic-based filter designs, errors corresponding to all the agents in the meta-heuristic framework are evaluated for every input sample considered. In an online application like NANC, this method is not appropriate, as only a single active loudspeaker and error microphone can be activated to achieve noise cancellation at any given instant.

In order to make this framework online, Rout et al. [16] proposed a de-multiplexer/multiplexer arrangement in a meta-heuristic-based ANC system. At any given time, only one agent is in the active electronic path between the reference

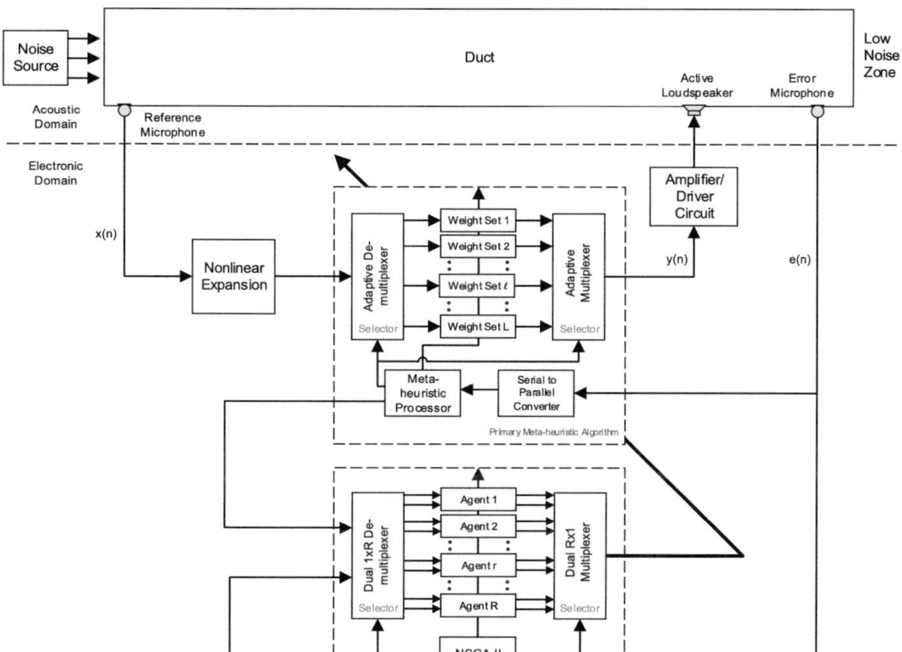

Fig. 23.2 Schematic diagram of the DNANC system

microphone and the secondary path. After evaluating Q errors corresponding to the first agent, the fitness of the agent is evaluated as the value of the mean-square-error. After an agent has finished processing Q samples of the input sound, the meta-heuristic engine switches the sound input to the next agent in the population. This process is continued for LQ sound input samples after which the meta-heuristic engine updates the weight vectors corresponding to each agent using the meta-heuristic algorithm. This process is repeated for rest of the primary noise samples.

It is to be noted that, the switching quotient Q and the agent count L in this framework remain constant throughout. However, this may lead to a sub-optimal design of the meta-heuristic controller, owing to the fact that different performance is achieved by this controller in dynamically-changing sound and path environments. Moreover, controller performance also depends heavily on the algorithm type supported by the controller engine. In an endeavour to overcome this limitation of the meta-heuristic ANC scenario, a novel NANC scheme with dynamic Q and L is designed in the next section.

Dynamic Nonlinear Active Noise Control System

The noise mitigation performance of an evolutionary-computing-based NANC system is largely dependent on the proper selection of the switching quotient Q and the agent count L. To the best of our knowledge, all the works reported in the field of evolutionary-computing-based ANCs use fixed values for Q and L, which are experimentally determined for a particular ANC scenario. In a practical implementation of an ANC system, fixed values of Q and L may not lead to optimal noise mitigation in a meta-heuristic-based NANC system. This is attributed to the dynamic nature of the ANC environment in terms of the transfer functions of the primary and secondary paths, as well as in terms of the reference noise.

In order to cope with a dynamically changing ANC environment, it is desirable for the meta-heuristic framework to be dynamic in terms of Q and L. A higher value of L can improve the noise cancellation capability of the ANC system. However, this improvement is achieved at the cost of increased computational load. It is desirable to have an optimal combination of Q and L that can simultaneously maximize noise cancellation and minimize computational load. This goal can be achieved by formulating a meta-heuristic-based NANC system as a multi-objective optimization problem. In this chapter, we have considered computational time (τ) and ensemble mean-square error (ξ) as the two conflicting objectives, with Q and L as the controlling variables. In contrast to a conventional meta-heuristic-based NANC system, this approach continuously updates Q and L for improving performance.

The basic form of the dynamic nonlinear ANC system proposed in this chapter is shown in Fig. 23.2. In the figure, the primary meta-heuristic block refers to the meta-heuristic controller in a conventional meta-heuristic-based NANC system as discussed in Sect. 1.2. To allow dynamicity, the traditional de-multiplexer/multiplexer arrangement has been replaced with an adaptive de-multiplexer/multiplexer system. Multi-objective optimization is achieved in the DNANC system by using a secondary meta-heuristic block, containing an evolutionary multi-objective optimization algorithm. The detailed mechanism of the interaction between the primary and secondary meta-heuristic blocks in the DNANC system is discussed below.

The primary meta-heuristic block shown in Fig. 23.2 combines a $1 \times L$ de-multiplexer and an $L \times 1$ multiplexer system. The de-multiplexer/multiplexer arrangement switches the non-linearly expanded sound samples $\boldsymbol{X}(n)$ between the component weight sets (agents for the primary meta-heuristic algorithm) after Q sound samples. As the proposed NANC system is dynamic in nature, the Q and L values are updated according to the existing acoustic scenario. The Q and L update is achieved using a secondary meta-heuristic block, which essentially contains a multi-objective optimization algorithm engine. The meta-heuristic processor should ensure that Q and L belong to a set of natural numbers by approximating the updated Q and L to the next highest integer. In this work, PSO, CSA, and DE have been considered as candidates for the primary meta-heuristic algorithm.

The secondary meta-heuristic algorithm employed in this chapter is the non-domination sorting genetic algorithm II (NSGA-II) proposed by Deb et al. [9]. NSGA-II, which falls under the category of multi-objective optimization algorithms have found successful applications in diverse fields of science and technology. Let $f_1(x_1, x_2)$ and $f_2(x_1, x_2)$ be two objective functions that are conflictive in nature, with x_1 and x_2 as the controlling variables. In a typical NSGA-II-based optimization problem, which aims to simultaneously optimize $f_1(x_1, x_2)$ and $f_2(x_1, x_2)$, the solution set gives rise to a Pareto-optimal front of solutions [9]. A Pareto- optimal front is a curve connecting all the Pareto-optimal solutions [8].

In this chapter, the two objective functions used are the computation time (τ) and the ensemble mean-square error (ξ), and the control variables are Q and L. In a real-time application like ANC, the evaluation of the objective functions for the various agents of NSGA-II requires a modification in the conventional NSGA-II framework, leading to the introduction of a dual $1 \times R$ de-multiplexer and a dual $R \times 1$ multiplexer. The two inputs to the de-multiplexer are the objective function values τ and ξ. For each agent, the two values that the rth agent passes through the multiplexer are $Q(r)$ and $L(r)$. The switching coefficient for the secondary meta-heuristic block corresponding to the rth agent is

$$\phi(r) = Q\,(r-1)\,L\,(r-1)\,\beta_p \tag{23.4}$$

where $Q(r-1)$, $L(r-1)$ and β_p correspond to the switching coefficient, the agent count, and the iteration count, respectively, for the primary meta-heuristic block due to the previous agent of the secondary meta-heuristic algorithm. A flowchart of the proposed NSGA-II based secondary meta-heuristic block is shown in Fig. 23.3 and the schematic diagram of the complete DNANC system is depicted in Fig. 23.2.

Fig. 23.3 Flowchart of NSGA-II based secondary meta-heuristic block

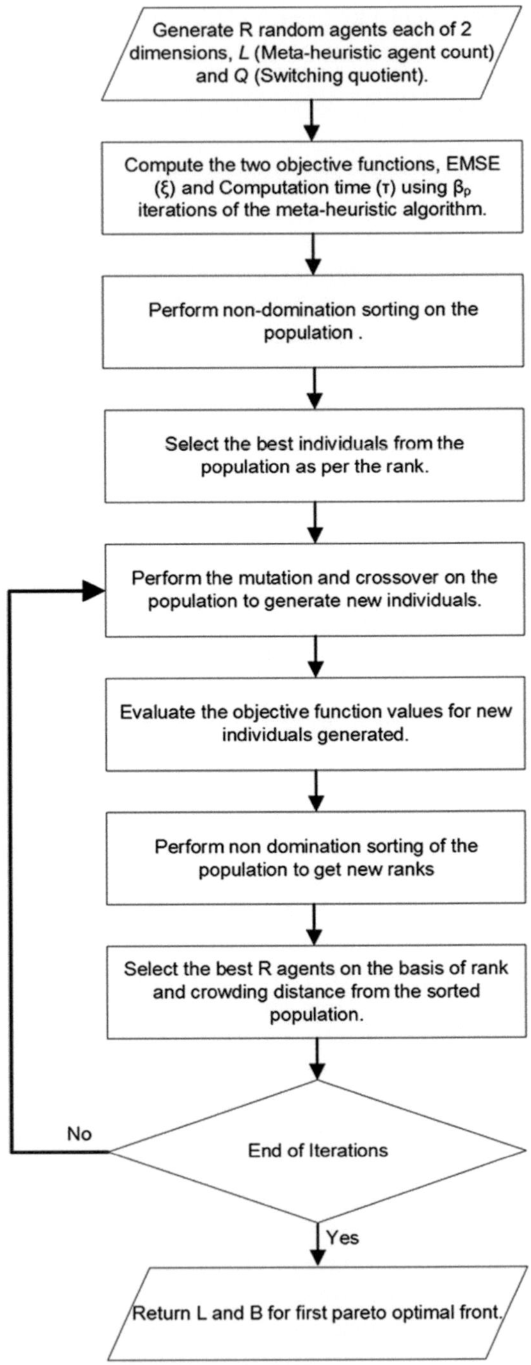

Simulation Study

In an endeavour to evaluate the noise cancelling performance of the proposed DNANC system, an extensive simulation study has been carried out in this section. The NSGA-II-based multi-objective optimization algorithm in the DNANC system aims to simultaneously improve noise cancellation and reduce computational load. The level of noise cancellation, which is the first objective function that needs to be minimized, is measured in terms of ensemble mean-square error (EMSE), which is defined as

$$\xi = 10\log_{10} E\left[e^2(n)\right] \tag{23.5}$$

where E [.] is the expectation operator. The second objective function is the computational time (τ) measured in seconds, which is the time required for the meta-heuristic ANC block to process the given set of input sound samples. Both objective functions have been modelled as functions of two variables: meta-heuristic algorithm switching quotient (Q) and the agent count (L). The switching quotient is the number of samples of sound processed through an agent of the meta-heuristic algorithm before the signal is routed to the next agent of the meta-heuristic framework.

In order to exemplify the robustness of the proposed DNANC system, a total of 9 different testing experiments have been included. Each experiment is a combination of different primary paths, secondary paths, and type of input sound. The various sound types considered are random noise, tonal noise, and logistic chaotic noise. In addition, primary and secondary paths characterized by low-nonlinearity as well as high-nonlinearity have also been used. All the experiments have been carried out in a MATLAB environment on an Intel Core i5 PC, having a 4-GB RAM and 2.60-GHz processor. The algorithms used are CSA, DE, and PSO for the primary meta-heuristic and NSGA-II for the secondary meta-heuristic. The tap length M for the FLANN filter is taken as 2 and the order of the FLANN, α is also 2 for all the cases studied in this chapter. For all the subsequent experiments considered in this chapter, the scaling factor and the crossover rate are taken to be 0.3 and 0.6, respectively, for DE. Similarly, in the case of CSA, the probability of discovering an alien egg is set to the value 0.15 and the step size is chosen to be 0.01, while the Levy flight parameter is chosen to be equal to 2.5. In PSO, the behavioural parameters c_1 and c_2 are set to the values 0.4 and 0.9, respectively, while linearly varying the inertia weight from 0.9 to 0.3 [11].

Case A: Random Input Noise

A set of three experiments is carried out in with a random input noise, which is uniformly distributed in the range $[-0.5, 0.5]$.

Experiment 1

In this experiment, the primary noise $d(n)$ reaching the error microphone is given by

$$\begin{aligned} d(n) = x(n) + 0.8x\,(n-1) + 0.3x\,(n-2) + 0.4x\,(n\text{–}3) \\ - 0.8x(n)x\,(n-1) + 0.9x(n)x\,(n-2) + 0.7x(n)x\,(n-3) \end{aligned} \tag{23.6}$$

where $x(n)$ is the input noise [18]. Similarly, the cancelling sound sensed at the error microphone is given by

$$\begin{aligned} \hat{d}(n) = y(n) + 0.35y\,(n-1) + 0.09y\,(n-2) - 0.5y(n)y\,(n-1) \\ + 0.4y(n)y\,(n-2)\,. \end{aligned} \tag{23.7}$$

Figure 23.4 shows the variation of EMSE with respect to time for the proposed DNANC system, implemented using CSA, DE, and PSO. The dynamic behaviour of the proposed method, which is attributed to the usage of NSGA-II, is evident from the variation of the meta-heuristic algorithm switching quotient (Q) and the agent count (L) plotted in Fig. 23.5. Both Q and L have been rounded to the next

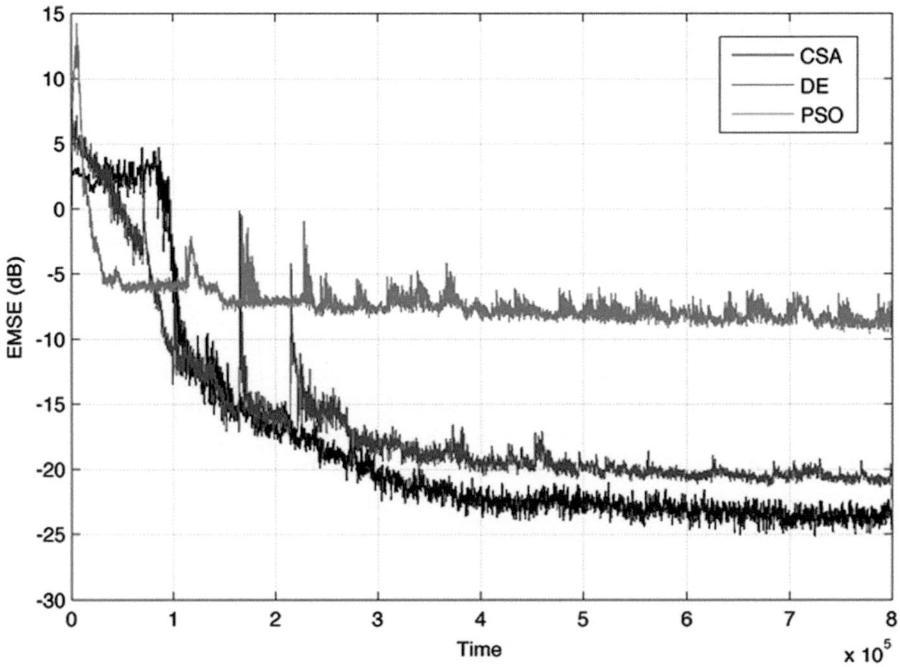

Fig. 23.4 Experiment 1: Variation with respect to time for a DNANC system with CSA, DE, and PSO as the primary meta-heuristic algorithm and random input noise. All results plotted have been averaged over ten independent trials

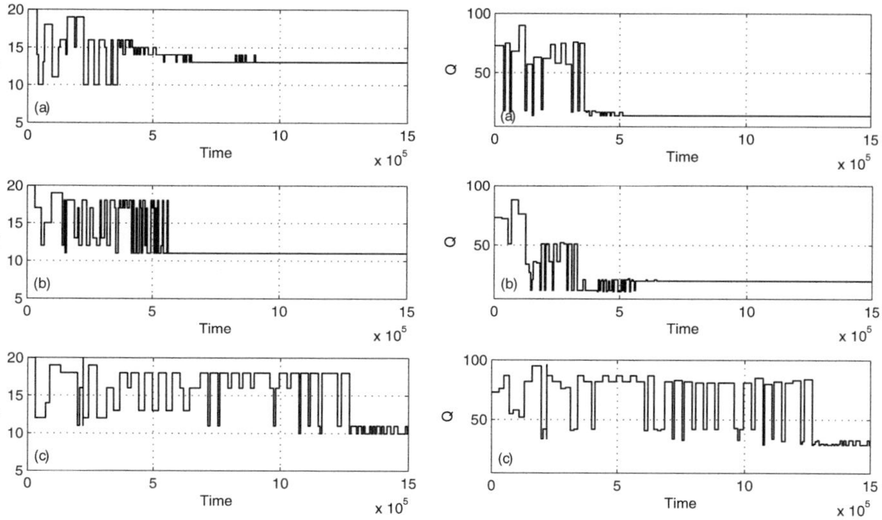

Fig. 23.5 Experiment 1: Variation of L and Q with respect to time for a DNANC system with (**a**) CSA, (**b**) DE and (**c**) PSO as the primary meta- heuristic algorithm and random input noise. NSGA-II is used as the secondary meta-heuristic algorithm

highest integer in cases where the algorithm returns a fractional value. It can be observed that the three meta-heuristic algorithms employed in this work converge to different values of Q and L. Thus, the DNANC scheme offers the freedom in NSGA-II to select the best Q and L, which minimizes EMSE and computational time simultaneously. The final average EMSE values obtained are −24.04, −20.25, and − 8.07 dBs for the DNANC system with the primary meta-heuristic algorithm taken as CSA, DE, and PSO, respectively. The best Pareto front obtained using NSGA-II with CSA as the meta-heuristic algorithm is shown in Fig. 23.6.

Experiment 2

The primary path employed in this experiment is given by

$$\begin{aligned} d(n) &= x(n) + 0.8x\,(n-1) + 0.3x\,(n-2) + 0.4x\,(n-3) \\ &- 0.8x(n)x\,(n-1) + 0.9x(n)x\,(n-2) + 0.7x(n)x\,(n-3) \\ &- 3.9x^2\,(n-1)\,x\,(n-2) - 2.6x^2\,(n-1)\,x\,(n-3) \\ &+ 2.1x^2\,(n-2)\,x\,(n-3)\,. \end{aligned} \tag{23.8}$$

The secondary path [18], as well as all the other simulations parameters used in this experiment, are the same as that of Experiment 1. All other simulation parameters are also similar to that of experiment one. Figure 23.7a shows the variation of EMSE with respect to time for the proposed DNANC system with

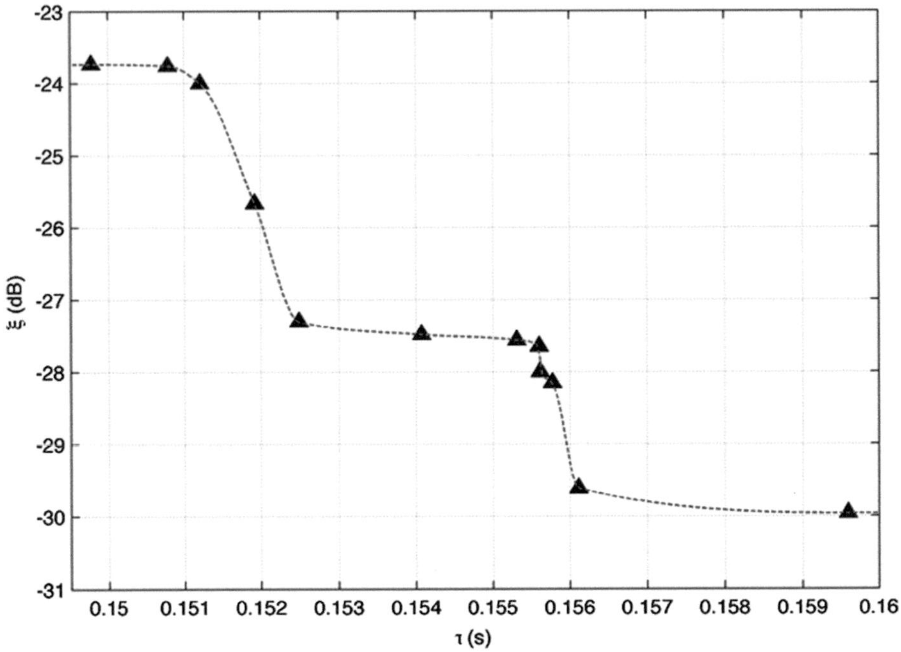

Fig. 23.6 Experiment 1: The best Pareto front obtained using NSGA-II with CSA as the primary meta-heuristic algorithm and random input noise

CSA, DE, and PSO as the primary meta-heuristic algorithm and NSGA-II as the secondary meta-heuristic algorithm. The final average EMSE values obtained are −9.48, −19.27, and − 8.86 dBs for the DNANC system with the primary meta-heuristic algorithm taken as CSA, DE, and PSO respectively.

Experiment 3

In this experiment, a primary path same as that of experiment two is considered. The secondary path considered in this experiment, which is a Hammerstein filter, is given by

$$\hat{d}(n) = w(n) + 0.2w\,(n-1) + 0.05w\,(n-2) \tag{23.9}$$

where

$$w(n) = \tanh\left[y(n)\right]. \tag{23.10}$$

All the other simulation parameters employed in this experiment are same as that used in experiment two. The change in EMSE with time is depicted in Fig. 23.7b. A

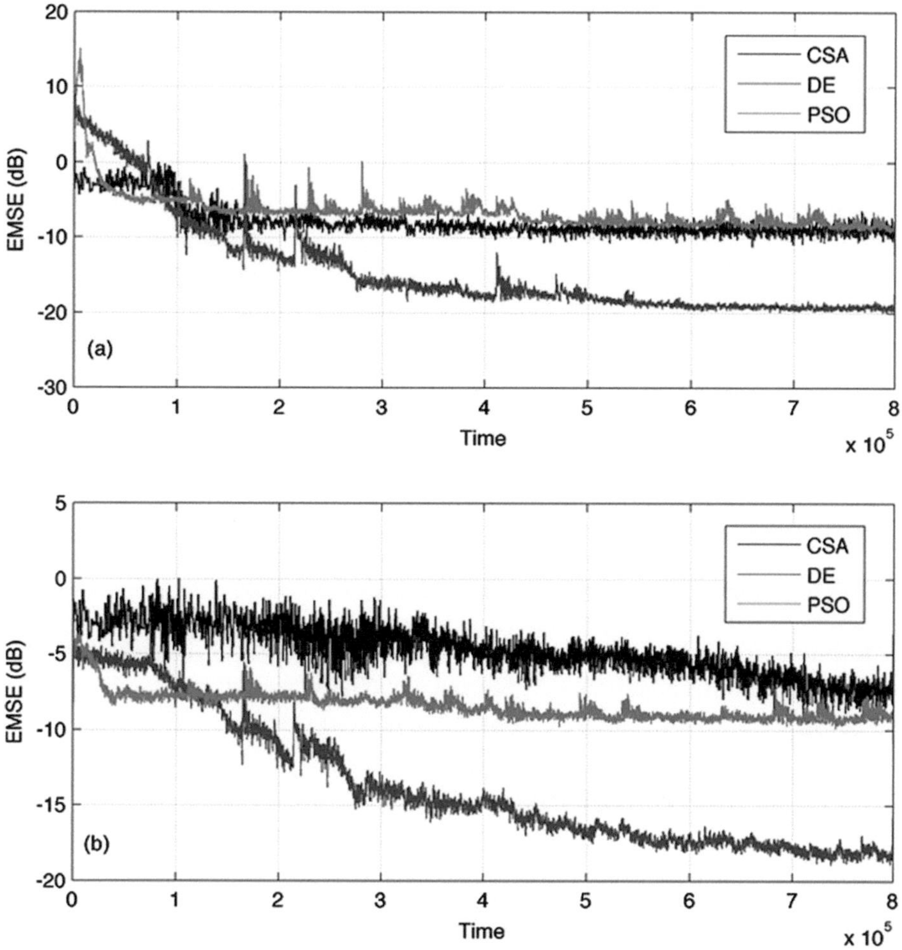

Fig. 23.7 Experiments 2 and 3: Variation of EMSE with respect to time for a DNANC system with CSA, DE, and PSO as the primary meta-heuristic algorithm and random input noise. (**a**) Experiment 2, (**b**) Experiment 3. All results plotted have been averaged over ten independent trials

final average EMSE of −7.32, −17.74, and − 8.88 dBs have been obtained for the DNANC system with the primary meta-heuristic algorithm taken as CSA, DE, and PSO, respectively.

Case B: Tonal Input Noise

Two sets of experiments have been conducted considering a sinusoidal signal given by

$$x(n) = \sin\left(\frac{2\pi f_n}{f_s}\right), \tag{23.11}$$

where $f_s = 10{,}000$ Hz is the sampling frequency and f_n is taken as 500 Hz [5].

Experiment 4

The primary and secondary paths, as well as the other simulation parameters applied in this experiment, are same as that in Experiment 1. The variation of EMSE with time for the DNANC system with CSA, DE and PSO as the primary meta-heuristic algorithms is shown in Fig. 23.8a, where the final average EMSE values are −12.24, −19.27, and − 8.86 dB, respectively. The improved noise mitigation performance of CSA- and DE-based DNANC systems are evident from the results.

Experiment 5

In this experiment, the primary path, the secondary path, and the simulation parameters used are same as that of Experiment 2. Figure 23.8b shows the variation of EMSE with time for the three primary meta-heuristic algorithms considered, with final average EMSE of −7.19, −11.98, and − 2.69 dB, respectively.

Case C: Logistic Chaotic Input Noise

A logistic chaotic input noise is used in the next two experiments, given by

$$x(n) = \gamma x\,(n-2)\ [1 - x\,(n-2)] \ \text{ for } n = 2, 3, \ldots \tag{23.12}$$

where $x(n) = 0.9$ for $n = 0$, 1 and $\gamma = 4$ [5].

Experiment 6

The primary path and the secondary path transfer functions used in this experiment are same as that of Experiment 1. The time behaviour of EMSE for CSA, DE, and PSO is depicted in Fig. 23.9a, while the final average EMSE obtained is −20.40, −18.74 and − 7.08 dB respectively. The CSA- and DE-based DNANC systems have been shown to outperform the PSO-based DNANC system for the logistic chaotic input noise considered.

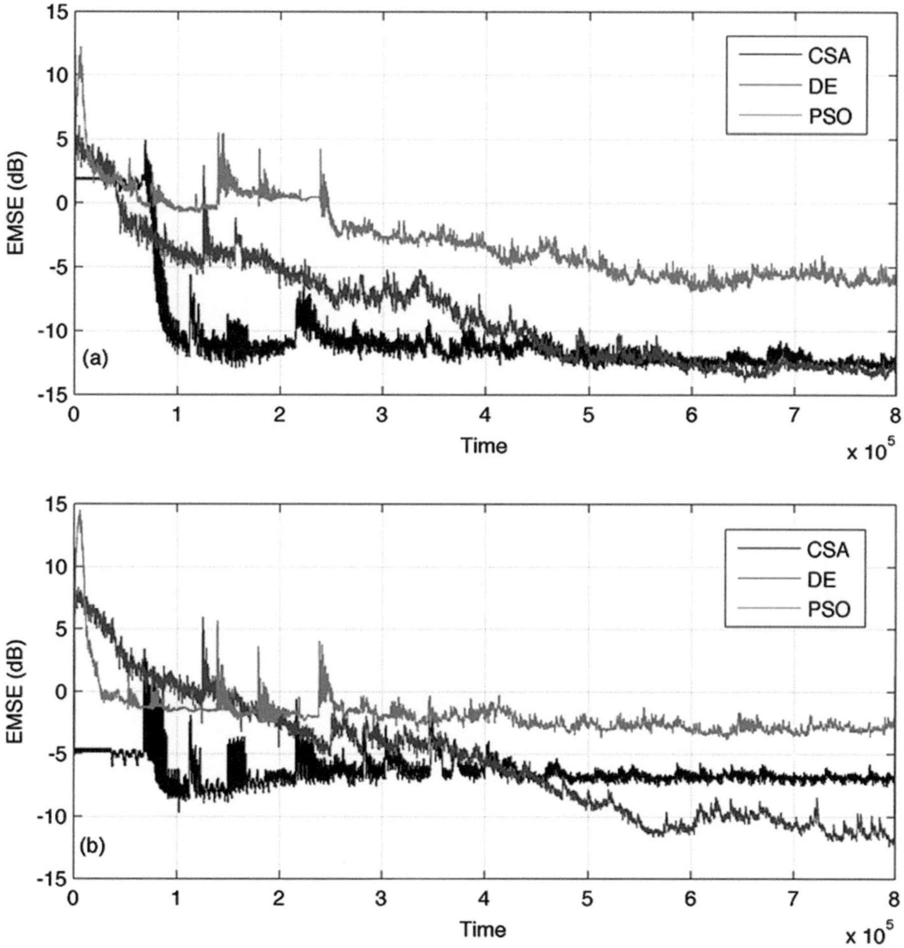

Fig. 23.8 Experiments 4 and 5: Variation of EMSE with respect to time for a DNANC system with CSA, DE, and PSO as the primary meta-heuristic algorithm and tonal input noise. (**a**) Experiment 4, (**b**) Experiment 5. All results plotted have been averaged over ten independent trials

Experiment 7

Another experiment has been carried out using the logistic chaotic noise given by (12). The primary and secondary paths used in this experiment is same as that used in Experiment 3. The improved noise cancellation achieved using CSA and DE primary meta-heuristic algorithms is evident from the variation of EMSE plotted in Fig. 23.9b, with a final average EMSE of −20.40, −18.74 and − 7.08 dBs respectively.

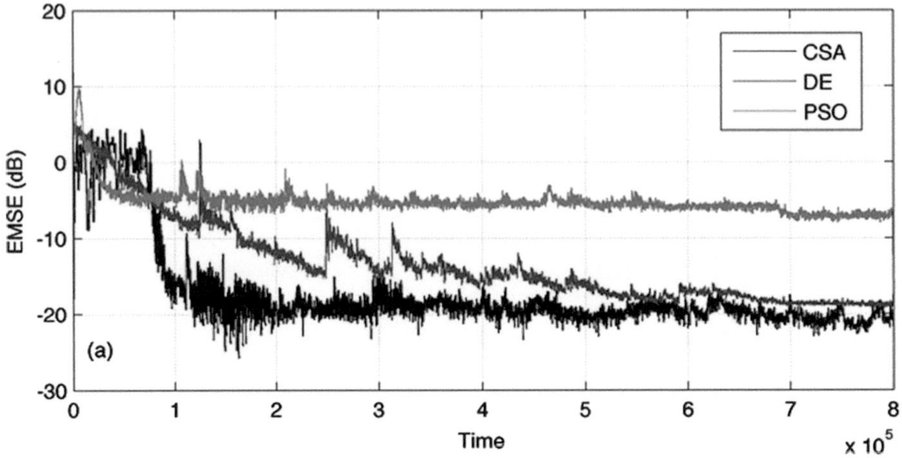

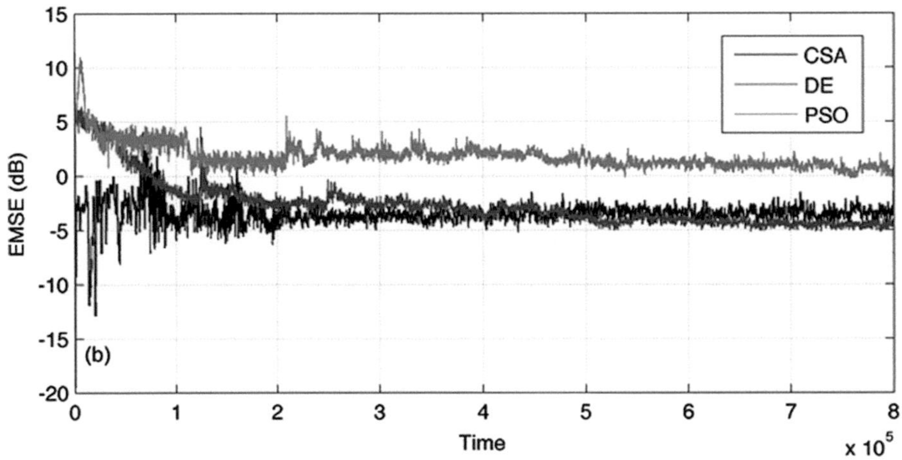

Fig. 23.9 Experiments 6 and 7: Variation of EMSE with respect to time for a DNANC system with CSA, DE, and PSO as the primary meta-heuristic algorithm and logistic chaotic input noise. (**a**) Experiment 6, (**b**) Experiment 7. All results plotted have been averaged over ten independent trials

Case D: Dynamically Changing Environment

In most of the real-life implementations of an ANC system, the type of the noise, as well as the characteristics of the primary and secondary paths, are subject to change. Two experiments have been conducted to evaluate the performance of the DNANC system under such scenarios.

Experiment 8

In this experiment, three different primary and secondary path combinations have been considered for a random noise uniformly distributed between [−0.5, 0.5]. In the first one-third of the primary sound samples, the paths used are same as that in Experiment 2. In the next one-third, the primary and secondary path transfer functions employed in Experiment 1 have been considered. The paths used in Experiment 3 have been used in the last third. The EMSE values are plotted against time in Fig. 23.10a. The dynamic behaviour of the proposed scheme is evident from

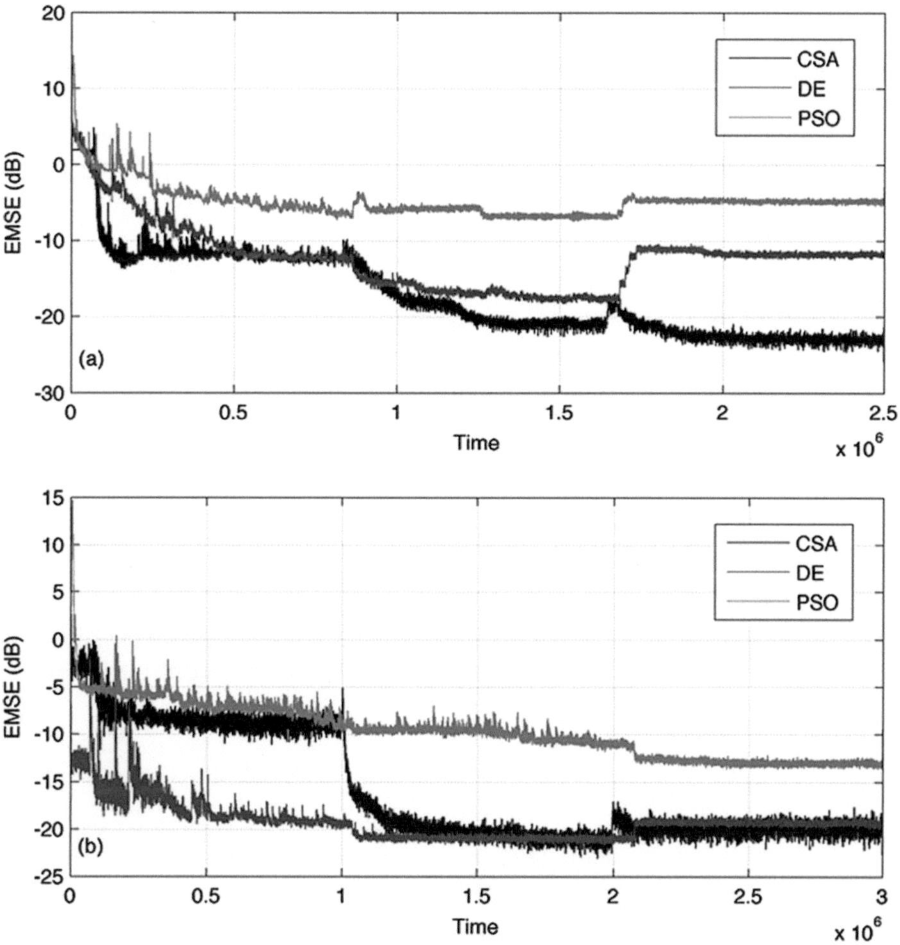

Fig. 23.10 (**a**) Experiment 8: Variation of EMSE with respect to time for a DNANC system with CSA, DE, and PSO as the primary meta-heuristic algorithm and FsLMS algorithm with online secondary path modelling (using Eriksson's method) for time-varying primary and secondary paths. (**b**) Experiment 9: Variation of EMSE with respect to time for a DNANC system and time-varying input noise with CSA, DE, and PSO as the primary meta-heuristic algorithm. All results plotted have been averaged over ten independent trials

the results. In order to verify the effectiveness of the proposed scheme for noise control without secondary path modeling, the results obtained in this experiment has been compared with that obtained using an FsLMS algorithm, with online secondary path modeling using the Eriksson's method [10]. The improvement in noise mitigation as well the enhanced adaptability of the proposed scheme is clear from Fig. 23.10a.

Experiment 9

The input noise considered in this experiment consists of tonal noise given by (11) followed by logistic chaotic noise given by (12), which is further followed by random noise uniformly distributed between $[-0.5, 0.5]$. Each of the noise type considered in this experiment exists for an equal duration during the experiment. The variation of EMSE with time for a DNANC system is shown in Fig. 23.10b. The robustness of the DNANC system to time-varying nature of the primary sound is clear from the results.

Concluding Remarks

A DNANC system, which is robust to dynamically changing sound inputs as well as to dynamically changing acoustic system environments, is proposed in this chapter. The dynamic behaviour of the proposed scheme is attributed to the integration of a multi-objective evolutionary algorithm like NSGA-II to an evolutionary-based NANC system. The DNANC framework is a step towards achieving an effective sound cancelling mechanism for ANC systems without the need for secondary path modeling.

Acknowledgements This work was supported by the Department of Science and Technology, Government of India under the Fast Track Scheme for Young Scientists (SERB/ET-0018/2013).

References

1. Akhtar MT, Abe M, Kawamata M (2005) A new structure for feed forward active noise control systems with improved online secondary path modeling. IEEE Trans Speech Audio Proc 13:1082–1088
2. Akhtar MT, Abe M, Kawamata M (2006) A new variable step size LMS algorithm-based method for improved online secondary path modeling in active noise control systems. IEEE Trans Audio Speech Lang Process 14:720–726
3. Ardekani IT, Abdulla WH (2012) Effects of imperfect secondary path modeling on adaptive active noise control systems. IEEE Trans Control Syst Technol 20:1252–1262

4. Aslam MS, Raja MAZ (2015) A new adaptive strategy to improve online secondary path modeling in active noise control systems using fractional signal processing approach. Signal Process 107:433–443
5. Behera SB, Das DP, Rout NK (2014) Nonlinear feedback active noise control for broadband chaotic noise. Appl Soft Comput 15:80–87
6. Chang CY, Chen DR (2010) Active noise cancellation without secondary path identification by using an adaptive genetic algorithm. IEEE Trans Actions Instrum Measur 59:2315–2327
7. Das DP, Panda G (2004) Active mitigation of nonlinear noise processes using a novel filtered-s LMS algorithm. IEEE Trans Speech Audio Proc 12:313–322
8. Deb K (2001) Multi-objective optimization using evolutionary algorithms. Wiley, Chichester
9. Deb K, Pratap A, Agarwal S, Meyarivan T (2002) A fast and elitist multi-objective genetic algorithm: NSGA-II. IEEE Trans Evol Comput 6:182–197
10. Eriksson LJ, Allie MC (1989) Use of random noise for on-line transducer modeling in an adaptive active attenuation system. J Acoust Soc Am 85:797–802
11. George NV, Panda G (2012a) A particle swarm optimization based de- centralized nonlinear active noise control system. IEEE Trans Instrum Meas 61:3378–3386
12. George NV, Panda G (2012b) A robust filtered-s LMS algorithm for nonlinear active noise control. Appl Acoust 73:836–841
13. Gholami-Boroujeny S, Eshghi M (2012) Non-linear active noise cancellation using a bacterial foraging optimisation algorithm. IET Signal Proc 6:364–373
14. Kuo SM, Morgan DR (1999) Active noise control: a tutorial review. Proc IEEE 87:943–975
15. Patel V, Gandhi V, Heda S, George NV (2016) Design of adaptive exponential functional link network-based nonlinear filters. IEEE Trans Circuit Syst I Regul Papers 63:1434–1442
16. Rout NK, Das DP, Panda G (2011) Particle swarm optimization based active noise control algorithm without secondary path identification. IEEE Trans Instrum Meas 61:554–563
17. Rout NK, Das DP, Panda G (2016) Particle swarm optimization based nonlinear active noise control under saturation nonlinearity. Appl Soft Comput 41:275–289
18. Sicuranza GL, Carini A (2011) A generalized FLANN filter for nonlinear active noise control. IEEE Trans Audio Speech Lang Process 19:2412–2417
19. Spiriti E, Morici S, Piroddi L (2014) A gradient-free adaptation method for nonlinear active noise control. J Sound Vib 333:13–30
20. Storn R, Price K (1997) Differential evolution-A simple and efficient heuristic for global optimization over continuous spaces. J Glob Optim 11:341–359
21. Tan LZ, Jiang J (1997) Filtered-X second-order Volterra adaptive algorithms. Electron Lett 33:671–672
22. Tyagi S, Katre V, George NV (2014) Online estimation of secondary path in active noise control systems using Generalized Levinson Durbin algorithm. In: Proceedings of the 19th international conference on digital signal processing, pp 552–555
23. Yang XS, Deb S (2009) Cuckoo search via L´evy flights. In: Proceedings of IEEE world congress on nature & biologically inspired computing, pp 210–214
24. Zhang M, Lan H, Ser W (2001) Cross-updated active noise control system with online secondary path modeling. IEEE Trans Speech Audio Proc 9:598–602
25. Zhang M, Lan H, Ser W (2005) On comparison of online secondary path modeling methods with auxiliary noise. IEEE Trans Speech Audio Proc 13:618–628
26. Zhao H, Zeng X, Zhang J (2010) Adaptive reduced feedback FLNN filter for active control of nonlinear noise processes. Signal Process 90:834–847

Chapter 24
Scheduling of Jobs on Dissimilar Parallel Machine Using Computational Intelligence Algorithms

Remya Kommadath and Prakash Kotecha

Abstract The success of Computational Intelligence (CI) techniques to solve combinatorial scheduling problem critically depends on efficient modelling of the problem. In this chapter, we study the scheduling of a set of jobs with different release and due dates on a set of dissimilar parallel machine problems to minimize the processing cost. The performance of five recent CI techniques *viz.,* Artificial Bee Colony, Dynamic Neighborhood Learning based Particle Swarm Optimizer, Genetic Algorithm, Multi-Population Ensemble Differential Evolution (MPEDE) and Sanitized Teaching-Learning based Optimization is evaluated on problems reported in the literature. It was observed from 750 unique trials (5 problems $\times$ 2 datasets $\times$ 5 algorithms $\times$ 15 runs) that MPEDE showed superior performance to the other four algorithms for larger problems.

Keywords Scheduling · Artificial Bee Colony · Particle Swarm Optimization · Genetic algorithm · Differential evolution · Teaching-learning based optimization

Introduction

Scheduling involves the allocation of tasks to the available resources in order to accomplish certain objectives within the specified time. In an increasingly competitive environment with several uncertainties, it becomes critical for organizations to optimally employ their limited resources for completing a given set of tasks to increase their profitability. Scheduling problems are combinatorial in nature and these have received significant attention from researchers working in the area of operations research. In a typical machine-scheduling problem, the jobs represent the tasks that are to be performed whereas the machines indicate the available resources.

R. Kommadath · P. Kotecha (✉)
Department of Chemical Engineering, Indian Institute of Technology Guwahati, Guwahati, Assam, India
e-mail: remya@iitg.ac.in; pkotecha@iitg.ac.in

F. Bennis, R. K. Bhattacharjya (eds.), *Nature-Inspired Methods for Metaheuristics Optimization*, Modeling and Optimization in Science and Technologies 16,
https://doi.org/10.1007/978-3-030-26458-1_24

The aim is to sequence and schedule the processing of tasks by the set of available resources to achieve the optimal solution with respect to certain objective(s). Some of the common objectives include minimization of makespan [17], minimization of the total processing cost [11, 12], and minimization of inventory cost [7]. These problems are predominantly solved using mathematical programming techniques by explicitly posing the problem in a Mixed-Integer Linear Programming (MILP) or Mixed-Integer Non-Linear Programming (MINLP) framework. Some other techniques such as Constraint Programming (CP), heuristic methods as well as stochastic optimization techniques have also been used in relatively fewer instances. The scheduling of a set of tasks with release and due date has been considered in the literature [12] and has been solved using MILP, CP, combined MILP-CP OPL model and a hybrid MILP/CP model. This work was further studied [10] with the development of advanced Benders cuts for the objective of minimization of makespan for the sub-problems. This work also relied on hybridizing CP and MILP. Minimization of makespan for parallel batch processing machines with unequal ready time and arbitrary job sizes have been solved using a combination of MILP and compound algorithm along with three different heuristic techniques [4]. Several surveys are available for the scheduling problems that discuss various aspects of the problem [1, 27].

Traditional mathematical programming techniques have been reported to be inefficient to solve large-scale scheduling problems due to an almost exponential increase in the number of binary variables and constraints. Computational Intelligence techniques possess several advantages such as (i) ability to handle nonlinearities, (ii) accommodates conflicting objectives, (iii) provides value-added solutions, (iv) solves black-box optimization problems, (v) does not require information about gradients, (vi) can be tuned to accommodate problem specific operators, (viii) does not require the problem to be postulated in the conventional equality and inequality form, (ix) the number of decision variables and constraints do not exponentially increase with an increase in the problem size, and (x) can be easily integrated with parallel and GPU computing. The success of these techniques for real parameter optimization has been significantly larger than combinatorial optimization problems. Despite their advantages, it should be noted that these techniques need to be judiciously applied to combinatorial optimization problems. These might exhibit poor performance if these are used to solve such problems posed in the conventional MILP or MINLP models that use a large number of artificial variables. In many instances, their success is critically dependent on appropriate choice of decision variables [2, 3] and their bounds, modification or incorporation of specific operators [9] that exploit the structure of the problem. Even though these techniques do not explicitly guarantee the global optima, these still can determine multiple reasonable schedules in a lower time as compared to the traditional methods [21]. These multiple solutions can provide flexibility to the users for selecting the appropriate schedule as per their requirements.

A hybrid version of Genetic Algorithm (GA) was utilized for minimizing the makespan for scheduling parallel batch processing machines with arbitrary job sizes [14]. The scheduling of jobs with flexibility in assigning machines to the different operations of each job has been solved [26] using a modified GA that employs rule-based assignment methods for creating initial population and also incorporates multiple strategies for reproduction and variation of the solutions. A combination of GA and Simulated Annealing (SA) has been proposed [30] to solve the multi-objective job shop scheduling with minimization of makespan, total work load of machines and critical machine work load as the objective functions. The parallel machine scheduling problem has also been solved with SA [16] and a Discrete Particle Swarm Optimization along with a hybrid variant [13]. A memetic algorithm based on Particle Swarm Optimization (PSO), SA, Nawaz-Enscore-Ham heuristic has also been devised to solve the flowshop scheduling problem [19]. Another hybrid version with PSO and a local search technique has also been reported [24] to solve the multi-objective job shop scheduling.

A large number of CI techniques [8, 23], Punnathanam and Kotecha [28] are developed every year but some of the proposed techniques are demonstrated on trivial optimization problems and their performance is not critically evaluated on benchmark optimization problems. In many cases, there have been issues in independently implementing the algorithms due to incomplete description and conflicting results reported in the literature [5, 22] including unfair comparisons. Some of these techniques have shown to be significantly better than the traditional techniques such as GA, SA and PSO on real parameter optimization problems but they have not been used or benchmarked with other algorithms for solving scheduling problems. In this work, we attempt to address this lacuna by solving the job shop scheduling problem with five recently proposed optimization techniques on various instances of a combinatorial scheduling problem. It should be noted that it has been reported that neither MILP nor CP were able to solve this problem on a standalone basis [12, 15].

The chapter is organized as follows: In the succeeding section, we provide a detailed description of the problem statement and the solution strategy. In Sect. 24.3, we briefly describe the five techniques which have been considered in this work. Section 24.4 presents the experimental settings, whereas the results and its detailed analysis is provided in Sect. 24.5. We conclude by summarizing the developments in this work and suggest possible future work.

Problem Statement

In this problem, a set of independent orders (I) need to be processed on a set of dissimilar parallel machines (M). A machine can process multiple orders whereas an individual order cannot be processed on multiple machines. It should be noted that a machine could process only one order at a given point of time. The processing cost and time depend on the order as well as the machine used for processing the order. The notations c_{im} and p_{im} denote the processing cost and processing time respectively of order i on machine m. Every order is associated with a release date (r_i) and due date (d_i). Pre-emption or extension of processing of orders is not permitted i.e., the processing of any order must be performed on or after the release date and has to be completed on or before the due date. The objective is to determine the minimum cost schedule such that all the orders are completed before their due dates subject to the satisfaction of all the constraints. Typically, for a MILP formulation, this problem is modelled with the inclusion of binary decision variables, say x_{im} which will be assigned a value of 1 if order i is processed on machine m and zero otherwise. Another binary variable y_{ij} is used to sequence the orders on a given machine and assumes a value of one if and only if the ith and j^{th} order are processed by the same machine with the ith order preceding the jth order. Another set of continuous variables are employed to determine the start and end times. The number of discrete variables can exponentially increase with an increase in the number of order and machines. Such a modeling strategy may not yield successful results with respect to CI techniques. The fact that CI techniques do not require an explicit optimization formulation has been exploited to efficiently model this problem [15] as described below.

Decision Variables A careful analysis of the problem would indicate that there are only two decision variables associated with every order, i.e., (i) the machine used to process a particular order and (ii) the start time of the order on the machine. Hence the problem can be efficiently modelled with $2I$ decision variables, which is not only independent of the number of machines but does not exponentially increase with an increase in the number of orders. For a problem with I orders, the decision variables can be divided into two distinct sets as in Fig. 24.1.

The first set indicates the machine that is assigned to each order whereas the second set indicates the starting time of each order. The first decision variable indicates the machine to which the first order is assigned, the second decision variable indicates the machine to which the second order is assigned and so on until the Ith decision variable. The $(I+1)^{\text{th}}$ variable indicates the starting time of the first order, the $(I+2)^{\text{nd}}$ decision variable indicates the starting time of the second order and so on till the starting time of the Ith order in the $2I$ position. For example, Eq. (24.1) shows a potential solution for the assignment of four machines to five orders.

$$X = \begin{bmatrix} \overbrace{2\;3\;1\;4\;3}^{Machines} & \underbrace{1\;3\;5\;2\;7}_{Starting\ time} \end{bmatrix} \tag{24.1}$$

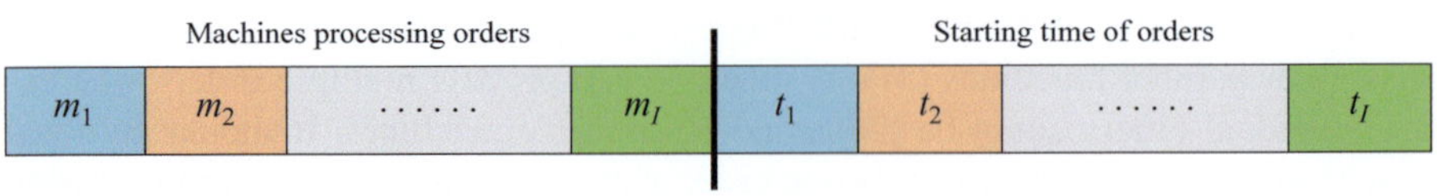

Fig. 24.1 Example of a solution vector

In the above example, order I_1 is assigned to machine M_2, order I_2 is assigned to machine M_3, order I_3 is assigned to machine M_1 whereas orders I_4 and I_5 are assigned to machine M_4 and machine M_3 respectively. This selection of the representation of the decision variables inherently satisfies the condition that an order is not assigned to more than one machine. The second set of decision variable indicates that I_1 which is assigned to M_2 is started at time 1. Similarly, I_2 (assigned to M_3) starts at time 3 whereas I_3, I_4 and I_5 start at time 5, 2, and 7 respectively on their specified machines.

Bounds The lower and upper bound is identical for all the decision variables in the first set with 1 being the lower bound and the maximum number of machines available being the upper bound. The lower bound for the second set of decision variables is the release date whereas the upper bound is the difference between the due date and the minimum processing time required for the order among all the machines. The lower bound of starting time ensures the processing of orders does not begin before the release date. Thus, unlike the first set of decision variables, the variables in the second set of decision variables need not have an identical lower and upper bounds.

Constraints The following three constraints are incorporated to ensure that a feasible schedule is obtained.

(i) **Due date:** The completion time of order i is the sum of the starting time and the processing time of the order on its assigned machine as in Eq. (24.2) and it should be less than or equal to the due date of the order.

$$end_i = t_i + p\left(i, m_i\right) \quad \forall i \in I \tag{24.2}$$

As per the decision variable in Eq. (24.1), $m_1 = 2$ (since order 1 is assigned to machine 2) $m_2 = 3$; $m_3 = 1$; $m_4 = 4$ and. Thus the violation in the due date constraint can be determined as Eq. (24.3).

$$P_i^{due} = \begin{cases} 0 & if\ \ d_i \geq end_i \\ |end_i - d_i| & otherwise \end{cases} \quad \forall i \in I \tag{24.3}$$

(ii) **Overlapping of Orders:** If more than a single order is processed on the same machine, then the start time of the subsequent order should not be lower than the end time of the preceding order. The penalty for violation of this requirement is calculated as Eq. (24.4).

$$P_i^m = \begin{cases} 0 & if\ \ end_i < t_{i+1} \\ |end_i - t_{i+1}| & otherwise \end{cases} \quad \forall i \in I_m, m \in M : t_i < t_{i+1} \tag{24.4}$$

It should be noted that the penalty is incurred if there is an overlapping with respect to the immediate succeeding order on that machine. In order to understand the above constraint, consider the following example, which has nine orders and three machines.

$$X = \left[\overbrace{3\ 1\ 2\ 1\ 2\ 1\ 2\ 1\ 3}^{Machines} \quad \underbrace{8\ 4\ 6\ 14\ 9\ 7\ 11\ 5\ 2}_{Starting\ \ time} \right]$$

As per the encoding scheme explained earlier, it can be observed that I_2, I_4, I_6 and I_8 are processed on M_1 whereas I_3, I_5 and I_7 are processed on M_2 with the remaining two orders processed on M_3. Without loss of generality and for ease in understanding, let us consider that the processing time of all the nine orders on their respective machines is 3 time units. As per Eq. (24.4),

- I_2 begins at 4 and will end at 7. However, I_8 on M_1 begins at 5. Thus there is overlapping of orders and hence $P_1^1 = 2$.
- I_8 begins at 5 and will end at 8. However, I_6 on M_1 begins at 7. Thus there is overlapping of orders and hence $P_2^1 = 1$.
- I_6 begins at 7 and will end at 10. However, I_4 on M_1 begins at 14. Thus there is no overlapping of orders and hence $P_3^1 = 0$.

Thus the total penalty due to the overlapping of orders on M_1 is 3. A similar analysis will yield that the total penalty is 1 for M_2 and 0 for M_3.

(iii) **Maximum Processing Time:** The difference between the earliest release date and the maximum due date among all the orders assigned to a particular machine provides an upper bound on the maximum time available to a machine and is determined as per Eq. (24.5).

$$H_m = \max_{i \in I_m} (d_i) - \min_{i \in I_m} (r_i) \quad \forall m \in M \tag{24.5}$$

The total amount of time required for completing all the orders on a machine can be determined as Eq. (24.6).

$$T_m = \sum_{i=1}^{I_m} p\,(i, m_i) \quad \forall m \in M \tag{24.6}$$

All the orders assigned to a particular machine should be completed within the available time. The amount of penalty for the violation of this constraint is calculated using Eq. (24.7).

$$P_m = \begin{cases} 0 & if \;\; H_m > T_m \\ T_m - H_m & otherwise \end{cases} \quad \forall m \in M \tag{24.7}$$

The total penalty for violation in the various constraints can be calculated as Eq. (24.8).

$$Penalty = \sum_{i=1}^{I} P_i^{due} + \sum_{m=1}^{M} \sum_{i=1}^{I_m} P_i^m + \sum_{m=1}^{M} P_m \tag{24.8}$$

Fitness Function The objective function is the determination of the least cost schedule for processing all the orders. The cost associated with a potential schedule is calculated using Eq. (24.8)

$$f = \begin{cases} \sum_{i=1}^{I} c\left(i, m_i\right) & if \;\; Penalty = 0 \\ \sum_{i=1}^{I} c\left(i, m_i\right) + Penalty + \sum_{i \in I} \max_{m \in M} \left\{c\left(i, m\right)\right\} & otherwise \end{cases} \tag{24.9}$$

For a feasible solution, the sum of all penalty must be equal to zero. An infeasible solution is penalized by the amount of violation of constraints as well as the maximum cost associated to process an ith order. This ensures that an infeasible solution will not have a better fitness value in comparison with a feasible solution. It should be noted that the above strategy can be easily incorporated in CI techniques but cannot be incorporated in mathematical programming technique as m_i is a decision variable that is used as an index to access the corresponding data. The p-codes of the objective functions and Gantt charts can be downloaded from https://goo.gl/4eVNL5.

Algorithm Description

In this section, we provide a brief description of the algorithms and additional details can be obtained from the relevant cited literature. These algorithms have not only been selected in view of their popularity but also because of the availability of their exact implementations. This helps in independent verification of the results and avoids the criticism that the poor performance of an algorithm is due to its improper implementation.

Artificial Bee Colony

ABC is a swarm intelligence based CI technique which imitates the intelligent behavior of honey bee swarm in determining the food sources. A potential solution of the optimization problem is referred to as food source and the fitness of the solution is referred as the nectar amount in the food source. Three types of bees are used in ABC to search the optimal solution (i) employed bees, (ii) the onlooker bees and (iii) the scout bees. In the first stage, the employed bees determine new food sources based on the food sources that it had visited previously and on the information received from other employed bees. The employed bee changes its position if it is able to determine a better food source. In the second stage, the onlooker bees determine new food sources similar to the employed bees but also use the information about the nectar content of food sources from the employed bees. The onlooker bees update their positions if they are able to determine a better food source than their previous food source. The solution, which repeatedly fails a specified number of times to determine a better solution in the employed bee phase or the onlooker phase, is replaced by a random solution in the scout phase of the algorithm. These three stages are repeatedly performed until the termination criterion is achieved. The scout bees are reported to explore the search space, as it does not utilize the information about the previous position whereas the onlooker and employed bees are reported to perform the exploitation of the search space. The food source discovered with the largest nectar content is considered as the best solution. The critical issues in the implementation of ABC are discussed in [22]. In this work, we have used the implementation of ABC that is publicly available (https://goo.gl/zqeujq).

Dynamic Neighborhood Learning Based Particle Swarm Optimizer (DNLPSO)

PSO is one among the most popular swarm intelligence technique and a large number of variants have been proposed to improve its performance. In PSO, the solutions are termed as particles and new potential solutions are determined using the historical best of a particular particle, (known as personal best), as well as the global best of all the particles. Comprehensive Learning PSO was proposed [18] to increase the explorative nature of the algorithm by (i) removing the dependency on the global best and (ii) relying on the personal best of other solutions rather than its own personal best. However, it has been reported that this leads to a reduction in the convergence rate. Dynamic Neighborhood Learning based Particle Swarm Optimizer (DNLPSO) [25] is a modified version of CLPSO in which each particle updates its velocity with the help of its personal best or the exemplar particle that is selected from its neighborhood and also the global best. The formation of the neighborhood with a ring topology has been recommended though other topologies can also be implemented. The learning from the neighborhood solution is implemented only if the personal best does not improve for a specified number (refreshing gap) of consecutive iterations. In order to encourage exploration, a dynamic neighborhood is employed which is formed after a specified number of iterations. In this work, we have employed the implementation of DNPLSO provided by the authors (http://bit.ly/30mLciT).

Genetic Algorithm (GA)

GA [6] is probably the most popular and one of the earliest evolutionary techniques. It is based on the working principles of natural genetics and natural selection. The search for an optimal solution begins with randomly generated solutions within the search space that constitutes the population. Each solution in the population is termed as chromosome whose fitness is represented by the value of the objective function. In every generation, reproduction operators are used to form the mating pool by choosing multiple copies of better chromosomes. Variation operator such as crossover and mutation are employed to generate offspring which are potential solutions. The occurrence of crossover or mutation of a chromosome is decided based on the values of user-defined parameters such as crossover and mutation probability. The reproduction, crossover and mutation operators perform the exploration and exploitation of the search space. At the end of every generation, the best chromosomes among the parents and offspring progress to the next generation. In this work, we have used the inbuilt *ga* function of MATLAB2016a.

Multi-population Ensemble Differential Evolution (MPEDE)

DE is also a popular evolutionary technique that has found applications in diverse areas. It can employ multiple mutation strategies and it is not always possible to select the best mutation strategy for an arbitrary problem. A variant of DE, MPEDE [29] embeds three mutation strategies in its working.

In particular, the population is divided into four subgroups in every generation, three of which are equal in size and are known as the indicator population whereas the remaining members constitute the reward population that is comparatively larger. Each indicator population is assigned a different mutation strategy (Current-to-pbest/1" and "rand/1" with the uniform binary crossover and Current-to-rand/1 without any crossover). It also employs an adaptive strategy to tune various parameters required in the mutation strategies. After periodic intervals, the best performing strategy, determined with respect to the improvement in fitness value and the number of functional evaluations utilized, is assigned to the reward population. The implementation of MPEDE provided by the authors is used in this work (http://bit.ly/33LktPc).

Sanitized Teaching-Learning Based Optimization (s-TLBO)

TLBO is a population based stochastic technique and has been proposed multiple times in literature. It is inspired by the concept of knowledge transfer in a classroom environment through the teaching and learning process. A solution is termed as student and the marks scored in various subjects offered to the student is analogous to the values of the design variables in the optimization problem. A set of solutions that constitutes the population is known as the class. The solution that has the best fitness is termed as teacher. Each iteration of TLBO involves two phases, viz., teacher phase and learner phase. In the teacher phase, a potential solution is generated for every member of the class based on the teacher, marks of the particular member, teaching factor and mean of the class. In the student phase, a potential solution is generated for every member of the class depending on the marks of the particular member and marks of a randomly selected learning partner. In the teacher phase as well as the learner phase, a new solution is accepted if it is better than the solution undergoing the teaching-learning process. In view of the various discrepancies reported in the literature [5], we have developed an in-house code which does not remove any duplicates and provides a deterministic relation between the maximum number of objective function evaluations and the number of iterations in s-TLBO (https://goo.gl/RGsbNM).

Experimental Settings

We have considered ten instances of job shop scheduling problems that have been widely used in literature [12, 15, 20] These instances arise from five problems with the smallest problem containing 3 orders and 2 machines and the largest problem comprises of 20 orders and 5 machines. The details of the five problems are provided in Table 24.1. Each of the five problems has two datasets arising from the difference in processing time. The ten instances are labeled as P1S1, P1S2 and so on till P5S2 in which 'P1' corresponds to Problem 1 and 'P5' corresponds to Problem 5 whereas 'S1' and 'S2' indicate to the two datasets of a particular problem. For the sake of brevity, the due date, release date, processing time and cost are provided in Table 24.2 for Problem 1. For the rest of the eight instances, the data can be obtained from the literature (https://goo.gl/QPM0G4). The first instance indicated by 'S1' have longer processing times and have been reported to be more complex as these contain fewer feasible schedules than the second instance S2.

Very often, there are conflicting claims about the performance of CI techniques, which are primarily attributed to improper implementation of the algorithm and the absence of the details regarding the values of algorithm-specific parameters. The details of the various algorithm-specific parameters are provided in Table 24.3 and are primarily based on the recommendations in the literature by their authors. In view of the stochastic nature of the metaheuristic techniques, 15 independent runs are executed for each of the ten instances. Hence a total of 750 unique trials (5 Problems × 2 datasets × 15 runs × 5 algorithms) are executed to evaluate the performance of the five algorithms. All the simulations were performed on MATLAB 2016a on an Intel i7 processor (3.6 GHz) PC with 16 GB RAM.

The population size N_p for each algorithm is a function of the number of decision variables of the problem and is set as 20*D*, where *D* is the number of decision variables of the problem (given in Table 24.1). For a fair comparison between the algorithms, the termination criterion for all the algorithms is set as the maximum allowed functional evaluations (*MaxFE*). The inbuilt *ga* function of MATLAB requires N_p (N_g + 1) + 1 objective function evaluation for performing N_g generations where N_p is the population size. The *ga* function is used for 500 generation and hence the maximum number of functional evaluation for all the algorithms is given by

$$MaxFE = 501N_p + 1 \tag{24.10}$$

CI techniques are primarily designed to solve the optimization problem with continuous decision variables. However, the decision variable corresponding to the machine used by a particular order has to be an integer. In order to accommodate these integer variables, the rounding scheme is employed as shown in Eq. (24.11)

$$\begin{aligned} y_i &= \lfloor y_c \rfloor \quad if\ \ y_c - \lfloor y_c \rfloor < 0.5, \\ y_i &= \lceil y_c \rceil \quad otherwise \end{aligned} \tag{24.11}$$

where y_c is the continuous variable determined by the algorithm and y_i is the integer value determined using this continuous variable. However, in the case of GA, the inbuilt function of MATLAB permits the use of integer variables and the same has been used to handle integer variables.

Results and Discussions

In this section, we evaluate the performance of five different stochastic population based CI techniques for solving the combinatorial job shop scheduling problems.

Table 24.1 Details of orders and machines for all the problems

Problem	Number of orders	Number of machines	Number of decision variables
P1	3	2	6
P2	7	3	14
P3	12	3	24
P4	15	5	30
P5	20	5	40

Table 24.2 Data for problem 1 [12]

					Durations			
			Cost on machine		Dataset 1		Dataset 2	
Order(i)	r_i	d_i	M1	M2	M1	M2	M1	M2
1	2	16	10	6	10	14	5	7
2	3	13	8	5	6	8	3	4
3	4	21	12	7	11	16	5	7

Table 24.3 Parameter settings for various algorithms

Algorithms	Parameters	Values
ABC	Number of onlooker bees	Population size
	Abandonment limit parameter	*round(0.6 × population size × problem dimension)*
	Upper bound of acceleration coefficient	1
DNLPSO	Acceleration coefficients	1.49445
	w_0	0.9
	w_1	0.4
	Refreshing group	3
	Regrouping gap	5
	Learning probability	$0.45 - 0.4\frac{(Population\ \ Index-1)}{(Population\ \ size-1)}$
	Parameter for determination of velocity bound	2
MPEDE	Fraction of indicator population for each strategy	0.2
	Best mutation strategy determining period	20

The performance of the algorithms in the individual runs is shown in Fig. 24.2. In P1S1, it can be seen that all the algorithms except DNLPSO are able to determine the best solution. In P1S2, it can be seen that the algorithms are able to either obtain the objective function value of 18 or 21. In P5S1, most of the runs of the algorithms are unable to reach the best solution and are clustered around the value of 400, while a few runs of MPEDE being able to reach the best solution. It can also be observed that DNLPSO is consistently unable to determine the best solution for any of the ten instances. Figure 24.3 shows the number of runs in which an algorithm was unable to determine a feasible solution. It can be observed that the number of infeasible runs in the dataset 1 is higher than for dataset 2. This can be attributed to the reason that the processing time in the second dataset is shorter and thus the number of feasible solutions is higher. In P1S2 and P2S2, all the algorithms are able to determine a feasible solution in all the runs. MPEDE has the lowest number of infeasible runs among all the algorithms followed by ABC.

The best, mean and standard deviation values of the best objective function value obtained in the fifteen runs by each algorithm are shown in Tables 24.4

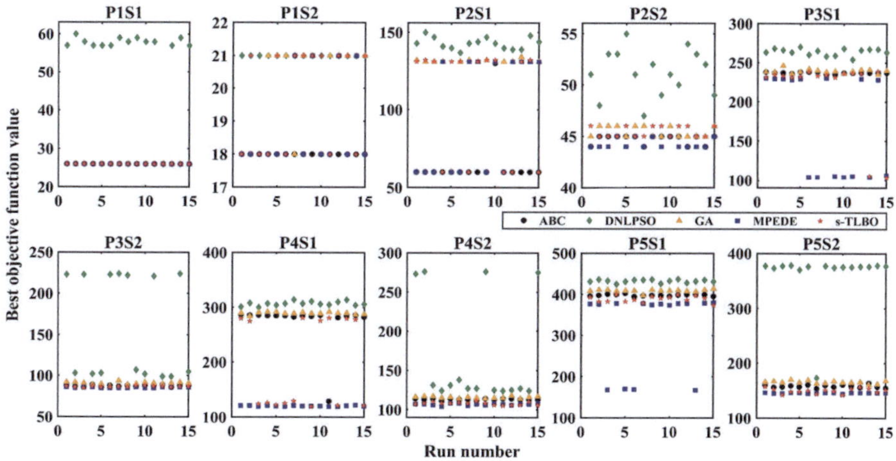

Fig. 24.2 Performance of the algorithms in the various runs

and 24.5. Table 24.4 also shows the results of Grey-Wolf Optimization and the JAYA algorithm which have been reported in recent literature [20] with identical termination conditions as used in this work. In Table 24.4, the values shown for GA (MATLAB) have been generated as described earlier whereas the value indicated against GA correspond to those provided in the literature [15] which does not clearly specify the termination criteria. It can be observed that the best value as well as the mean value obtained by all the algorithms for P1S1 are identical except for DNLPSO. Moreover, the value of zero to standard deviation implies that the algorithms are able to determine the global optima of this problem in all the runs. In the case of P1S2, all the algorithms except DNLPSO are able to determine the best value. Moreover, ABC is able to determine the solution in all the runs whereas s-TLBO showed the largest standard deviation. Thus in the case of P1, ABC seems to outperform all the other algorithms as it is able to consistently determine the best solution. The convergence curve of each algorithm for the best run of P1 is shown in Fig. 24.4. The values in y-axis show the best solution discovered until the completion of the specified number of functional evaluations denoted by the value in x-axis. It can be observed that all the algorithms are able to determine an identical solution in both instances except DNLPSO in P1S2. For all the problems, the Gantt chart is shown for the best solution that has been determined across the five algorithms. Figure 24.5 shows the Gantt chart for the two instances of P1. It can be observed that the processing time of the orders in the second instance are longer in P1S1 as compared to P1S2.

Similar to P1, ABC outperforms the other algorithms with its ability to determine the better solution in both the instances of P2. Moreover, as reflected in the mean values, it is able to determine better solutions consistently. MPEDE is also able to determine the best solution reported by ABC and has a better mean than all algorithms except ABC. Across all algorithms, the standard deviation is high in

Table 24.4 Statistical analysis of best objective function value obtained in problem 1 to problem 3

Problems	Algorithms	Best	Worst	Mean	Median	SD
P1S1	ABC	26	26	26	26	0
	DNLPSO	26	60	55.8	58	8.3
	GA (MATLAB)	26	26	26	26	0
	MPEDE	26	26	26	26	0
	TLBO	26	26	26	26	0
	GWO	26	–	26	26	0
	JAYA	26	–	26	26	0
	GA	26	26	26	–	0
P1S2	ABC	18	18	18	18	0
	DNLPSO	21	21	21	21	0
	GA (MATLAB)	18	21	20	21	1.46
	MPEDE	18	21	18.8	18	1.37
	TLBO	18	21	19.8	21	1.52
	GWO	18	–	18	18	0
	JAYA	18	–	18	18	0
	GA	18	21	18.72	–	1.28
P2S1	ABC	60	130	64.67	60	18.07
	DNLPSO	137	150	143	143	3.76
	GA (MATLAB)	60	134	126.67	131	18.46
	MPEDE	60	131	97.87	131	36.66
	TLBO	60	132	107.6	131	34.84
	GWO	60	–	117.9	132	28.9
	JAYA	60	–	117.9	132	28.9
	GA	–	–	–	–	–
P2S2	ABC	44	45	44.73	45	0.46
	DNLPSO	47	55	51.2	51	2.27
	GA (MATLAB)	45	46	45.47	45	0.52
	MPEDE	44	45	44.33	44	0.49
	TLBO	45	46	45.47	45	0.52
	GWO	44	–	45.7	46	1.3
	JAYA	44	–	45.7	46	0.6
	GA	–	–	–	–	–
P3S1	ABC	235	238	236.93	237	0.96
	DNLPSO	254	270	263.8	265	4.43
	GA (MATLAB)	235	246	239.13	239	2.75
	MPEDE	104	230	170.93	228	64.26
	TLBO	103	239	216.8	233	45.87
	GWO	102	–	217.9	233	42.3
	JAYA	233	–	237.4	237	2.5
	GA	–	–	–	–	–

(continued)

Table 24.4 (continued)

Problems	Algorithms	Best	Worst	Mean	Median	SD
P3S2	ABC	86	89	87.53	88	0.99
	DNLPSO	99	224	158.67	107	62.19
	GA (MATLAB)	87	94	90.73	91	1.67
	MPEDE	85	86	85.53	86	0.52
	TLBO	85	88	86.67	87	0.82
	GWO	86	–	90.1	90	1.8
	JAYA	88	–	90.2	90	1.3
	GA	–	–	–	–	–

P2S1. The performance of GA and s-TLBO are almost identical and these have not been able to determine the best solution of 44 even in one of their runs. The performance of DNLPSO is not satisfactory. The time required for completing all the orders in both the instances is 28 time units and corresponds to the due date of order 4. The convergence curve corresponding to the best run of every algorithm for P2 is shown in Fig. 24.6. It can be observed that in P2S1, ABC requires larger number of objective function evaluations to determine the best solution. In both instances, MPEDE shows the fastest convergence to the best solution. The Gantt chart for P2 is shown in Fig. 24.7.

In P3S1, the best value reported by GWO is better than all the five algorithms considered in this work. It is observed that s-TLBO reports the best solution of 103 for P3S1. Though the best solution reported by MPEDE is 104, it has a very impressive mean of 170.93 against the mean of 216.80 determined by s-TLBO and 217.90 determined by GWO. The performance of the rest of the three algorithms is not satisfactory as their best value is more than twice that of s-TLBO and MPEDE. In the case of P3S2, MPEDE and s-TLBO are able to determine an identical best solution with MPEDE being able to determine a better mean and a lower standard deviation. Figure 24.8 shows the convergence curve of the algorithms for P3. In both the instances, s-TLBO is able to converge faster than MPEDE to a better solution. The Gantt chart for P3 is given in Fig. 24.9. For P3S1, it can be observed that orders 4, 5, 6, 8 and 10 are processed on M1, the orders 2, 3, 9 and 12 are processed on M2 whereas the remaining three orders are processed on M3. In the case of P3S2, M1 is not used to process any order whereas eight orders are processed on M2 and four orders are processed on M3.

In P4S1, it can be observed that GWO was able to determine a better solution (best) compared to the identical solution of 119 determined by s-TLBO and MPEDE. However, the mean and standard deviation of MPEDE is significantly better than that of GWO and TLBO. Among the other algorithms, ABC performs better than GA and DNLPSO. In P4S2, MPEDE is able to determine the best solution and has a better mean than s-TLBO. However, the difference in performance for this problem is not as prominent as P4S1. In Fig. 24.10, it can be observed that all the algorithms have converged, with MPEDE and s-TLBO providing an identical solution for P4S1 whereas MPEDE converges to a better solution than s-TLBO in

Table 24.5 Statistical analysis of best objective function value obtained in the problem 4 to problem 5

Problems	Algorithms	Best	Worst	Mean	Median	SD
P4S1	ABC	129	287	274.47	285	40.26
	DNLPSO	300	314	306.93	307	4.08
	GA (MATLAB)	284	292	289.53	290	2.13
	MPEDE	119	122	120.2	120	0.94
	TLBO	119	281	195.87	129	80.27
	GWO	118	–	147	123	57.1
	JAYA	134	–	282.7	286	21.4
	GA	120	283	185.67	–	75.73
P4S2	ABC	111	114	113.2	113	0.86
	DNLPSO	124	276	166.87	127	67.6
	GA (MATLAB)	114	118	115.67	116	1.23
	MPEDE	104	108	106.73	107	1.16
	TLBO	105	111	107.73	108	1.75
	GWO	108	–	111.9	112	2.2
	JAYA	110	–	116.2	116	2
	GA	105	117	109.96	–	1.94
P5S1	ABC	395	403	398.47	398	2.23
	DNLPSO	425	437	432.27	433	3.67
	GA (MATLAB)	398	414	408.87	409	3.48
	MPEDE	167	380	321.6	377	95.58
	TLBO	374	400	388.93	391	7.75
	GWO	167	–	352.2	380	73.8
	JAYA	392	–	399.4	400	3.6
	GA	165	389	349.26	–	75.43
P5S2	ABC	154	164	158.2	157	2.68
	DNLPSO	173	378	362.13	376	52.36
	GA (MATLAB)	161	170	165.53	165	2.47
	MPEDE	144	146	145.67	146	0.62
	TLBO	142	157	149.67	150	4.7
	GWO	151	–	157.8	158	3.5
	JAYA	158	–	164.2	164	2.8
	GA	146	167	151.63	–	3.60

P4S2. The Gantt chart for P4 is shown in Fig. 24.11. For the first instance, all the machines are used whereas only three machines are used in the second instance. In both instances, all the orders are completed at 38 with order 9 being the last processed order.

In P5S1, it can be observed that the performance of MPEDE is significantly better than all the algorithms including s-TLBO which was otherwise performing reasonably similar to MPEDE in the other instances. It should be noted that the best solution determined by MPEDE and GWO are identical. Though the standard

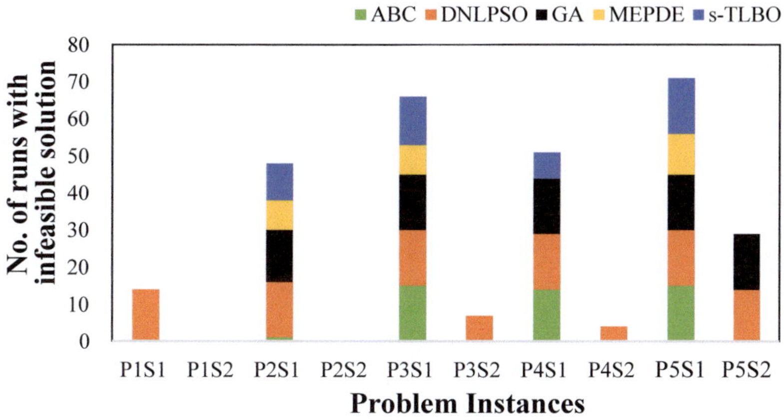

Fig. 24.3 Details of infeasible runs

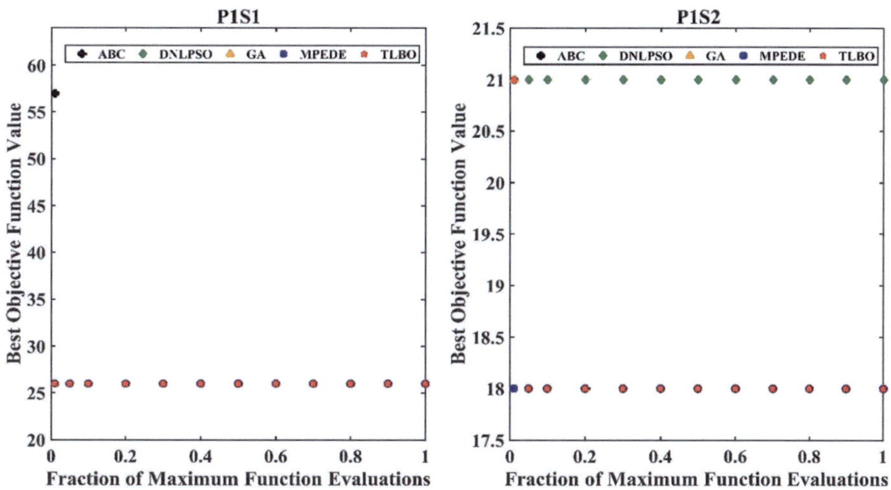

Fig. 24.4 Convergence curves for problem 1

deviation of MPEDE is higher than s-TLBO and GWO, the mean of MPEDE is better than that of both algorithms. In P5S2, the best solution determined by s-TLBO is better than that determined by MPEDE but the mean determined by MPEDE is better than that of s-TLBO. As in all the other problems, the performance of DNLPSO is the least satisfactory. Figure 24.12 shows the convergence curve for P5 and it can be observed that some of the algorithms have not converged but have been terminated due to the utilization of the maximum number of objective function evaluations. Unlike in other problems, MPEDE utilizes more than 80% of the objective function evaluations to determine the best solution. Figure 24.13 shows the Gantt chart for P5 and it can be observed that all the 20 orders are scheduled

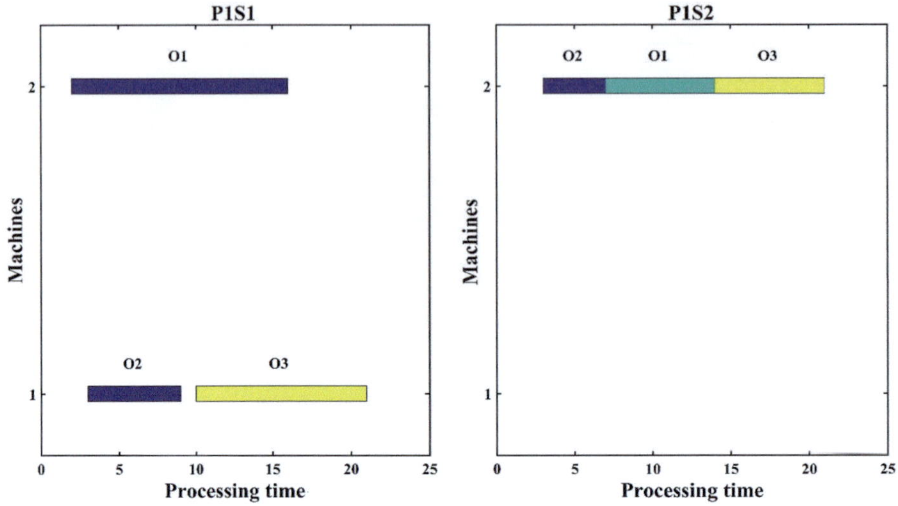

Fig. 24.5 Gantt chart for problem 1 (3 orders and 2 machines)

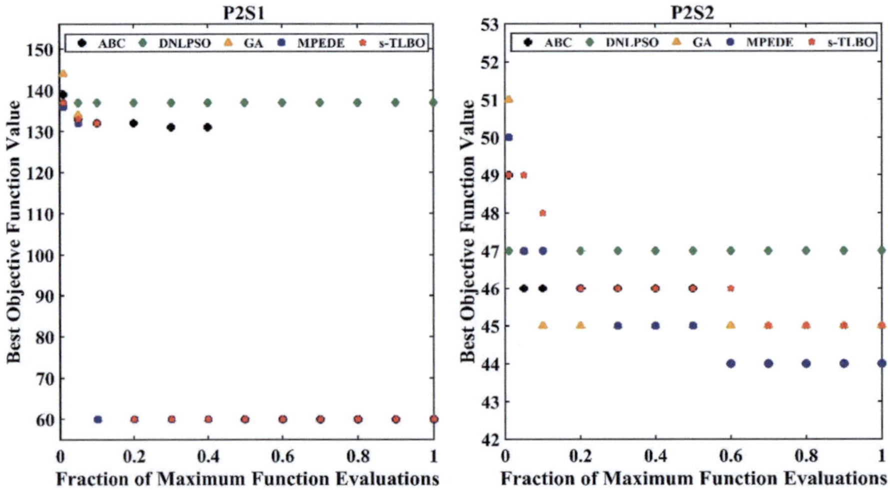

Fig. 24.6 Convergence curves for problem 2

without any conflict on any of the machines. Moreover, all the orders are processed in according to their release date and are completed by the due date. On analyzing the overall performance of algorithms on this scheduling problem, it is observed that MPEDE has consistently determined better solutions in majority of runs except in two smaller problems (P1S2 and P2S1).

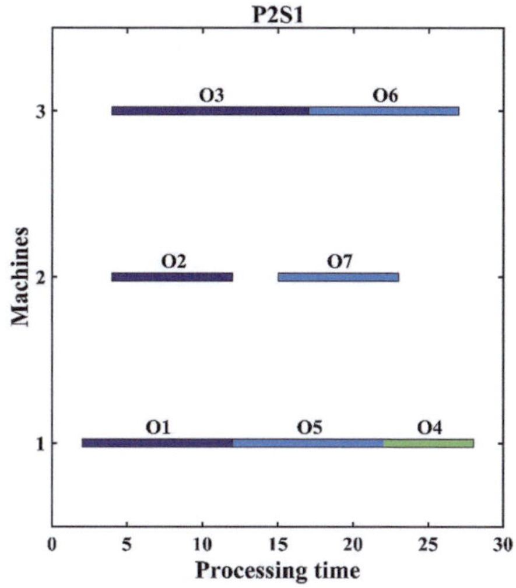

Fig. 24.7 Gantt chart for problem 2 (7 orders and 3 machines)

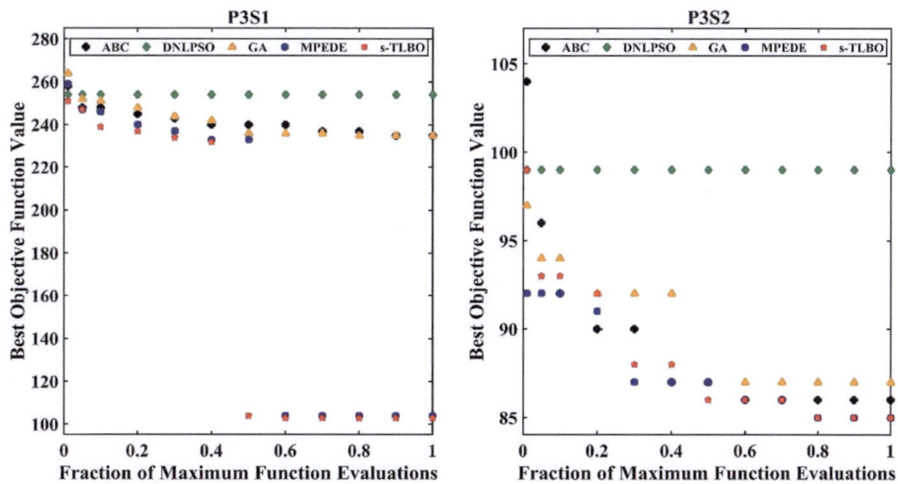

Fig. 24.8 Convergence curves for problem 3

Time Complexity

Time complexity provides a measure of the time required by an algorithm to solve the optimization problem. In this work, we determine the time complexity of the algorithm on every problem instance as per the following procedure that is predominantly based on the procedure used to benchmark computational intelligence techniques in the competitions held at the Congress on Evolutionary Computation.

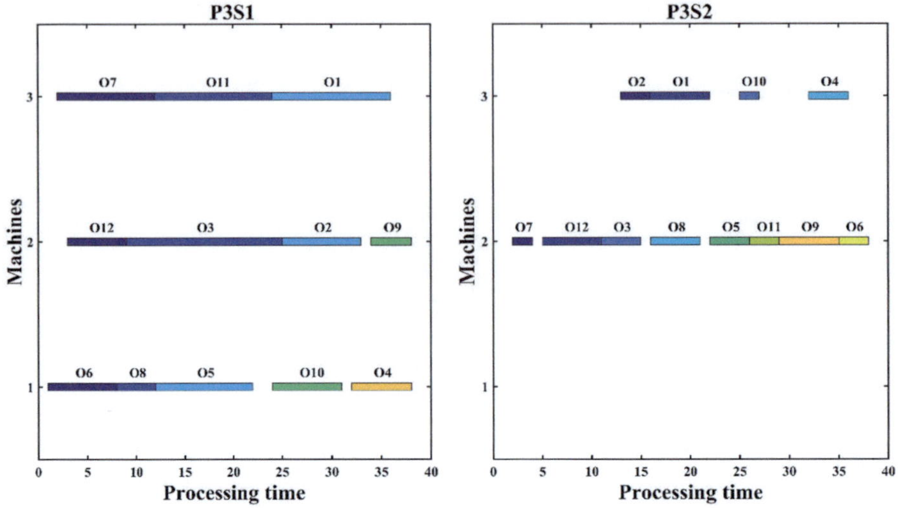

Fig. 24.9 Gantt chart for problem 3 (12 orders and 3 machines)

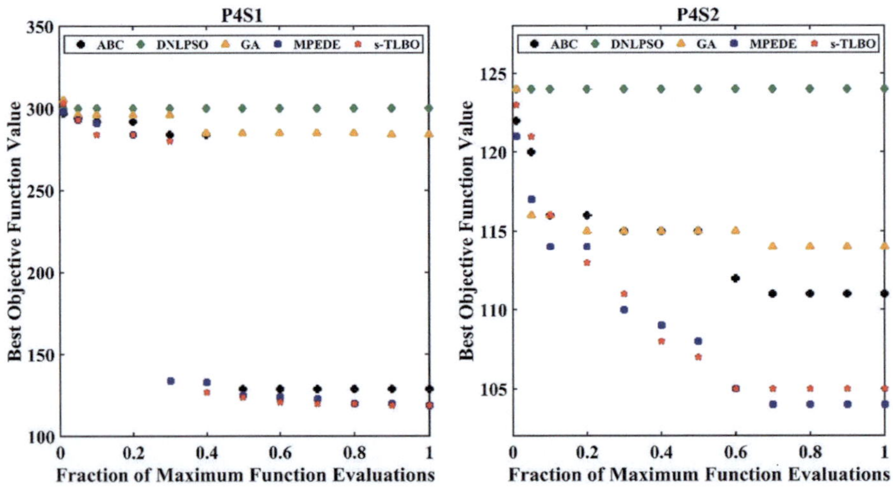

Fig. 24.10 Convergence curves for problem 4

Step 1: Determine the time required (T_M) for computing certain basic mathematical operations for 10^6 times.

Step 2: Determine the time required (T_O) to evaluate the objective function for MaxFe times.

Step 3: Determine the time required (T_A) for an algorithm to solve a particular instance (with MaxFe as the termination criterion). In view of the stochastic

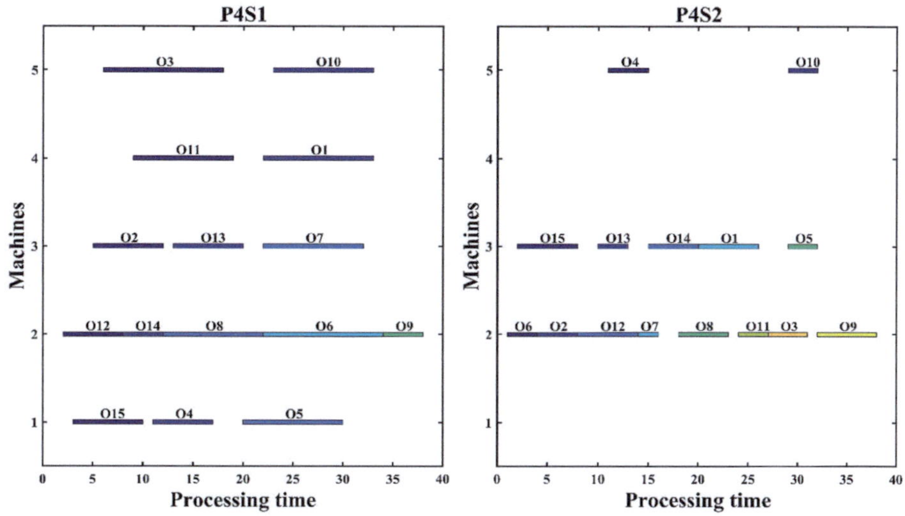

Fig. 24.11 Gantt chart for problem 4 (15 orders and 5 machines)

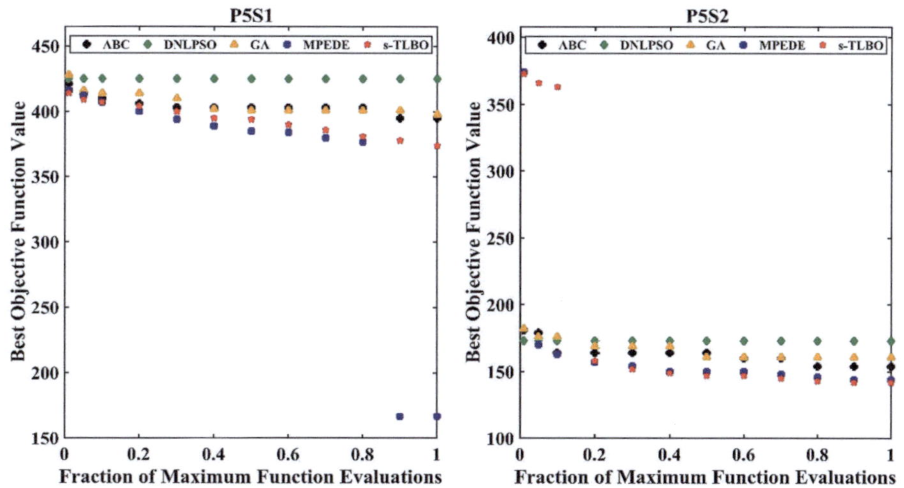

Fig. 24.12 Convergence curves for problem 5

nature of the algorithms, T_A is determined for five runs and its mean (T_{Av}) is used to evaluate time complexity.

Step 4: The time complexity of an algorithm is given by $(T_{Av} - T_O)/ T_M$.

It should be noted that Step 3 is to be performed for every combination of the algorithm and problem instance whereas Step 2 is independent of the algorithm and is to be executed for every problem instance. The time complexity for the 50 combinations (5 algorithms × 2 datasets × 5 problems) is depicted in Fig. 24.14.

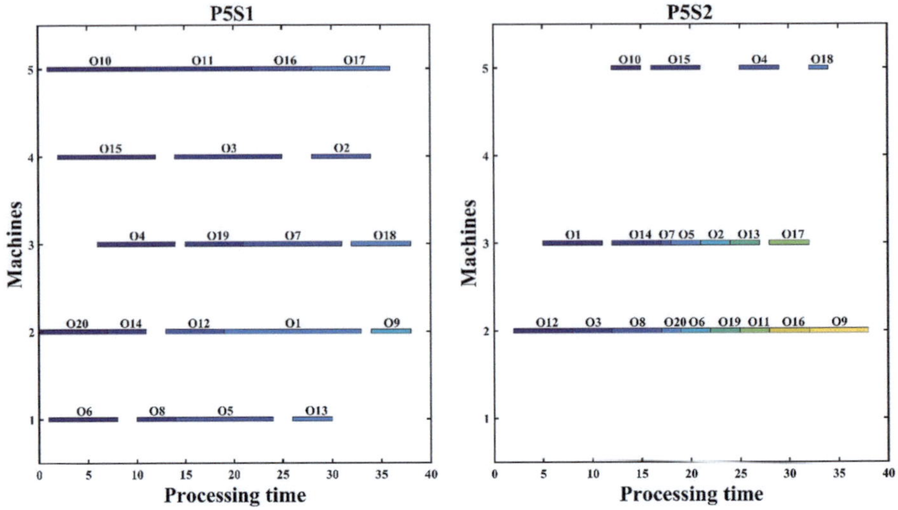

Fig. 24.13 Gantt chart for problem 5 (20 orders and 5 machines)

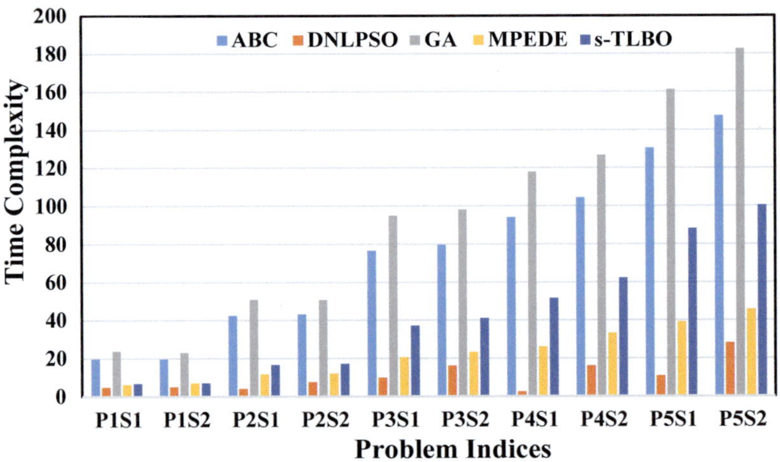

Fig. 24.14 Time complexity of the algorithms

Across all algorithms, it can be observed that the time complexity is higher for the second dataset than the first dataset. Among the algorithms, DNLPSO possesses the lowest time complexity whereas the inbuilt *ga* function of MATLAB possesses the highest time complexity. However, it should be noted that DNLPSO was not able to determine the best solution. The best solution is usually determined by either MPEDE or s-TLBO and it can be observed that MPEDE has a lower time complexity than s-TLBO. Moreover, the increase in the time complexity with an increase in the problem dimension is relatively lower for MPEDE.

Conclusion

The performance of five recent optimization techniques has been evaluated on the dissimilar job shop scheduling problems and it was observed that MPEDE was able to consistently determine the best solution followed by s-TLBO even for the larger problems. It was also observed that the first dataset (S1) which involves higher processing time are comparatively challenging to solve due to the presence of only a few feasible solutions. DNPLSO showed relatively poor performance but was able to generate better solutions in its initial stage. The time complexity of MPEDE does not significantly scale up with an increase in the size of the problem. Future work can involve the design of efficient schemes as well as hybridization among various algorithms to efficiently determine the optimal solution.

References

1. Allahverdi A, Ng CT, Cheng TCE, Kovalyov MY (2008) A survey of scheduling problems with setup times or costs. Eur J Oper Res 187(3):985–1032
2. Chauhan SS, Kotecha P (2018) An efficient multi-unit production planning strategy based on continuous variables. Appl Soft Comput 68:458–477
3. Chauhan SS, Sivadurgaprasad C, Kadambur R, Kotecha P (2018) A novel strategy for the combinatorial production planning problem using integer variables and performance evaluation of recent optimization algorithms. Swarm Evol Comput, 43:225–243
4. Chung SH, Tai YT, Pearn WL (2009) Minimising makespan on parallel batch processing machines with non-identical ready time and arbitrary job sizes. Int J Prod Res 47(18):5109–5128
5. Črepinšek M, Liu SH, Mernik L (2012) A note on teaching learning-based optimization algorithm. Inf Sci 212:79–93
6. Deb K (2001) Multi-objective optimization using evolutionary algorithms. Wiley, New York. ISBN 978-81-265-2804-2
7. Dobson G, Arai Yano C (1994) Cyclic scheduling to minimize inventory in a batch flow line. Eur J Oper Res 75(2):441–461
8. Eskandar H, Sadollah A, Bahreininejad A, Hamdi M (2012) Water cycle algorithm – a novel metaheuristic optimization method for solving constrained engineering optimization problems. Comput Struct 110–111:151–166
9. Hasda RK, Bhattacharjya RK, Bennis F (2017) Modified genetic algorithms for solving facility layout problems. Int J Interact Des Manuf 11(3):713–725
10. Hooker JN (2005) A hybrid method for the planning and scheduling. Constraints 10(4):385–401
11. Hooker JN, Ottosson G, Thorsteinsson ES, Kim HJ (1999) On integrating constraint propagation and linear programming for combinatorial optimization. In: Proceedings of 16th national conference on artificial intelligence. Orlando, MIT Press, pp 136–141
12. Jain V, Grossmann IE (2001) Algorithms for hybrid MILP/CP models for a class of optimization problems. INFORMS J Comput 13(4):258–276
13. Kashan AH, Karimi B (2009) A discrete particle swarm optimization algorithm for scheduling parallel machines. Comput Ind Eng 56(1):216–223
14. Kashan AH, Karimi B, Jenabi M (2008) A hybrid genetic heuristic for scheduling parallel batch processing machines with arbitrary job sizes. Comput Oper Res 35(4):1084–1098

15. Kotecha PR, Bhushan M, Gudi RD (2011) Constraint programming and genetic algorithm. Stoch Glob Optim World Sci 2:619–675. https://doi.org/10.1142/9789814299213_0018
16. Lee WC, Wu CC, Chen P (2006) A simulated annealing approach to makespan minimization on identical parallel machines. Int J Adv Manuf Technol 31(3):328–334
17. Lenstra JK, Shmoys DB, Tardos É (1990) Approximation algorithms for scheduling unrelated parallel machines. Math Program 46(1):259–271
18. Liang JJ, Qin AK, Suganthan PN, Baskar S (2006) Comprehensive learning particle swarm optimizer for global optimization of multimodal functions. IEEE Trans Evol Comput 10(3):281–295
19. Liu B, Wang L, Jin YH (2007) An effective PSO-based memetic algorithm for flow shop scheduling. IEEE Trans Syst Man Cybern B Cybern 37(1):18–27
20. Maharana D, Kotecha P (2019) Optimization of Job Shop Scheduling Problem with Grey Wolf Optimizer and JAYA Algorithm. In: Panigrahi B., Trivedi M., Mishra K., Tiwari S., Singh P. (eds) Smart Innovations in Communication and Computational Sciences. Advances in Intelligent Systems and Computing, vol 669. Springer, Singapore
21. Mastrolilli M, Gambardella LM (2000) Effective neighbourhood functions for the flexible job shop problem. J Sched 3(1):3–20
22. Mernik M, Liu SH, Karaboga D, Črepinšek M (2015) On clarifying misconceptions when comparing variants of the artificial bee Colony algorithm by offering a new implementation. Inf Sci 291:115–127
23. Mirjalili S (2015) Moth-flame optimization algorithm: a novel nature-inspired heuristic paradigm. Knowl-Based Syst 8:228–249
24. Moslehi G, Mahnam M (2011) A Pareto approach to multi-objective flexible job-shop scheduling problem using particle swarm optimization and local search. Int J Prod Econ 129(1):14–22
25. Nasir M, Das S, Maity D, Sengupta S, Halder U, Suganthan PN (2012) A dynamic neighborhood learning based particle swarm optimizer for global numerical optimization. Inf Sci 209:16–36
26. Pezzella F, Morganti G, Ciaschetti G (2008) A genetic algorithm for the flexible job-shop scheduling problem. Comput Oper Res 35(10):3202–3212
27. Pfund M, Fowler JW, Gupta JND (2004) A survey of algorithms for single and multi-objective unrelated parallel-machine deterministic scheduling problems. J Chin Inst Ind Eng 21(3):230–241
28. Punnathanam V, Kotecha P (2016) Yin-Yang-pair optimization: a novel lightweight optimization algorithm. Eng Appl Artif Intell 54:62–79, http://bit.ly/31SmmYE
29. Wu G, Mallipeddi R, Suganthan PN, Wang R, Chen H (2016) Differential evolution with multi-population based ensemble of mutation strategies. Inf Sci 329:329–345
30. Xia W, Wu Z (2005) An effective hybrid optimization approach for multi-objective flexible job-shop scheduling problems. Comput Ind Eng 48(2):409–425

Chapter 25
Branch-and-Bound Method for Just-in-Time Optimization of Radar Search Patterns

Yann Briheche, Frederic Barbaresco, Fouad Bennis, and Damien Chablat

Introduction and Context

Set covering is a well-known problem in combinatorial optimization. The objective is to cover a set of elements, called the universe, using a minimum number of available covers. The theoretical problem is known to be generally NP-difficult to solve [1], and is often encountered in industrial processes and real-life problem. In particular, the mathematical formulation of the set cover problem is well-suited for radar search pattern optimization of modern radar systems.

Electronic scanning and numerical processing allow modern radars to dynamically use bi-dimensional beam-forming, giving them great control on the radar search pattern. While traditional rotating radars search the space sequentially along the azimuth axis and reproduce at each azimuth the same pattern along the elevation axis, modern radars can optimize the search pattern along both axis simultaneously (Fig. 25.1). Those new possibilities requires a more sophisticated formulation for the optimization problem of the radar search pattern. Approximation of radar search pattern optimization as a set cover problem offers a flexible yet powerful formulation [2], capable of accounting radar specific constraints (localized clutter,

Y. Briheche (✉)
Thales Air Systems, Voie Pierre-Gilles de Gennes, Limours, France

Laboratoire des Sciences du Numérique de Nantes, UMR CNRS 6004, Nantes, France
e-mail: yann.briheche@thalesgroup.com

F. Barbaresco
Thales Air Systems, Voie Pierre-Gilles de Gennes, Limours, France

F. Bennis · D. Chablat
Laboratoire des Sciences du Numérique de Nantes, UMR CNRS 6004, Nantes, France

F. Bennis, R. K. Bhattacharjya (eds.), *Nature-Inspired Methods for Metaheuristics Optimization*, Modeling and Optimization in Science and Technologies 16,
https://doi.org/10.1007/978-3-030-26458-1_25

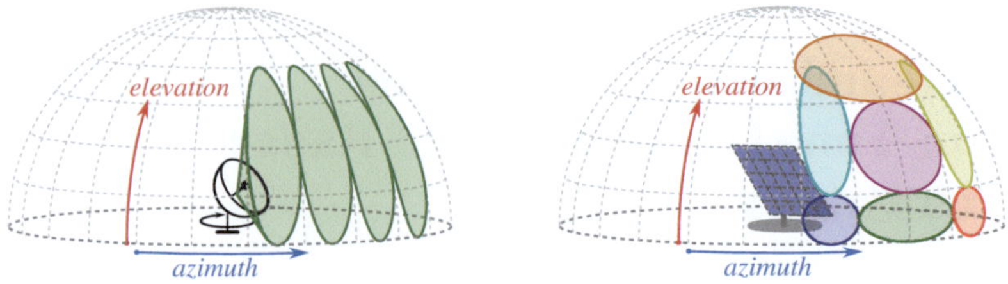

Fig. 25.1 Radar search pattern for a rotating radar (left) and a modern fixed-panel radar (right)

adaptive scan-rate updates, multiple missions) without changing the underlying mathematical structure of the optimization problem.

Various optimization algorithms have been proposed for solving the set cover problem: exact methods such as branch-and-bound combined with relaxations methods, and approximation algorithms such as greedy algorithm, simulated annealing and genetic algorithms (see [3] for a recent survey of those methods). In practice, problems of reasonable size can be efficiently solved by branch-and-bound exploration using linear relaxation for lower bound estimation. More importantly, branch-and-bound features interesting characteristics, making it particularly fit for producing just-in-time solutions, for example for radars in operational situation.

Problem Statement

Definition

Let $G = \{g_{m,n}\}$ be the set representation of a finite bi-dimensional M-by-N regular grid with

- each element $g_{m,n}$ representing a cell indexed by $(m, n) \in [0, M[\times[0, N[\subset \mathbb{N}^2$. The grid contains MN cells.
- each couple (m, n) representing the node at the intersection of the m-th horizontal line and the n-th vertical line with $(m, n) \in [0, M] \times [0, N] \subset \mathbb{N}^2$. The grid has $(M + 1)(N + 1)$ nodes.

On the grid, a rectangular cover is a subset of elements included in a rectangle, uniquely defined by its upper left corner node (m_0, n_0) and its lower right corner node (m_1, n_1), such that $0 \leq m_0 < m_1 \leq M$ and $0 \leq n_0 < n_1 \leq N$. The set representation of a cover defined by corners (m_0, n_0) and (m_1, n_1) is:

$$C = \{g_{m,n},\ (m, n) \in [m_0, m_1[\times[n_0, n_1[\}$$

For example, in cover C_7 (Fig. 25.2), the corners are $(m_0, n_0) = (0, 1)$ and $(m_1, n_1) = (1, 2)$.

Let $\mathcal{C} = \{C_1, \ldots, C_D\}$ be a collection of D rectangular covers on G.

Let $T_C \in \mathbb{R}^+$ be the associated cost of cover $C \in \mathcal{C}$ (also noted T_i for cover C_i).

Find a minimum cost sub-collection $\mathcal{S} \subset \mathcal{C}$ covering all cells of grid G.

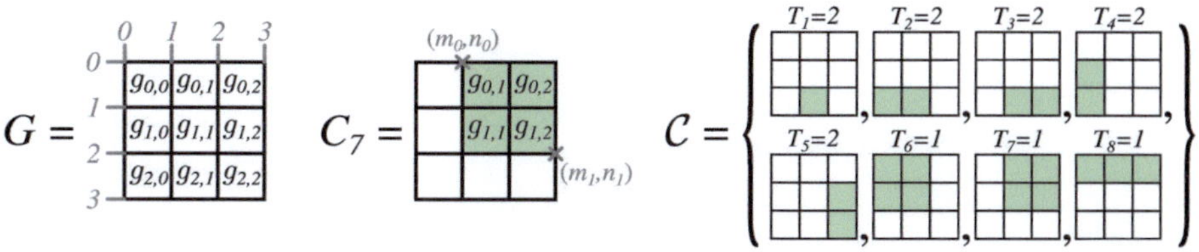

Fig. 25.2 Grid G to cover (left), a cover (center) and the collection $\mathcal{C}$ of available covers (right)

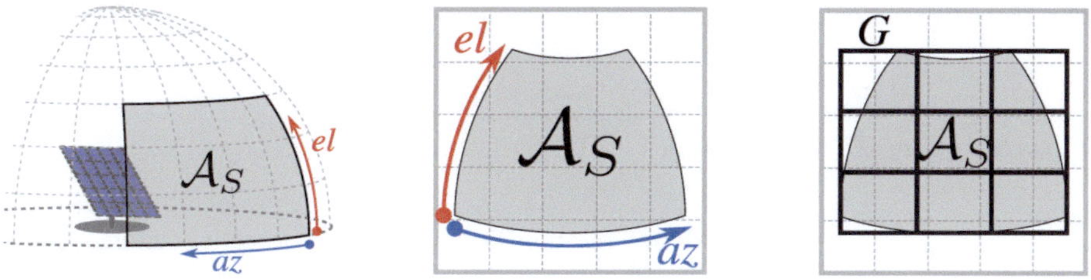

Fig. 25.3 Surveillance area $\mathcal{A}_S$ (left), its projection in direction cosines (center) and the surveillance grid (right)

Example

There are six available covers such as in Fig. 25.2 to cover G:

- $\mathcal{S}_1 = \{C_1, C_4, C_5, C_6, C_7\}$ is a valid sub-optimal covering collection with total cost $T_1 + T_4 + T_5 + T_6 + T_7 = 8$.
- $\mathcal{S}_2 = \{C_2, C_3, C_6, C_7\}$ is a valid optimal covering collection with total cost $T_2 + T_3 + T_6 + T_7 = 6$, as there are no solution with total cost 5 or less.
- $\mathcal{S}_3 = \{C_2, C_5, C_6, C_8\}$ is another optimal covering collection, thus an optimal solution is not necessarily unique.

The optimization formulation of this set cover problem can be written as:

$$\begin{aligned} &\min \textstyle\sum_{C \in \mathcal{S}} T_C \\ &s.t.\ \forall g_{m,n} \in G, \exists C \in \mathcal{S},\ g_{m,n} \in C \\ &\mathcal{S} \subset \mathcal{C} \end{aligned} \tag{25.1}$$

In the case of radar search pattern application, the grid G represents the surveillance area (Fig. 25.3), while each cover $C \in \mathcal{C}$ represents a radar beam and its detection area on the grid G (Fig. 25.4). The associated cost T_C is the duration required to emit the radar signal, then receive and process the echo. The total cost of collection of radar beams is the time required to emit all beams in sequential order, as the radar cannot emit simultaneously several beams.

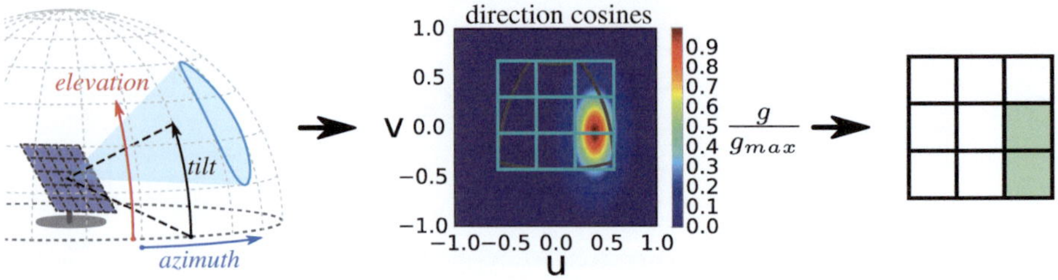

Fig. 25.4 Radar detection beam (left), its radiation pattern (center) and the associated cover (right)

Combinatorial Complexity

A rectangle cover is uniquely define by its upper left and lower right corners. Those corners are mathematically defined by choosing two values m_0 and m_1 among the $M+1$ horizontal lines, and two values n_0 and n_1 among the $N+1$ vertical lines on the grid. Thus there are at most

$$\binom{M+1}{2}\binom{N+1}{2} = \frac{MN(M+1)(N+1)}{4} = O(M^2N^2)$$

possible distinct rectangles on a M-by-N grid. And so the maximum number of possible sub-collections of rectangular covers on the grid is $2^{MN(M+1)(N+1)/4}$.

Even for a 10-by-10 grid, which is relatively small, the number of possible sub-collections is approximately 10^{900}, which is far too big to allow the use of brute-force exploration.

Integer Programming

Problem Formulation

The set cover problem can be written as an integer program by using matrix formulations. We represent each cover $C \in \mathcal{C}$ as a binary M-by-N matrix noted $\mathbf{C}$, or as a binary vector of length MN noted $\mathbf{c}$ (Fig. 25.5):

$$\mathbf{C}(m,n) = \mathbf{c}(m+Mn) = \begin{cases} 1 \text{ if } g_{m,n} \in C \\ 0 \text{ otherwise} \end{cases}$$

For each cover $C_i \in \mathcal{C}$, let $x_i \in \{0,1\}$ be the binary selection variable of cover C_i, such that the vector $\mathbf{x} = (x_1, \ldots, x_D) \in \{0,1\}^D$ represents the sub-collection $\mathcal{S} = \{C_i \in \mathcal{C}\ s.t.\ x_i = 1\}$, containing the chosen covers.

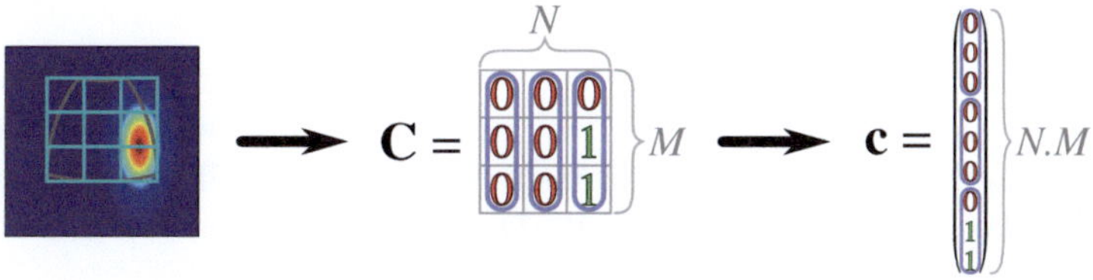

Fig. 25.5 Radar detection beam (left), its binary matrix representation (center) and its binary vector representation (right)

Let $\mathbf{T} = (T_1 \cdots T_D)^T$ be the cost vector and let

$$\mathbf{A} = \begin{pmatrix} \mathbf{c}_1 & \cdots & \mathbf{c}_D \end{pmatrix} = \begin{pmatrix} \mathbf{C}_1(0,0) & \cdots & \mathbf{C}_D(0,0) \\ \mathbf{C}_1(1,0) & \cdots & \mathbf{C}_D(1,0) \\ \vdots & \ddots & \vdots \\ \mathbf{C}_1(m,n) & \cdots & \mathbf{C}_D(m,n) \\ \vdots & \vdots & \vdots \end{pmatrix}$$

be the cover matrix.

Then the set cover problem (25.1) can be written as the following integer program:

$$\begin{aligned} &\min \mathbf{T}^T.\mathbf{x} \\ &s.t.\ \mathbf{A}\cdot\mathbf{x} \geq \mathbf{1} \\ &\quad \mathbf{x} \in \{0, 1\}^D \end{aligned} \tag{25.2}$$

where $\mathbf{1}$ is the vector $(1 \cdots 1)$ of length MN. As an example, the set cover problem represented in Fig. 25.2 can described by the following Equation:

$$\mathbf{A} = \begin{pmatrix} 0&0&0&0&0&1&0&1 \\ 0&0&0&1&0&1&0&0 \\ 0&1&0&1&0&0&0&0 \\ 0&0&0&0&0&1&1&1 \\ 0&0&0&0&0&1&1&0 \\ 1&1&1&0&0&0&0&0 \\ 0&0&0&0&0&0&1&1 \\ 0&0&0&0&1&0&1&0 \\ 0&0&1&0&1&0&0&0 \end{pmatrix}, \mathbf{T} = \begin{pmatrix} 2&2&2&2&2&1&1&1 \end{pmatrix}^T \text{ and } \mathbf{x} = \begin{pmatrix} x_1 \\ x_2 \\ x_3 \\ x_4 \\ x_5 \\ x_6 \\ x_7 \\ x_8 \end{pmatrix} \tag{25.3}$$

In their general form, integer programs are NP-hard to solve. Intuitively, this means that solving those problems is difficult, and requires some form of exhaustive enumeration of all possible solutions, whose number is often exponential in respect to the problem size.

Linear Relaxation

The linear relaxation of an integer program can be obtained by relaxing the integrality constraint of (25.2) into a positivity constraint, allowing the variables $(x_i)_{1\leq i\leq D}$ to take continuous values in [0, 1]:

$$\begin{aligned} &\min \mathbf{T}^T.\mathbf{x} \\ &s.t.\ \mathbf{A}\cdot\mathbf{x}\geq\mathbf{1} \\ &\quad\ \mathbf{0}\leq\mathbf{x}\leq\mathbf{1} \end{aligned} \tag{25.4}$$

Any valid solution of the integer program is also a valid solution of its linear relaxation. Consequently the optimal value of the linear relaxation is inferior to the optimal value of the integer program, since an optimal solution of the integer program is a valid solution of the linear relaxation.

Note that the constraint $\mathbf{x}\leq\mathbf{1}$ is in fact unnecessary, since the problem

$$\begin{aligned} &\min \mathbf{T}^T.\mathbf{x} \\ &s.t.\ \mathbf{A}\cdot\mathbf{x}\geq\mathbf{1} \\ &\quad\ \mathbf{0}\leq\mathbf{x} \end{aligned} \tag{25.5}$$

has the same optimal solutions as (25.4). Intuitively, in the linear relaxation, a cell is going to be covered by a sum of "fractional" covers (with $x_i < 1$), or as at least one integer cover (with $x_i = 1$) and thus has no need for covers with $x_i > 1$.

We formalize mathematically this idea in the following proof:

- Let $\mathbf{x}^b = (x_1^b\cdots x_D^b)$ be an optimal solution of (25.5). Let $\mathbf{y} = (y_1\cdots y_D)$ with $y_i = \min\{x_i^b, 1\}$.
  Immediately we have $\mathbf{T}^T\mathbf{y}\leq\mathbf{T}^T\mathbf{x}^b$. Since $\mathbf{x}^b$ is a valid solution of (25.5):

$$\forall(m,n), \sum_i x_i^b C_i(m,n) = \sum_{x_i^b\leq 1} x_i^b C_i(m,n) + \sum_{x_i^b>1} x_i^b C_i(m,n)\geq 1$$

  - Case 1: $\sum_{x_i^b>1} x_i^b C_i(m,n) = 0$

$$\sum_i y_i C_i(m,n)\geq\sum_{x_i^b\leq 1} x_i^b C_i(m,n) = \sum_{x_i^b\leq 1} x_i^b C_i(m,n)+\sum_{x_i^b>1} x_i^b C_i(m,n)\geq 1$$

  - Case 2: $\sum_{x_i^b>1} x_i^b C_i(m,n) > 0$
    $\exists j$ s.t. $x_j^b > 1$ and $C_j(m,n) > 0$, and since $C_j(m,n)\in\{0,1\}$, $C_j(m,n) = 1$, thus:

$$\sum_i y_i C_i(m,n) \geq y_j C_j(m,n) \geq C_j(m,n) \geq 1$$

So $\mathbf{y}$ is a valid solution of (25.5), and since $\mathbf{T}^T\mathbf{y} \leq \mathbf{T}^T\mathbf{x}^b$, $\mathbf{y}$ is also an optimal solution of (25.5).

From this, we deduce $\mathbf{T}^T\mathbf{y} = \mathbf{T}^T\mathbf{x}^b$ and with $\mathbf{T} > \mathbf{0}$, we deduce $\mathbf{x}^b = \mathbf{y}$, so $\mathbf{x}^b$ is a valid solution for (25.4).

- Let $\mathbf{x}^a$ be an optimal solution of (25.4), $\mathbf{x}^a$ is also a valid solution of (25.5), so $\mathbf{T}^T\mathbf{x}^b \leq \mathbf{T}^T\mathbf{x}^a$. We just showed that if $\mathbf{x}^b$ is an optimal solution of (25.5) then it is also a valid solution of (25.4), so we also have $\mathbf{T}^T\mathbf{x}^a \leq \mathbf{T}^T\mathbf{x}^b$, and thus we can conclude $\mathbf{T}^T\mathbf{x}^b = \mathbf{T}^T\mathbf{x}^a$.

  So any optimal solution of (25.4) is an optimal solution of (25.5) and reciprocally, thus both problems have the same set of optimal solutions.

Furthermore, the positivity constraints $\mathbf{0} \leq \mathbf{x}$ can be integrated in the matrix formulation with

$$\mathbf{R} = \begin{pmatrix} \mathbf{A} \\ \mathbf{I} \end{pmatrix} \text{ and } \mathbf{d} = \begin{pmatrix} \mathbf{1} \\ \mathbf{0} \end{pmatrix}$$

by rewriting the linear program as

$$\begin{aligned} &\min \mathbf{T}^T.\mathbf{x} \\ &s.t.\ \mathbf{R} \cdot \mathbf{x} \geq \mathbf{d} \end{aligned} \tag{25.6}$$

And the three formulations of the linear relaxation (25.4), (25.5) and (25.6) are equivalent.

The integer program representing our set cover problem and its linear relaxation have two more interesting properties:

- Easily-checked feasibility: an integer program is feasible if there is at least one solution validating all constraints. It is possible that no valid solution exists if some constraints are conflicting, or if one constraint is impossible. In our case, feasibility is easy to check: the integer program as well as its linear relaxation are feasible if and only if $\mathbf{x}_F = (1 \cdots 1)$ is a feasible solution, i.e. $\mathbf{A} \cdot \mathbf{x}_F = \sum_{i=1}^{D} \mathbf{c}_i \geq \mathbf{1}$:
  - if $\mathbf{x}_F$ is a valid solution, then the problem is feasible by definition.
  - if $\mathbf{x}_F$ is an invalid solution, then there is an invalidated constraint for $\mathbf{x}_F$, i.e.:
    $\exists (m,n)$ s.t. $\sum_{i=1}^{D} \mathbf{C}_i(m,n) < 1$, and since $\forall (i,m,n), \mathbf{C}_i(m,n) \in \{0,1\}$,
    $\exists (m,n)$ s.t. $\forall i, \mathbf{C}_i(m,n) = 0 \Rightarrow \exists (m,n)$ s.t. $\forall (x_i)_{1 \leq i \leq D}, \sum_{i=1}^{D} x_i \mathbf{C}_i (m,n) = 0 < 1$
    In other words, A has its $(m + Mn)$-th row filled with zeros, corresponding to a constraint which can be satisfied by no solution.

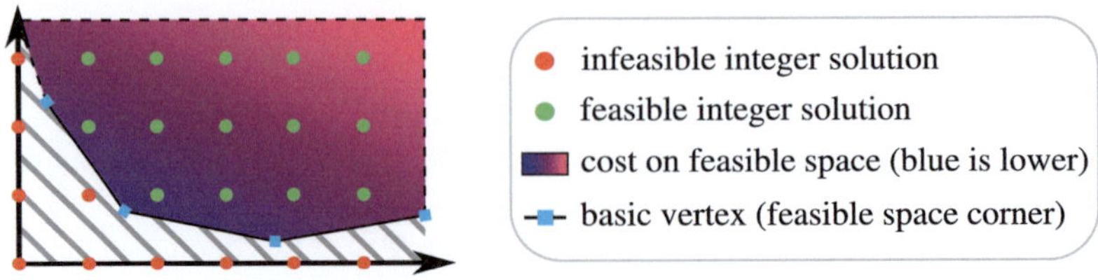

Fig. 25.6 Decision space for 2D linear and integer programs

Intuitively, $\mathbf{x}_F$ represents C, the collection of all available covers itself, and if it is an invalid solution, then there is a cell which cannot be covered. This can happen in a real system if there is a cell which cannot be scanned, because of an obstacle or because the radar has not enough power to achieve the desired detection range.

- Boundedness: a recurring question for linear programs is whether they are bounded, that is whether the cost function is bounded (below for minimization) for valid solutions. In our case the cost function is positive and thus bounded below by 0.

Linear Programming

There are three important geometrical aspects describing the decision space of the integer and linear programs (Fig. 25.6):

- $\mathbf{T}$ is the cost function gradient. The cost function is linear and its gradient is constant. $-\mathbf{T}$ is the direction of maximum decrease of the cost function.
- $\mathbf{A}$ is the cover matrix. Each row of $\mathbf{A}$ correspond to a detection constraint on a cell of G. In the decision space, each constraint corresponds to an hyperplane, the limit between the halfspace of solutions validating the constraint and the halfspace of solutions violating the constraint. The intersection of those halfspace forms a convex polyhedron.
- The positivity constraint of the linear relaxation $\mathbf{0} \leq \mathbf{x} \leq \mathbf{1}$ bounds the values of the valid solutions in the hypercube $[0, 1]^D$.

  For the integer program, the integrality constraint $\mathbf{x} \in \{0, 1\}^D$ further reduces the set of valid solutions to the vertices of the hypercube $[0, 1]^D$.

So the set of valid solutions for the linear relaxation is the intersection of the valid halfspaces for all constraints, and the hypercube $[0, 1]^D$. Geometrically, it is a bounded convex polyhedron in $\mathbb{R}^D$, and can be described by its vertices ("corners"). Each vertex of this polyhedron is a point where at least D hyperfaces of the polyhedron intersect, in other words, a point where D constraints are tight.

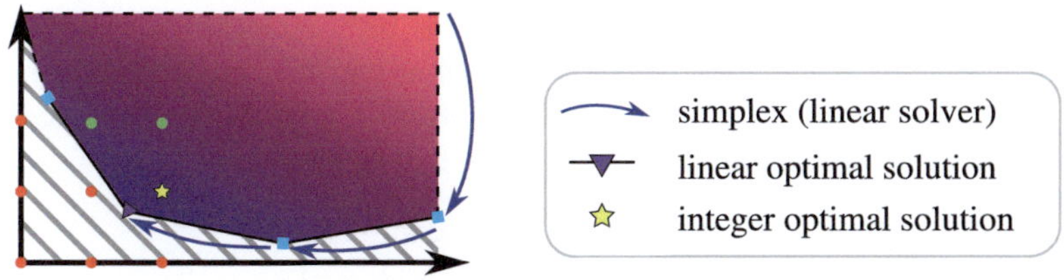

Fig. 25.7 Illustration of Dantzig's simplex method for solving linear programs

Such a point is called a basic solution (or basic vertex) of the linear program. It has been proved that if a linear program is bounded and feasible, then it has a basic optimal solution [4].

For a linear program, the polyhedron convexity allows the use of descent methods, such as Dantzig's simplex method, represented in Fig. 25.7, which moves from vertex to vertex on the feasible polyhedron until it reaches an basic optimal solution, i.e. a vertex with no decreasing neighbor. However, this type of method generally cannot be used to solve integer programs, for which solutions are isolated points.

So for a given basic optimal solution, we have D tight constraints. Let $B \leq MN$ be the number of tight detection constraints. If $D > B$ then we have $Z = D - B$ tight bound constraints. By considering formulation (25.5), we know that those Z tight bound constraints are of the form $x_i \leq 0$, thus $x_i = 0$. The corresponding Z variables are called *non-basic* variables and are zeroes. The other $D - Z = B$ variables are called *basic* variables and can be non-zero values. We reorganize the variables as $\mathbf{x}^T = (\mathbf{x}_B^T \ \mathbf{x}_Z^T)$, where $\mathbf{x}_B$ are the basic variables, and $\mathbf{x}_Z$ are the non-basic variables. Thus we have B tight detection constraints in $\mathbf{A}$, such that

$$\mathbf{A}_B \mathbf{x}_B = \mathbf{1}$$

where $\mathbf{A}_B$ is the square B-by-B submatrix of $\mathbf{A}$ linking the basic variables $\mathbf{x}_B$ to the tight detection constraints. Furthermore, $\mathbf{A}_B$ is necessarily non-singular: since the hyperplanes of all constraints intersect, the constraints are linearly independent.

Integral Program and Total Unimodularity

So for any basic optimal solution of the linear program, there is a (possibly non-unique) square non-singular submatrix $\mathbf{A}_B$ of $\mathbf{A}$ such that $\mathbf{A}_B \mathbf{x}_B = \mathbf{1}$ (note that the reverse is not true, the condition is necessary, but not sufficient). Thus, $\mathbf{x}_B = \mathbf{A}_B^{-1} \mathbf{1}$ and $\mathbf{x}^T = (\mathbf{x}_B^T \ \mathbf{x}_Z^T) = (\mathbf{x}_B^T \ \mathbf{0})$.

So if $\mathbf{A}_B^{-1}$ is an integral matrix (i.e. contains only integral values), then $\mathbf{x}$ is also integral. This means that the linear program and the integer program share an optimal solution. The integrality of an invert matrix is determined by the determinant of its forward matrix:

$$\det(\mathbf{A}_B) \in \{-1, +1\} \Rightarrow \mathbf{A}_B^{-1} = \frac{\mathrm{com}(\mathbf{A}_B)}{\det(\mathbf{A}_B)^{-1}} \text{ is integral}$$

A matrix $\mathbf{A}$ is said to be *unimodular* if $\det(\mathbf{A}) \in \{-1, 0, +1\}$. A matrix is said to be *totally unimodular* if all its square submatrices are unimodular. An integer program whose constraint matrix is totally unimodular can be directly solved by linear relaxation, because the integer program and its linear relaxation have the same basic optimal solutions. In this case, the problem and its associated convex polyhedron are said to be integral. Geometrically, this means that all vertices of the polyhedron are integral points.

Total unimodularity is an important concept in combinatorial optimization, because it reduces integer programming to linear programming, which is theoretically an "easier" problem. Linear programming is solvable in polynomial time, and very efficient practical algorithms exist.

One-Dimensional Cover Problem

For example, let us consider the one-dimensional case of our problem, with $M = 1$:

- $G = \{g_n,\ n \in [0, N[\}$ is a vector with N cells to covers.
- a cover $C = \{g_n,\ n \in [n_0, n_1[\}$ is a contiguous subset of G, uniquely defined by a starting element n_0 and an ending element n_1 such that $0 \leq n_0 < n_1 \leq N$. Each cover can be represented by a binary vector

$$\mathbf{c}(n) = \begin{cases} 1 \text{ if } n_0 \leq n < n_1 \\ 0 \text{ otherwise} \end{cases}$$

  which contains contiguous "ones", i.e. there is no "zero" between two "ones".
- $\mathcal{C} = \{C_1, \ldots, C_D\}$ is a collection of covers on G.

An example of this problem is presented in Fig. 25.8.

For the one-dimensional cover problem, the cover matrix $\mathbf{A} = \begin{pmatrix}\mathbf{c}_1 & \cdots & \mathbf{c}_D\end{pmatrix}$ is an *interval matrix*, i.e. each column $\mathbf{c}_i$ of $\mathbf{A}$ has its ones "consecutively". Interval matrices are known to be unimodular [5], and thus totally unimodular, since every submatrix is also an interval matrix. Every basic optimal solution of the linear relaxation is an integral solution, and a valid solution for the integer program. Solving the linear relaxation of the one-dimensional cover is sufficient to solve the problem itself, making it an "easy" problem.

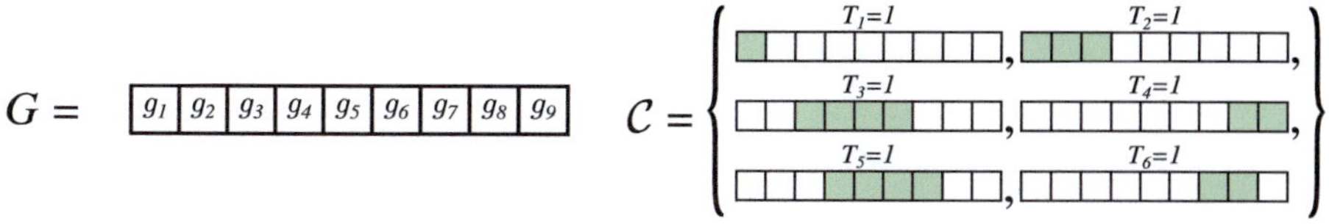

Fig. 25.8 Instance of the one-dimensional cover problem

However this not the case for the two-dimensional cover problem. For the problem represented in Fig. 25.2 and described by Eq. (25.3), the linear basic optimal solution is $\mathbf{x}_L = (0\ \frac{1}{2}\ \frac{1}{2}\ \frac{1}{2}\ \frac{1}{2}\ \frac{1}{2}\ \frac{1}{2}\ \frac{1}{2})^T$. A possible $\mathbf{A}_B$ submatrix for this solution is

$$\mathbf{A}_B = \begin{pmatrix} 0 & 0 & 0 & 0 & 0 & 1 & 0 & 1 \\ 0 & 0 & 0 & 1 & 0 & 1 & 0 & 0 \\ 0 & 1 & 0 & 1 & 0 & 0 & 0 & 0 \\ 0 & 0 & 0 & 0 & 0 & 1 & 1 & 1 \\ 0 & 0 & 0 & 0 & 0 & 1 & 1 & 0 \\ 1 & 1 & 1 & 0 & 0 & 0 & 0 & 0 \\ 0 & 0 & 0 & 0 & 0 & 0 & 1 & 1 \\ 0 & 0 & 0 & 0 & 1 & 0 & 1 & 0 \\ 0 & 0 & 1 & 0 & 1 & 0 & 0 & 0 \end{pmatrix}$$

which has a determinant of -2, thus explaining the appearance of $\frac{1}{2}$ values in the linear basic solution. This counterexample disproves the total unimodularity of the two-dimensional cover.

Integrality Gap

For the two-dimensional cover problem, the linear optimal solution might not be a valid solution for the integer problem and may have fractional values. The optimal cost of the linear relaxation described by Eq. (25.3) is $\mathbf{T}^T\mathbf{x}_L = \frac{11}{2}$. The optimal cost of the integer program cannot be fractional and is necessarily integer (in this case, it is 6). The difference between the optimal cost of the integer program and its linear relaxation is called the *integrality gap*.

Dynamic Programming

Interestingly, the difficulty gap between the one-dimensional and two-dimensional cover problems can be found by a completely different, algorithmic approach using dynamic programming. Dynamic programming is a method for solving an

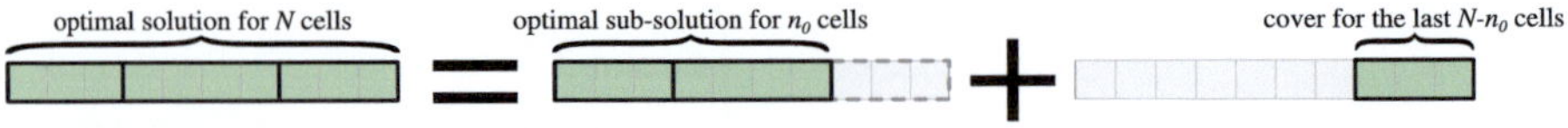

Fig. 25.9 Optimal substructure of the one-dimensional cover problem

optimization problem by recursively solving smaller sub-problems. Problems solved by dynamic programming usually possess an *optimal substructure*, which means that an optimal solution can be constructed by combining optimal solutions of its sub-problems. The method is particularly efficient if this substructure can be broken down recursively in a polynomial number of sub-problems.

This the case for the one-dimensional cover (Fig. 25.9). An optimal subcollection $\mathcal{S}$ covering G for this problem is going to be the combination of:

- a cover $\mathbf{c}$ including the last cell, so such that $n_0 \leq n < N(= n_1)$. Thus $\mathbf{c}$ covers the last cell but might also covers some previous cells.
- a subcollection covering optimally the first n_0 cells (which are not covered by $\mathbf{c}$)

Then it's possible to recursively define the solution covering optimally the first K cells as the union of a one cover including the K-th cell and a solution covering optimally the first k cells with $k < K$. This is formalized by the following equation, called the *Bellman recursion*:

$$\mathbf{x}_{[1,K]} = \begin{cases} \mathbf{0} & \text{if } K = 0 \\ \mathbf{x}_{[1,K-1]} & \text{if } \mathbf{A}\mathbf{x}_{[1,K-1]} \geq \mathbf{1}_{[1,K]} \\ \text{argmin}\{\mathbf{T}^T\mathbf{x} \text{ s.t. } \mathbf{x} = & \\ (\mathbf{x}_{[1,k]} + \mathbf{x}_i),\ k < K,\ \mathbf{A}\mathbf{x} \geq \mathbf{1}_{[1,K]}\} & \text{otherwise} \end{cases}$$

where

- $\mathbf{x}_{[1,K]}$ is the solution covering optimally at least the first K cells
- $\mathbf{1}_{[1,K]}$ is the vector of length N starting with K ones and ending with $N - K$ zeroes
- $\mathbf{x}_i$ is the vector of length D with a one at position i and zeroes elsewhere, representing the addition of the cover $\mathbf{c}_i$.
- k is the first non-zero element of the given cover $\mathbf{c}_i$

And the recursion can be described by the following steps

- $\mathbf{x}_{[1,0]}$ is the solution covering zero cells, initialized at $\mathbf{0}$
- If the solution covering $K - 1$ cells also covers the K-th cell, we keep it.
- Otherwise, search the optimal solution covering the first K cells as a combination of:
  - a cover $\mathbf{c}_i$ containing the K-th cell
  - the sub-solution $\mathbf{x}_{[1,k]}$ optimally covering the first $k < K$ cells not covered by $\mathbf{c}_i$

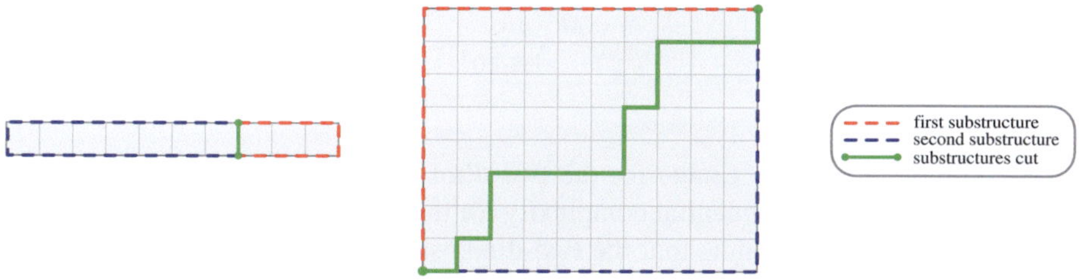

Fig. 25.10 Substructure decomposition of the one-dimensional cover problem (left) and the two-dimensional cover problem (middle)

So the dynamic programming method computes optimal sub-solutions for all sub-problems of covering the first K cells for $K \leq N$. So we must solve N sub-problems. For each sub-problem, the solution is computed as a combination of a cover and a smaller substructure optimal solution, so requires $O(D)$ steps to search through all covers. Thus, the algorithmic complexity of the dynamic programming algorithm is $O(ND)$, which is polynomial.

A natural question would be whether this approach can be generalized to the two-dimensional cover problem. Let us consider an optimal solution for the two-dimensional cover problem. It can be viewed as a combination of a rectangular cover C including the last bottom-right cell and an optimal cover for the substructure of cell not covered by C.

So the optimal substructure of the two-dimensional cover problem is the cover sub-problem of a first "top-left" half of the M-by-N grid G. The number of sub-problems is equal to the number of way of cutting G into two substructures: a top-left part and a bottom-right part, as presented in Fig. 25.10. Equivalently, this is equal to the number of paths between the top-right corner and the bottom-left corner of G.

A cut is constituted by $N + M$ edges on the grid, with M vertical edges and N horizontal edges. Any cut can be defined uniquely by choosing the N vertical edges (or equivalently M horizontal edges) among the $N + M$ edges. So the number of possible paths between two opposite corners of G, and thus the number of cover sub-problems on G is $\binom{N+M}{N} = \binom{N+M}{M}$.

Let $K = \min\{N, M\}$, then the number of possible cuts can be bound below by the following approximation using Stirling's formula

$$\binom{N+M}{N} \geq \binom{2K}{K} \simeq \frac{\sqrt{2\pi 2K}(2K)^{2K}}{e^{2K}} \left(\frac{e^K}{\sqrt{2\pi K}K^K}\right)^2 = \frac{2^{2K}}{\sqrt{\pi K}}$$

Thus, the number of sub-problems to solve grows exponentially with the grid size: an increase by 10 of the grid size increase the number of sub-problems by approximately $2^{2\cdot 10} \approx 10^6$. Even for small values, the number of sub-problems explodes (Table 25.1):

Table 25.1 Number of sub-problems

$N = M$	10	20	30	40	50
$\binom{2N}{N}$	$\simeq 10^5$	$\simeq 10^{11}$	$\simeq 10^{17}$	$\simeq 10^{23}$	$\simeq 10^{29}$

So while theoretically usable for the two-dimensional cover problem, dynamic programming has an exponential complexity for this problem, making the approach rather unpractical. This hints that the two-dimensional cover problem is computationally harder than the one-dimensional cover problem. The two-dimensional cover problem is in fact NP-difficult to solve [6]. To efficiently solve the two-dimensional cover problem, we need to use a more general optimization method.

Branch&Bound

Integer programs are generally NP-hard optimization problems: there is currently no know algorithm capable of finding quickly an optimal solution. Informally, exact algorithms "have to" search through the solution space.

The space of all possible solutions can be represented as a finite binary tree with depth p, each node representing the value choice of an integer variable (Fig. 25.11). Each end leaf represents a solution for the integer program. The number of possible solutions is finite, but grows exponentially and is usually huge: in our case, 2^D possible solutions.

Exploring the entire tree is computationally unfeasible in reasonable time. However it is possible at each node to estimate a lower bound of the node subtree best solution, by solving its linear relaxation. Knowing their lower bound, it is possible to avoid exploring certain subsets. This method is known as the branch-and-bound method [7]:

- Branching: Each branch at the current node (with depth $i - 1$) correspond to a chosen value, 0 or 1, for the next variable x_i. In each branch, x_i is no longer a variable but a parameter. The current problem is thus divided into 2 smaller sub-problems, each considering a different value for x_i and each having one less variable.
- Bounding: The current problem is relaxed into a linear program, whose solution is a lower bound of the current problem best solution. Depending on the lower bound value, the node sub-tree will be explored next (if it is the most promising branch), later (if there is a more promising branch), or never (if a better solution has already be found in another branch).

Defining what a promising branch is a difficult question, a lower bound is not necessarily better since deeper nodes may have higher bounds while being closer to optimal solutions. Integer programming solvers usually rely on various heuristics to define the exploration strategy and improve bound estimations.

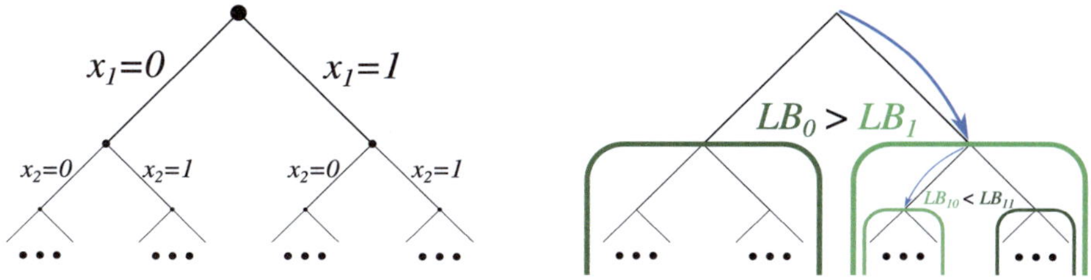

Fig. 25.11 Finite tree of solutions (left) and branch-and-bound method (right)

Description

We present in this section a pseudo-code describing a basic implementation of the branch-and-bound method in Algorithm 1. Each node in the tree can be described by the sequence of choices leading to it:

$$N = (x_1, x_2, \ldots, x_n)$$

From a given node, we can compute its children $N_0 = (x_1, \ldots, x_d, 0)$ and $N_1 = (x_1, \ldots, x_d, 1)$. At each node N explored, $(x_1, x_2, \ldots, x_n)$ are set, and we solve a linear relaxation of the problem with respect to the variables $(x_{n+1}, \ldots, x_D)$ using the simplex method, then add N to the list of nodes to explore.

The algorithm can be summarized by the following steps:

0. Initialization:
Initialize the list of node to explore with the root node.

1. Exploration:
Pop next node to explore from the list of nodes and compute its linear relaxation.

2. Bounding:
If the current node relaxation value is less than the current best solution found, proceed to Step 3, otherwise, drop current node and go back to Step 1.

3. Update:
If the current node relaxation is an integral solution, then its an improving solution (note that an end leaf always yield an integral solution). Update best current solution and proceed to Step 1.
Otherwise:

4. Branching:
Compute the current node children. For each child, check if the descendants contains a valid solution (this can be done by summing covers already used by the parent, the cover of the child node if used, and covers available to the descendants). If the child node is valid, add it to the list of node to explore. Proceed to Step 1.

Algorithm 1 Branch-and-bound

```
% LP_SOLVE is the relaxation subroutine called during branching
function LP_SOLVE(N)
    (x_1, ..., x_{d-1}) := N                          ▷ At node N, the first d − 1 variables are set
    (x_d, ..., x_D) := argmin{∑_{j=d}^{D} T_j x_j, s. t. A · x ≥ 1}   ▷ Optimization of non-set variables
    return x_L := (x_1, ..., x_d, x_{d+1}, ..., x_D)
end function

% Initialization
N_root = ()
𝒩 := {N_root}                                       ▷ Start with root node
x_best := x_F = (1 ··· 1)     ▷ Best solution found so far (by default, x_F is a valid solution)

% Exploration
while 𝒩 is not empty do
    N := pop(𝒩)                                      ▷ Take next node in 𝒩
    x_L := LP_SOLVE(N)                                ▷ Solve node relaxation

    % Bounding
    if T^T · x_L < T^T · x_best then       ▷ Explore node N only if it can improve best solution

        % Update
        if x_L ∈ {0, 1}^D then                        ▷ Check if x_L is an integral solution
            x_best := x_L else

        end
        (x_1, ..., x_d) := N

            % Branching
            for x ∈ {0, 1} do                         ▷ Compute children of node N
                N_c := (x_1, ..., x_d, x)
                if ∑_{j=1}^{d} x_j c_j + x c_{d+1} + ∑_{j=d+2}^{D} c_j ≥ 1 then   ▷ Check child feasibility
                    𝒩 := 𝒩 ∪ {N_c}                   ▷ Add child to the candidate list
                end if
            end for
        end if
    end if
end while
return x_best
```

This very generic description is just a presentation of the general idea of the method. Efficient implementations of the branch-and-method usually combined several techniques such as cutting planes, diving heuristics and local branching to improve bounds estimation and speed.

Application Example

In this section, we apply and describe the behavior of the branch-and-bound method on the example represented in Fig. 25.2 and described in Eq. 25.3.

- $\mathcal{N} = \{\}$, $\mathbf{x}_{best} = (1\ 1\ 1\ 1\ 1\ 1\ 1\ 1)$, $f_{best} = \mathbf{T}^T \cdot \mathbf{x}_{best} = 13$:
  Solving the root relaxation yields the linear solution $(0\ \frac{1}{2}\ \frac{1}{2}\ \frac{1}{2}\ \frac{1}{2}\ \frac{1}{2}\ \frac{1}{2}\ \frac{1}{2})$ with cost $\frac{11}{2} \leq 13$. Root node children (0) and (1) are feasible, and thus added to the exploration list $\mathcal{N} := \{(0), (1)\}$
- $\mathcal{N} = \{(0), (1)\}$, $\mathbf{x}_{best} = (1\ 1\ 1\ 1\ 1\ 1\ 1\ 1)$, $f_{best} = \mathbf{T}^T \cdot \mathbf{x}_{best} = 13$:
  Relaxation of (0) yields the same linear solution $(0\ \frac{1}{2}\ \frac{1}{2}\ \frac{1}{2}\ \frac{1}{2}\ \frac{1}{2}\ \frac{1}{2}\ \frac{1}{2})$ with cost $\frac{11}{2}$. We add the children $(0, 0)$ and $(0, 1)$ to the exploration list $\mathcal{N} := \{(1), (0, 0), (0, 1)\}$
- $\mathcal{N} = \{(1), (0, 0), (0, 1)\}$, $\mathbf{x}_{best} = (1\ 1\ 1\ 1\ 1\ 1\ 1\ 1)$, $f_{best} = \mathbf{T}^T \cdot \mathbf{x}_{best} = 13$:
  Relaxation of (1) yields the linear optimal solution $\mathbf{x}_L = (1\ 0\ 1\ 1\ \frac{1}{2}\ \frac{1}{2}\ \frac{1}{2}\ \frac{1}{2})$ with cost $\frac{15}{2} < 13$. We add the children $(0, 0)$ and $(0, 1)$ to the exploration list $\mathcal{N} := \{(1, 0), (1, 1)\}$
- $\mathcal{N} = \{(0, 0), (0, 1), (1, 0), (1, 1)\}$, $\mathbf{x}_{best} = (1\ 1\ 1\ 1\ 1\ 1\ 1\ 1)$, $f_{best} = \mathbf{T}^T \cdot \mathbf{x}_{best} = 13$:
  Relaxation of $(0, 0)$ yields the linear optimal solution $\mathbf{x}_L = (0\ 0\ 1\ 1\ 0\ 0\ 1\ 1)$ with cost $6 < 13$. $\mathbf{x}_L$ is an integral solution, thus we update the best current solution $\mathbf{x}_{best} := \mathbf{x}_L$; $f_{best} := 6$.

At this point, we can already deduce than we have found an integer optimal solution. The root relaxation has linear optimal cost $\frac{11}{2}$. Since any integer solution is a valid linear solution, it has an integer cost greater than the linear optimal cost $\frac{11}{2}$, so greater than 6. This suffices to prove the optimality of $\mathbf{x}_{best} = (0\ 0\ 1\ 1\ 0\ 0\ 1\ 1)$ for the integer program described by Eqs. (25.2) and (25.3).

Multiple Solutions Enumeration

While we could terminate the exploration once we have found an optimal solution, we also have the possibility to pursue the exploration in order to found alternative optimal solutions.

In engineering applications, multiple solutions are a desirable feature for engineers and operators who can select a solution among multiple candidates based on their expertise. This choice in turn can be analyzed to define preferences, to add secondary selection criterion to the method or even refined the model into a multi-objective optimization problem.

Multiple solutions enumeration can be done by slightly modifying steps **2.** and **3.** of the branch-and-bound method:

2. **Bounding**:
   If the current node relaxation value is less than or equal to the current best solution found, proceed to Step 3, otherwise, drop current node and go back to Step 1.
3. **Update and Enumerate**:
   If the current node relaxation is an integral solution, then its an improving solution. If it is strictly better than the current solution, empty the set of best solutions and update best current solution. Otherwise, update the set of best solutions. Proceed to Step 4 (as there could be other optimal solutions among the children of the current node).

This result in modifications to Algorithm 1 pseudo-code as described in Algorithm 2.

If we pursue the method application to the numerical example previously described:

- $\mathcal{N} = \{(0,0), (0,1), (1,0), (1,1)\}$, $\mathbf{x}_{best} = (1\ 1\ 1\ 1\ 1\ 1\ 1\ 1)$, $f_{best} = \mathbf{T}^T \cdot \mathbf{x}_{best} = 13$:

Algorithm 2 Multiple solutions enumeration branch-and-bound

```
% Initialization
...
x_best := x_F = (1···1)        ▷ Best solution found so far (by default, x_F is a valid solution)
X_best := {x_F}                                          ▷ Set of best solutions found so far

% Exploration
while N is not empty do
  ...

  % Bounding
  if T^T · x_L ≤ T^T · x_best then      ▷ Explore N if its relaxation is at least as good as x_best

    % Update and Enumerate
    if x_L ∈ {0, 1}^D then                             ▷ Check if x_L is an integral solution
      if T^T · x_L < T^T · x_best then
        x_best := x_L
        X_best := {x_L} else
    end
    X_best := X_best ∪ {x_L}
      end if
    end if

    % Branching
    for x ∈ {0, 1} do
      ...
    end for
  end if
end while
return X_best
```

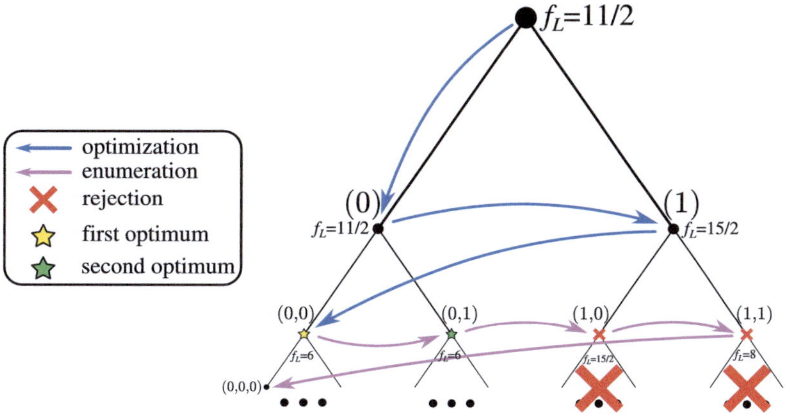

Fig. 25.12 Graphical representation of the branch-and-bound application example

Relaxation of $(0, 0)$ yields the linear optimal solution $\mathbf{x}_L = (0\ 0\ 1\ 1\ 0\ 0\ 1\ 1)$ with cost $6 \leq 13$. $\mathbf{x}_L$ is an integral solution, thus we update the best current solution $\mathbf{x}_{best} := \mathbf{x}_L$; $f_{best} := 6$.
We add the children $(0, 0, 0)$ and $(0, 0, 1)$ to the exploration list $\mathcal{N}$.

- $\mathcal{N} = \{(0, 1), (1, 0), (1, 1), (0, 0, 0), (0, 0, 1)\}$, $\mathbf{x}_{best} = (0\ 0\ 1\ 1\ 0\ 0\ 1\ 1)$, $f_{best} = \mathbf{T}^T \cdot \mathbf{x}_{best} = 6$:
  Relaxation of $(0, 1)$ yields the linear optimal solution $\mathbf{x}_1 = (0\ 1\ 1\ 0\ 0\ 1\ 1\ 0)$ with cost $6 \leq 6$. $\mathbf{x}_1$ is an integral solution, thus added to $\mathcal{X}_{best} := \{\mathbf{x}_{best}, \mathbf{x}_1\}$. We add the children $(0, 1, 0)$ and $(0, 1, 1)$ to the exploration list $\mathcal{N}$.
- $\mathcal{N} = \{(1, 0), (1, 1), (0, 0, 0), \ldots\}$, $\mathbf{x}_{best} = (0\ 0\ 1\ 1\ 0\ 0\ 1\ 1)$, $f_{best} = \mathbf{T}^T \cdot \mathbf{x}_{best} = 6$:
  Relaxation of $(1, 0)$ yields the linear optimal solution $\mathbf{x}_L = (1\ 0\ 1\ 1\ \frac{1}{2}\ \frac{1}{2}\ \frac{1}{2}\ \frac{1}{2})$ with cost $\frac{15}{2} > 6$. We drop node $(1, 0)$ and proceed with the next node.
- $\mathcal{N} = \{(1, 1), (0, 0, 0), \ldots\}$, $\mathbf{x}_{best} = (0\ 0\ 1\ 1\ 0\ 0\ 1\ 1)$, $f_{best} = \mathbf{T}^T \cdot \mathbf{x}_{best} = 6$:
  Relaxation of $(1, 1)$ yields the linear optimal solution $\mathbf{x}_L = (1\ 1\ 1\ 0\ 0\ 0\ 1\ 1)$ with cost $8 > 6$. We drop node $(1, 1)$ and proceed with the next node.

This numerical example is graphically represented in Fig. 25.12, with the optimization phase, the enumeration phase and some nodes rejection.

Just-in-Time Criteria

One of the most interesting features of the branch-and-bound method from an operational point of view is the possibility to use a "just-in-time" criteria. For example, a radar system with an embedded computer must optimize its cover just before a mission start. However, it only has 5 min to perform the optimization.

Algorithm 3 Just-in-time branch-and-bound

```
% Exploration
current_time := time()                                         ▷ Get current time
while 𝒩 is not empty AND current_time ≤ time_limit do
    ...
end while
return 𝒳_best, B_𝒩
```

A "just-in-time" is a time limit condition that would ensure that even if the optimum has not been reached, the algorithm will return the best solution it found in the available lapse of time. Another strength of the method is the fact that linear relaxation provides a lower bound of the optimal solution value:

$$B_{\mathcal{N}} = \min\{\mathbf{T}^T \cdot \mathbf{x}_L \ : \ \mathbf{x}_L = \text{LP\_SOLVE}(N), N \in \mathcal{N}\}$$

thus during the computation of the method, we always have an interval of confidence for the optimal solution value, above the lower bound but below the current best value:

$$B_{\mathcal{N}} \leq \mathbf{T} \cdot \mathbf{x}_{opt} \leq \mathbf{T} \cdot \mathbf{x}_{best}$$

Knowing the lower bound, we can compute the *(worst-case) relative optimality gap* as:

$$\Delta_{opt} = \frac{\mathbf{T} \cdot \mathbf{x}_{best} - B_{\mathcal{N}}}{B_{\mathcal{N}}}$$

which give as a percentage the best gain we can hope from the optimal solution relatively to the current best solution. The pseudo-code modifications required to account a time limit and provided the current lower bound are described in Algorithm 3.

In practice, if the algorithm has a broad choice of available covers, it will find very quickly a good quality solution. Typically within $\leq 10\%$ of relative optimality gap. However closing those last percents to reach the optimal solution can be difficult. Because the decision space is often huge, the algorithm spends a long time crossing out possibilities. In some case even, the algorithm finds quickly the optimal solution, and spends a long time proving its optimality.

Application to Radar Engineering

In this section, we give a study case example of radar search pattern optimization and its simulation results. We present first a quick informal description with intuitive and quantitative insight on our mathematical model of the radar system.

Radar Model

An active radar is a system capable of detecting distant metallic objects, by sending electromagnetic waves and listening to reflected echoes. To perform detection in a given azimuth-elevation direction (az, el), the radar antenna is electronically controlled to focus power in direction (az, el), maximizing the radiation pattern in that direction. A signal containing a series of impulses is then sent through the radar. Upon reception, the reflected signal is filtered to detect echoes. A longer signal is more energetic and easier to filter out. The energy received by the radar from a target at distance R is

$$E_r = K \frac{g^2 T}{R^4} \tag{25.7}$$

where K is a constant accounting for the radar emitting power, internal losses, target reflectability, etc., g is the antenna radiation pattern in direction (az, el), and T is the signal duration.

Equation (25.7) is a simpler version of the *radar equation* [8], a fundamental concept in radar theory. It formalize the intuitive idea that the reflected energy increases with antenna directivity and signal duration, but decreases with the target distance. The radar has a certain detection threshold E_t, and detects a target only if its reflected energy is above this threshold.

For a given radar system, we have a set of feasible rectangular radiation patterns for the antenna and a set of available signals. The combination of a radiation pattern and a time signal is called a *dwell*.

Simulation Parameters

The desired detection range is defined by a minimum distance $D_{\min}$ and a minimum altitude $H_{\min}$. We want to detect targets within that range, closer than $D_{\min}$ and below altitude $H_{\min}$, so the desired detection range is defined as:

$$R_c(az, el) = \begin{cases} D_{\min} & \text{if } el \leq \operatorname{asin}\left(\frac{H_{\min}}{D_{\min}}\right) \\ \frac{H_{\min}}{\sin(el)} & \text{otherwise} \end{cases}$$

Informally, the volume defined by the detection range resembles a sliced cylinder (Fig. 25.13). The radar can use two different type of signals: a short signal and a long signal.

In our simulation, the detection grid G is a 20×20 lattice with 326 valid cells. We computed 866 feasible dwells in our study case, with 815 dwells using a short time signal with duration T_s and 51 dwells using a long time signal with duration T_l. We compute the cost vector $\mathbf{T}$ with size 866 associating each dwell to its signal time duration.

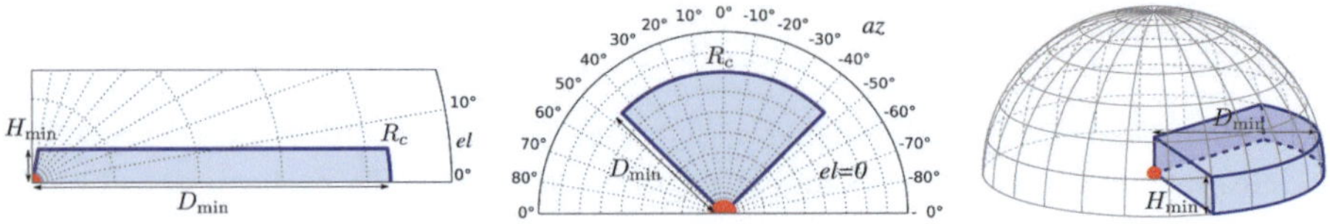

Fig. 25.13 Desired detection range: elevation cut (left), azimuth cut (center) and 3D view (right)

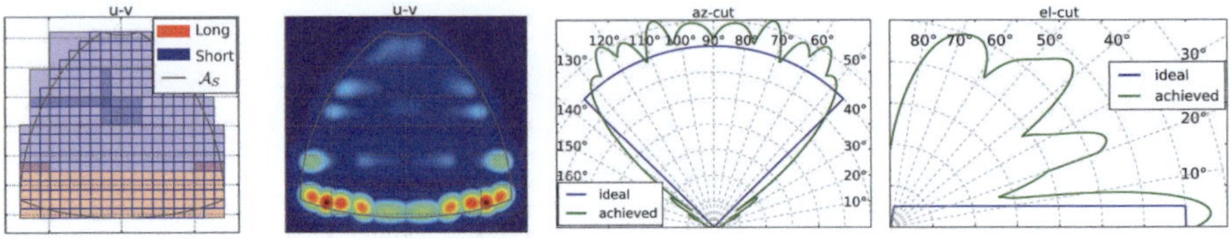

Fig. 25.14 From left to right: computed radar search pattern, emission pattern and detection range

We want to find a optimal radar search pattern, i.e. a sub-collection of dwells among the 866 available dwells covering all 326 valid detection cells with minimal total time-budget. For each of the 866 dwells, we use Eq. (25.7) to compute the dwell detection cover on the 326 cells. From the detection covers, we can compute the cover matrix **A** with shape 326×866.

Having computed **T** and **A**, we can use the branch-and-bound method described previously to search an optimal radar search pattern. The corresponding integer program has 866 variables and 326 detection constraints.

The optimization is done through the CPLEX solver [9], which implements an improved version of the branch-and-bound. The total time required to find the solution is 5 s on an i7-3770@3.4GHz processor.

Optimal Solution

The returned optimal solution is shown in Fig. 25.14: the left sub-figure shows the discrete covers of the 20 dwells used in the pattern. Ten dwells use a short signal (represented in blue) and cover high elevation, and ten dwells use a long signal (represented in red) and cover low elevations.

This result is explained by the fact that a radar must usually achieve high detection range near the horizon (where targets are located) and low detection range at high elevation (since most aircrafts have limited flying altitude). It makes sense to use longer, and thus “more energetic” dwells at low elevations than at high elevations.

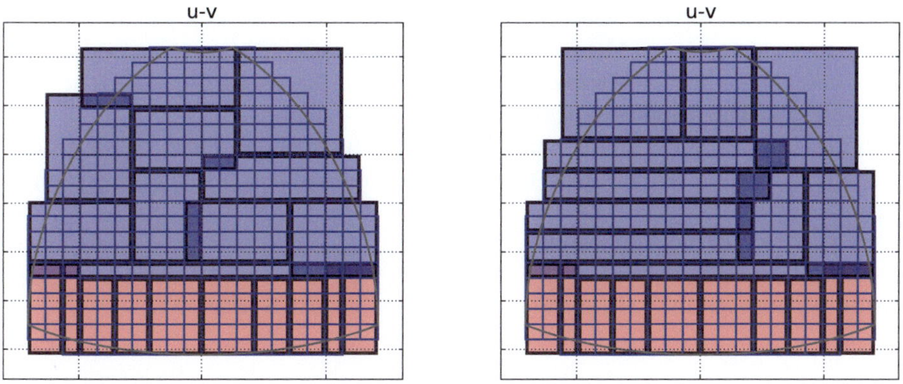

Fig. 25.15 Two other possible optimal solutions

Enumeration

As we have seen before, there may be multiple optimal solutions. In this simulation we managed to found 3500 different optimal solutions in 5 min. However, this search is unlikely to be exhaustive: due to the wide choice of possible dwells, there is often an extremely high number of possible alternative optimal solutions. Finding all solutions is unfeasible in practice.

However optimal solutions share certain characteristics: all solutions have ten shorts signal dwells at high elevation and ten long signal dwells at low elevation. The long-signal dwells (in red) are mostly the same for all optimal solutions found, and form an *optimality invariant*. The short-signal dwells (in blue) are however different for each solution.

Intuitively, the low-elevation area is more "energetically demanding"; thus low-elevation detection constraints are the "hardest constraints" of the problem, and do not leave a lot of choice for covering the low-elevation area. High-elevation detection constraints are in comparison "easier" and can be validated by different covers (Fig. 25.15).

Conclusion

The branch-and-bound method is a practical and powerful technique. It can be used as an exact algorithm if exploration is pushed to its completion, when there is no more branch left with a potentially better solution. It can also be used as an heuristic, with stopping criterion based on an time limit or a optimality gap threshold. This is especially useful in operational situations with broad choices, when finding a good solution is easy, but proving optimality is difficult.

The method is very generic, and can be used to solve a lot of different combinatorial problems. Many of those problems have evident practical values and important applications in various industries, such as the set cover problem in radar applications. The versatility of the method and its various "flavors" can be used for different purposes: enumeration permits analysis of the radar "possibilities" during conception, while just-in-time criteria improves resources management in operational situations. Branch-and-bound is an extremely efficient tool for a broad variety of engineering applications.

Acknowledgements This work is partly supported by a DGA-MRIS scholarship.

References

1. Vazirani VV (2001) Approximation algorithms. Springer, New York
2. Briheche Y, Barbaresco F, Bennis F, Chablat D, Gosselin F (2016) Non-uniform constrained optimization of radar search patterns in direction cosines space using integer programming. In: 2016 17th International Radar Symposium (IRS)
3. Yelbay B, Birbil Şİ, Bülbül K (2015) The set covering problem revisited: an empirical study of the value of dual information. J Ind Manag Optim 11(2):575–594
4. Matouek J, Gärtner B (2006) Understanding and using linear programming (universitext). Springer, New York/Secaucus
5. Nemhauser GL, Wolsey LA (1988) Integer and combinatorial optimization. Wiley-Interscience, New York
6. Briheche Y, Barbaresco F, Bennis F, Chablat D (2018) Theoretical complexity of grid cover problems used in radar applications. J Optim Theory Appl 179(3):1086–1106 [Online]. https://doi.org/10.1007/s10957-018-1354-x
7. Conforti M, Cornuejols G, Zambelli G (2014) Integer programming. Springer Publishing Company, Incorporated
8. Skolnik M (2008) Radar handbook, 3rd edn. McGraw-Hill Education, New York
9. IBM ILOG CPLEX Optimization Studio, v12.6 (2015) http://www-03.ibm.com/software/products/en/ibmilogcplestud/

Chapter 26
Optimization of the GIS-Based DRASTIC Model for Groundwater Vulnerability Assessment

Sahajpreet Kaur Garewal, Avinash D. Vasudeo, and Aniruddha D. Ghare

Abstract Groundwater vulnerability assessment is an essential tool for identification, classification and analysis of factors affecting groundwater and thus to take necessary steps for reduction of adverse environmental consequences. DRASTIC is a widespread approach for assessment of groundwater vulnerability. It is a region specific method and a major challenge is to manage its intrinsic subjectivity for evaluating the approximate parameters and assigning rates and weight to the parameters, depending on the regional factors affecting groundwater. In order to minimize the subjectivity of DRASTIC, in the present study DRASTIC methodology was optimized by addition of Land use (Lu) parameter and rebuilding the parameters rates and weight. The criteria for the optimization of rates were mean nitrate concentration and weight of the parameters was Analytical Hierarchy Process (AHP). The modified DRASTICLu model formed by optimizing the conventional DRASTIC by applying various modification shows good correlation with field quality parameters.

Keywords Groundwater · Vulnerability · DRASTIC · Land use · Nitrate · AHP

Introduction

Groundwater is globally important and necessary environmental element for sustainable and economical development. Increasing urbanization has induced tremendous pressure on existing natural resources. Last few decades have witnessed overexploitation and quality degradation of surface and ground water, which has affected the living creature and surrounding environment [9]. Contamination of water is a serious issue in India, 80% of the surface water resources and various

S. K. Garewal (✉) · A. D. Vasudeo · A. D. Ghare
Department of Civil Engineering, Visvesvaraya National Institute of Technology, Nagpur, Maharashtra, India

F. Bennis, R. K. Bhattacharjya (eds.), *Nature-Inspired Methods for Metaheuristics Optimization*, Modeling and Optimization in Science and Technologies 16,
https://doi.org/10.1007/978-3-030-26458-1_26

groundwater zones are found to be contaminated by organic, inorganic and physical contaminants [21]. Over exploitation of the groundwater has been observed mainly in the areas where municipal water supply is limited or the other available surface water resources are polluted. While making efforts to meet the required quantity of water as per demand, it is essential to maintain the specific quality of water. Above scenario calls for the necessary step to make suitable strategies for effective planning, management and development of available water resource.

The effort to enhance the groundwater condition is limited due to its invisibility in nature and large cost and time requirement [23]. With increasing awareness about the significance of groundwater resources, attempts are being made to reduce, prevent and eliminate the groundwater contamination [13]. At the national and regional levels, some progress is being made for estimating the quantity and quality of water available and coordinating efforts to manage its use. Assessment of groundwater vulnerability has become an effective tool, which identifies the area less or more susceptible to contamination, allow the better understanding of local hydrogeology of the area and helps in making effective policies for the regional development of groundwater [23].

The methodology adopted for vulnerability assessment of groundwater is basically classified as statistical methods, overlay and index methods and process-based simulation models, which is selected on the basis of data availability. A GIS based DRASTIC method is a well-known approach used for assessment of groundwater vulnerability. It has been applied successfully over the world such as India [5, 13, 17, 19], Japan [3], China [6, 7, 20, 22], Turkey [16], Iran [4, 8, 12, 15] and so on.

Though worldwide usage of DRASTIC methodology for groundwater vulnerability assessment it has been criticized by various researchers due to its subjectivity towards selected parameters and assigned rates and weights [15]. DRASTIC is a region specific method, the parameters can be added or subtracted and modification can be applied to assigned rates and weights depending upon the local hydrogeology of the area [3]. The map showing groundwater vulnerable zones generated using DRASTIC approach is unique can only be used for a specific study area. On the basis of aforesaid literature, it can be concluded that optimization of the DRASTIC methodology by applying various modification helps in better assessment of groundwater vulnerability.

In the present study, groundwater vulnerability of Nagpur city is evaluated using the conventional DRASTIC methodology and for better prediction of groundwater vulnerable zone, an optimization to DRASTIC methodology was attempted based on the:

(i) Addition of land use parameter along with the intrinsic DRASTIC parameters.
(ii) Revising the rates assigned to the parameters using mean nitrate concentration.
(iii) Revising the parameters weight using AHP.

Study Area

Nagpur urban is located at geographical centre of India, indicated by zero mile stone (Fig. 26.1). It is situated between longitude 79°00′ – 79°15′ and latitude 21°00′ – 21°15′ at 310 m a.s.l. The city occupies approx 218 sq. km of area within Nagpur Municipal Corporation (NMC) limit with the population of about 2,405,665. The stratigraphy of the city is comprised by massive Deccan trap and Archeans crystalline in the western zone of the city. Sandstone aquifers are located in North-East zone and, central and North-South is occupied by lameta formation. The major water bodies within the city limits are Nag River and Pilli River. Nag

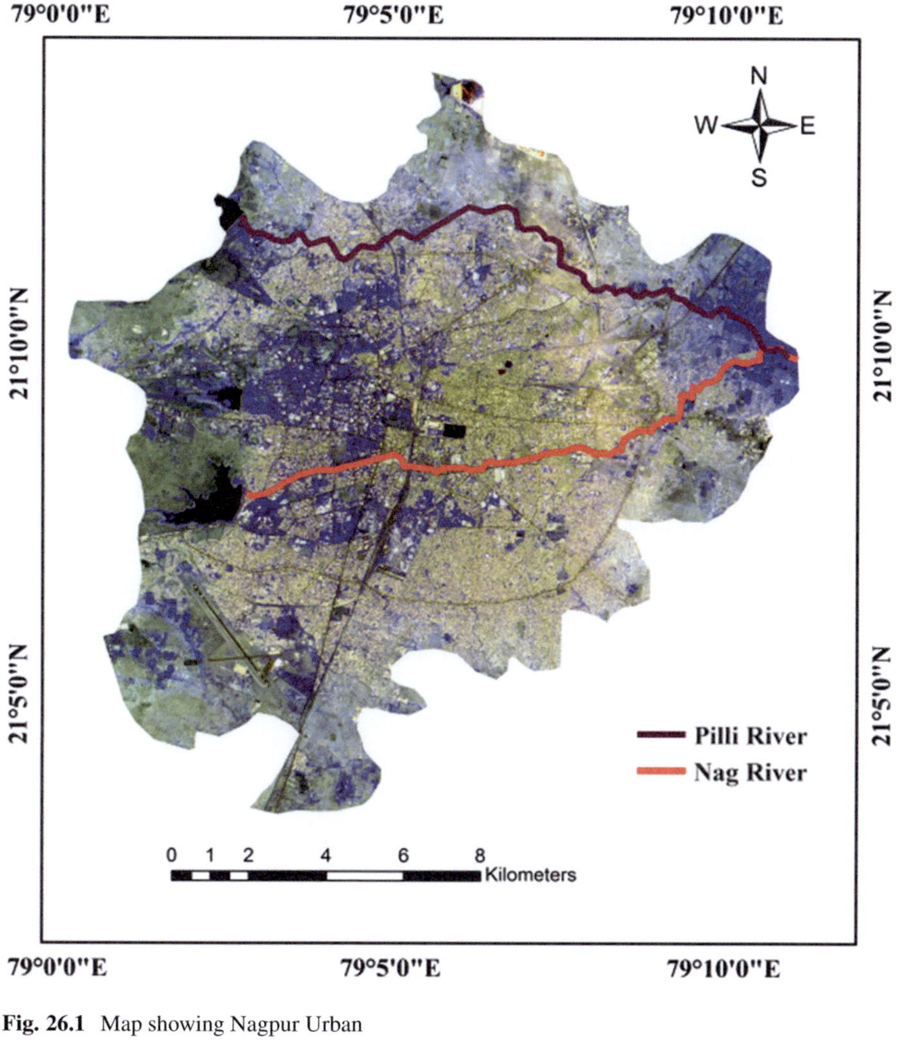

Fig. 26.1 Map showing Nagpur Urban

River is originated from Ambazari Lake and runs towards the east of the city meets Pilli River before joining the Kanhan River. The city named after the Nag River. Pilli River is originated from Gorewada lake travels approx 17 km within the city. The average temperature in winter is 12 °C and 45 °C during summer and mean precipitation is 1100 mm.

Methodology

DRASTIC (Conventional Method)

The intrinsic vulnerability of groundwater is evaluated including DRASTIC seven hydro-geological parameters such as Depth to water table (D), Recharge (R), Aquifer media (A), Soil media (S), Topography (T), Impact of vadose zone (I) and Hydraulic conductivity (C) [1]. The parameters are characterized into different sub-parameters and rated on the scale from 1 to 10 based on their effect on vulnerability of groundwater. Weight is assigned to all the parameters from 1 to 5 showing their importance in overall vulnerability assessment. The Intrinsic Vulnerability Index (IVI) is evaluated by addition of all the parameters using Eq. (26.1).

$$\mathrm{IVI} = \mathrm{D_rD_W} + \mathrm{R_rR_W} + \mathrm{A_rA_W} + \mathrm{S_rS_W} + \mathrm{T_rT_W} + \mathrm{I_rI_W} + \mathrm{C_rC_W} \tag{26.1}$$

Where,

w and r are the weight and rates assign to individual parameters in the overlay analysis.

The groundwater vulnerability Index is evaluated by assigning the rates and weight to the parameters which are already documented by [1] using Delphi technique. The resultant vulnerability map is classified in different vulnerable zones showing the area under very high to very low vulnerability index using Natural breaks (jerks) classification tool of GIS. Figure 26.2 shows the complete methodology adopted in the present study. The intrinsic vulnerability evaluated using DRASTIC parameters are calibrated and validate using Field quality data.

Optimization of Conventional DRASTIC

The major constraint of the DRASTIC approach is its intrinsic subjectivity in selecting the appropriate parameters and determining the weighting coefficients and rating scales of the parameters. An attempt has been made to optimize the DRASTIC methodology with the aim of better prediction of groundwater vulnerable zone of Nagpur city by:

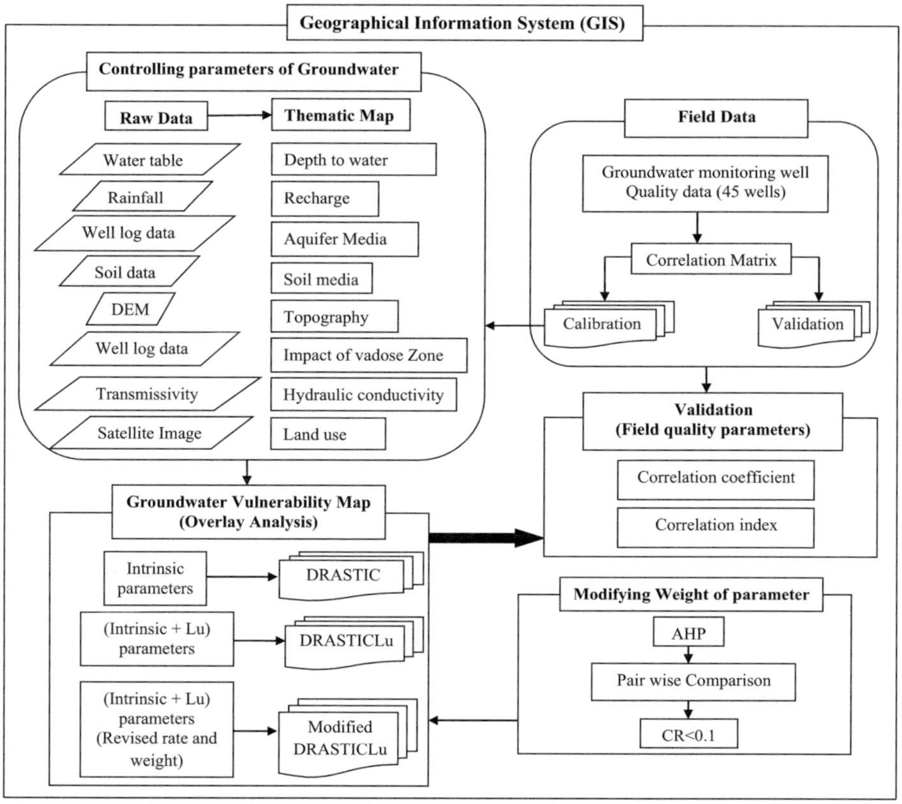

Fig. 26.2 Methodology for the present study

Addition of Land Use Parameter with Conventional DRASTIC Parameters (DRASTICLu)

Groundwater vulnerability map evaluated using DRASTIC methodology involves the intrinsic parameters of aquifer responsible for transportation and attenuation of contamination, irrespective of any actual site specific data or source of the contaminant. There is no data or methodology available to identify whether the presence of the contaminant in the groundwater is due to the properties of the material or by the anthropogenic activities at the land surface [5]. In the present study, Land Use (Lu) parameter is incorporated with the intrinsic parameter of the aquifer to involve the effect of different land occupied by various activities affecting the potential of groundwater. The land use parameter classification is rated on the basis, that the area which contributes more to groundwater contamination were assigned higher value. The Vulnerability Index (VI) is calculated by adding the land use parameter with IVI as formulated in Eq. (26.2)

$$\mathrm{VI} = \mathrm{IVI} + \mathrm{Lu_r Lu_w} \tag{26.2}$$

Where,

w and r are the weight and rate assigned to additional Land use (Lu) parameter.

Revising the Rates of the Parameters Using Quality Data

The ratings of the DRASTIC parameters are revised using mean nitrate concentration (quality data). The presence of source at land surface and presence of the contaminant in groundwater at the particular location is not correlative. The contaminant travels from the land surface to groundwater is affected by various elements such as properties of media through which it is traveling, the slope of the area, the concentration of contaminant and so on. The mean concentration of nitrate is observed under various parameter classifications to revise the rating based on the presence of the actual contaminant in the area. Nitrate is selected for the modification of parameter rate because they are not present in the groundwater under the natural condition it's mainly due to anthropogenic activities at the land surface.

Revising the Weights of the Parameters Using AHP (Modified DRASTICLu)

The weight assigned to DRASTIC parameters proposed by [1] can give unrealistic results sometimes due to the changing hydrogeology of the area [16]. Analytic Hierarchy Process (AHP) is a Multi Decision Making Analysis (MCDA) tool proposed by [14], which assist the decision maker to make the best decision by setting the priorities between the parameters. AHP is incorporated in the DRASTIC method by [18] for revising the rate and weight of DRASTIC parameters and followed by various researches [11, 16]. In AHP, pair wise comparison matrix between the parameters are formed by comparing the parameters in a standard nine level scale developed by Saaty [14]. In the scale, 9 signifies extremely more important parameter and 1/9 signifies extremely less important parameter. The principal Eigen vector and value evaluated by pair-wise comparison matrix is helpful in decision making. For judgment of the comparison matrix, Consistency Ratio (CR) is evaluated which should be within threshold value 0.1(10%). CR is the ratio of Consistency Index (CI) calculated using Eq. (26.3) to Random Index (RI), which is already defined for 'n' order matrix [14]. If CR fails to be within the range than the answer of comparison is revised.

$$CI = \frac{\lambda_{max} - n}{n - 1} \tag{26.3}$$

Where,

CI is Consistency Index, λ_{max} is largest eigen value of respected 'n' order matrix.

Results

Preparation of DRASTIC Thematic Map

The data used for preparing the thematic map of DRASTIC parameters are collected from the different government organization, research institutes and previous research papers. All the operation used for obtaining the maps is performed in ArcGIS version 10. The map of Depth to water table is generated using data of 45 monitoring wells collected from CGWB, Nagpur department. The available data is in the statistical discrete form, Kriging interpolation tool is used to generate a surface map of groundwater table showing different level within the city limit. The depth to water level in the city varies from 0.7 to 15.2 m (Table 26.1). Greater the depth to water level, lesser will be the chance of contamination as the contaminant will take more time to reach aquifer and attenuation of contaminant will also take place. Rainfall gives a major contribution for groundwater recharge; in the study

Table 26.1 Rates are weight of the parameters

Parameter	Sub parameter	DRASTIC [1]		Average nitrate concentration	Modified DRASTIC	
		Rating	Weights		Rating (Average No3 concentration)	Weight (AHP)
Depth of water (m)	0.7–2.61	10	5	21	No changes	0.035
	2.62–3.87	9		46.78		
	3.87–4.69	8		41.36		
	4.69–5.95	6		84.12		
	5.95–7.86	4		71.96		
	7.86–10.77	2		54.14		
	10.77–15.2	1		52		
Recharge (mm)	423–433	9	4	67.86	10	0.160
	417–422	8		63.24	9	
	411–416	7		61	8	
	405–410	5		25.5	4	
	396–404	3		22.19	3	
Aquifer media	Amgaon-Gneiss-complex	8	3	87.9	10	0.198
	Unclassified-Gneiss	7		63.16	8	
	Massive basalt	4		49.86	6	
	Intertrapean	1		21	3	
Soil media	Sand	8	2	86.33	10	0.047
	Clayey	7		72.4	7	
	Clay loam	3		57.21	8	

(continued)

Table 26.1 (continued)

Parameter	Sub parameter	DRASTIC [1]		Average nitrate concentration	Modified DRASTIC	
		Rating	Weights		Rating (Average NO_3 concentration)	Weight (AHP)
Topography (%)	<2.69	10	1	89.67	No changes	0.089
	2.69–5.01	9		67.89		
	5.01–7.90	7		30.67		
	7.90–11.56	5		57.80		
	11.56–16.19	4		39.40		
	16.19–22.93	3		60.45		
	>22.93	1		12.33		
Impact of vadose zone (m)	0.60–3.16	8	5	43.41	No changes	0.037
	3.16–3.88	7		58.42		
	3.88–4.48	6		89.58		
	4.48–5.08	5		89.22		
	5.08–5.88	4		74.33		
	5.88–7.08	3		71.83		
	7.08–10.79	2		52		
Hydraulic conductivity (m/s)	10^{-3}–10^{-4}	9	3	95.33	10	0.172
	10^{-4}–10^{-5}	8		63.16	8	
	10^{-5}–10^{-6}	6		51.98	6	
	$>10^{-6}$	5		21	3	
Land use	Agriculture	9	5	70.72	10	0.262
	Built-Up	7		66.925	8	
	Water bodies	5		65.22	6	
	Wasteland	3		51.56	5	
	Forest	2		15.99	2	

30–40% of rainfall is considered as recharge [2, 5]. Recharge in the study varies from 396 to 433 mm which is nearly constant (Table 26.1), greater the recharge more amount of contaminants will infiltrate to the groundwater. Aquifer media is acquired using litho log data collected from CGWB, Nagpur. Mainly the city is comprises of hard rock aquifer tabulated in Table 26.1. Soil media restrict the movement of recharge from the surface to groundwater. The soil media in the city varies from sand, clay to clayey loam (Table 26.1). The topography is specified in the form of slope generated from DEM (Table 26.1). The slope is flat to mild in the eastern zone and steeped in the western zone of the city. Impact of vadose zone is given the form of a thickness of the zone calculate using DEM and depth to the water level of the city and methodology proposed by [10] (Table 26.1). Lesser the thickness of vadose zone more will be the contamination and vice versa. Hydraulic conductivity depends on the characteristic of media through which the groundwater travels. Table 26.1 shows the hydraulic conductivity range within the city; Greater the hydraulic conductivity more will be the contamination migration. Land use is obtained from the satellite image (LISS III) in the current study, comprising of built-up area, agriculture, water bodies, forest and waste lands (Table 26.1). Each classification has an individual impact on groundwater contamination.

DRASTIC (Conventional Method)

Intrinsic vulnerability index (DRASTIC) map evaluated by overlay analysis of seven thematic maps using Eq. (26.1) is shown in Fig. 26.3. The vulnerability map of groundwater is classified in five vulnerable zones on the basis of the vulnerability index value, higher the index value more will be the chance of contamination. From the observation of the resultant intrinsic vulnerability map of the study area, South and South-West region is found to safe from contamination having least vulnerability index. Center to south region is having moderate vulnerability index and East and North-East region is at high risk having higher vulnerability index value. The areas covered under different vulnerable zones are shown in Fig. 26.4.

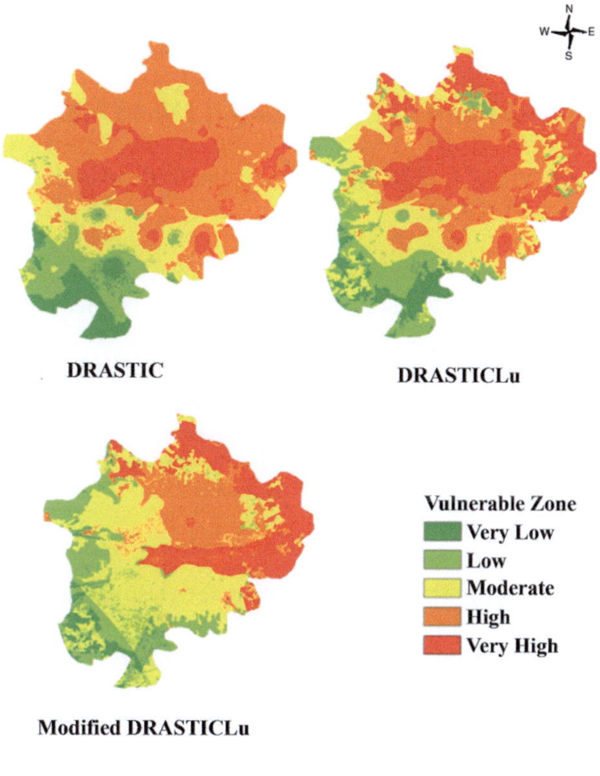

Fig. 26.3 Groundwater Vulnerability maps of Nagpur city

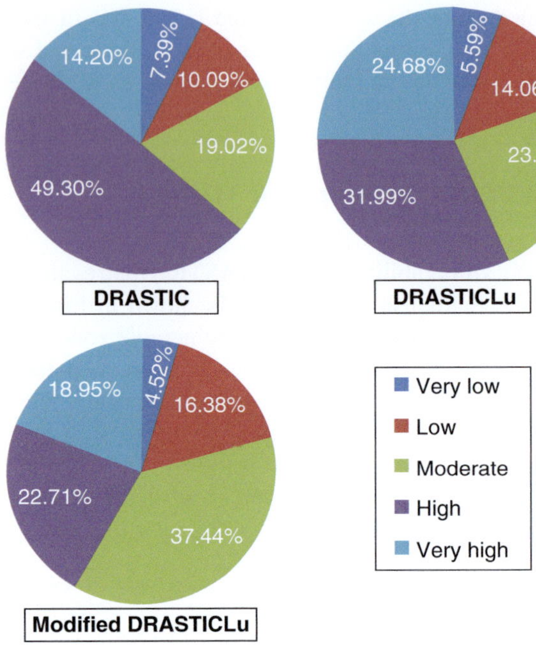

Fig. 26.4 Areas covered under different vulnerable zone

Optimization of Conventional DRASTIC

Addition of Land Use Parameter with Conventional DRASTIC Parameters (DRASTICLu)

The conventional DRASTIC approach is enhanced by the addition of land use parameter. Vulnerability index map is obtained using Eq. (26.2) from overlay analysis eight parameters is shown in Fig. 26.3. The resultant vulnerability map (DRASTICLu) shows that the East and North-East zone is occupied by higher vulnerability index except few isolated pocket which lies under moderate vulnerability index. The Central south zone is covered with the moderate vulnerability index and South-West is safe having least vulnerability index. The area occupied by different vulnerable zone is shown in Fig. 26.4.

Revising the Rates of the Parameters Using Quality Data

The rates of the parameters used in DRASTIC methodology are revised using mean nitrate concentration of the city. The concentration of nitrate is observed in different parameters classification. For each classification, an average value of nitrate is calculated by using the data of monitoring wells lying under the same classification. The mean nitrate value of all the classes is observed and the classification carrying the maximum average nitrate concentration is rated as 10 and the other classification is calibrated accordingly to the scale of 10 (Table 26.1). It was observed from the analysis that depth to water level, topography and impact of vadose zone are not showing any trend, so the ratings of these parameters remain unchanged.

Revising the Weights of the Parameters Using AHP (Modified DRASTICLu)

The groundwater vulnerability map is generated using all the eight parameters rated using mean nitrate concentration and weighted using AHP technique. The weights of all the parameters are modified using pair wise comparison matrix using AHP approach. The resultant groundwater vulnerability (Modified DRASTICLu) map shows different vulnerable zones within the city limit (Fig. 26.3). From the close observation of the Modified DRASTICLu groundwater vulnerability map, it was observed that South-West zone is under least vulnerability index, Central and Center west is under moderate vulnerability and East-North is under high to very high vulnerability index. The area under different vulnerable zone according to Modified DRASTICLu is shown in Fig. 26.4.

Validation

The groundwater vulnerability maps evaluated using DRASTIC approach and various modifications are region specific generated using regional hydrogeology of the study area, which needs to be validated. To evaluate the effectiveness of obtained vulnerability maps, correlation coefficient is evaluated between the resultant maps and nitrate concentration of the city using different regression techniques tabulate in Table 26.2. The results shows that Modified DRASTICLu shows a better correlation with the nitrate concentration followed by DRASTICLu and DRASTIC.

As DRASTIC methodology has been already modified using mean nitrate concentration, other quality parameters such EC, TH and Cl are used to find the effectiveness of the resultant vulnerability map. The Correlation Index (CI) method suggested by [15] is used in the study to find correlation between vulnerability maps and quality parameters. The vulnerability map and the field quality parameter are classified in five classes showing very high to very low vulnerability. The classification of the quality parameters is based on the permissible limit (drinking water) of the parameters. To find a correlation, the wells data are observed in the vulnerability maps. If the well contamination value and vulnerable zone show convergence, mean both lies in same vulnerable class for example very high contamination lies in a very high vulnerable zone than the numbers of wells are

Table 26.2 Correlation between Nitrate concentration and DRASTIC models

Correlation	DRASTIC	DRASTICLu	Modified DRASTICLu
Pearson coefficient	0.249	0.310	0.561
Spearman coefficient	0.174	0.215	0.521
Kendall's coefficient	0.130	0.154	0.377

Table 26.3 Correlation Index between quality parameters and groundwater vulnerability map

Map	Zones	Electrical conductivity (EC)					Chlorine (Cl)					Total Hardness (TH)				
		VL	L	M	H	VH	VL	L	M	H	VH	VL	L	M	H	VH
X	VL	0	0	0	0	0	2	4	3	7	7	0	0	0	1	1
	L	0	2	2	2	3	1	2	3	1	3	0	1	0	1	2
	M	3	4	6	8	8	0	0	1	2	1	3	5	7	7	7
	H	0	0	3	1	2	0	0	4	1	1	0	0	1	1	0
	VH	0	0	0	1	0	0	0	0	1	1	0	0	3	2	3
		CI = 170					CI = 140					CI = 177				
Y	VL	0	0	0	0	0	0	4	7	6	6	0	0	0	2	0
	L	0	2	2	2	3	1	2	3	1	3	0	1	0	0	3
	M	1	4	9	8	7	0	0	1	2	1	0	6	11	6	6
	H	0	0	3	1	2	0	0	3	1	2	0	0	1	1	0
	VH	0	0	0	1	0	0	0	0	1	1	0	0	2	3	3
		CI = 176					CI = 141					CI = 178				
Z	VL	0	0	0	0	0	0	5	13	6	0	0	0	2	0	0
	L	0	1	5	3	2	0	1	6	2	0	0	1	1	2	0
	M	0	4	13	5	4	0	0	1	3	0	0	4	13	7	4
	H	0	2	2	2	1	0	0	2	1	1	0	0	1	0	1
	VH	0	0	0	0	1	0	0	1	1	2	0	0	3	3	3
		CI = 184					CI = 157					CI = 186				

VH = Very High; *H* = High; *M* = Moderate; *L* = Low; *VL* = Very Low
X = DRASTIC; Y = DRASTICLu; Z = Modified DRASTICLu

multiplied by 5. If the difference is 1 like very high contamination value wells lies in a high vulnerable zone or high contamination value wells lies in very high or moderate vulnerable zone than the numbers of wells are multiplied by 4. Similarly for the difference 2, 3 and 4 the numbers of wells are multiplied by 3, 2 and 1 respectively. Table 26.3 shows the wells lying in the different vulnerable zone and calculated CI. The result of the analysis shows higher correlation index of Modified DRASTICLu with field quality parameters in comparison with DRASTICLu and DRASTIC.

Discussion and Conclusion

The groundwater vulnerability of Nagpur city is evaluated using the conventional DRASTIC method and by applying optimization to the conventional approach by adding a land use parameter and modifying the rates using mean nitrate concentration and weight using AHP approach. The resultant vulnerability map shows that the Modified DRASTICLu method shows a good correlation with field quality parameters followed by DRASTICLu and DRASTIC. DRASTIC includes only hydro-geological parameters of the aquifer, whereas DRASTICLu includes the effect of anthropogenic activities of land surface by involving the land use parameter in the analysis along with hydro-geological parameters, which increase the effectiveness of DRASTICLu method in the study area. The groundwater vulnerability map evaluated using optimization, by applying various modification to DRASTIC shows the best result in the study area in comparison to other applied approach as it includes the source of contamination (land use parameter), presence of contamination in groundwater by modifying the rates using mean nitrate concentration and revising the weight using pair wise comparison of involved parameter (AHP technique). All this modification helps in better assessment of groundwater vulnerability of Nagpur city.

From the resultant groundwater vulnerability maps, it can be concluded that East zone of the city is more vulnerable to contamination, while the south zone is safe having least vulnerability index. Various groundwater zones and monitoring wells are affected by the higher contaminant concentration, which needs effective remedial measure for the further prevention of groundwater contamination. The vulnerability map obtained using Modified DRASTICLu approach can be used for making effective policies for regional groundwater development.

References

1. Aller L, Lehr J, Petty R, Bennett T (1987) A standardized system to evaluate groundwater pollution using hydrogeologic setting. J Geol Soc India 29(1):23–37
2. Baalousha H (2006) Vulnerability assessment for the Gaza Strip, Palestine using DRASTIC. Environ Geol 50(3):405–415
3. Babiker IS, Mohamed MA, Hiyama T, Kato K (2005) A GIS-based DRASTIC model for assessing aquifer vulnerability in Kakamigahara Heights, Gifu Prefecture, central Japan. Sci Total Environ 345(1–3):127–140
4. Barzegar R, Moghaddam AA, Baghban H (2015) A supervised committee machine artificial intelligent for improving DRASTIC method to assess groundwater contamination risk: a case study from Tabriz plain aquifer, Iran. Stoch Env Res Risk A 30:1
5. Gupta N (2014) Groundwater vulnerability assessment using DRASTIC method in Jabalpur District of Madhya Pradesh. Int J Recent Technol Eng 3(3):36–43
6. Huan H, Wang J, Teng Y (2012) Assessment and validation of groundwater vulnerability to nitrate based on a modified DRASTIC model: a case study in Jilin City of Northeast China. Sci Total Environ 440:14–23
7. Jang C-S, Lin C-W, Liang C-P, Chen J-S (2015) Developing a reliable model for aquifer vulnerability. Stoch Env Res Risk A 30(1):175–187
8. Javadi S, Kavehkar N, Mousavizadeh MH, Mohammadi K (2011) Modification of DRASTIC model to map groundwater vulnerability to pollution using nitrate measurements in agricultural areas. J Agric Sci Technol 13:239–249
9. Karamouz M, Ahmadi A, Akhbari M (2011) Groundwater hydrology engineering, planning and Managment. CRC press, London
10. Li R, Zhao L (2011) Vadose zone mapping using geographic information systems and Geostatistics a case study in the Elkhorn River basin, Nebraska, USA. In: International symposium on water resource and environmental protection, IEEE, pp 3177–3179
11. Neshat A, Pradhan B, Dadras M (2014) Groundwater vulnerability assessment using an improved DRASTIC method in GIS. Resour Conserv Recycl 86:74–86

12. Neshat A, Pradhan B, Pirasteh S, Mohd Shafri HZ (2013) Estimating groundwater vulnerability to pollution using a modified DRASTIC model in the Kerman agricultural area, Iran. Environ Earth Sci 71(7):3119–3131
13. Rahman A (2008) A GIS based DRASTIC model for assessing groundwater vulnerability in shallow aquifer in Aligarh, India. Appl Geogr 28(1):32–53
14. Saaty T (1980) The analytic hierarchy process. McGraw-Hill, New York
15. Sadeghfam S, Hassanzadeh Y, Nadiri AA, Zarghami M (2016) Localization of groundwater vulnerability assessment using catastrophe theory. Water Resour Manag 30:4585–4601
16. Sener E, Davraz A (2013) Assessment of groundwater vulnerability based on a modified DRASTIC model, GIS and an analytic hierarchy process (AHP) method: the case of Egirdir Lake basin (Isparta, Turkey). Hydrogeol J 21(3):701–714
17. Sinha MK, Verma MK, Ahmad I, Baier K, Jha R, Azzam R (2016) Assessment of groundwater vulnerability using modified DRASTIC model in Kharun Basin, Chhattisgarh, India. Arab J Geosci 9(98):1–22
18. Thirumalaivasan D, Karmegam M, Venugopal K (2003) AHP-DRASTIC: software for specific aquifer vulnerability assessment using DRASTIC model and GIS. Environ Model Softw 18(7):645–656
19. Umar R, Ahmed I, Alam F a (2009) Mapping groundwater vulnerable zones using modified DRASTIC approach of an alluvial aquifer in parts of central ganga plain, Western Uttar Pradesh. J Geol Soc India 73(2):193–201
20. Wang J, He J, Chen H (2012) Assessment of groundwater contamination risk using hazard quantification, a modified DRASTIC model and groundwater value, Beijing plain, China. Sci Total Environ 432:216–226
21. WaterAid (2016) FSM – urban wash: an assessment of Faecal sludge management policies and programmes at the national and select states level. WaterAid, New Delhi
22. Wu H, Chen J, Qian H (2016) A modified DRASTIC model for assessing contamination risk of groundwater in the northern suburb of Yinchuan, China. Environ Earth Sci 75(483):1–10
23. Yu C, Yao Y, Hayes G, Zhang B, Zheng C (2010) Quantitative assessment of groundwater vulnerability using index system and transport simulation, Huangshuihe catchmentment, China. Sci Total Environ 408:6108–6116

MIX
Papier aus verantwortungsvollen Quellen
Paper from responsible sources
FSC® C105338

If you have any concerns about our products,
you can contact us on
ProductSafety@springernature.com

In case Publisher is established outside the EU,
the EU authorized representative is:
Springer Nature Customer Service Center GmbH
Europaplatz 3, 69115 Heidelberg, Germany

Printed by Libri Plureos GmbH
in Hamburg, Germany